Vinod Wadhawan

Understanding Natural Phenomena

Self-Organization and Emergence in Complex Systems

Understanding Natural Phenomena

Self-Organization and Emergence
in Complex Systems

Vinod Wadhawan

CreateSpace Independent Publishing Platform
Charleston, SC, USA
2018

For my grandchildren Rhea, Richa, Nishka, and Arjun.

For all those who respect and love science and technology.

First edition published in July 2017
Second edition: September 2017
Second print: January 2018

FRONT COVER. A flower is a work of art, but there is no artist involved. The flower evolved from lesser things which, in turn, evolved from still lesser things, and so on, all the way down. For example, the symmetry of a flower is the end result of a long succession of spontaneous processes and events, as also of some simple 'local rules' in operation, all constrained, even aided, by the infallible second law of thermodynamics for 'open' systems. In fact, the second law is the mother of all organizing principles, leading to the enormous amounts of cumulative self-organization, structure, symmetry, and 'emergence' we see in Nature.

As science turns to complexity, one must realize that complexity demands attitudes quite different from those heretofore common in physics. Up to now, physicists looked for fundamental laws true for all times and all places. But each complex system is different: apparently there are no general laws for complexity. Instead, one must reach for 'lessons' that might, with insight and understanding, be learned in one system and applied to another. Maybe physics studies will become more like human experience.

Goldenfeld and Kadanoff (1999)

I think the next century will be the century of complexity.

Stephen Hawking (2000)

Successful ecosystems are complex adaptive systems, as are successful cities and societies. According to the scientist James Lovelock's Gaia concept, the Earth as a whole is a complex adaptive system. One of its long-term adaptations that should be of concern to all of us may well be to get rid of our species to protect itself. Whether that happens or not could come down to whether we are able to understand the rules that govern its complexity, and whether we have the wisdom to adapt ourselves and conform to those rules.

Len Fisher (2009)

OTHER BOOKS BY VINOD WADHAWAN

Introduction to Ferroic Materials
Gordon & Breach Science Publishers, Amsterdam (2000)
ISBN 90-5699-286-4

Smart Structures
Blurring the Distinction between the Living and the Nonliving
Oxford University Press, Oxford (2007)
ISBN 978-0-19-922917-8

Complexity Science
Tackling the Difficult Questions We Ask about Ourselves
and about Our Universe
LAP Lambert Academic Publishing, Saarbrücken (2010)
ISBN 978-3-8383-7754-4

Nauka Zlozonosci
Trudne pytania, ktore zadajemy o sobie i o naszym Wszechswiecie
(In Polish. Translation by Malgorzata Koraszewska.)
Racjonalista.pl, Wroclaw (2010)
ISBN 978-83-62503-02-5

Latent, Manifest, and Broken Symmetry
A Bottom-up Approach to Symmetry,
with Implications for Complex Networks
Printed by CreateSpace, Charleston, SC, USA (2011)
ISBN 978-1463766719 / 1463766718

Contents (abridged list)

Foreword — xvii

Preface — xvii

1. Overview 3
2. The Philosophical and Computational Underpinnings of
    Complexity Science 9
3. The Second Law of Thermodynamics 25
4. Dynamical Evolution 37
5. Relativity Theory and Quantum Mechanics 47
6. The Nature of Information 57
7. Darwinian Evolution, Complex Adaptive Systems, Sociobiology . . . 75
8. Symmetry is Supreme 83
9. The Standard Model of Particle Physics 95
10. Cosmology Basics 101
11. Uncertainty, Complexity, and the Arrow of Time 117
12. The Cosmic Evolution of Complexity 123
13. Why Are the Laws of Nature What They Are? 127
14. The Universe is a Quantum Computer 131
15. Chaos, Fractals, and Complexity 135
16. Cellular Automata as Models of Complex Systems 143
17. Wolfram's 'New Kind of Science' 149
18. Swarm Intelligence 157
19. Nonadaptive Complex Systems 163
20. Self-Organized Criticality 169
21. Characteristics of Complex Systems 175

22. Evolution of Structure and Order in the Cosmos 183
23. The Primary and Secondary Chemical Bonds 187
24. Cell Biology Basics 193
25. Evolution of Chemical Complexity 197
26. What is Life? 207
27. Models for the Origins of Life 211
28. Genetic Regulatory Networks and Cell Differentiation 219
29. More Ideas on the Origins of Species: From Darwin to Margulis . . . 223
30. Coevolution of Species 231
31. The Various Energy Regimes in the Evolution of Our Ecosphere . . . 241

32. Evolution of Niele's Energy Staircase After the Emergence of Humans . . 249
33. Computational Intelligence 261
34. Adaptation and Learning in Complex Adaptive Systems 273
35. Smart Structures 281
36. Robots and Their Dependence on Computer Power 287
37. Machine Intelligence 295
38. Evolution of Language 303
39. Memes and Their Evolution 307
40. Evolution of the Human Brain, and the Nature of Our Neocortex . . . 311
41. Minsky's and Hawkins' Models for how Our Brain Functions . . . 319
42. Inside the Human Brain 325

43.   Kurzweil's Pattern-Recognition Theory of Mind 331
44.   The Knowledge Era and Complexity Science 337
45.   Epilogue 347

A1.   Equilibrium Thermodynamics and Statistical Mechanics 357
A2.   Probability Theory 365
A3.   Information and Uncertainty 369
A4.   Thermodynamics and Information 375
A5.   Systems Far from Equilibrium 379
A6.   Quantum Theory and Particle Physics 389
A7.   Theory of Phase Transitions and Critical Phenomena 401
A8.   Chaos Theory 413
A9.   Network Theory and Complexity 421
A10.   Game Theory 439

Bibliography. 453

Index 481

Acknowledgements 491

About the Author 492

Contents (detailed list)

Foreword   xvii

Preface   xxxi

I. Complexity Basics

1. Overview **3**
1.1   Preamble 3
1.2   A whirlpool as an example of self-organization 5
1.3   Spontaneous pattern formation: the Bénard instability 6
1.4   Recent history of investigations in complexity science 8
1.5   Organization of the book 8

2. The Philosophical and Computational Underpinnings of Complexity Science **9**
2.1   The scientific method for understanding natural phenomena 9
2.2   Reductionism and its inadequacy for dealing with complexity 12
2.3   The Laplace demon 13
2.4   Holism 15
2.5   Emergence 16
2.6   Scientific determinism, effective theories 17
2.7   Free will 18
2.8   Actions, reactions, interactions, causality 21
2.9   The nature of reality 23

3. The Second Law of Thermodynamics **25**
3.1   The second law for isolated systems 25
3.2   Entropy 26
3.3   The second law for open systems 27
3.4   Nucleation and growth of a crystal 29
3.5   The second law is an emergent law 32
3.6   Emergence, weak and strong 33
3.7   Nature abhors gradients 33
3.8   Systems not in equilibrium 34
3.9   Thermodynamics of small systems 35

4. Dynamical Evolution **37**
4.1   Dynamical systems 37
4.2   Phase-space trajectories 37
4.3   Attractors in phase space 38
4.4   Nonlinear dynamical systems 40
4.5   Equilibrium, stable and unstable 40
4.6   Dissipative structures and processes 42
4.7   Bifurcations in phase space 43
4.8   Self-organization and order in dissipative structures 44

5. Relativity Theory and Quantum Mechanics **47**
5.1   Special theory of relativity 47
5.2   General theory of relativity 49
5.3   Quantum mechanics 52

5.4   Summing over multiple histories 55

6.   The Nature of Information **57**
6.1   Russell's paradox 57
6.2   Hilbert's formal axiomatic approach to mathematics 58
6.3   Gödel's incompleteness theorem 59
6.4   Turing's halting problem 60
6.5   Elementary information theory 63
6.6   Entropy means unavailable or missing information 65
6.7   Algorithmic information theory 66
6.8   Algorithmic probability and Ockham's razor 69
6.9   Algorithmic information content and effective complexity 70
6.10   Classification of problems in terms of computational complexity 70
6.11   'Irreducible complexity' deconstructed 71

7.   Darwinian Evolution, Complex Adaptive Systems, Sociobiology **75**
7.1   Darwinian evolution 75
7.2   Complex adaptive systems 77
7.3   The inevitability of emergence of life on Earth 79
7.4   Sociobiology, altruism, morality, group selection 81

8.   Symmetry is Supreme **83**
8.1   Of socks and shoes 83
8.2   Connection between symmetry and conservation laws 83
8.3   Why so much symmetry? 84
8.4   Growth of a crystal as an ordering process 85
8.5   Broken symmetry 86
8.6   Symmetry aspects of phase transitions 88
8.7   Latent symmetry 89
8.8   Latent symmetry and the phenomenon of emergence in complex systems 90
8.9   Broken symmetry and complexity 91
8.10   Symmetry of complex networks 92

9.   The Standard Model of Particle Physics **95**
9.1   The four fundamental interactions 95
9.2   Bosons and fermions 96
9.3   The standard model and the Higgs mechanism 98

10.   Cosmology Basics **101**
10.1   The ultimate causes of all cosmic order and structure 101
10.2   The Big Bang and its aftermath 102
10.3   Dark matter and dark energy 105
10.4   Cosmic inflation 108
10.5   Supersymmetry, string theories, M-theory 109
10.6   Has modern cosmology got it all wrong?111

11.   Uncertainty, Complexity, and the Arrow of Time **117**
11.1   Irreversible processes, and not entropy, determine the arrow of time 117
11.2   Irreversible processes *can* lead to order 117
11.3   The arrow of time and the early universe 118
11.4   When did time begin? 119
11.5   Uncertainty and complex adaptive systems 120

12. The Cosmic Evolution of Complexity 123
12.1 Our cosmic history 123
12.2 We are star stuff 124

13. Why Are the Laws of Nature What They Are? 127
13.1 The laws of Nature in our universe 127
13.2 The anthropic principle 128

14. The Universe is a Quantum Computer 131
14.1 Quantum computation 131
14.2 Quantum entanglement 132
14.3 The universe regarded as a quantum comphter 133

15. Chaos, Fractals, and Complexity 135
15.1 Nonlinear dynamics 135
15.2 Extreme sensitivity to initial conditions 136
15.3 Chaotic rhythms of population sizes 137
15.4 Fractal nature of the strange attractor 139
15.5 Chaos and complexity 141

16. Cellular Automata as Models of Complex Systems 143
16.1 Cellular automata 143
16.2 Conway's Game of Life 143
16.3 Self-reproducing automata 145
16.4 The four Wolfram classes of cellular automata 146
16.5 Universal cellular automata 147

17. Wolfram's 'New Kind of Science' 149
17.1 Introduction 149
17.2 Wolfram's principle of computational equivalence (PCE) 150
17.3 The PCE and the rampant occurrence of complexity 151
17.4 Why does the universe run the way it does? 152
17.5 Criticism of Wolfram's NKS 153

18. Swarm Intelligence 157
18.1 Emergence of swarm intelligence in a beehive 157
18.2 Ant logic 159
18.3 Positive and negative feedback in complex systems 160

19. Nonadaptive Complex Systems 163
19.1 Composite materials 163
19.2 Ferroic materials 163
19.3 Multiferroics 164
19.4 Spin glasses 165
19.5 Relaxor ferroelectrics 166
19.6 Relaxor ferroelectrics as vivisystems 167

20. Self-Organized Criticality, Power Laws 169
20.1 The sandpile experiment 169
20.2 Power-law behaviour and complexity 170
20.3 Robust and nonrobust criticality 173

21. Characteristics of Complex Systems 175

II. Pre-Human Evolution of Complexity

22.   Evolution of Structure and Order in the Cosmos 183
22.1   The three eras in the cosmic evolution of complexity 183
22.2   Chaisson's parameter for quantifying the degree of complexity 183
22.3   Cosmic evolution of information 184
22.4   Why so much terrestrial complexity? 186

23.   The Primary and Secondary Chemical Bonds 187
23.1   The primary chemical bonds 187
23.2   The secondary chemical bonds 189
23.3   The hydrogen bond and the hydrophobic interaction 190

24.   Cell Biology Basics 193

25.   Evolution of Chemical Complexity 197
25.1   Of locks and keys in the world of molecular self-assembly 197
25.2   Self-organization of matter 199
25.3   Emergence of autocatalytic sets of molecules 202
25.4   Positive feedback, pattern formation, emergent phenomena 204
25.5   Pattern formation: the BZ reaction 205

26.   What is Life? 207
26.1   Schrödinger and life 207
26.2   Koshland's 'seven pillars of life' 209

27.   Models for the Origins of Life 211
27.1   The early work 211
27.2   The RNA-world model for the origin of life 213
27.3   Dyson's proteins-first model for the origins of life 215
27.4   Why was evolution extremely fast for the earliest life? 218

28.   Genetic Regulatory Networks and Cell Differentiation 219
28.1   Circuits in genetic networks 220
28.2   Kauffman's work on genetic regulatory networks 221

29.   Ideas on the Origins of Species: From Darwin to Margulis 223
29.1   Darwinism and neo-Darwinism 223
29.2   Biological symbiosis and evolution 225
29.3   What is a species 227
29.4   Horizontal gene transfer in the earliest life forms 228
29.5   Epigenetics 229

30.   Coevolution of Species 231
30.1   Punctuated equilibrium in the coevolution of species 231
30.2   Evolutionarily stable strategies 232
30.3   Of hawks and doves in the logic of animal conflicts 234
30.4   Evolutionary arms races and the life-dinner principle 236

31.   The Various Energy Regimes in the Evolution of Our Ecosphere . . .241

31.1   The thermophilic energy regime 242
31.2   The phototrophic energy regime 244
31.3   The aerobic energy regime 245

III. Humans and the Evolution of Complexity

32.   **Evolution of Niele's Energy Staircase After the Emergence**
     of Humans 249
32.1   The pyrocultural energy regime 249
32.2   The agrocultural energy regime 251
32.3   The carbocultural energy regime 252
32.4   The green-valley approach to System Earth 253
32.5   The imperial approach to System Earth 254
32.6   A nucleocultural energy regime? 256
32.7   A possible 'heliocultural' energy regime 258

33.   **Computational Intelligence** 261
33.1   Introduction 261
33.2   Fuzzy logic 262
33.3   Neural networks, real and artificial 263
33.4   Genetic algorithms 265
33.5   Genetic programming: Evolution of computer programs 267
33.6   Artificial life 271

34.   **Adaptation and Learning in Complex Adaptive Systems** 273
34.1   Holland's model for adaptation and learning 273
34.2   The bucket brigade in Holland's algorithm 274
34.3   Langton's work on adaptive computation 276
34.4   The edge-of-chaos existence of complex adaptive systems 278

35.   **Smart Structures** 281
35.1   The three main components of a smart structure 281
35.2   Reconfigurable computers and machines that can evolve 283

36.   **Robots and Their Dependence on Computer Power** 287
36.1   Behaviour-based robotics 287
36.2   Evolutionary robotics 288
36.3   Evolution of computer power per unit cost 290

37. Machine Intelligence 295
37.1   Artificial distributed intelligence 295
37.2   Evolution of machine intelligence 296
37.3   The future of intelligence and the status of humans 298

38.   **Evolution of Language** 303

39.   **Memes and Their Evolution** 307

40.   **Evolution of the Human Brain, and the Nature of Our Neocortex** . . . 311
40.1   Evolution of the brain 312
40.2   The human neocortex 313
40.3   The history of intelligence 315

41.   Minsky's and Hawkins' Models for how Our Brain Functions 319
41.1   Marvin Minsky's 'Society of Mind' 319
41.2   Can we make decisions without involving emotions? 320
41.3   Hawkins' model for intelligence and consciousness 323

42. Inside the Human Brain 325
42.1   Probing the human Brain 325
42.2   Peering into the human brain 327

43.   Kurzweil's Pattern-Recognition Theory of Mind 331

44.   The Knowledge Era and Complexity Science 337
44.1   The wide-ranging applications of complexity science 337
44.2   Econophysics 338
44.3   Application of complexity-science ideas in management science 341
44.4   Cultural evolution and complexity transitions 343
44.5   Complexity leadership theory 345
44.6   Complexity science in everyday life 345

45.   Epilogue 347

IV. Appendices

A1.   Equilibrium Thermodynamics and Statistical Mechanics 357
A1.1   Equilibrium thermodynamics 357
A1.2   Statistical mechanics 360
A1.3   The ergodicity hypothesis 360
A1.4   The partition function 361
A1.5   Tsallis thermodynamics of small systems 361

A2.   Probability Theory 365
A2.1   The notion of probability 365
A2.2   Multivariate probabilities 365
A2.3   Determinism and predictability 367

A3.   Information and Uncertainty 369
A3.1   Information theory 369
A3.2   Shannon's formula for a numerical measure of information 370
A3.3   Shannon entropy and thermodynamic entropy 371
A3.4   Uncertainty 372
A3.5   Algorithmic information theory 373

A4.   Thermodynamics and Information 375
A4.1   Entropy and information 375
A4.2   Kolmogorov-Sinai entropy 376
A4.3   Mutual information and redundancy of information 377

A5.   Systems Far from Equilibrium 379
A5.1   Emergence of complexity in systems far from equilibrium 379
A5.2   Nonequilibrium classical dynamics 380
A5.3   When does the Newtonian description break down? 383

A5.4   Generalization of Newtonian dynamics 384
A5.5   Pitchfork bifurcation 386
A5.6   Extension of Newton's laws 386

A6.   Quantum Theory and Particle Physics **389**
A6.1   Introduction 389
A6.2   The Heisenberg uncertainty principle 389
A6.3   The Schrödinger equation 390
A6.4   The Copenhagen interpretation 391
A6.5   Time asymmetry 391
A6.6   Multiple universes 391
A6.7   Feynman's sum-over-histories formulation 392
A6.8   Quantum Darwinism 393
A6.9   Gell-Mann's coarse-graining interpretation 393
A6.10 Poincaré resonances and quantum theory 394
A6.11 Model-dependent realism, intelligence, existence 396
A6.12 The principle of conservation of quantum information 397
A6.13 Particle physics 398

A7.   Theory of Phase Transitions and Critical Phenomena **401**
A7.1   A typical phase transition 401
A7.2   Liberal definitions of phase transitions 401
A7.3   Instabilities can cause phase transitions 402
A7.4   Order parameter of a phase transition 403
A7.5   The response function corresponding to the order parameter 404
A7.6   Phase transitions near thermodynamic equilibrium 404
A7.7   The Landau theory of phase transitions 405
A7.8   Spontaneous breaking of symmetry 407
A7.9   Field-induced phase transitions 407
A7.10 Ferroic phase transitions 408
A7.11 Prototype symmetry 409
A7.12 Critical phenomena 409
A7.13 Universality classes and critical exponents 410

A8.   Chaos Theory **413**
A8.1   The logistic equation 413
A8.2   Lyapunov exponents 416
A8.3   Divergence of neighbouring trajectories 417
A8.4   Chaotic attractors 419

A9.   Network Theory and Complexity **421**
A9.1   Graphs 421
A9.2   Networks 425
A9.3   The travelling-salesman problem 426
A9.4   Random networks 427
A9.5   Percolation transitions in random networks 428
A9.6   Small-world networks 429
A9.7   Scale-free networks 431
A9.8   Evolution of complex networks 432
A9.9   Emergence of symmetry in complex networks 433
A9.10 Chua's cellular nonlinear networks as a paradigm for
       emergence and complexity 435

A10. Game Theory **439**
A10.1 Introduction 439
A10.2 Dual or two-player games 442
A10.3 Noncooperative games 449
A10.4 Nash equilibrium 450
A10.5 Cooperative games 450

Bibliography **453**

Index **481**

Acknowledgements **491**

About the Author **492**

Foreword

In medieval times, our understanding of the world around us was primarily in the realm of religion and magic. However, it was in the 15th century that a more rational approach to the study of nature began to appear, followed in the early 18th century by the so-called period of enlightenment. Nonetheless, the role of religion continued to dominate thought right through the 19th and twentieth centuries. Yet, here we are in the 21st century, when one would have thought that rationality would be the order of the day, we are nonetheless still surrounded by irrationality, and partly religious magical thinking. You have only to type the word "crystal" into Google to see page after page on the magical healing of crystals. For we scientists, such beliefs make no sense at all and even can be seen as an attack against the scientific method itself. No doubt, Nature is observed to be complex and at times may seem to be mysterious, but that does not mean that we should give up and substitute the concept of "belief" for true scientific examination. This is why the material described in this book is so useful and important to understand today.

Vinod Wadhawan has been a crusader for rationality in thinking and public discourse for many years. Though this book has been designed as a comprehensive textbook on complexity science, it serves many other purposes as well. He explains how the known processes and understandings of complex systems can develop from often simple beginnings. While such happenings may often seem to the layman to be strange or even magical, they are generally susceptible to scientific reasoning. For example, consider the appearance in a fluid of regular hexagonal-shaped cells when the fluid is under a large temperature gradient. This beautiful phenomenon is called Bénard convection and is fully understood once one appreciates the underlying thermal convection currents in the fluid. As Vinod quotes from others, "Nature abhors gradients".

Here you will read about a mixture, or better a fusion, of philosophical and scientific ideas, in a rather accessible language. After all the field of physics was, and is to this day in Scotland, known as Natural Philosophy. This soon gets us into a discussion of determinism and whether free will exists, subjects that have before them centuries of discussion. One of the means of rationalising the ways of nature is through the now generally well-accepted ideas inherent in thermodynamics, especially the Second Law for open systems. The law itself is not provable, but as with so many examples in science, leads to conclusions that can be tested. Despite this lack of direct proof, the laws of thermodynamics have stood the test of time and we do not know of exceptions. Wadhawan makes considerable use of this Law in explaining the phenomena associated with changes from simple to complex behaviour.

An important message that suffuses the book is that most complex systems are far too complex to be understandable in terms of the usual reductionistic approach of conventional science. One just cannot set up and solve a tractable number of differential equations for catching the essence of most of the complex systems. One has to look beyond reductionism, and attempt a holistic approach. Very often, difference equations come to the rescue. The useful tip in this book seems to be: work with difference equations if you cannot work with differential equations in a meaningful way for trying to comprehend a complex system. The book gives a pride of place to the subject of cellular automata for this reason.

So, in the first part of the book the reader is treated to a whole range of topics from concepts of evolution, relativity, quantum theory through the fundamental ideas of symmetry, particle physics, chaos theory, and causes of complexity in nature. Vinod then takes us on a tour of pre-human evolution of complexity, addressing knotty questions such as the meaning of life (but not 42 as in the Hitchhiker's Guide to the Galaxy!), and the fundamental basis of the Darwinian view of the evolution of species. Darwinian evolution is a subset of dynamical evolution. Dynamical evolution, controlled as well as aided by the second law of thermodynamics for open systems, is at the heart of what the science of complex systems is all about. This fact is brought out very clearly in the book.

In the next part of the book we meet the evolution of complexity during human existence, including the founding of various algorithms, robotics and functions of the human brain. Many of these problems today remain unresolved, of course, but such is the nature of the scientific method that constant progress is actually being made in their understanding. In this we are currently living through a remarkable period of rapid developments of ideas. Appearance of humans on the scene has led to a rapid increase in the rate of evolution of complexity. Even more significantly, our remarkable progress in the field of artificial intelligence has brought up a critical situation in which our robots are already getting better than us in more and more aspects. As pointed out by Vinod, the self-evolution of robots can occur exponentially rapidly, whereas we humans are hardly evolving on that time scale. There remains the question as to whether the human brain itself can be reproduced artificially. We are now making considerable progress in understanding how the brain works and one can argue that surely there will come a time when science will enable a complete artificial intelligence to be built, complete with the ability to reason, to think and perhaps even develop a conscience. Precisely what this means is a hot topic of current debate. Perhaps the relatively new field of quantum computing will open this door; but, of course, prediction of the future is difficult and likely to be wrong! What is certain though is that artificial intelligence is advancing at such a remarkable rate that the old science fiction view of robots is beginning to look seriously realistic. The other day I watched a small machine running around independently mowing the lawn in a neighbour's garden. I saw with astonishment how it carefully manoeuvred itself around objects such as a chair on the lawn. Look at mobile phones. My first computer had 8K store on a magnetic drum, but today's mobile phones are several thousand times more powerful and are capable, for instance, of allowing photographs to be taken at phenomenal resolution. This has been a triumph of the development of many fields, including lens design, new materials and new software techniques, let alone the ability to make telephone calls. Who would have thought of such phenomena outside of the world of scientific fiction a few years ago?

Clearly the future belongs to robots. If they turn out, for instance, to be made mainly of inorganic materials, they will outlast all humans, and this even raises the question as to whether humans as a species will continue to exist or even if they need to exist. These are deep, possibly troubling, but certainly exciting prospects to consider, both as a matter of practicality and of ethics. While we still have some control on robots, we should apply our minds to what kind of a future we want for ourselves. And good decisions in this regard require a basic minimum understanding of the science of complex systems by a wide cross section of society. We are living through a very special time. This is where this book comes in.

It can be seen that Vinod Wadhawan has set himself a momentous and daunting task in putting together into a single book so many apparently diverse concepts and ideas that might at first seem to be so disparate as to be intractable. But in fact, we see that there are common threads, often called simply the Laws of Nature by some. These laws are rapidly becoming ever more understood and a careful reading of this book will help us with our observations of the world around us, so that though we may continue to ask "why?", sometimes we will come up with a rational explanation.

This book is epic in the sense that it covers so much ground that one is left somewhat dizzy. And yet, it all makes sense once one realizes how it is possible for something that is complex, for example a flower, to evolve via natural processes from humble beginnings. After all, starting with single-cell creatures such as amoebae we follow a complicated but rational evolutionary path to arrive at the most complex organizations that we know of – ourselves. So, if you follow the logic of this book, starting with the basic concepts of thermodynamics, symmetry, quantum theory and so on, you will be treated to many many thought-provoking ideas, which will likely challenge your own preconceptions and leave you thirsting for more.

Now a few words about the author. I have personally known Vinod for a long time, ever since he came to work for a while in my laboratory. At the time he was working on a phenomenon known as ferrogyrotropy, wherein certain crystals that show chiral ("handedness" if you prefer) properties, the chiral properties can be switched by application of an external stimulus. I think he was the only person in the world then studying this phenomenon. After he left Oxford in 1980 we kept in constant contact, with Vinod playing an important role as one of my regional editors with the international journal, Phase Transitions, for which I was the general editor. I noted that every paper sent to the journal from Indian authors had been closely edited by Vinod beforehand, and so I knew that I could rely entirely automatically on his personal skill and judgement. Vinod's ability at writing in English is commendable: he obviously has had the benefit of a classical education. Since returning to India, he has produced several books, starting with topics related to ferroic materials and smart structures, and eventually moving on to the more philosophical concepts that have to do with the science of complexity. So, we come to this his latest book, where Vinod has supplied us with many nice examples of complexity and how it arises, and as a result the reader will finish the book much more informed than at the beginning. That, after all, is the purpose of a book like this.

A. M. Glazer
Emeritus Professor of Physics and Emeritus Fellow of Jesus College Oxford
Former Vice President, International Union of Crystallography
January 2018

Preface to the First Edition

I am a scientist and I take pride in the fact that we humans have invented and perfected the all-important *scientific method* for investigating natural phenomena. Wanting to understand natural phenomena is an instinctive urge in all of us. In this book I make a case that taking the complexity-science route for satisfying this urge can be a richly rewarding experience. Complexity science enables us (fully or partially) to find answers to even the most fundamental questions we may ask about ourselves and about our universe. We call them *the Big Questions*: How did our universe emerge out of 'nothing' at a certain point in time; or is it that it has been there always? Why and how has structure arisen in our universe: galaxies, stars, planets, life forms? How did life emerge out of nonlife? How does intelligence emerge out of nonintelligence? These are difficult questions. But, as Mark Twain is said to have said, 'there is something fascinating about science. One gets such wholesale of conjecture out of such a trifling investment of fact'. As you will see in this book, the Big Questions, as also many others, can be answered with a good amount of credibility by using just the following 'trifling investment of facts':

1. *Gradients tend to be obliterated spontaneously.* Concentration gradients, temperature gradients, pressure gradients, etc. all tend to decrease spontaneously, till a state of equilibrium is reached, after which the gradients cannot fall any further. This is actually nothing but a nonstatistical-mechanical version of the *second law of thermodynamics*. [Why do gradients arise at all, at a cosmic level? The original cause of all gradients in the cosmos is the continual expansion and cooling of our universe. At the local (terrestrial) level, the energy impinging on our ecosphere from the Sun is the main factor creating gradients.]

2. *It requires energy to prevent a gradient from annulling itself, or to create a new gradient.* A refrigerator works on this principle, as also so many other devices.

3. *Left to themselves, things go from a state of less disorder to a state of more disorder, spontaneously.* This is the more familiar version of the second law of thermodynamics. Examples abound. Molecules in a gas occupy a larger volume spontaneously if the larger volume is made available to them; but there is practically no way they would occupy the smaller volume again, on their own.

4. *If a system is not left to itself, i.e., if it is not an isolated system and can therefore exchange energy and/or matter with its surroundings, then a state of lower disorder can sometimes arise locally.* [This is in keeping with the second law of thermodynamics, as generalized to cover 'thermodynamically open' systems also.] Growth of a crystal from a fluid is an example. A crystal has a remarkably high degree of order and design, even though there is no designer involved. To borrow a phrase from Stuart Kauffman, this is 'order for free'.

5. *If a sustained input of energy drives a system far away from equilibrium, the system may develop a structure or tendencies which enable it to dissipate energy more and more efficiently.* This is called *dissipation-driven adaptive organization*. England (2013) has shown that all dynamical evolution is more likely to lead to structures and systems which get better and better at absorbing and dissipating energy from the environment.

6. *The total energy of the universe is conserved.* This is known as the energy-conservation principle. Since energy and mass are interconvertible, the term 'energy' used here really means 'mass plus energy'.

7. *Natural phenomena are governed by the laws of quantum mechanics.* Classical mechanics, though adequate for understanding many day-to-day or 'macroscopic' phenomena, is only a special, limiting, case of quantum mechanics.

8. *There is an uncertainty principle in quantum mechanics, one version of which says that the energy-conservation principle <u>can</u> be violated, though only for a very small, well-specified duration.* The larger the violation of energy conservation, the smaller this duration is.

9. *It can be understood fully in terms of the second law of thermodynamics that in a system of interacting entities, entirely new (unexpected) behaviour or properties can arise if the interactions are appropriate and strong enough.* 'More is different' (Anderson 1972). The technical term for this occurrence is *emergence.* Complexity science is mostly about self-organization and emergence, and we shall encounter many examples of them in this book. To mention a couple of them here: the emergence of life out of nonlife; and the emergence of human intelligence in a system of nonintelligent entities, namely the neurons. Interestingly, the second law of thermodynamics is itself an emergent law. The motion of a molecule is governed by classical or 'Newtonian' mechanics, which has time-reversal symmetry, meaning that if you could somehow reverse the direction of time, the Newtonian equations of motion would still hold. And yet, when you put a large number of these molecules together, there are interactions among them and there emerges a *direction* of time: Time increases in the direction in which irreversible processes occur. As I shall discuss later in the book, even the causality principle is an emergent principle.

10. *The dynamics of evolution of a complex system of interacting entities is mostly through the operation of 'local rules'.* Chua (1998) has introduced the important notion of cellular nonlinear networks (CNNs), and enunciated a *local-activity dogma.* According to it, in order for a 'nonconservative' system or model to exhibit any form of complexity, the associated CNN parameters must be such that either the cells or their couplings are *locally active.*

11. *The most adaptable are the most likely to survive and propagate.* Any species, if it is not to become extinct, must be able to survive and propagate, in an environment in which there is always some intra-species and/or inter-species competition because different individuals may all have to fight for the same limited resources like food or space or mates. The fittest individuals or groups for such tasks (i.e., the most *adaptable* ones) stand a greater chance of winning the game and, as a result, the population gets better and better (more adapted) at survival and propagation in the prevailing conditions: the more adaptable or 'fitter' ones are not only more likely to survive, but also stand a greater chance to pass on their genes through mating to the next generation.

It is remarkable that an enormous number and variety of natural phenomena can be understood in terms of just these few 'commonsense' facts, by adopting the complexity-science approach. Complexity science helps us understand, to a small or large extent, even those natural phenomena which fall outside the scope of conventional reductionistic science.

What is complexity science, and how is its operational space different from that of conventional science? Let us begin by answering the question: What does the phrase 'system under investigation' mean in conventional science? Strictly speaking, since everything interacts with everything else, the entire cosmos is one big single system. But such an approach cannot take us very far because it is neither tractable nor useful. So, depending on our interest, we define a subsystem which is a *'quasi-isolated system'*. A quasi-isolated system is an imaginary construct, such that what is outside it can be, to a good approximation, treated as an unchanging (usually large) 'background', or 'heat bath' etc. This approach is so common in conventional science that we just say 'system' when what we really mean is a carefully identified quasi-isolated system. An example from rocket science will illustrate the point. For predicting the initial trajectory of a rocket, we can assume safely that a truck moving an adequate distance away from the launching site will not affect the trajectory *significantly*. Conventional science deals mostly with such 'simple' or 'simplifiable' systems. Complexity science, by contrast, deals with systems which must be treated in their totality; for them it is mostly not possible to identify a 'quasi-isolated subpart'.

By definition, a complex system is one which comprises of a large number of 'members', 'elements' or 'agents', which interact substantially with one another and with the environment, and which have the potential to generate qualitatively new collective behaviour. That is, there can be an *emergence* of new (unexpected) spatial, temporal, or functional structures or patterns. Different complex systems have different 'degrees of complexity', and the amount of information needed to describe the structure and function of a system is one of the measures of that degree of complexity (Wadhawan 2010).

'Complexity' is something we associate with a complex system (defined above). It is a technical term, and does not mean the same thing as 'complicatedness'.

The idea of writing this book took shape when I was working on my book *Smart Structures: Blurring the Distinction between the Living and the Nonliving* (Wadhawan 2007). Naturally, there was extensive exposure to concepts from complexity science. Like the subject of smart structures, complexity science also cuts across various disciplines, and highlights the basic unity of all science. The uneasy feeling grew in me that, in spite of the fact that complexity is so pervasive and important, it is not introduced as a well-defined subject even to science students. They are all taught, say, thermodynamics and quantum mechanics routinely, but not complexity science. Even among research workers, although a large number are working on one complex system or another (and not just in physics or chemistry, but also in biology, brain science, computational science, economics, etc.), not many have learnt about the basics of complexity science in a coherent manner at an early stage of their career. I have tried to write a book on complexity that takes this subject to the classroom at a fairly introductory but comprehensive level. There is no dumbing down of facts, even at the cost of appearing 'too technical' at times.

Here are some examples of complex systems: beehives; ant colonies; self-organized supramolecular assemblies; ecosystems; spin-glasses and other complex materials; stock markets; economies of nations; the world economy; the global weather pattern. The origin and evolution of life on Earth was itself a series of emergent phenomena that occurred in highly complex systems. Evolution of complexity is generally a one-way traffic: The new emergent features may (in principle) be deducible from, but are not reducible to, those operating at the next lower level of complexity. Reductionism stands discounted.

As I said earlier, emergent behaviour is a hallmark of complex systems. Human intelligence is also an emergent property: Thoughts, feelings, and purpose result from the interactions among the neurons. Similarly, even memories are emergent phenomena, arising out of the interactions among the large number of 'unmemory-like' fragments of information stored in the brain.

What goes on in a complex system is essentially as follows: There is a large number of interacting agents, which may be viewed as forming a *network*. In the network-theory jargon, the agents are the 'nodes' of the network, and a line joining any two nodes (i.e., an 'edge') represents the interaction between that pair of agents. Any interaction amounts to communication or exchange of information. The action or behaviour of each agent is determined by what it 'sees' others doing, and its actions, in turn, determine what the other agents may do. Further, the term *game-playing* is used for this mutual interaction in the case of those complex systems in which the agents are 'thinking' organisms (particularly humans). Therefore a partial list of topics covered in this book is: information theory; network theory; cellular automata; game theory.

Exchange of information in complex systems, controlled like other macroscopic phenomena by the second law of thermodynamics, leads to self-organization and emergence. In particular, biological evolution is a natural and inevitable consequence of such ongoing processes, an additional factor here being the cumulative effects of *mutations* and *natural selection*. This book has chapters on evolution of complexity of all types: cosmic, chemical, biological, artificial, cultural.

Networked or 'webbed' systems have the all-important *nonlinearity* feature. In fact, nonlinear response, in conjunction with substantial departure from equilibrium, is the crux of complex behaviour. There are many types of nonlinear systems. The most important for our purposes in this book are those in which, although the output (y) is indeed proportional to the input (x), the proportionality factor (m) is not independent of the input; i.e., m is not a constant factor, but rather varies with what x is. For a linear system we have $y = m x$, with m having a fixed value, not varying with x. But for a nonlinear system, the equation becomes $y = m(x) x$; now m is not a constant. This has far-reaching consequences for the (always networked) complex system. In particular, its future progression of events is very sensitive to conditions at any particular point of time (the so-called 'initial conditions'). This sensitivity to initial conditions is also the hallmark of *chaotic systems*. In fact, there is a well-justified viewpoint that it is impossible to discuss several types of complex systems without bringing in concepts from chaos theory. And, what is more, complex systems tend to evolve to a configuration wherein they can operate near the so-called *edge of chaos* (neither too much order, nor too much chaos). There is a chapter on chaos which elaborates on these things.

Inanimate systems can also be complex. Whirlpools and whirlwinds are familiar examples of dynamic nonbiological complex systems. Even static physical systems like some nanocomposites may exhibit properties that cannot always be deduced from those of the constituents of the composite. A particularly fascinating class of complex materials are the so-called *multiferroics*. A multiferroic is actually a ferroic crystalline material (a 'natural' composite) which just refuses to be homogeneous over macroscopic length scales, so that the same crystal may be, say, ferroelectric in some part, and ferromagnetic in another. In a multiferroic, two or all three of the electric, magnetic and elastic interactions compete in a delicately balanced manner, and even a very minor local factor can tilt the balance in favour of one or the other. This class of materials offers great scope for basic research and for device applications, particularly in smart structures.

The current concern about ecological conservation and global warming points to the need for a good understanding of complex systems, particularly their holistic nature. Mother Earth is a single, highly complex, system, now increasingly referred to as *the System Earth*.

A better understanding of complexity may well become a matter of life and death for the human race. And the subject of complexity science is still at the periphery of science. It has not yet become mainstream, in the sense that it is not taught routinely even at the college level. That cannot go on.

There are already a substantial number of great books on complexity science, and I have drawn on them. But I believe that this book is student-friendly and teacher-friendly, and it brings home the all-pervasive nature of the subject. Here are its salient features:

1. It provides a comprehensive update on the subject.

2. It can serve as introductory or supplementary reading for an undergraduate or graduate course on any branch of complexity science.

3. Practically all the mathematical treatment of the subject has been pushed to the appendices at the end of the book, so the main text can be comprehended even by those who are not too comfortable with equations. This is important because a large fraction of the educated public must get the hang of the nature of complexity, so that we can successfully meet the challenges posed to our very survival as a species.

4. Both among scientists and nonscientists there is a large proportion of people who are insufficiently trained about the explaining power of complexity science when it comes to some of the deepest puzzles of Nature and, hopefully, this book would help remedy the situation to some extent.

5. The book has a certain all-under-one-roof character. The topics covered are so many and so diverse that it would be well-nigh impossible for a reader, specializing in a particular branch of complexity science, not to get exposed to what is going on in the rest of complexity science! This is important, because using the insights gained in one complex system for trying to understand another complex system is the hallmark of complexity science.

6. A proper understanding of what complexity science has already achieved will also help discredit many of the claims of mystics, supernaturalists, and pseudoscientists.

Bengaluru   Vinod Wadhawan
July 2017

Preface to the Second Edition

A number of minor corrections and other improvements have been incorporated. The font size has been reduced by 10%. New information has been added, and some less relevant material has been removed.

Bengaluru   Vinod Wadhawan
September 2017

Preface to the Second Print

This print includes a Foreword by Prof. A. M. Glazer of the University of Oxford. A former Vice President of the International Union of Crystallography, he is a veteran crystallographer and a great teacher. I am grateful to him for his kind words and also many other useful inputs.

Bengaluru   Vinod Wadhawan
January 2018

I. Complexity Basics

This part of the book introduces the reader to the basic concepts used in trying to understand complexity. Practically no equations are used in this and the next two parts. Most of the mathematics has been pushed to the fourth part, namely the appendices. The aim is to make the essence of complexity accessible to a large and diverse readership.

Complexity science is an effort to discern and theorize common patterns in complex systems from multiple scientific perspectives. Many scientific disciplines are already associated with powerful models and theories: in biology, for example, there is the theory of evolution, in economics there is utility maximization and game theory, and in engineering mathematics there is Alan Turing's theory of computation.

Complexity science seeks to connect these theories, to find explanatory and predictive frameworks that allow us to, for example, describe biological mechanisms in computational terms or social structures in energetic terms.

For the last few decades we have been steadily surveying the landscape of complex phenomena, and it is gratifying that along the way we find that complex systems nominally unrelated bear strong family resemblances. These similarities include how the mathematical structure of evolutionary adaptation looks a lot like the mathematics of learning, that the distribution of energy within a body made of tissues and fluids follows rules similar to those governing the distribution of energy in a society, that networks within cells adhere to the geometric principles we find on the internet, and that the rise and fall of ancient civilizations follow a sequence similar to the extraordinary growth and contraction of urban centres we see in our own millennium.

David Krakauer, Santa Fe Institute (2015)

http://www.csmonitor.com/Science/Complexity/2015/1115/Complexity-Worlds-hidden-in-plain-sight

1. Overview

Life differs not in kind from non-life, rather only in degree – in fact, degree of complexity – and therefore, however humbling, does not deserve the adjective 'special'.

Chaisson, *Cosmic Evolution*

1.1 Preamble

Let us begin with the first Big Question (BQ1) I mentioned in the Preface: Does our universe have a beginning and an end, or has it been there forever? Although doubts are being raised increasingly, the majority view among experts continues to be that our universe has probably emerged from 'nothing' at a certain specific moment, ~13.72 billion years ago. This was the so-called *Big Bang* moment (Krauss 2012, 2017).

In my opinion we may, sooner or later, end up accepting the idea that our universe has no beginning or end, although we would still have to pay due attention to the fact that something very drastic and critical happened 13.72 billion years ago, as indicated by the vast amount of accumulated cosmological data. Any alternative better theory will still have to accommodate the large number of facts that have given support and substance to the Big Bang model.

For our purpose in this book, so far as the evolution of complexity is concerned, the most relevant fact is that there has been, and continues to be, a sustained creation of gradients (thermal gradients, spatial gradients, etc.) in our part of the universe. And, as will become increasingly clear as we proceed, complexity evolves and order emerges as these gradients tend to annihilate themselves (for achieving greater stability for the system as whole). In the Big Bang model the gradients arise because the universe keeps expanding and cooling. Any better model of the universe will still have to have the feature that spatial and thermal gradients keep getting created, at least in our part of the universe. So let us just assume that, from the vantage point of evolution of complexity, the Big Bang model holds. But I shall return to BQ1 in a later chapter after introducing some concepts and jargon.

According to the Big Bang model, there was practically no structure or order in our universe in the beginning. There was just radiation, and gravity. As time passed, more and more structure and order emerged spontaneously, leading in due course to the evolution of life also. A living being embodies an enormous amount of order and organization, compared to anything inanimate. Why and how did this happen? That is the second Big Question (BQ2).

A short answer is possible for the question '*Why* has so much order emerged in the universe spontaneously, when there was none to start with?' Or, to paraphrase Stuart Kauffman (1995b), '*Why do we get order for free?*' The short answer is that the second law of thermodynamics, as applied to '*open*' systems, permits (in fact, encourages) the spontaneous creation of order locally, as long as there is no decrease of disorder globally. The growth of a crystal (a highly ordered system) from a fluid (a highly disordered system) is a common example of how order can emerge out of disorder. Such '*self-organization*' is rampant in Nature.

Admittedly, this short answer to the 'why' part of BQ2 is somewhat simplistic: The reference to the second law must be supplemented by invocation of the fact that our universe has been expanding and cooling all the time, and this expansion and cooling has been creating gradients, thus providing the '*free energy*' needed for the relentless evolution of order and complexity.

The other part of BQ2 is: '*how* do we get order for free?' Naturally the answer to that depends on what system we are talking about. In this book I shall first discuss the evolution of complexity in the *cosmos* as a whole, from the Big Bang onwards. That narrative will bring us to the epoch when the atoms got formed. I would then discuss the *chemical* evolution of complexity (atoms to molecules, molecules to macromolecules and biomolecules, and so on). Evolution of *biological* complexity will be taken up next, followed by the evolution of *artificial* or man-made complexity, including cultural complexity etc.

The third Big Question (BQ3) we would like to have an answer for is: how did intelligence emerge from non-intelligent beginnings? As we shall see, the answer comes from first recognizing the fact that this can happen only in what are called '*complex adaptive systems*' (CASs). CASs are dynamical complex systems that not only evolve with time (like any other dynamical system), but also *learn* from the information they acquire. They also have the proclivity to strike an optimum balance between order and disorder; in other words, they thrive best near the so-called *edge of chaos*. For answering BQ3 properly a lot of basic complexity science must be introduced first, particularly the 'swarm intelligence' paradigm. Therefore, more on this later.

Many formal definitions have been attempted for complex systems (Wadhawan 2010). A complex system consists of a large number of *connected* members, elements or agents, which have the potential to generate qualitatively *new* collective behaviour. As we shall see later in the book, the emergence of life out of nonlife is a striking example of that. Per Bak (1996), acknowledging the difficulties in defining complex systems uniquely, settled for a very general definition: *Complex systems are systems with a large variability*. In the context of this definition, an example of a *non*-complex system is that of a gas; another is that of a crystal. Both are epitomes of uniformity or sameness, with hardly any variability; all portions are the same.

We define complexity as something we associate with a complex system.

Among other things, the science of complex systems exposes the excessive dependence of traditional science on reductionism and constructionism (Anderson 1972). But the self-organization feature of complex systems leads to the emergence of properties and phenomena which *cannot* always be anticipated from an application of the laws of physics. There are surprises, and there is *perpetual novelty* in the time evolution of complex systems.

It turns out that there is a strong degree of *interdependence* among various levels and types of complexity in a system, and there is considerable merit in the idea that our entire universe is one big complex system. Thus *holism* has a certain degree of validity, although we must always guard against the unscientific usages of this term.

Nearer home and at a more practical level, our ecosphere is one single complex system, or *Gaia* (Lovelock 1979). Terrestrial complexity is of major interest to all of us because understanding it adequately is a matter of survival for us. We humans have the unique ability to *influence* the course of many types of evolution, including our own. Therefore it is important that humanity understands the intricacies of complexity science, and plans the broad outlines of its future in a sensible way.

We humans and our interactions with one another, and with our biosphere, are among the most complex systems imaginable. What is our future going to be like? Although we cannot make definite predictions, even probabilistic statements about the more likely scenarios can have a salutary effect on how we conduct our affairs (e.g. regarding the management of climate change) to achieve high levels of sustainability.

Let us discuss some examples to get a feel for how self-organization and pattern-formation occur in complex systems (Renard 2000).

1.2 A whirlpool as an example of self-organization

Imagine a bathtub filled with water. Suppose you suddenly pull out the stop. An interesting vortex structure develops soon, as the water drains out (Fig. 1.1). The vortex is an *ordered dynamic structure* that has emerged spontaneously out of stagnant water. So this is *self-*organization. There is, of course, a *driving force*, namely gravity. And although there is emergence of order locally, there is no violation of the second law of thermodynamics globally (there cannot be). We should consider the bathtub and the drainage system together. Under the action of the driving force (gravity), the system is in a *far-from-equilibrium condition* in which, although there is creation of order locally, there is an overall increase of disorder (entropy) as the water is accelerated in an irreversible manner down the drain.

The whirlpool is an *energy-dissipating structure*. It is also an example of *energy-driven organization* (Niele 2005), popularly known as self-organization, resulting in emergent or unexpected properties or patterns.

Niele (2005) made a distinction between driving forces and *shaping forces* in the emergence and evolution of complex dissipative structures. In the whirlpool, the driving force sustaining the energy-dissipating structure results from an *energy gradient*, whereas the shaping forces come from interactions within the whirlpool and with the surrounding bathtub etc. For example, the shape of the bathtub and the shape of the drainage hole influence the shape of the vortex. The shaping force within the whirlpool is encoded in the structure of water molecules and in the interactions among them. No water molecule has any embedded information or instructions about how to construct the vortex. Yet '*strings of synchronized interactions*' among the water molecules do the shaping of the complex vortex structure.

Fig. 1.1 A schematized whirlpool.
Image credit: Free Images – Andy's Whirlpool, at openclipart.org.
https://openclipart.org/detail/16148/whirlpool

In the context of complexity, the important point here is that it is impossible to work backwards from the observed whirlpool structure, and calculate the details of the positions and velocities of all the atoms and the interactions among them that have given rise to the observed complex behaviour. This is generally true of all '*dissipative systems*' (Prigogine 1996). And most real-life systems *are* dissipative systems. Similarly, except in a broad macroscopic or hydrodynamic sense, the observed complexity of the whirlpool cannot be predicted in detail from the underlying simplicity of the shapes of the molecules and the interactions among them. The gross features of order in the whirlpool are on a scale millions of times larger than the features of the interactions causing them. In any case, one cannot perform computations at infinite speed, and a system is said to be *computationally irreducible* if the simplicity underlying it cannot be worked out or computed in reasonable time. Both reductionism and constructionism stand discounted (Anderson 1972).

Another kind of self-organization can occur in systems in which the members are *living* organisms. Globally coherent patterns can emerge in them out of *local interactions*. Flocking behaviour of birds is an example of this. Simple local rules like *separation* (avoidance of crowding, or short-range repulsion), *alignment* (steering towards the average heading of neighbours), and *cohesion* (steering towards the average position of neighbours, or long-range attraction), result in well-organized flock patterns, even though nobody is in command (Fisher 2009).

Many other such examples of self-organization can be seen in Nature: shoals of fish; swarms of insects; bacteria colonies; herding behaviour of land animals. Experiments carried out on humans showed something similar: When just 5% of the 'flock' changed direction, others followed suit (Fisher 2009).

1.3 Spontaneous pattern formation: the Bénard instability

Beyond a certain threshold of departure from equilibrium, entirely new features can emerge in a complex system. A well-known example of this is the emergence of the so-called Bénard instability, and the resulting highly ordered pattern of circular vortices.

Consider a thin layer of a fluid constrained between two flat, horizontal plates. The lower surface is at a higher temperature $T + \Delta T$ than the upper surface, which is at a temperature T. In Fig. 1.2 the two horizontal plates and the fluid between them are depicted edge-on. When the temperature difference is small, an elementary volume of the liquid close to the bottom plate tends to rise because its density is a little less than that of the liquid above it. But this tendency towards upward movement is countered by the local dissipation of heat due to conduction, viscosity, diffusion, etc. The hotter droplet simply loses its excess heat, continuously, to the surrounding liquid by these processes, and nothing *organized* happens (Fig. 1.2a).

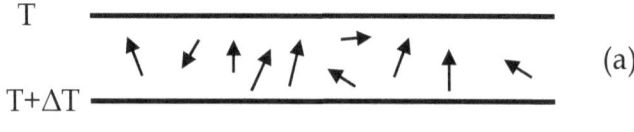

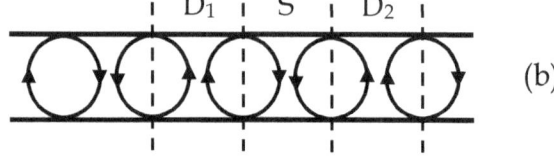

Fig. 1.2 The Bénard experiment as an example of self-organization and pattern formation. A horizontal liquid layer (a) subjected to a sufficient (but not too large) vertical temperature gradient develops a complex pattern of circular vortices (b).

But when the temperature gradient exceeds a certain critical value, the difference in the density of the liquid at the top and the liquid at the bottom is so substantial that now convection becomes the *dominant (more efficient) mode of heat dissipation*. What is more, the liquid confined between the two plates begins to *organize* itself into a coherent, highly ordered, spatial pattern (Bodenschatz 1991; Renard 2000) (Fig. 1.3). How?

Consider two neighbouring droplets near the bottom plate (marked D_1 and D_2 in Fig. 1.2b), and the small space S between them. Both D_1 and D_2 rise by convection, and then fall after reaching the top plate. In the intervening space between them the motion of the liquid is downwards, this liquid having come from both the droplets. Thus in the vertical cross-section there are cylindrical or near-cylindrical vortices. The top view exhibits hexagonal symmetry (Fig. 1.3), rather like what we get when we have a close packing of equal cylinders.

This is self-organization into a pattern of what are called *Bénard cells*, a pattern having hexagonal symmetry, even though no such pattern of symmetry has been superimposed from the outside on the configuration of Fig. 1.2a (Atkins 1994; Prigogine 1996). The molecules move *coherently*, forming hexagonal-shaped cells of convection having cylindrical symmetry.

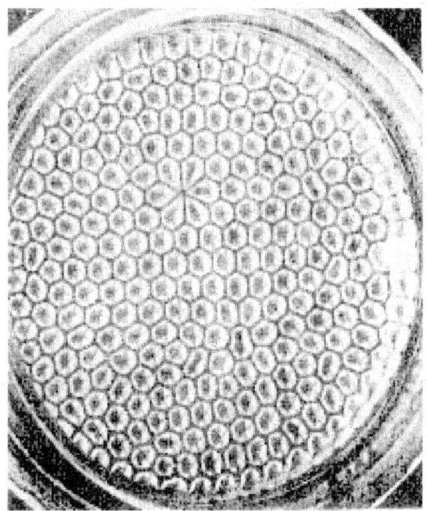

Fig. 1.3 A top view of the Bénard pattern. Image credit:
http://www.viten.com/nyviten/renard.htm

As will be explained later in the book, this is also a case of a *generalized phase transition* (or a *bifurcation* in 'phase space'), occurring spontaneously at a certain critical value of the temperature gradient (see Solé and Goodwin 2000). The formation of the pattern by the Bénard instability provides a *more efficient* way of dissipating heat than the mechanisms of conduction etc. operative before the phase transition took place.

Another way of interpreting what has happened is to say that *Nature abhors gradients*. Pattern formation is a result of the tendency of all systems to find *efficient* ways of annulling gradients of all kinds (Chaisson 2001; Margulis and Sagan 2002; Michaelian2011; England 2013).

If the temperature difference between the two plates is brought to a small value again, the ordered structure, which had *emerged* out of the chaotic motion, disappears.

The Bénard system is an example of an *energy-dissipating structure* (Kauffmann 1995a;

Prigogine 1996). It is also an example of energy-flow-driven organization, or self-organization, which results in *emergent* or new behaviour.

1.4 Recent history of investigations in complexity science

Simon (1996) identified three bursts of activity on research on complex systems in the 20th century. The first was after World War I (WWI), when terms like 'holism', 'Gestalts', and 'creative evolution' were introduced and debated. This epoch was strongly anti-reductionistic. It focused on the assertion that *the whole is more than the sum of its parts.*

The second major eruption of interest in complexity science began after WWII. In this phase, terms like 'information', 'feedback', 'cybernetics', and 'general systems' were bandied around. This epoch was rather neutral on the question of reductionism. Instead, it highlighted the importance of feedback and self-stabilization (*homeostasis*) for the sustenance of a complex system.

Finally, the current eruption of activity started in the early 1990s, wherein complexity is often linked to 'chaos', 'adaptive systems', 'genetic algorithms', and 'cellular automata'. The current tendency is to focus more, if not mainly, on the *mechanisms* that sustain a complex system, and also on the analytical tools for describing and analysing complexity.

I think the most important thing about complexity science is that it helps us understand how *design can arise even when there is no designer involved.* A flower certainly looks like a work of art, but the fact is that there is no artist behind that work of art: It is just a result of a whole succession complexity-evolution processes.

1.5 Organization of the book

The book is in four parts. Part I introduces the basics of complexity science.

Complexity has been evolving ever since the beginning of our universe (or whatever it was that happened at the Big Bang moment). At some stage we humans emerged on the scene. Our importance compared to that of other creatures is that we are *aware* that complexity (including biological complexity) has been evolving. What is more, we have the intelligence and the ability to *influence* the course of its evolution. This fact has made a qualitative change to the course and speed of evolution of complexity. Therefore Part II is about the *pre-human* evolution of complexity, starting with the Big Bang; and Part III deals with humans and their influence on the evolution of complexity. Network theory and game theory play an important role in the contents of Part III. For example, intelligence emerges from the interactions among the network of neurons, even though each neuron is a non-intelligent entity. And game theory is about the games that humans and certain other creatures play when they interact with one another.

Part IV comprises of a rather large set of appendices. My effort has been to push almost all the material involving mathematical equations to the appendices. This should make the rest of the book accessible to even those readers who are not too comfortable with equations.

2. The Philosophical and Computational Underpinnings of Complexity Science

In this chapter we take a brief look at the philosophical and computational basis of science in general, and complexity science in particular. Let us begin by paying tribute to the celebrated '*scientific method*' invented by us humans for a systematic and rational investigation of natural phenomena.

2.1 The scientific method for understanding natural phenomena

In questions of science, the authority of a thousand is not worth the humble reasoning of a single individual.

Galileo Galilei

I am a scientific pantheist, credulous in my own way. The culture of science is a distinct one and certainly mine. I believe that the latest discoveries in biology, chemistry, and physics are true, or at least true for the moment, for science is a method, not a destination. I believe we live in the body of the world and that we are compelled to know the world ... I believe that science is about connection and complexity, harmony and surprise. Science is about beauty. The more I see – the more I know – the more beautiful the world seems.

Sharman Apt Russell

The Wikipedia describes the scientific method as follows: 'The scientific method is a body of techniques for investigating phenomena, acquiring new knowledge, or correcting and integrating previous knowledge. To be termed scientific, a method of inquiry must be based on empirical and measurable evidence subject to specific principles of reasoning'. And according to the *Oxford English Dictionary* the scientific method is: 'a method or procedure that has characterized natural science since the 17th century, consisting in systematic observation, measurement, and experiment, and the formulation, testing, and modification of hypotheses'.

The basic scientific approach is as follows (Sagan 1995; Popper 2005). Suppose there is a set of observations about a natural phenomenon that we want to understand. The scientific method for doing this is the following (I call it the *8-fold way*):

1. A minimum necessary set of axioms. There is an agreed, minimum necessary, set of axioms which are taken as givens (their validity is either a matter of assumption, or has been established already).

2. Rules of logic. There is an agreed set of rules for logical reasoning.

3. Hypotheses. The logical rules for reasoning, as well as the axioms, are used along with a hypothesis (or model) for describing and interpreting the observations we humans have made about the natural phenomenon under investigation. It is not important how the hypothesis is arrived at, because it is anyway going to be tested thoroughly and repeatedly. And there can even be more than one competing hypotheses for explaining the same set of observations or empirical evidence.

4. Agreed meaning of each word. Every word used for making any statement in science should have the same agreed meaning for everybody. This requirement becomes particularly important when concepts like 'consciousness' are discussed or investigated. Further, in the scientific method an important general practice is to define words or concepts in terms of things that are *observable* or, better still, *measurable*.

5. Verification by objective and reproducible observations. A hypothesis must be able to explain the observations in a logically consistent way, and it must successfully stand the test of *repeated* experimental verification. If its success is only partial, we try to modify and improve it, and then check against the observations again. That is how we gradually arrive at the best, i.e., the most successful, hypothesis *at a given point of time in our history*. A pre-requisite to the verifiability requirement is the easy and widespread availability of detailed information about the experiments conducted and the conclusions drawn. This is ensured by the practice of publishing this information in peer-reviewed scientific journals. Such journals also act as repositories of authentic information for future generations. Thus scientific knowledge gets archived in a systematic manner, and becomes available to anybody at any time for learning and for scrutiny. The result is that over 300 years or more, humanity has nurtured and accumulated a formidable body of knowledge called science. The spectacular progress of our science and technology is a result of all this.

6. Predictive capability of the selected hypothesis. A validated hypothesis is an example of '*induction*', i.e., inference of a general or universal conclusion from a number of singular or individual observations. Our confidence in its validity grows if it not only explains what has been already observed, but also enables us to '*deduce*' correctly some predictions about what more can be expected to be observed about the natural phenomenon. Thus both *induction* and *deduction* are parts of the scientific method.

7. Elevation of a hypothesis to the status of a theory. A hypothesis (or a whole set of related, mutually consistent, *set* of hypotheses) that has repeatedly stood the test of experiment, and that can successfully predict and explain a whole range of experimental observations, gradually acquires the status of a theory. There is nothing permanent about a theory. Any theory gets abandoned in the light of new data and interpretations which result in a better theory. *The beauty of the scientific method is that it is self-correcting*. No wrong theory can hold sway for long.

8. Falsifiability. During the entire process of: (i) statement of the research problem; (ii) use of logical reasoning; and (iii) drawing of conclusions from the data and the reasoning, an important constraint usually put in by the scientific method is that only *falsifiable* statements can be made. The term 'falsifiable statement' was introduced by Karl Popper (2005). I explain its meaning with the help of an example.

Consider the following statement (Wudka 1998):

S1: 'The moon is populated by little green men who can read our minds and will hide whenever anyone on Earth looks for them, and will flee sufficiently quickly into deep space whenever a spacecraft comes near'. This statement is so worded that no one can ever observe the postulated green men and demonstrate that the statement is false; so the statement is *unfalsifiable* (and therefore not permitted in scientific discourse).

Next, consider the following statement:

S2: 'There are no little green men on the moon'. This is clearly a falsifiable statement. All you have to do to prove it false is to show material evidence for the existence of even one green man. Berry (2010) attributes the following famous statement to Einstein: '*Many experiments may prove me right, but it takes only one to prove me wrong'.*

Only falsifiable statements are permitted in the scientific method. Therefore S1 is an unscientific statement or theory, and S2 is a scientific statement or theory.

In work beginning in the 1930s, Popper gave falsifiability a renewed emphasis as a criterion for acceptable statements in science. He also pointed out that *not all unfalsifiable claims are fallacious; they are just unfalsifiable.* As long as proper skepticism is retained and proper evidence is given, even an unfalsifiable claim can be a legitimate form of reasoning (but not of what *finally* becomes a part of science). We should never assume that we *must* be right simply because we cannot be proved wrong.

Why did Popper emphasize the falsifiability requirement? It was a part of the effort to tackle what he called '*the problem of induction*'. As stated above, the process of doing science involves generalization from individual observations, and this is always fraught with uncertainty. How many observations or measurements should we make so as to be able to generalize correctly? Generally, all we can say is: the larger the number, the better. But there is always the possibility that the next observation (which we did not make) may go against the generalization. So we can only have low or high *probabilities*, but not certainties, in the induction process. The larger the number of observations which agree with the generalization, the more likely it is that the generalization is valid.

Similarly, the greater the variety of conditions in which the observations and measurements are made, the greater the probability that the inductive generalization is true. The question arises: Which variations in the conditions of observation and measurement are considered significant and relevant, and which ones are not. This is decided by the theory we believe in for the domain of investigation. If the theory is wrong, we are likely to be led astray, till somebody comes up with a better theory.

Thus, because of 'the problem of induction', strong or weak likelihood, rather than complete certainty, is what the inferred laws of science are all about. *Popper emphasized the falsifiability requirement in an effort to minimize the chances of inductivism going wrong.* At the centre of the scientific method is the act of making statements based on existing theories. By restricting ourselves strictly to making only falsifiable statements, we are ensuring that even a single observation or measurement that disagrees with the pre-supposed hypothesis or theory is enough to dismiss the generalization, namely the theory, we inferred by the process of induction.

Notice the intellectual humility of the scientist. Scientific spirit means an ever-present willingness to give up even our pet theories and opinions if the evidence demands so. Contrast this with what is said in many of the organized religions. In them, certain statements cannot be questioned, and there are statements or beliefs in them which are unfalsifiable.

Votaries of faith may be quick to point out that the choice for axioms, mentioned in the 8-fold way above, is also a matter of blind faith. No, it is not. To understand why, let us consider the example of quantum theory.

All natural phenomena are governed by the laws of quantum mechanics. Why the laws of Nature are what they are is something I shall discuss later in the book. The laws of quantum mechanics are highly counter-intuitive for us humans. The quantum theory is based on certain assumed axioms, like any theory is. But the most important thing here is that *the quantum theory is the most repeatedly and the most thoroughly tested theory ever*. It is the best theory we have *at present* for understanding the world around us. If anybody does not agree, he/she is most welcome to come up with another theory, with its own assumed set of axioms and logical structure. If the new theory is better supported by experimental evidence than the present quantum theory, science and scientists will have no compunctions whatsoever in abandoning the existing theory, and accepting the new one. This is not faith and reverence; in fact it is the negation of all that.

[I discussed inductive and deductive modes of reasoning in this section. Are they always used in complexity science also, in their original form? Not always. Complex systems usually comprise of a large number of units, agents, or actors, and that too with changing patterns of interactions. Their investigation therefore gets too difficult for the usual kinds of mathematical analyses. Computer simulation is often the only method available then. One specifies the simple local rules of interaction, and then observes (on a computer) how the overall system evolves with time. This is agent-based modelling or bottom-up modelling. This way of doing science is in contrast to the induction and deduction methods, but with a twist. I quote Axelrod (2006): '*Like deduction, it starts with a set of explicit assumptions. But unlike deduction, it does not prove theorems. Instead, an agent-based model generates simulated data that can be analysed inductively. Unlike typical induction, however, the simulated data come from a rigorously specified set of rules rather than direct measurement of the real world. Whereas the purpose of induction is to find patterns in data and that of deduction is to find consequences of assumptions, the purpose of agent-based modelling is to aid intuition*'.]

2.2 Reductionism and its inadequacy for dealing with complexity

> *With reductionism comes the conviction that a court proceeding to try a man for murder is "really" nothing but the movement of atoms, electrons, and other particles in space, quantum and classical events, and ultimately to be explained by, say, string theory.*
>
> S. Kauffman

Complexity science compels us to take a fresh look at how we have been doing much of our science so far, namely by the reductionistic approach. Complexity science dares to look at research problems (the 'Big Questions') too difficult to handle by conventional science. Naturally, a fresh look at how Nature has to be investigated is needed. Reductionism is a case in point.

Reductionism is the philosophy that the explanation of all phenomena can be *reduced to* their simplest components. The basic premise is that everything can be *ultimately* explained in terms of the bottom-level laws of physics. The spirit behind this approach is that the universe is governed by natural laws which are fixed and comprehensible at the fundamental level. This works well when we are dealing, for example, with *linear* systems. Such systems obey the *principle of linear superposition*, which says that, if x_1 and x_2 are two solutions of the dynamical equation describing a system, then $C_1 x_1 + C_2 x_2$ is also a solution, where C_1 and C_2 are constants. This is not true for nonlinear dynamical systems, so reductionism breaks down for them, unless the nonlinearity is of a mild and manageable nature. One class of systems wherein this is certainly not true are what are called *chaotic systems* (to be discussed later); they are strongly nonlinear. Similarly, it is impossible to understand the functioning of a brain reductionistically.

Reductionism has excellent validity where it is applicable, namely for simple or *simplifiable*, rather than complex, systems. As I explained in the Preface, for simple or simplifiable systems, it is realistic to identify a quasi-isolated subsystem for investigation, assuming that what is not a part of the subsystem is just some (large) 'heat bath' or 'background'. It is also presumed that

the principle of causality holds for the subsystem, meaning that the effect-cause-effect-cause-
... chain of events can be traced or *reduced* all the way down to the 'ultimate cause(s)'. It is
implicitly assumed that 'the whole is *equal* to the sum of its parts'. Reductionistic science has
had remarkable successes in predicting, for example, the occurrence of solar eclipses with a
very high degree of precision in both space and time (Fig. 2.1).

Fig. 2.1 The solar eclipse of 15 January 2010, as recorded from Dhanushyakodi (near Rameshwaram),
Tamil Nadu, India.
Image credit: Vinayak Kolvankar, BARC, Mumbai.

An approach related to reductionism is *constructionism* (Anderson 1972), which says that we
can start from the laws of physics and *predict* all that we see in the universe. Both reductionism
and constructionism assume the availability of data of infinite precision and accuracy, as well
as unlimited time and computing power at our disposal, and are therefore unrealistic
assumptions for application to many of the complex systems we encounter in practice. What is
more, random events at critical junctures in the evolution of complex systems can make it
impossible for us to make meaningful predictions. The 'Laplace demon' construct helps us
understand this.

2.3 The Laplace demon

> *Occurrences in this domain are beyond reach of exact prediction because of*
> *the variety of factors in operation, not because of any lack of order in nature.*
> Albert Einstein

Imagine a superintelligent and superhuman creature (often called the Laplace demon) who
knows at one instant of time the position and momentum of every particle in our universe, as
also the forces acting on each particle. Assuming the availability of a good enough
supercomputer, is it possible for the Laplace demon to predict the future in every detail? The
answer would be 'yes' according to the tenets of constructionism, provided *unlimited*
computational power and time are available. In reality, there are limits on the speeds of
computation, as well as on the extent of computation one can do. These limits are set by the
laws of physics, and also by the limited nature of the resources available in the universe
(Anderson 1972; Williams 1997; Lloyd and Ng 2004; Davies 2005; Chaitin 2006; Lloyd 2006).
Here are some of the reasons for this:

- The *bit* is the basic unit of information, and the *bit-flip* the basic operation of information processing. It costs energy to process information. Energy and time are related through the Heisenberg uncertainty principle of quantum mechanics. This principle puts a lower limit on the time needed for processing a given amount of energy and information.

- The finite speed of light puts an upper limit on the speed with which information can be exchanged among the constituents of a processor. [It is an experimental fact that no object, wave, or signal can move faster than light. This means that information cannot be exchanged or processed at speeds greater than that of light.]

- A third limit is imposed by entropy which, as we shall see later in the book, is a measure of unavailable information: One cannot store more bits of information in a system than permitted by its entropy.

- The universe is believed to have begun with the Big Bang, ~13.7 billion years ago. Therefore, light cannot have traversed distances greater than 13.7 billion light years [a light year is the distance that light travels in one year]. Regions of space separated by larger distances than that cannot have a causal relationship. And the Laplace demon cannot have a supercomputer larger than the size of the universe (Lloyd 2006).

- Most of the interactions underlying natural phenomena are nonlinear, rather than linear. This, as we shall see in a later chapter, can make the dynamics of even deterministic systems unpredictable.

- In the time-evolution of many systems, there are events which are necessarily random, and therefore cannot be predicted, except in probabilistic terms.

Thus there are limits on available computational power. Predictions based on the known laws of physics but requiring larger computational power than the limits stated above are not possible. In any case, predictions (which are always based on data of *finite* precision) cannot have unlimited precision, not even for otherwise deterministic situations (e.g., if the system is in the so-called *chaotic regime*). The implication of this conclusion for complex systems is that, beyond a certain level of *computational complexity*, new and unexpected organizing principles can arise (see Duke 2006): If the fundamental-level physical laws cannot apply for completely determining the future states of a complex system, then higher-level *laws of emergence* may need to be recognized and used for rationalizing what is observed.

A striking example of this type of *strong emergence* is the origin and evolution of life. From the deterministic point of view, it is a computationally intractable problem. Therefore new, higher-level laws, different from the bottom-level laws of physics and chemistry, might have played a role in giving to the genes and the proteins the functionality they possess at present.

Complexity beyond a certain threshold can lead to the emergence of new principles (Prigogine 1996).

Complex systems usually have a *hierarchical structure*. The new principles and features observed at a given level of complexity may sometimes be deducible from those operating at the previous lower level of complexity (constructionism). Similarly, starting from a given observed level of complexity, one can sometimes work backwards and infer the previous lower level of complexity (reductionism). But reductionism and constructionism have only limited (localized) ranges of applicability for complex systems. Chaotic systems (discussed later in the book) provide a particularly striking example of this. The Laplace demon is not capable of predicting *on a long-term basis*, for example, the weather of a chosen region, nor can he start from the observed weather pattern at a given instant of time and work out the positions and momenta of all the molecules.

Anderson (1972), in his famous '*More is Different*' paper, emphasized the emergence of complexity in a variety of condensed-matter systems. As happens even now, scientists sometimes tend to take the validity of both reductionism and constructionism for granted. Anderson pointed out that:

. . . the reductionist hypothesis does not by any means imply a constructionist one: The ability to reduce everything to simple fundamental laws does not imply the ability to start from those laws and reconstruct the universe. In fact, the more the elementary particle physicists tell us about the nature of the fundamental laws, the less relevance they seem to have to the very real problems of the rest of science, much less to those of society. The constructionist hypothesis breaks down when confronted with the twin difficulties of scale and complexity. The behaviour of large and complex aggregates of elementary particles, it turns out, is not to be understood in terms of a simple extrapolation of the properties of a few particles. Instead, at each level of complexity entirely new properties appear.

A common thread running through the behaviour of all complex systems is the breakdown of the principle of linear superposition: Because of the nonlinearities involved, a linear superposition of two solutions of an equation describing a complex system is not necessarily a solution. This fact lies at the heart of the failure of the reductionistic approach when it comes to understanding complex systems.

Reductionism and constructionism may be relevant, even highly useful, when applied to two neighbouring or contiguous levels of complexity. We can understand quite well how elementary particles come together to form atoms, or how atoms come together to form molecules, or how molecules come together to form supramolecular assemblies, and so on. But, starting from elementary particles and their known properties, we cannot predict the formation of supramolecular assemblies or life forms. The whole is more than the sum of its parts.

2.4 Holism

> *You see, "Mu" is an ancient Zen answer which, when given to a question, UNASKS the question. Here, the question seems to be, "Should the world be understood via holism, or via reductionism?" And the answer of "Mu" here rejects the premises of the question, which are that one or the other must be chosen. By unasking the question, it reveals a wider truth: that there is a larger context into which both holistic and reductionistic explanations fit.*
>
> Douglas Hofstadter, *Gödel, Escher, Bach*

> *Zen is holism, carried to its logical extreme. If holism claims that things can only be understood as wholes, not as sums of their parts, Zen goes one further, in maintaining that the world cannot be broken into parts at all. To divide the world into parts is to be deluded, and to miss enlightenment.*
>
> Douglas Hofstadter, *Gödel, Escher, Bach*

The term 'holism' was coined in 1929 by J.C. Smuts from the Greek word 'holos', meaning *all* or *whole*. Smuts defined holism as '*The tendency in nature to form wholes that are greater than the sum of the parts through creative evolution*'. According to Smuts, '[Holism] regards natural objects as wholes. . . . It looks upon nature as consisting of discrete, concrete bodies and things . . . [which] are not entirely resolvable into parts; and . . . which are more than the sums of their parts, and the mechanical putting together of their parts will not produce them or account for their characters and behaviour'.

Holism is the belief that, since everything interacts with everything else, the whole is generally greater than the sum of its parts, and it is impossible to understand a system as simply a superposition of its parts. The entire system must be considered as a whole. Thus, holism really means 'wholism'. The system as a whole determines how the parts behave (see Wahl: https://medium.com/age-of-awareness/understanding-complexity-a-prerequisite-for-sustainable-design-fd45990e3bd6.) Everything interacts with everything else, and therefore *two plus two can be different from four.*

Holism has a *weak* interpretation and a *strong* interpretation (Simon 1996), depending on whether the related *emergence* of new phenomena is weak or strong (see below).

2.5 Emergence

> Given the idea that neo-Darwinism-derived complexities might be superposed on those more conventionally generated by energy alone, perhaps we can agree that biology is physics with added features. That may be why, in physics, the simplest model for any particular phenomenon is usually the right one, whereas in biology historical accidents and accumulating complexity often invalidate the influence of Ockham's razor. That added degree of complexity might also explain why biology has no natural laws per se, only a set of coarse guidelines.
>
> Chaisson, *Cosmic Evolution*

The word *emergence* was coined in 1875 by the psychologist George Henry Lewes in his book *Problems of Life and Mind*, to refer to the appearance of novel (unpredicted) phenomena in a complex system. He wrote: 'Every resultant is either a sum or a difference of the co-operant forces; their sum, when their directions are the same – their difference, when their directions are contrary. Further, every resultant is clearly traceable in its components, because these are homogeneous and commensurable. It is otherwise with emergents, when, instead of adding measurable motion to measurable motion, or things of one kind to other individuals of their kind, there is a co-operation of things of unlike kinds. The emergent is unlike its components insofar as these are incommensurable, and it cannot be reduced to their sum or their difference'.

In a complex system, fundamental units get self-organized on many distinct *hierarchical levels*. Unpredictable or unexpected new properties can arise when the fundamental units or constituents in a lower level become organized into higher levels or higher order. This is when we speak about emergent phenomena.

Weak emergence

Consider a single isolated object. It has certain properties which give it an identity. But certain other properties of it become meaningful (or arise) only when there is another object around, with which it can interact. An example is gravitational phenomena. We need two objects to speak of a gravitational interaction between them. Gravitation is an example of a weak form of emergence: It is something new in the sense of being not present when only one object was there, but it does not negate the applicability of reductionism here. Weak emergence simply means that the parts of a complex system have mutual relations that do not exist for the parts in isolation.

An enzyme is a protein molecule that catalyses the linking of two molecules. Naturally, it has to be in an environment in which the molecules to be linked are present. Thus, although the enzymatic catalytic action is an emergent property, it can be given a reductionistic explanation in terms of the basic physico-chemical processes among the molecules involved. This is another example of weak emergence; it does not pose any threat to the reductionistic way of doing science.

It is convenient to introduce new terms like 'voltage' that are defined by relations among observable quantities (think of Ohms law). Such terms are introduced to avoid reference to the details of the component subsystems, referring only to the aggregate properties (Simon 1970). There is often a near-independence of hierarchical systems from the details of their component subsystems. Also, there may be a short-run independence of the subsystems from the slower movements of the total system. By adopting this weak-emergence approach one can build nearly independent theories for each successive level of complexity. What is more, one can also build bridging theories that explain how each higher level can be accounted for in terms of the elements and relations of the next level below. Reductionism is often applicable between two contiguous hierarchical levels of complexity.

Strong emergence and holism

If the emergent phenomena in a higher level of complexity cannot be reduced to (cannot be explained in terms of) the properties of the lower level(s) of organization, one speaks of strong emergence.

Strong emergence is associated with the strong interpretation of holism, and has had even some run-away unsavoury unscientific consequences, particularly in the context of living systems. For living systems, some proponents of the strong interpretation of holism claim that 'the putting together of their parts will not produce them or account for their characters and behaviours'. This implies a *vitalism*, that is totally unacceptable in modern molecular biology.

The debate becomes even more shrill when strong holism is used by some people for understanding the nature of the mind. They use it to support both the claim that machines cannot think and the claim that thinking involves more than the arrangement and behaviour of neurons.

When applied to complex systems in general, proponents of this special (unscientific) brand of strong holism postulate new properties and relations among the subsystems that did not exist in the components of the system components. They reject what they call the 'mechanistic' explanations of strong emergence, and invoke a 'creative' principle. But it so happens that the so-called *Ockham's razor* (Chapter 6) is always there in science to cut and slash such mumbo jumbo.

2.6 Scientific determinism, effective theories

> *Quantum physics might seem to undermine the idea that nature is governed by laws, but this is not the case. Instead it leads us to accept a new form of determinism: given the state of a system at some time, the laws of nature determine the* probabilities *of various futures and pasts rather than determining the future and the past with certainty.*
>
> Hawking and Mlodinow (2010)

The discussion in the previous section has the underpinnings of 'scientific determinism', and points to the repeated use of what are called 'effective theories'. Laplace was perhaps the first to clearly enunciate the basic tenets of scientific determinism, according to which, *given the state of the universe at one instant of time, a complete set of natural laws fully determines both the future and the past* (see Hawking and Mlodinow 2010). That was *classical* scientific determinism. Since then we humans have made the fundamental discovery that all natural phenomena are governed by the laws of quantum mechanics, and that laws of classical mechanics are subservient to those of quantum mechanics, applicable in certain special situations only. This fact has necessitated a sea change in our outlook: We now speak the language of probabilities, rather than certainties, as also of *several* futures and pasts, rather than *the* future and *the* past. Scientific determinism now has a quantum flavour.

Effective theories

As we have seen above, real-life situations are so complex that it is not enough to have knowledge of the 'complete set of fundamental laws of physics' for understanding all natural phenomena. Knowledge of the laws of physics may, *in principle*, be adequate for understanding physical phenomena, but *in practice* it is found essential to formulate many an additional (empirical) law as an *effective theory*.

An example is the gravitational force experienced by a macroscopic object on the surface of the Earth. The gravitational interaction is present between any two atoms, but we cannot formulate and solve exactly the equations governing the gravitational interaction between every atom in the macroscopic object and every atom in the Earth. Instead, an effective theory is formulated in terms of the object's mass and a few other numbers like the value of the gravity constant at the surface of the Earth.

Similarly, in chemistry we cannot hope to formulate and solve the totality of equations describing the electric interactions among all the positive and negative charges in a system. Instead, an effective theory involving concepts like valence deals with how chemical reactions occur.

This approach continues as we go up the ladder of increasing complexity. Details at one hierarchical level of complexity are 'summarized' or 'integrated over' to generate some *effective parameters* which are used for describing the details of the next higher level: From particle physics to macroscopic physics and chemistry; from chemistry to biology; and so on. An effective theory is essentially a framework we create for modelling certain observed phenomena, without describing in detail all the underlying processes.

2.7 Free will

> *Common sense inclines, on the one hand, to assert that every event is caused by some preceding events, so that every event can be explained or predicted ... On the other hand, .. common sense attributes to mature and sane human persons ... the ability to choose freely between alternative possibilities of acting.*
>
> Karl Popper

The computational limit to understanding the reality of our complex universe manifests itself in a dramatic way when we ponder over the notion, or probably the *illusion*, of free will. Michael Shermer (2006) has described it very succinctly:

As with the God question, scholars of considerable intellectual power for many millennia have failed to resolve the paradox of feeling free in a determined universe. One provisional solution *is to think of the universe as so complex that the number of causes and the complexity of their interactions make the predetermination of human action pragmatically impossible. We can even assign a value to the causal net of the universe to see just how absurd it is to think we can get our minds around it fully. It has been calculated that in order for a computer in the far future of the universe to resurrect in a virtual reality every person who ever lived or could have lived (that is, every possible genetic combination to create a human), with all the causal interactions between them and their environment, it would need . . . . 10^{23} bits of memory. Suffice it to say that no computer in the conceivable future will achieve this level of power; likewise, no human brain even comes close.*

The enormity of this complexity leads us to feel as though we were acting freely as uncaused causers, even though we are actually causally determined. Since no set of causes we select as the determiners of human action can be complete, the feeling of freedom arises out of this ignorance of causes. To that extent, we may act as though we were free. There is much to gain, little to lose, and personal responsibility follows.

The handy notion of free will is actually one more effective theory, like the many mentioned in the previous section. Even though free will is, in all probability, only an illusion, one can often make good progress in cataloguing and understanding psychological phenomena by *pretending* that people have free will (Hawking and Mlodinow 2010).

One goes a step further when modelling economics. The effective theory we use there is that (i) people have a free will; (ii) their behaviour may or may not be rational; and (iii) their decisions may be sometimes based on a defective analysis of the limited data at their disposal.

Although the free-will notion serves as a useful and convenient effective theory, we should not lose sight of the fact that biological processes, including those occurring in the brain, are governed by (and only by) the laws of physics and chemistry. Our actions are therefore governed only by the laws of physics and chemistry. Although assessments and opinions differ, a number of experiments done on humans under controlled conditions have demonstrated this. For example, electrical stimulation of appropriate regions of the brain can make a person '*want*' to move the hand, arm, or foot. Such findings are in line with the scientific idea of physical causes leading to physical effects, and go against the genuineness or fundamental validity of the concept of free will (however, see Hameroff 2012).

In philosophy, two opposite positions have been in vogue: *compatibilism vs. incompatibilism*. The former accepts the possibility of *both* determinism and free will. The philosopher Daniel Dennett subscribes to it (see *Stanford Encyclopaedia of Philosophy* 2009). And incompatibilism says that determinism and free will are incompatible. This leads to three possibilities: (i) Choose determinism (*hard determinism*). (ii) Choose free will (*metaphysical libertarianism*). (iii) Reject both determinism and free will (*hard indeterminism*).

According to Smith (2011): 'Haynes, a neuroscientist at the Bernstein Centre for Computational Neuroscience in Berlin, put people into a brain scanner in which a display screen flashed a succession of random letters. He told them to press a button with either their right or left index fingers whenever they felt the urge, and to remember the letter that was showing on the screen when they made the decision. The experiment used functional magnetic resonance imaging (fMRI) to reveal brain activity in real time as the volunteers chose to use their right or left hands. The results were quite a surprise. . .. The conscious decision to push the button was made about a second before the actual act, but the team discovered that a pattern of brain activity seemed to predict that decision by as many as seven seconds. Long before the subjects were even aware of making a choice, it seems, their brains had already decided.'

Smith argues that consciousness of a decision may be a mere *biochemical afterthought*, with no influence on a person's actions, so free will is an illusion.

The neuroscientist V. S. Ramachandran is rather ambivalent on the question of free will being an illusion. In his 2010 book *The Tell-Tale Brain* he talks about a patient suffering from 'ideomotor apraxia', which is an inability to perform suggested skilled actions. This is what he writes on page 131: 'What he lacks is the ability to conjure up a mental picture of the required action – in this case combing – which must precede and orchestrate the actual execution of the action. These are functions one would normally associate with mirror neurons'. Perhaps it is the *conjuring-up* part (a *semi-conscious* action?) which is showing up in the recorded neural activity *before* the actual action in the above-mentioned experiment by Haynes.

Ramachandran also makes the point that, since we have a huge number of neurons in the cortex, there is always a lot of activity going on there, although we are conscious of only a very small part of it. Our information is being processed by the brain all the time, and many alternative decisions are being thrown up into our consciousness. Our conscious brain selects one or the other of these alternatives. Therefore it is more a case of 'free won't' rather than 'free will'.

What is more, since a lot of information processing occurs (unconsciously and then consciously), and that takes time, it is wrong to associate only the last stage with the moment of decision-making. Kurzweil (2012) gives an analogy to explain this:

'Consider the analogy to a military campaign. Army officials prepare a recommendation to the president. Prior to receiving the president's approval, they perform preparatory work that will enable the decision to be carried out. At a particular moment, the proposed decision is presented to the president, who approves it, and the rest of the mission is then undertaken. Since the "brain" represented by this analogy involves the unconscious processes of the neocortex (that is, the officials under the president) as well as its conscious processes (the president), we would see neural activity as well as actual actions taking place prior to the official decision's being made. We can always get into debates in a particular situation as to how much leeway the officials under the president actually gave him or her to accept or reject a recommendation, and certainly American presidents have done both. But it should not surprise us that mental activity, even in the motor cortex, would start before we were aware that there was a decision to be made'.

Sam Harris (2012) has argued at length about why free will is just an illusion. Here are some further points made in his book, which are of practical importance:

• The realization that free will is an illusion makes us more humane when it comes to reacting to crime, hatred, etc. How can you hate a person when you know that he/she is not *fully* in control of the bad or undesirable behaviour?
• Nevertheless, when somebody commits a crime, society must still react effectively, not in the spirit of punishment or retribution or hatred, but for providing deterrence against future crime.
• Being aware of the contribution of the unconscious to our apparently 'freely chosen' actions can make us more responsible and sensible about how we react to situations and take decisions.

The free-will example is typical of what happens in many complex systems. Consider the following three questions:

(i) What will be the weather pattern of our world exactly six months from now?

(ii) What will be the values of all the shares in a stock market exactly six months from now?
(iii) What deliberate action Mr. X will be taking exactly six months from now?

These are all questions about complex systems involving a huge number of, respectively, interacting matter and forces, interacting people, and interacting neurons. Only probabilistic answers can be given, if at all, and this is in spite of the fact that entirely deterministic processes may be involved at every micro-step in the time and space evolution of the system. The power and relevance of the rather young science of complex systems is that we can learn about certain general trends, commonalities, and broad features by investigating a variety of complex systems. This is a highly worthwhile exercise because, after all, the questions being examined are some of the most profound and relevant, the kind of questions normally left unattended to by conventional reductionistic science.

2.8 Actions, reactions, interactions, causality

> *Simple cause and effect relationships only exist in theory, not in reality.*
> Frederic Vester

The question of free will, discussed above, is linked to that of causality. The *causality principle* says that every phenomenon or event is an 'effect' for which there must be a 'cause' that precedes it. Actually this principle is only an effective theory, although very useful in many day-to-day situations, particularly when it comes to understanding simple (or simplifiable) *macroscopic* systems. But an effective theory may not always have a strict logical basis, and may not be universally applicable.

We generally tend to subscribe to the idea that any action is followed by a reaction. In fact, we have 'Newton's 3rd law of motion' which says that to every action there is an equal and opposite reaction. But can we always talk in terms of actions and reactions, or causes and effects? No.

Consider two protons (identical positively charged particles), moving towards each other (Fig. 2.2a). The force of repulsion between them is not much when they are far apart, but increases as they approach each other, resulting in a bending or reversal of their trajectories. They cannot get too close to each other because of the repulsion, and go their separate ways after coming as close as they can (Fig. 2.2b). Both proton trajectories have been affected by such an encounter. Can you tell which proton is the cause and which is the effect? No. Instead of cause and effect, or action and reaction, it makes better sense here to talk only of an *interaction*. Bring in a third proton and it would be even easier for you to agree with me.

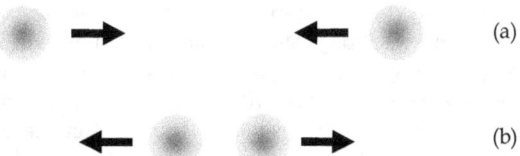

Fig. 2.2 Two protons approaching each other (a). They cannot come too close because of the Coulombic repulsion between them. Depending on their relative velocities, they come as close as possible, and then reverse their direction of movement (b).

Cut to our solar system. Does the Sun go around the Earth, or does the Earth go around the Sun? There are complications because of the Moon and other planets etc. (and other celestial bodies), so imagine a simpler situation in which we have only the Sun and the Earth. Most people will say that it is the Earth that goes around the Sun. The psychology of such an answer is that the Sun is much heavier than the Earth. But the reality is that the two go around each other. There is a *centre of mass* (c.m.) for the Sun-plus-Earth system taken as whole, and they both go around that point. [Of course, it is also true that the Sun has so much more mass compared to the Earth that the c.m. of the Sun is very close to the c.m. of the composite Sun-plus-Earth system; therefore for many practical purposes it is a good approximation to assume that the two coincide.]

The human tendency is that the larger object is generally taken as causing the effect on the smaller object(s). The real thing is that there are only interactions, rather than actions and reactions always. Philosophers have been tying themselves into knots by carrying the action-reaction or causality idea too far. The absurdity of it all becomes palpable when they even talk of 'downward causality'. While the causality principle is very useful (even valid in most situations) as an effective theory, it is a good idea to remember that, deep down under, we just have interactions, rather than actions (causes) and reactions (effects).

The cause-effect notion is convenient (even valid) to use in a large number of practical situations, with the proviso that the effect never precedes the cause. And the last part of this statement is further subject to what the special theory of relativity (to be introduced in a later chapter) demands, namely that a signal from the cause cannot travel faster than the speed of light. What this implies is that the meaning of the word 'simultaneous' is observer-dependent. The cause precedes the effect for all so-called 'inertial' observers. The cause and the effect are separated by a 'timelike' interval, and the effect is in the future of the cause.

And according to the general theory of relativity, the effect must belong to the future 'light cone' of its cause, even when the 'spacetime' is curved. More on this later.

Consider a beehive. It is a complex system. It has *swarm intelligence*. No one is in command, not even the queen bee. Each bee follows some very simple *local rules*, and interacts with other bees in the hive. The effect here is the *emergent* property of swarm intelligence. What is the 'ultimate' cause of this intelligence? Not the action of any one bee. The beehive is the archetypal example of a system in which it is often meaningless to talk about causes and effects, or actions and reactions. It is *interactions*, through and through. This is not an isolated example. Complex systems are generally like that.

The causality idea is well-entrenched in the human psyche, in spite of the above-mentioned limiting provisos. In fact, much of our conventional science is based on it. Logical reasoning in conventional science is one big chain of cause-effect-cause-effect- . . . . interpretations: An effect becomes the cause for the next event or process, and so on. But conventional science is often quite unfit for tackling complexity-related, highly nonlinear, problems. Radically new thinking is needed when the situation is such that any simplifying assumption can destroy the very essence of the complex system being investigated, or when it is impossible to model a system in terms of an adequate number of differential equations (try modelling life processes in terms of an adequate number of differential equations!). But some of the unconventional, alternative, approaches formulated in complexity science have met with strong resistance from many practitioners of conventional analytical science. Mindsets do not change easily.

We should be prepared to think in terms of interactions and correlations when necessary, rather than actions and reactions all the time. Such an approach will help us better understand a large number of natural phenomena, and keep us away from absurdities like 'downward causality'.

Time to make a technical point here, and it is regarding emergence: The point I make is that *the causality principle is an emergent principle*. We have seen above that causality does not exist when we have just two interacting protons or other elementary particles. When the number of interacting entities is increased more and more, a stage can come when cause and effect can be distinguished, at least in a practical effective sense, and the causality principle is then said to have *emerged* as a new operative principle.

2.9 The nature of reality

Before the advent of quantum physics, we had classical physics, and the view of reality was as follows: Objects exist and have well-defined attributes like speed and mass which are absolute, in the sense that they are properties of the objects and are independent of who is observing them, or even whether somebody is observing them at all. This was classical deterministic realism, or *objectivism*.

This view of reality underwent a sea change when classical physics was superseded by quantum physics. At present we all agree that, in accordance with the principles of quantum mechanics, an elementary particle does not have a definite position and velocity: We cannot tell its position and velocity with arbitrarily high precision. Therefore we can no longer say that the particle 'really' has specific attributes.

A lucid discussion of the meaning of 'reality' has been given by Hawking and Mlodinow (2010). They take the view that '*there is no picture- or theory-independent concept of reality*'. They argue in support of what they call '*model dependent realism*', and define it as the idea that 'a physical theory or world picture is a *model* (generally of a mathematical nature) and a set of rules that connect the elements of the model to observations'. Science should be interpreted in this framework only. Any deeper or absolute meaning of reality is not possible.

One may argue that, although in the microscopic world of elementary particles it is necessary to abandon the classical picture, classical physics is fine when we are dealing with the macroscopic world. For example, if there is a fruit-laden tree in a garden with no birds etc. around, the tree is there no matter who is watching it, or even if nobody is watching. It is futile to say that the tree may or may not be there when nobody is watching it. Suppose a bird came when no observer was around and ate up some of the fruits and dropped some more on the ground. When a human sees the aftermath, he/she can hardly think of saying that the tree does not exist if nobody is watching. Opinion is sharply divided on such philosophical issues and, according to Hawking and Mlodinow (2010), one can salvage the situation by adopting the model, or *effective theory*, that the tree has an existence which is an absolute reality *for all practical purposes*. This is reminiscent of David Hume's statement that, although we have no rational grounds for believing in objective reality, we also have no choice but to act as if it is true.

According to this model-dependent realism, all that matters is whether or not a model agrees with observation. There may be more than one models which agree with what we have observed. It is not always as if one model is better at describing 'reality' than the other, only more convenient and economical in content in a chosen frame of reference. For example, consider the question: Does something or somebody exist when we are not viewing it? There are two opposite models (materialism *vs.* idealism) for answering this. Which model of reality is correct? Naturally the one that is *simpler*, *self-consistent*, and *most successful* in terms of its predicted consequences. I think this is where materialism wins hands down. The materialistic model is that the entity exists even when nobody is observing it. This model is far more successful in explaining reality than the opposite model. And we can do no better than build models of whatever there is to observe, understand, and explain.

Suppose 100 persons are asked to describe an object, including its colour, and all of them say that it is a chair. Further, suppose 98 of them say that it is a *red* chair, but the other two disagree about the colour seen by the majority. If further investigation shows that these two persons have a colour-blindness problem, the model of reality we humans build is that the object is a red chair.

But suppose it turns out that these two persons are *not* colour blind, and no matter what tests we carry out, we are unable to explain why they do not see or describe the chair as red. We then go (tentatively) by the majority view, or *consensus*. Of course, any model of reality can change in the light of new data and insights. This is the approach we adopt in science for building up our knowledge. We build models and theories of reality, and we accept those which are most successful in explaining what we humans observe collectively *at that time*.

A model is a good model if it is elegant and self-consistent; it contains no or only a few arbitrary or adjustable parameters; it explains most or all of the existing observations; and it makes detailed and falsifiable predictions (Hawking and Mlodinow 2010).

Reality is nothing deeper than the best available scientific model for it. Often a phenomenon or entity is so complex that no sensible model has been formulated yet. In such a case, we have to wait till science makes more progress.

Reality is not time-independent. Consider our current understanding of the origin of our universe. Scientists agree, by and large, that the current epoch of our universe began with a Big Bang, ~13.72 billion years ago. This model of reality stands on three pillars: (I) the observed 'Hubble expansion' of the universe; (ii) the observed cosmic-wave background; and (iii) the good agreement between the predicted and observed relative abundances of the light elements hydrogen, helium, and lithium. Lawrence Krauss (2012) has discussed them in detail in his book *A Universe from Nothing*. I shall consider only the first of them here, as explained by Krauss (2012).

Observations show that the rate of expansion of our universe is increasing. What is more, the observable universe is expanding at present at a rate that is not much lower than the speed of light. Galaxies which we can see today will one day recede from us at a speed greater than the speed with which even the fastest possible signals (electromagnetic radiation) from them can reach us. *They will then become invisible to us, permanently*. This will happen ~2 trillion years into the future.

We and our solar system will die out in ~5 billion years. But other civilizations and advanced science can emerge elsewhere in the future. Suppose you are one of those astronomers in the future. When you look around the cosmos with your highly advanced telescopes, you would see *none* of the ~400 billion galaxies we humans see today. What will be your model of reality then? Certainly not the same as ours at present. *There is no such thing as absolute, unique, invariant reality*.

We are 'lucky' to be living in a period when evidence for the Big Bang is available. Astronomers far enough into the future will have *no evidence* and *no reason* to believe that the Big Bang occurred at all, and that there were galaxies other than their own:

We live at a very special time . . . the only time when we can observationally verify that we live at a very special time! (Krauss 2012).

3. The Second Law of Thermodynamics

In thermodynamics we aim to describe the bulk behaviour of systems in terms of a few measurable parameters like pressure, volume, and temperature. A clear distinction is made between *isolated systems* and *open systems*. The former are those which cannot exchange matter or energy with the surroundings, and the latter are those which can.

3.1 The second law for isolated systems

A basic familiarity with the second law of thermodynamics is absolutely essential for understanding the evolution of complexity in Nature. To get a feel for what the law says, let us first consider the 'free expansion' of a gas. Imagine some gas in a chamber (labelled A in Fig. 3.1(a)), separated by a partition from another chamber (B) of the same volume. This second chamber is empty to start with. So there is gas on one side of the partition and vacuum on the other. If the partition disappears, the molecules of the gas would be able to move into the right half of the enlarged chamber also (Fig. 3.1b), and soon the gas will occupy the entire (doubled) volume uniformly.

This expansion of the gas will occur even if there is no force of repulsion among the molecules. It is purely a consequence of their kinetic energies, random motions, and collisions with one another and with the walls of the chamber.

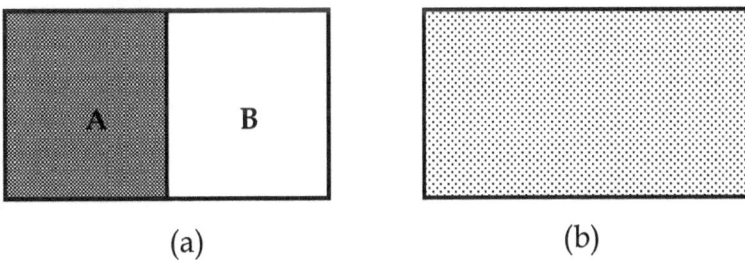

(a)   (b)

Fig. 3.1 Free expansion of a gas in vacuum. The region to the right of the partition in (a) is vacuum. When the partition is removed, as in (b), the gas expands spontaneously (*and irreversibly*) to occupy the additional volume available to it.

The free expansion of the gas is governed by random chance processes. Let us say that there are n molecules of the gas. Before the partition is removed, all the molecules are in the left half of the enlarged chamber (Fig. 3.1(a)). So the probability of finding any of the molecules in the left half is 100%, and it is zero for finding that molecule in the right half. After the partition has been removed (Fig. 3.1(b)), there is a 50% chance of finding a molecule in the left half, and 50% chance for finding it in the right half. It is like tossing a coin, and saying that we associate 'heads' with finding the molecule on the left, and 'tails' with finding it on the right. In both cases the chance is 50% or ½.

Next let us ask the question: What is the probability that all the n molecules of the gas will ever occupy the left half of the chamber again? This probability is the same as that of flipping a coin n times and finding 'heads' in each case, namely $\frac{1}{2}^n$.

Considering the fact that usually n is a very large number (typically of the order of the Avogadro number, i.e. $\sim 10^{23}$), the answer is very close to zero indeed. In other words, the free expansion of the gas is an *irreversible* process, for all practical purposes. On removal of the partition, the gas has *spontaneously* gone into a state of greater *disorder*, and it cannot spontaneously go back to the initial state of order. Why do we say 'greater disorder'? It is because the probability of finding any specified molecule at a particular location in the left half of the chamber is now only half its earlier value. Here, by 'order' we mean that there is a 100% chance that a thing is where we expect it to be, so 50% chance means a state of less order, or greater disorder.

So intuitively we have no trouble agreeing that, left to themselves (with no inputs from the outside), things are more likely to tend towards a state of greater disorder. This is all that the second law of thermodynamics says for isolated systems. It says that *if we have an isolated system, i.e.,. a system which cannot exchange energy or matter with the surroundings, then, with the passage of time, it can only go towards a state of greater disorder on its own, and not a state of lesser disorder.* This happens because, as illustrated by the free-expansion-of-gas example above, a more disordered state is *more probable* than a less ordered state.

The key phrase in the above statement is '*isolated* system'. I shall have much to say about what happens when the system is *not* an isolated one, but later. In fact, it is practically impossible to find a system which does not exchange energy and/or matter at all with the surroundings.

3.2 Entropy

> *(Entropy quantifies) how dispersed the energy is among the particles in a system, and how diffuse those particles are throughout space. It increases as a simple matter of probability: There are more ways for energy to be spread out than for it to be concentrated. Thus, as particles in a system move around and interact, they will, through sheer chance, tend to adopt configurations in which the energy is spread out. Eventually, the system arrives at a state of maximum entropy called "thermodynamic equilibrium," in which energy is uniformly distributed.*
>
> Emily Singer (2014)
>
> https://www.quantamagazine.org/a-new-thermodynamics-theory-of-the-origin-of-life-20140122/

There is a clear need to quantify the extent or degree of disorder, or rather the difference of disorder between one state and another. The concept of entropy (S) does just that. Technical details are given in Appendix A1. Here I shall just give you a feel for how it is done.

Let us go back to the free-expansion experiment depicted in Fig. 3.1. How much has the disorder of the gas increased on free expansion to twice the volume? Consider any molecule of the gas. After the expansion, there are twice as many positions at which the molecule may be found. And at any instant of time for any such position, there are twice as many positions at which a second molecule may be found, meaning that the total number of possibilities for the two molecules is now 2×2, or 2^2. Thus for n molecules there are 2^n ways in which the gas can fill the chamber after the free expansion. We say that in the double-sized chamber the gas has 2^n more *accessible states*, or *microstates*.

The symbol W is normally used for the number of microstates accessible to a system. This number doubled when the gas expanded to twice the volume. So, one way of quantifying the degree of disorder could be to say that entropy S, a measure of disorder, is proportional to W: $S \propto W$. Ludwig Boltzmann, who did pioneering work in establishing this field called *statistical thermodynamics*, did something even better than that for quantifying disorder. He defined entropy S as proportional to the *logarithm* of W ($S \propto \log W$). In his honour, the constant of proportionality (k_B) is now called the Boltzmann constant. Thus entropy is defined by the famous equation $S = k_B \log_2 W$ (or just $S = k \log W$). In the public mind, this 'Boltzmann equation' is just about as familiar as the Einstein equation $E = mc^2$.

To see the merit of introducing the logarithm in the definition of entropy, let us apply it to calculate the increase of entropy when the gas expands to twice the volume. Since $W = 2^n$, we get $S = n$. This makes sense. Introduction of the logarithm in the definition of entropy makes it, like energy or mass, a property *proportional* to the number of molecules in the system. Such properties are described as having the *additivity* feature. If the number of molecules in the gas is doubled from n to $2n$, it makes sense that the entropy also doubles in a linear fashion.

3.3 The second law for open systems

We need energy to fight against entropy.
Albert Szent-Györgyi

If disorder is always more likely than order, why should any order arise at all? *Globally* speaking, i.e., if we can regard the entire universe as one big *isolated* system, disorder must go on increasing with time because that is the most probable thing to happen. But *locally*, conditions can exist which favour the creation of order. [In fact, even globally, since our universe is expanding and cooling, and thus creating gradients, order can emerge because processes occur that tend to annul the gradients.]

A simple example makes the point. Take some water in a beaker and go on adding common salt to it gradually. In the beginning, when you add a little salt and stir the solution, all the salt dissolves. Go on adding more and more salt and keep stirring. A stage comes when all the salt cannot be dissolved, no matter how much or how long you stir. Some salt remains undissolved and settles at the bottom of the container. Now heat the container a bit. You see that at this higher temperature you are able to dissolve more salt than at room temperature. In fact, for every such temperature (at a fixed ambient pressure) there is a fixed amount of salt that would fully dissolve in a given amount of water. When we add only this much salt, all of it dissolves and we obtain a so-called *saturated solution*.

Take a saturated solution prepared at a temperature above room temperature, and just let it cool back to room temperature, on its own. What you see is that nice cube-shaped crystals of common salt appear spontaneously. It is understandable that since at room temperature the water can dissolve less of common salt than at an elevated temperature, the excess salt separates out. But why do we get *crystals*?

A crystal is a highly ordered material. The atoms are arranged very regularly on a lattice. By contrast, a fluid is highly disordered: Its molecules are moving around chaotically, and there is no underlying lattice structure. So, we seem to have got '*order for free*'. We did nothing except to cool the solution. How can the simple act of cooling create order (or design!)? Is it a violation of the second law of thermodynamics? No.

The important thing to realise here is that we are dealing with a thermodynamically *open* system, rather than an *isolated* system. Our container and the saturated solution in it can exchange heat with the surroundings; otherwise it will not cool down to room temperature. So it is not an isolated system.

In fact, two types of non-isolated systems are sometimes defined in thermodynamics. Those which can exchange both heat and matter with the surroundings are called *open* systems. And those which can exchange only heat, but not matter, with the surroundings are called *closed* systems. It is like closing the lid of the container containing the common salt solution, so that no matter can be exchanged but heat can still flow through the walls of the container and the lid. [Sometimes this distinction between 'open' and 'closed' is not made; for many purposes the more important distinction is between 'isolated' and 'nonisolated or open'.]

So the second law must to be generalized to encompass such situations also. For this, let us first see what has really happened during the crystallization of common salt (NaCl).

When we dissolve NaCl in water, the two atoms of NaCl separate, and move around as a positive ion (Na^+) and a negative ion (Cl^-). This is a disordered state in which the large number of positive and negative ions keep colliding and separating. They separate because, although there *is* a force of attraction between a positive ion and a negative ion, the thermal fluctuations in the solution are strong enough to rip them apart quite often. But as the temperature is lowered the thermal fluctuations weaken, making it possible for clusters of NaCl to persist for long or short durations. As we keep cooling the solution, there comes a temperature for which the average attractive force is stronger than the disruptive forces, so we get a few small crystals of NaCl (Wadhawan 2000). As we keep lowering the temperature, the crystals grow in size because the ions of Na^+ and Cl^- can get attached to ions of opposite charge on the surface of an existing crystal. The atomic structure of such a crystal is depicted in Fig. 3.2.

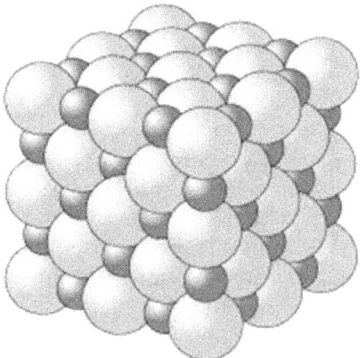

Fig. 3.2 The highly ordered structure of a crystal of NaCl. The small spheres depict Na^+ ions, and the bigger spheres the Na^- ions. Image credit:
http://commons.wikimedia.org/wiki/File:Natriumchloridtyp.jpg

In this closed (but not isolated) system, order emerges because, in a certain range of temperatures, the forces of attraction among the ions of NaCl are stronger than the local entropy-increasing tendencies.

To generalize the second law to include such situations, one introduces in thermodynamics the notion of *free energy*. [A derivation for the generalization of the second law is given in Section A1.1.] The free energy is the energy that is *free* or *available* for doing work. Remember, in general, not all energy is available for doing work; some part of it is trapped as heat energy, i.e., energy of chaotic motion of atoms or molecules. The free energy G (the so-called Gibbs free energy) is defined by the equation $G = H - TS$. Here H is the so-called *enthalpy*, and S is the entropy at temperature T. For the crystal-growth example described here, H is just the energy E that was released (the '*binding energy*') when the atoms of Na and Cl got bound together: This much energy must be expended on the crystal if we want to pull apart all the atoms again.

The generalized version of the second law says that *the free energy G cannot increase; it tends to decrease to its minimum possible value in any given situation.*

So, for nonisolated systems, there is an interplay of the H term and the TS term. An increase in entropy does mean a decrease in G (as demanded by the generalized second law). But now the possibility is available that TS can even *decrease* in magnitude, so long as there is a larger decrease in H, resulting in a net decrease in G. And a decreased S means a decrease in disorder. This is how an ordered crystalline state becomes possible, starting from a disordered fluid state.

This is also how all order or self-organization emerges spontaneously in Nature.

Conclusion: No 'order for free' is possible for an isolated system, but it is possible for a nonisolated system. A nonisolated system is ever exchanging matter and/or energy with the surroundings, and is therefore seldom in a state of equilibrium. Its approach to equilibrium is not always steady, and often leads to pockets of self-organization.

Let us dwell some more on the crystal-growth example mentioned above, so as to introduce the notion of *activation energy* or *activation barrier* early in this book.

3.4 Nucleation and growth of a crystal

As explained above, a crystal of NaCl grows from its saturated aqueous solution because the Gibbs free energy gets lowered in the process. But as every crystal-grower knows, there is mostly a two-step process in crystal growth: *Nucleation*, followed by *growth*. This is a feature of all so-called *first-order* phase transitions (to be discussed later). Let us see why.

There are two competing phases here: The crystalline phase (say Phase II) and the solution phase (Phase I). The free energy G of each phase has a certain dependence on temperature, as depicted schematically in Fig 3.3. Below a certain critical temperature T_c the free energy of Phase II is lower than that of Phase I. So, in accordance with the second law for non-isolated systems, the system lowers its free energy through the emergence of a crystal from the saturated solution.

All natural phenomena occur because their occurrence lowers the free energy.

Activation barrier

I have skipped an important detail in the above description of the growth of a crystal from its saturated solution. When the temperature is at T_c, or, equivalently, when the concentration is at the saturation value of the solution of NaCl for that temperature, it may perhaps be expected that perhaps a further lowering of the temperature by even an infinitesimal amount would result in the separation or crystallization of the entire solute from the solution. It does not happen that way for first-order phase transitions. The solution has to be *supercooled* substantially (or, equivalently, a substantial supersaturation has to be created for that temperature) to obtain a significant amount of crystallization. Here is why.

There are two competing factors involved here. One is that for any temperature below T_c, formation of the crystal will result in a lowering of the free energy by a certain amount (say ΔG), as demanded by what is shown in Fig. 3.3. So this factor favours crystallization. But there is another factor which *opposes* this tendency for the formation of a crystal from the solution.

It is referred to as the '*surface energy factor*': The average atomic structure of the solution is the same everywhere, and there are no surfaces in the solution. The solute atoms or molecules in the solution move around haphazardly, colliding with one another and with the walls of the container. For temperatures far above T_c the molecules just collide and then go their separate ways. The thermal agitation at such temperatures is so high that there are hardly any chances that two or more molecules will stick together for long.

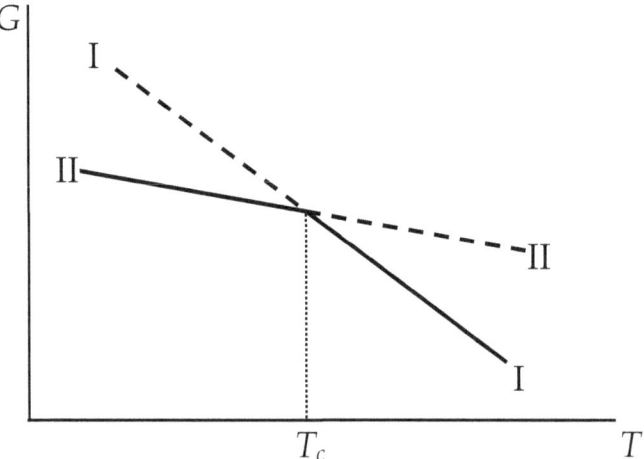

Fig. 3.3 Temperature variation of the free energy G for two competing phases I and II across a phase transition. At a certain critical temperature T_c the two phases have the same free energy. Above T_c Phase I has a lower free energy, so it is more stable than Phase II. Below T_c it is Phase II which has lower free energy. So a transition from Phase I to Phase II occurs when the system is cooled through the temperature T_c.

But as the temperature is lowered close to T_c, the chances of two or more molecules sticking together for substantial lengths of time in the form of small or large *clusters* increase. However, although clusters of molecules do exist above T_c, the chances of such clusters existing for long periods of time are not high. The reason is that any such cluster has a *surface* which separates it from the surrounding solution, and the creation and/or growth of this surface costs energy, the so-called *surface energy*.

Let us take a look at the contributions of these two factors to the total change of free energy G (Wadhawan 2000). Consider a cluster. For simplicity of argument, let us assume a spherical shape for it, with a radius r. There is a volume-dependent contribution to the change of free energy, and there is a surface-area-dependent contribution. The volume term lowers the free energy because the crystalline phase is more stable that the fluid phase at that temperature. By contrast, the surface term *increases* the free energy because the creation and growth of a surface between the fluid and the crystal costs energy. These two opposite trends are shown in Fig. 3.4. The volume of a sphere increases with r as r^3, and its surface area increases as r^2.

Curve 1 is for the volume term; it makes a decreasing contribution to free energy as r increases. And Curve 2 is for the surface term, showing an increasing contribution to the free energy with increasing r. The sum of these two contributions is shown in the curve labelled 1+2.

Because of thermal fluctuations, molecular clusters of various sizes form and dissolve back spontaneously. We see from Curve 1+2 that there is a *critical radius* r^*, above which an increase of r results in a lowering of the overall free energy G. So, if a cluster or crystallite of a radius larger than r^* somehow forms spontaneously, its growth into still larger sizes is favoured by the second law. By contrast, for cluster sizes below r^* there is an *increase* of free energy with increasing r. The free energy can decrease if r *decreases*. So such clusters of molecules just dissolve back into the solution.

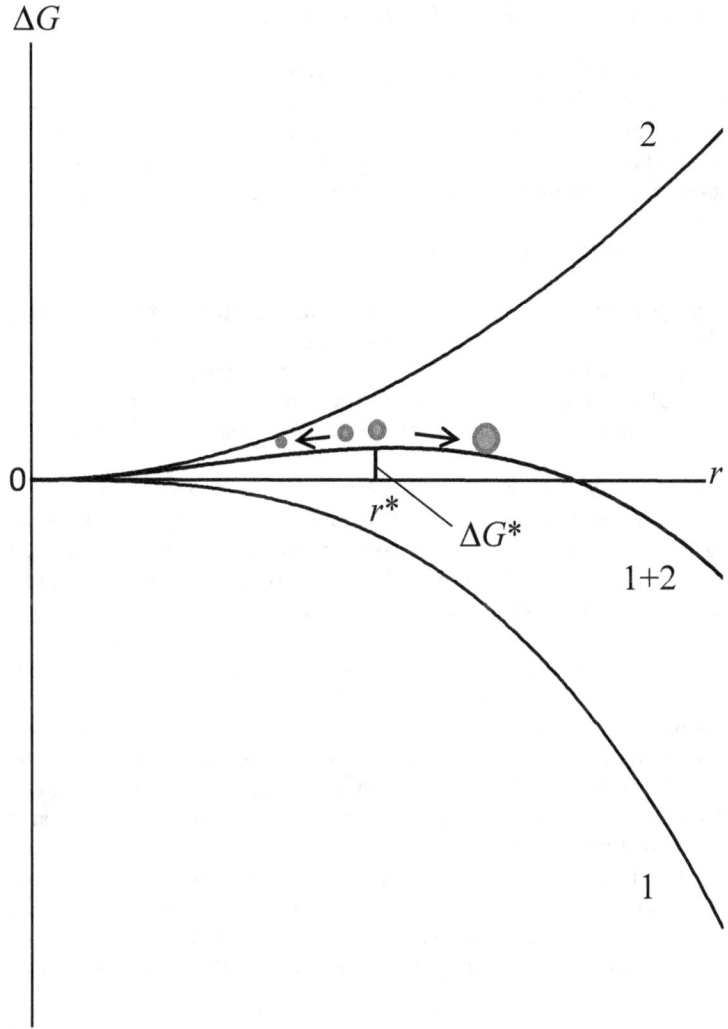

Fig. 3.4 Variation of bulk free energy (Curve 1), and surface free energy (Curve 2) with the radius r of a crystallite tending to emerge from the solution. If its starting radius is larger than r^* its growth will result is a lowering of G. But if the radius is less than r^* then the direction of decreasing G is that of decreasing r, i.e., a dissolution of the crystallite.

Seeding

There is another method available for overcoming the activation barrier. Instead of supercooling excessively for large-enough crystallites to form spontaneously, or waiting excessively for nucleation and growth to occur, we can dip into the solution an already grown crystal ('seed') of the same material. Now the activation barrier has been already crossed because the seed is the big nucleus, so growth can occur by deposition of molecules on the already existing surface of the seed crystal. This 'seeding' of an only slightly supersaturated solution is a good way of overcoming or reducing the activation barrier.

In general, any process or action that results in a lowering or overcoming of the activation barrier can be viewed as seeding or its equivalent.

[This has implications for how the gradual, many-step, emergence of life may have been aided by the presence of rocks submerged in sea water. Like crystal growth, chemical reactions can also be accelerated by the lowering of activation barriers. Some of the chemical reaction pathways, conducive to the ultimate emergence of life, could have got accelerated by the activation-barrier-lowering provided by appropriate rocks.]

3.5 The second law is an emergent law

The second law is a law of Nature which every macroscopic system must obey, in spite of the fact that the dynamics of the individual *microscopic* constituents of the system does not require that the law be obeyed. Because of this law, in the macroscopic world in which we live we take the direction of increasing time as that in which entropy increases and irreversibility occurs. All macroscopic phenomena have the property of irreversibility, or *time-asymmetry*.

The second law is an *emergent* law, because it is not deducible from the laws of mechanics applicable to the microscopic particles comprising any macroscopic system. The laws governing the dynamics of the microscopic particles are *time-symmetric*. Thus the time-asymmetry of the evolution of macroscopic systems (inherent in the content of the second law) is an emergent property, not present at the microscopic level.

The existence of emergent laws has bothered several scientists from times immemorial. In fact it continues to do so. Some of them are not able to reconcile to the very idea of emergence. Even Boltzmann, the originator of statistical thermodynamics, did try for a while to reconcile the time-symmetry of Newton's equations of motion with the time-asymmetry inherent in the second law. Starting from Newton's equations, Boltzmann succeeded in achieving time-asymmetry in the nonlinear transport equation derived by him. The solution of this equation was the famous *Boltzmann H function*. However, he soon realized that, although he had succeeded in deriving the occurrence of time-asymmetric behaviour in a macroscopic system, starting from time-symmetric dynamical equations for the individual molecules, he had introduced *probability considerations* for doing this, and this had serious connotations regarding the nature of *uncertainty*. [Elementary probability theory is introduced in Appendix A2.]

He then went to the other extreme, and approached the whole problem from the statistical angle alone. But the equation derived by him for what we now call *Boltzmann probability* has a *dynamical* flavour. It is the probability that a dynamical system will be in one of a given set of microstates, and is equal to the fraction of the total observation time that the system spends in that set of states.

Boltzmann's work met with stiff resistance and ridicule because his contemporaries just could not accept emergence. The second law is a statistical law. People in those days made a distinction between an exact law which is *never* violated and a statistical law which is *practically never* violated for a system of large number of constituents. Boltzmann was very depressed by the response to his profound work, and this possibly contributed to his suicide.

It has been suggested by some scientists that the time-asymmetric nature of the second law is partly a consequence of our inability to keep individual track of each of the very large number ($\sim 10^{23}$) of molecules in a macroscopic system, implying that the origin of its time-asymmetric nature is *statistical*. But a better rationalization can be achieved by probability-based arguments. Let us refer to the free-expansion-of-a-gas experiment and Fig. 3.1(a) again. All the molecules of the gas are in the left half, and their positions and velocities are randomly distributed. At the moment the partition is removed, there *happens to be* a particular set of positions and velocities at that moment. It is important to realize that any other set of random positions and velocities at that moment will still give an irreversible diffusion of the molecules to twice the volume. So this is a highly probable thing to happen. Now look at the final configuration in which the gas has, on its own, occupied twice the volume by diffusion (Fig. 3.1(b)). The motion of each molecule is governed by time-symmetric laws of dynamics. If we can somehow reverse the velocities of all the molecules at one particular moment of time, they would go back to occupying the left half of the chamber shown in Fig. 3.1(a) (assuming that there are no dissipative processes involved). But for this to happen the initial conditions, or the boundary conditions, for all the $\sim 10^{23}$ molecules will have to be one particular set from among the infinite such sets of initial conditions, and the probability for that to be the case is very very close to zero. That is why irreversibility and time-asymmetry arise at the macroscopic level.

Scrambling an egg is easy. Unscrambling it is impossible.

3.6 Emergence, weak and strong

The second law is an example of *weak* emergence. Although different parts of the macroscopic system considered above are interconnected, they are not necessarily *interdependent*. If we select a smaller part of such a macroscopic system, the second law still holds for it.

Similarly, the pressure and temperature of a gas are emergent properties, but only weakly emergent properties. There is uniformity, rather than variability.

Another example of weak emergence is the formation of a liquid (e.g. water (H_2O)) from gases (hydrogen and oxygen).

Strong emergence, by contrast, is something which is for a system as a whole. There is *interdependence* among the various subparts, and each part is indispensable for the overall emergent behaviour.

Strong emergence is global. Weak emergence may be local.

3.7 Nature abhors gradients of all types

The second law is normally stated in terms of entropy or free energy (which are concepts from statistical mechanics). But here is another version of the law, more relevant and important from the vantage point of the *dynamical* evolution of thermodynamic systems (Margulis and Sagan 2002):

Nature abhors gradients of all types, and tends to annul them.

Any gradient tends to get neutralized. For example, it requires a good deal of effort to maintain vacuum in a vessel. Vacuum is tantamount to gradient of pressure w.r.t. the surroundings, so Nature tends to fill the empty space with whatever atoms or molecules are available for doing so.

I used the 'free expansion of gas' example depicted in Fig. 3.1 to explain the genesis of the second law in terms of increase of entropy. What is equally true is that there was a pressure gradient when the partition was removed, and Nature destroyed this gradient irreversibly. The second law is nothing but a statement of the tendency for the spontaneous annulling of gradients like thermal gradients, pressure gradients, concentration gradients, etc.

[If Nature 'abhors' gradients, why does it 'love' creating them in the first place via the relentless expansion and cooling of our universe?! This conundrum exists not only for the gradients version of the second law stated here, but also for its entropy or information-theoretic version. The second law says that, for a closed system, entropy cannot decrease. As we shall see in Section 6.6, entropy means unavailable or missing information. When entropy increases, information is lost, so things are consistent and therefore understandable. But, in quantum mechanics, the basic thing we work with is the 'wavefunction' (Appendix A6). The wavefunction carries all the information about the system. The famous Schrödinger equation determines how the wavefunction evolves with time, and no matter what happens, the information encoded in the wavefunction can never get lost (see Section A6.12 for why this should be so). So how do we reconcile this conservation of information with the fact that entropy does increase and therefore information does get lost? That is the conundrum. The answer generally given is that information is indeed conserved, except that it has become practically irretrievable to us. It is out there somewhere! Notions such as 'quantum entanglement' are invoked for justifying this stance further (see Section A6.12 for more on this). I venture to offer another rationalization here. As our universe expands and cools, gradients get created, out of 'nothing'. Therefore Nature obliterates gradients to compensate for (or to make amends for!) having created them in the first place. This is similar to the argument given in cosmology for the continuous creation of energy/mass out of nothing, as our universe expands: We say that the creation of this (positive) energy is exactly balanced by the concomitant creation of an equal amount of negative energy locked up in the (attractive) gravitational potential energy of our universe.]

The second law for open systems is *the primary organizing principle for all natural phenomena*. The relentless expansion and cooling of our universe has been creating gradients of various types, which tend to get annulled as the blind forces of Nature take the local systems towards old or new equilibria. New patterns and structures, embodying more and more information, get created spontaneously when new equilibrium structures arise. This is already quite a profound statement, but we can extend it further and say that Nature tends to find *efficient* ways of neutralizing gradients of all types. When a system is moved away from a state of equilibrium by any influx of energy and/or matter, a gradient gets created. The Bénard instability depicted in Fig. 1.2 in Chapter 1 is a good example of how convection currents set in when conduction etc. are no longer *efficient* ways of destroying the large temperature gradient.

The natural tendency of complex systems to find more efficient ways of destroying gradients is very often the underlying cause for pattern formation, and also for the emergence of new structures or phenomena. The Bénard pattern is an example of this.

Thus, not only is the second law an emergent law, it is also instrumental, in turn, for the occurrence of many types of self-organization and emergence in Nature.

3.8 Systems not in equilibrium

A thermodynamic system is said to be in a *steady state* when its pressure, volume, temperature etc. do not change with time. If, in addition, there is no flow of energy or matter through the system, it goes into a state of *equilibrium*. Thus, in a system at equilibrium there is no change of its mass, energy, shape etc. with time.

It is for nonequilibrium systems that the second law allows for (even aids) the local creation of order and complexity. Nonequilibrium systems always have gradients of some kind or other. Natural phenomena are governed by the tendency to annul these gradients, as demanded by the second law. Consider osmosis as an example. Suppose there is a membrane, with salty water on one side, and pure water on the other. The second law tells us that the difference in salt concentration across the membrane will be reduced as time passes. This will happen spontaneously, not requiring the expenditure of any energy, as the system tries to reach a state of equilibrium (namely, no concentration gradients). But if we want *reverse* osmosis, it would require that energy be spent. This is how seawater desalination is done.

Forward and reverse osmosis are mirror-image processes, which reflect the time-symmetry or reversibility of the kinetics of motion at the microscopic level. A *reciprocity relation* connects the two processes, as demonstrated in the Nobel-Prize winning work of Lars Onsager in 1968.

The reciprocity relations are an example of how even nonequilibrium systems can be highly ordered, regular, or symmetric. The Bénard pattern is another example (see Section 1.3).

I discuss some more details of systems not in equilibrium in Appendix A5. It has been realized by researchers that even when a system is not in equilibrium, individual parts of it can be coherent and in equilibrium (Rubi 2008). This becomes possible if the forces acting on the system are not too strong, and if the properties do not change too much over short distances. Such systems (described as *linear* nonequilibrium systems) can be assigned a local temperature and a local entropy, which changes from one island of order to another.

3.9 Thermodynamics of small systems

Laws of thermodynamics need to be reformulated for 'small' systems. Some technical details are given in Appendix A1. The second law still holds, but the definition of entropy gets generalized for thermodynamically small systems. As discussed in Section A1.5, we speak of '*Tsallis thermodynamics*' in this context.

For extremely small systems it is meaningless to talk of an overall temperature, and thermal fluctuations dominate so much that it is meaningless to define a macrostate for stating the second law. Even the time scale at which a property or a process is to be described has to be chosen carefully and meaningfully for doing decent thermodynamics (Rubi 2008).

The second law, when stated in a language of meaningful terms, has a very long arm, and nothing escapes it.

4. Dynamical Evolution

At all levels - cosmology, geology, biology, and human society - we see a process of evolution in regard to instabilities and fluctuations.

Ilya Prigogine, *The End of Certainty*

In the public mind the word 'evolution' tends to be associated with Darwin. I shall describe Darwinian biological evolution in a later chapter, but it is important to realize that *any* dynamical system (living or nonliving) undergoes evolution as time passes. Emergence and self-organization are some of the consequences of the dynamical evolution of complexity. Darwinian evolution is a subset of dynamical evolution, with 'natural selection' as its distinctive feature.

4.1 Dynamical systems

A *static* system, as the name implies, is one which does not change with time. It is difficult to find a good example of a truly static system and the notion is, at best, an idealisation. What we have everywhere are *dynamical* systems. They are, by definition, systems which change with time, or *evolve* with time. Much of physical science concerns itself with studies of the time evolution of dynamical systems.

4.2 Phase-space trajectories

The concept of *phase space* (or *state space*) is a very powerful way of depicting the time evolution of dynamical systems. The concept is simple to explain and understand, and is almost essential for getting a good feel for the evolution of complexity in Nature.

Let us first consider the 'real' or actual 3-dimentional space in which we live. Suppose we want to specify the location of a particle in this space. We introduce three reference axes or coordinate axes, one for each dimension of the 3-dimensional space, and the coordinates of a point are specified as, say (x, y, z). If the point moves to a different location, it would have a different set of coordinates, say (x', y', z').

Now imagine an assembly of N particles. At any instant of time, any particle in this assembly is at a particular point in space, so we can specify its location in terms of three coordinates, say (x, y, z). At that instant of time, the particle also has some momentum. The momentum, being a vector quantity, can be specified in terms of its three components along the three coordinate axes, as say (p_x, p_y, p_z). Thus six parameters, (x, y, z, p_x, p_y, p_z), are needed for specifying the position and momentum of a particle at any instant of time. Therefore, for N particles, we need to specify $6N$ parameters for a complete description of the system at any instant of time. Generally, for real systems like molecules in a gas, the number N is very large, being typically of the order of the Avogadro number ($\sim 10^{23}$). So this is a very messy, in fact impossible, way of depicting graphically such a system of N particles. The concept of phase space solves this depiction and representation problem in a very elegant way.

Imagine a 6-dimensional '*hyperspace*' in which three of the axes are for specifying the position coordinates of a particle, and the other three are for specifying the momentum components of the same particle. In this imaginary space the position and momentum of a particle at any instant of time is represented by a *single* point. Similarly, for representing simultaneously the configuration of N particles, we can imagine a $6N$-dimensional space (called *phase space* or *state space*). A point in this space represents the state of the entire system of N particles at a particular instant of time. As time progresses, this representative point traces a trajectory, called the *phase-space trajectory*. Such a trajectory records the history or *time-evolution* of the entire dynamical system.

Fig. 4.1 illustrates this. In it we have introduced the simplification that, for depiction purposes, all the position coordinates ($3N$ in number) are given the generic symbol q, and only one axis is drawn to denote all the $3N$ axes. Similarly, all the $3N$ momentum components are given a representative symbol p, and only one axis is taken to represent all of them. In reality, there are a total of $6N$ axes in phase space, $3N$ for the position components, and $3N$ for the momentum components.

Some variations of the concept of such an imaginary phase space or state space are: *representation space*; *search space*; *configuration space*; *solution space*; etc. The basic idea is the same. One imagines an appropriate number of axes, one for each 'degree of freedom'. Any point in such a hyperspace represents the state of the entire system at an instant of time.

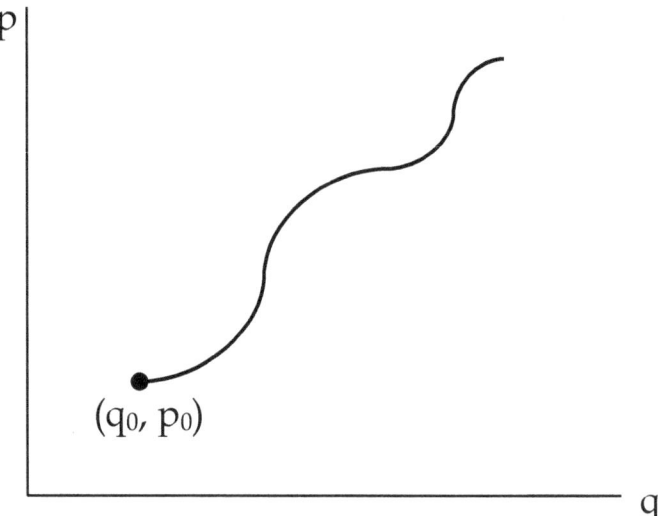

Fig. 4.1 A phase-space trajectory depicting the evolution of the dynamical state of a classical system, starting from the initial conditions represented by the point (q_0, p_0) in phase space.

4.3 Attractors in phase space

Generally speaking, dynamical systems lose energy through processes like friction, or emission of radiation, etc. We can say that practically all dynamical systems are '*dissipative systems*'. The phase-space idea is particularly useful for depicting this fact graphically.

As an example, let us consider a simple pendulum (a vertical string fixed at the top, and having a weight attached to its lower end). Suppose I pull the weight horizontally by a small distance x_0 along the x-axis, and then release it. The weight starts performing an oscillatory motion around the point $x = 0$. At the moment I released the weight it was at rest, so its momentum was zero, and it had only potential energy. On releasing it the potential energy starts decreasing as the weight moves towards the point $x = 0$, and its momentum starts increasing. This goes on till the point $x = 0$ is reached. At this moment the potential energy is zero (it got fully converted to kinetic energy corresponding to the momentum $-p_x$).

Because of this momentum, the weight overshoots the point $x = 0$ and moves in the opposite direction. When it has moved a certain distance $-x_0$ it stops, having spent all its kinetic energy for acquiring an equivalent amount of potential energy.

Then it starts moving towards the point $x = 0$ again. At $x = 0$ it has acquired the maximum (but oppositely directed) momentum p_x. And so on.

What is the phase-space trajectory of the system for this set of events? The answer is that it is a circle in a plane defined by the x-axis and the p_x-axis (Fig. 4.2(a)). The weight successively and repeatedly passes through a whole continuum of points in phase space, including the points $(-x,0)$, $(0,p_x)$, $(x,0)$, $(0,-p_x)$.

If there is no dissipation of energy, the phase-space trajectory in this experiment is a *closed* loop (a circle in this case) because the particle repeatedly passes through all the allowed (i.e., energy-conserving) position-momentum combinations again and again. Since the trajectory is fixed or constant, the area enclosed by the closed loop is also constant.

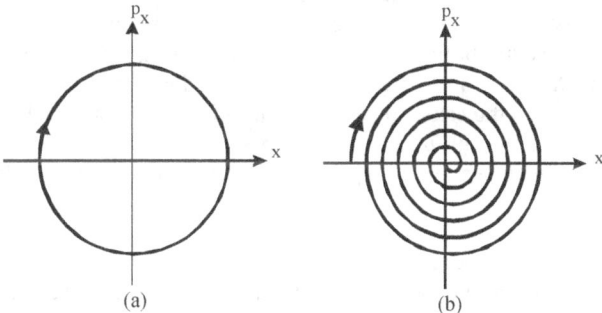

(a)   (b)

Fig. 4.2 (a) Phase-space trajectory for a 'conservative' system; i.e., a system that does not lose energy as time passes. (b) Phase-space trajectory for a system that loses energy with time, called a dissipative system.

But in reality, dissipative forces like friction are always present, and in due course all the energy expended in displacing the weight from its initial equilibrium position will be dissipated as heat. As the total energy decreases, the magnitude of the maximum value of the x-coordinate during the trajectory cycle, as also of the maximum value of p_x, would decrease with time, implying that the area enclosed by the trajectory in phase space will progressively decrease, till the particle finally comes to a state of rest or zero momentum. Thus, because of the gradual dissipation of energy, the phase-space trajectory *spirals* towards a state of zero area (Fig. 4.2(b)).

This final configuration corresponds to what is called an *attractor* in phase space: It is as if the dissipative dynamics of the system is 'attracted' by the point $(0,0,0,0,0,0)$ as its energy gets dissipated.

What is happening in phase space here is rather like a marble that is set rolling in a bowl, spiralling towards the bottom of the bowl: The bowl acts like a *basin of attraction*. The phase-space region around the attractor $(0,0,0,0,0,0)$ is the basin of attraction for the oscillator problem we have considered here.

Let us also note here that I had moved the weight by an *arbitrary* (though small) amount. The exact magnitude of this small amount of displacement is not important. In each such experiment (with different starting values of *x*), the dissipative system always approaches the same attractor. We say that, in this example, there is a *unique* basin of attraction around the *unique* attractor.

4.4 Nonlinear dynamical systems

In the above experiment with the simple pendulum, if I move the weight only by a small amount, its restorative force is *linearly proportional* to the displacement. If we plot this force f_x as a function of *x*, we get a straight line (which is a *linear* curve). But if the displacement is too large, the restorative force is not linearly proportional to the displacement *x*, and we are then dealing with a *nonlinear dynamical system*.

In our day-to-day life we tend to do 'linear' thinking. For example, if the number of guests expected for a party doubles, we plan for twice as much food. But the majority of natural phenomena are governed by nonlinear dynamics. Consider water at, say, 70°C. Heating it a little raises the temperature by a small amount, and it is still a liquid. Small cause, small consequence. Such linear behaviour continues till the temperature is close to 100°C, the so-called *phase-transition point* for water at atmospheric pressure. Now a small increase of temperature has a drastic effect: Water changes to steam. This happens because now the system is in a state of *unstable* equilibrium (like being on top of a hill, rather than being in a valley), and even a small perturbation is enough to make it 'role down the hill', so to speak, with no tendency to come back to the same (unstable) equilibrium state again. This 'no going back' situation leaves the system with no choice but *to seek out a new, different, state of stable equilibrium*. [In the next section I explain the difference between stable and unstable equilibrium.]

Many real-life systems are governed by nonlinear dynamics. In particular, most of the evolution of complexity on our Earth occurs because it receives a persistent and therefore cumulatively large amount of energy from the Sun, pushing it into nonlinear, nonequilibrium, regimes of dynamic behaviour, from which it keeps trying to 'roll downhill' in phase space, so as to approach stable equilibrium configurations.

4.5 Equilibrium, stable and unstable

> *Irreversibility can no longer be identified with a mere appearance that would disappear if we had perfect knowledge. Instead, it leads to coherence, to effects that encompass billions and billions of particles. Figuratively speaking, matter at equilibrium, with no arrow of time, is 'blind,' but with the arrow of time, it begins to 'see.' Without this new coherence due to irreversible, nonequilibrium processes, life on earth would be impossible to envision.*
>
> Ilya Prigogine (1996)

The term 'equilibrium' is used for the time-invariant state to which an isolated system tends to evolve. Uniformity of temperature is an important feature of a system at equilibrium. Such a system is in a state of maximum entropy; therefore there is no further production of entropy.

Now suppose the system is pushed away from its state of equilibrium by an influx of matter/energy. If the disturbance is small the system would, in all probability, tend to return to equilibrium. But if pushed too far away from equilibrium, it can really go 'over the hill', and is then unable to return to the old equilibrium configuration (see Fig. 4.3). If the influx of energy/matter pushing the system more and more away from equilibrium continues to be present, the system would seek new steady states. This is how much of *self-organized order* and *pattern formation* emerges. There is a local lowering of entropy (or increase of order), and a concomitant evolution of the degree of complexity.

One makes a distinction between stable and unstable equilibrium. Try balancing a pencil on its tip. When you release it, the direction in which it falls is purely a matter of chance. This is an example of unstable equilibrium leading to a new state of stable equilibrium (namely the pencil lying horizontally on the floor), and there is no spontaneous going back to the vertical configuration.

Fig. 4.3 depicts stable and unstable equilibria schematically. Complexity cannot emerge and increase in systems at stable equilibrium (Fig. 4.3a). If such a system is an isolated one, its entropy tends to its maximum value. If, on the other hand, the system is not isolated but is kept at a fixed temperature (i.e., if it is a *closed* system, but not an isolated system), then it is the free energy $F (= E - TS)$ which tends to the minimum value. In fact, a stable-equilibrium state is *defined* by the fact that the free energy is at its minimum value for that state. Thus the focus shifts to free energy, rather than entropy. In either case, fluctuations or perturbations tend to die out because the system returns to the stable equilibrium, i.e., back to its original configuration.

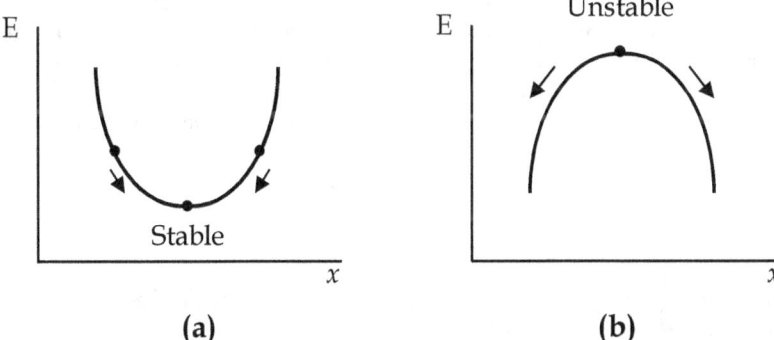

Fig. 4.3 Stable equilibrium (a); and unstable equilibrium (b). The plots show schematically the variation of internal energy E as a function of some relevant parameter, or coordinate x, in phase space.

For unstable equilibrium (Fig. 4.3(b)) the situation is totally different: The free energy is at its *maximum* value at the equilibrium point. Consequently any fluctuation or perturbation takes the system further away from equilibrium, because the system can lower its free energy by doing so (a tendency allowed and favoured by the second law of thermodynamics). *Nonequilibrium situations resulting from thermodynamic instabilities play a major role in the emergence of new structures and higher complexity.*

Any isolated system away from stable equilibrium tends to achieve the equilibrium configuration. The second law of thermodynamics says that an isolated system tends to attain a state of maximum entropy. It follows that the entropy at equilibrium is higher than that before the equilibrium was reached. Thus an isolated system away from stable equilibrium is a *lower-entropy* system (compared to the equilibrium value of the entropy).

Let us consider the Earth, inclusive of the atmosphere surrounding it. Suppose it gets completely isolated from the rest of the universe at any instant of time. Naturally, it will start approaching a state of equilibrium. A state of equilibrium will mean the end of all life processes, as it would be the state of highest entropy or disorder. *Equilibrium is death.* The reason why life is sustained on the Earth is that we are not living in an isolated system. There is a constant influx of energy from the Sun, which keeps the Earth in a lower-entropy, away-from-equilibrium, state. Moreover, the mean temperature of the Sun is much higher than that of the Earth. Since entropy, by definition, has an inverse dependence on temperature, what we receive from the Sun is low-entropy energy. This low-entropy energy from the Sun is converted into higher-entropy energy by a variety of *dissipative processes*, including all the processes that sustain life (Morrison 1964; Avery 2003; Niele 2005).

Nonequilibrium conditions must exist for order to emerge or entropy to decrease locally. A familiar example is the growth of a crystal from a fluid. Nonequilibrium conditions are created and maintained for crystal growth by imposing a thermal gradient or a concentration gradient. In the absence of such a gradient, the crystal cannot grow in size.

4.6 Dissipative structures and processes

> *Phase transitions often involve broken symmetries and the emergence of new entities. Spencer's idea that the evolution of life is characterized by increasing complexity can be made precise in the context of dissipative self-organization.*
>
> Klaus Mainzer, *Symmetry and Complexity*

Two types of nonequilibrium systems must be distinguished: linear and nonlinear. For the former the departure from equilibrium is small, and a steady-state nonequilibrium configuration is feasible. For such systems, fluctuations can dampen out.

Not so for the nonlinear type. They are fundamentally different from stable-equilibrium and linear-nonequilibrium systems (Glandsdorff and Prigogine 1971). *The fluctuations in them do not dampen.* In fact, persisting effects of fluctuations and instabilities are the norm here. There is no minimum of potential energy or free energy towards which the system may tend to gravitate with the passage of time.

'*Near-equilibrium laws of nature are universal, but when they are far from equilibrium, they become mechanism dependent*' (Prigogine 1996).

To illustrate this point, let us first consider a *closed* system in which there are chemical reactions occurring: $\{A\} \Leftrightarrow \{X\} \Leftrightarrow \{F\}$. Here $\{A\}$ denotes the initial set of chemicals, and $\{F\}$ the final products. $\{X\}$ stands for intermediate products. If the system has reached equilibrium, there are as many reactions from $\{A\}$ to $\{X\}$ as there are from $\{X\}$ to $\{A\}$. Similarly between $\{X\}$ and $\{F\}$. The ratio $\{A\}/\{F\}$ has a definite value corresponding to a maximum value for the entropy.

Let us next consider an *open* system, in which the input and output of chemicals can be adjusted. We start with a certain value for the ratio $\{A\}/\{F\}$ (say, the equilibrium value mentioned above), and progressively increase the value of this ratio. This amounts to taking the system further and further away from equilibrium. What happens to $\{X\}$ now?

For a given set of $\{A\}$ and $\{F\}$, more than one concentrations $\{X\}$ are possible. However, only one of these corresponds to thermodynamic equilibrium. Starting from this *thermodynamic branch*, one can extend the analysis to nonequilibrium situations. The work of Prigogine and coworkers showed that the thermodynamic branch becomes unstable at some critical 'distance' from equilibrium (Fig. 4.4). The point where this occurs as a function of some suitable control parameter is called a *bifurcation point* (Turing 1952). Beyond this point, *new phenomena arise* (e.g., oscillating chemical reactions, chemical waves, nonequilibrium spatial structures, etc.). The term *'dissipative structures'* was coined for structures which possess such spatial and/or temporal order.

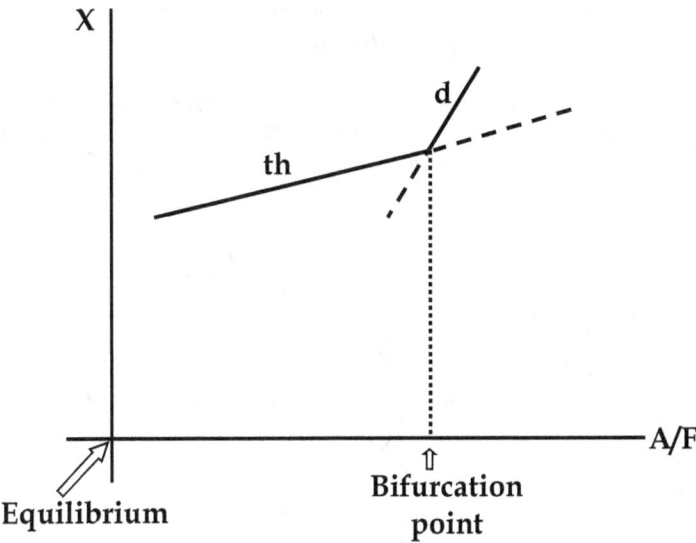

Fig. 4.4 The transition from 'stable' to 'unstable' for the thermodynamic branch of a far-from-equilibrium system. There can be two steady-state solutions ('th' and 'd') when we plot X as a function of A/F (see text for details). There is a *bifurcation point* such that 'th' is stable when A/F is below the bifurcation point, and 'd' is stable when A/F is above it (Prigogine 1996).

Two conditions are necessary for the occurrence of dissipative structures in chemical systems: (i) departure from equilibrium beyond a certain critical distance in phase space; and (ii) the availability of two-way catalytic steps $X \Leftrightarrow Y$ (Prigogine 1996).

4.7 Bifurcations in phase space

> *Thus, bifurcation mathematically only means the emergence of new solutions of equations at critical values. Actually, bifurcation and symmetry breaking is a purely mathematical consequence of the theory of nonlinear differential equations. But, bifurcations of final states as solutions of differential equations correspond to qualitative changes of dynamical systems and the emergence of new phenomena in nature and society . . .*
>
> Klaus Mainzer, *Symmetry and Complexity*

In physics the term 'phase' has a fairly well defined meaning. For example, liquid water and steam are two different (homogeneous) phases of H_2O. But when it comes to complex systems, the term 'phase transition' gets used loosely and liberally. For example, Ramachandran (2010) spoke of a *'mental phase transition'* in the evolutionary history of our brain: 'Then sometime about a hundred and fifty thousand years ago there was an explosive development of certain key brain structures and functions whose fortuitous combinations resulted in the mental abilities that make us special in the sense that I am arguing for. We went through a *mental* phase transition. All the same old parts were there, but they started working together in new ways that were far more than the sum of their parts.'

In such contexts, a better phrase than 'phase transition' would be *bifurcation in phase space*. As some control parameter changes in a sustained manner, a situation can arise wherein the system has moved so far away from equilibrium that its phase-space trajectory suddenly undergoes a bifurcation: There are two alternative trajectories which can result in a lowering of free energy, and the one actually taken is often a matter of pure chance: Even the tiniest of random fluctuations can make the system choose one trajectory over the other. This is how the unpredictability factor arises in the dynamical evolution of any complex system. *And this can happen even for an otherwise deterministic situation.*

Fig. 4.5 provides an example of what is commonly called *pitchfork bifurcation* in phase space. If we identify the variable on the *x*-axis with the concentration X discussed above, the three possible solutions of the dynamical equation can be depicted as in Fig. 4.5.

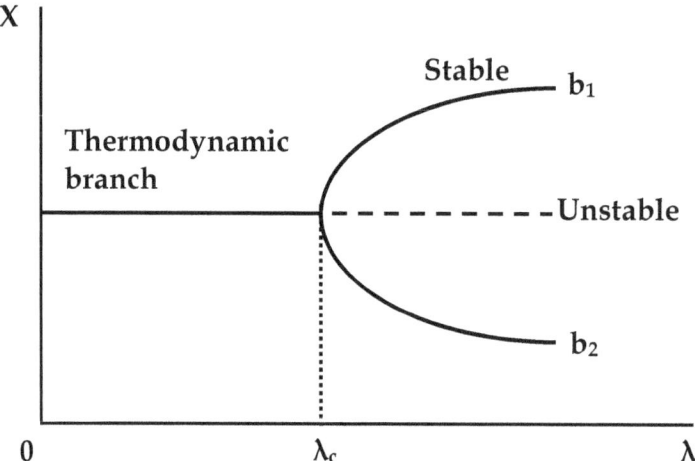

Fig. 4.5 Pitchfork bifurcation. The value of the control parameter λ is a measure of the 'distance' from the equilibrium configuration, and X is the chemical concentration considered in the text. At the bifurcation point λ_c, the thermodynamic branch (which is stable from $\lambda = 0$ to $\lambda = \lambda_c$) becomes unstable, and two possible new solutions b_1 and b_2 emerge (Prigogine 1996).

These results, though illustrated here by considering an example from chemistry, have universal applicability. The onset of lasing action is also an example of a bifurcation in phase space, or of a *generalized phase transition*. After a laser system has been assembled, the lasing action can start only when a certain fine-tuning is done. The fine tuning amounts to varying some control parameter, and at a critical value of the control parameter the lasing action is achieved. The onset of the lasing action is rather like the emergence of order (or a breaking of symmetry) at a phase transition (like crystalline ice emerging from liquid water). The 'ordered-state' characteristic of the laser is that in it a coherent emission of radiation occurs.

Lasing action is an *emergent phenomenon* arising in a complex system. Order emerges out of disorder (or less order) when the bifurcation in phase space occurs.

Symmetry breaking is a commonly occurring adjunct of bifurcations in phase space. Ever since the Big Bang, there has been a succession of symmetry-breaking bifurcations, each resulting in the emergence of new states of order, the most spectacular example being the emergence of life out of nonlife.

4.8 Self-organization and order in dissipative structures

Dissipative structures are characterized by the important role played by thermodynamic fluctuations. For near-equilibrium situations the fluctuations tend to die out. But a system driven beyond a bifurcation point has to make a choice between the two solutions b_1 and b_2

depicted schematically in Fig. 4.6.

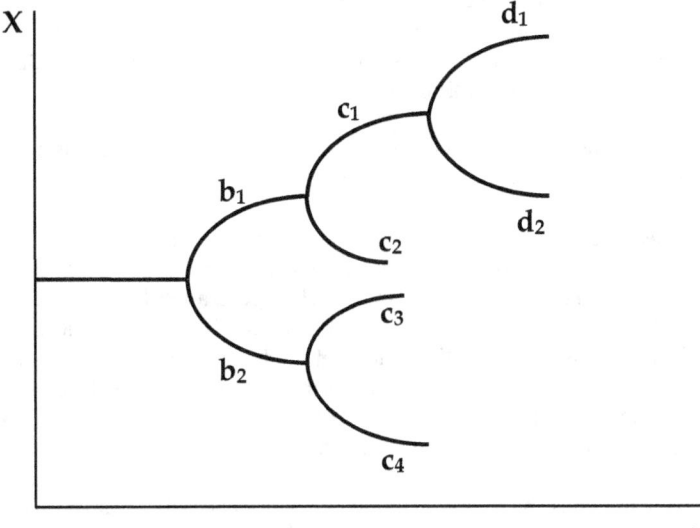

Fig. 4.6 Successive bifurcations as a system is taken further and further away from equilibrium.

This choice is arbitrary *as it depends on what happens to be the nature of the random fluctuation at the moment the system becomes unstable. This introduces an irreversible probabilistic element into the physics involved, even when no quantum effects are significant. Even if we know the initial conditions and the boundary conditions with infinite precision, we still cannot predict with certainty the final state.*

On further departure from equilibrium, a succession of bifurcation points may arise, and fluctuations determine randomly the actual state the system is in (Fig. 4.6).

Bifurcations can cause a breaking of symmetry. The configuration chosen by the system generally has a lower symmetry than the symmetry prior to the bifurcation (Nicolis and Prigogine 1989). And a lower symmetry means a state of higher order. Thus *an open system pushed sufficiently far from equilibrium generally self-organizes into a more ordered structure, with a lowered local entropy.*

As a function of time, there is a succession of *deterministic* processes (between bifurcations) and *probabilistic* processes (in the unpredictable choice of branches). Even the history of the system gets locked-in, to a certain extent. For example, suppose the present state of the system is found to be d_2. From this we can conclude that it must have taken the evolution path $b_1c_1d_2$. But if the system is at b_1 at some instant of time, we have no way of telling that it would evolve to d_2. It may as well evolve by the path b_1c_2 to c_2, or the path $b_1c_1d_1$ to d_1.

There is an important lesson in this. Life as we see it today evolved along a particular state-space trajectory sequence, from among a huge number of other possible trajectories. Chance played a role at the bifurcation points. Therefore, it is impossible to create in the laboratory those conditions from which the evolution would be *exactly* along the phase-space path that *happened* to have been chosen by the blind forces of Nature.

That we humans are what we are is a matter of chance. We may as well have evolved to be something very different. Of course, natural selection played its role, and natural selection is *not* a random process. But the overall chain of processes did have chance occurrences interspersed here and there. It is like wind *vs.* stagnant air. In both of them the molecules of air move in random directions, but in a wind there is an average velocity which is along a nonrandom direction. Biological natural selection superimposes an overall direction to the randomness inherent in most of the bifurcation events.

One can extrapolate such conclusions to other systems. 'Our universe has followed a path involving a succession of bifurcations' (Prigogine 1996).

A principle of self-organization was enunciated by the British cybernetician Ross Ashby (1962). According to it, *an open dynamical system tends to move towards the nearest attractor in phase space*. And what is the mechanism of self-organization by movement towards the nearest attractor? It is the deterministic, as also the probabilistic ('stochastic'), variations that occur in any dynamical system, enabling it to explore different regions in phase space until it reaches an attractor. Entering the attractor stops further variation outside the basin of attraction, and thus restricts the freedom of the components of the system to behave independently. There is an increase of coherence, or decrease of local entropy.

A distinction is sometimes made between self-assembly and self-organization. The latter is characterized by a continuous exchange of energy with the surroundings, whereas the former is simply a consequence of minimization of free energy of a thermodynamically open system.

5. Relativity Theory and Quantum Mechanics

All natural phenomena are governed by the laws of quantum mechanics. In our day-to-day experience it is usually not necessary to invoke the laws of quantum mechanics, as its limiting case, classical mechanics, suffices. Quantum effects usually become prominent only at very small length scales or very small masses.

Classical mechanics is a limiting case, not only of quantum mechanics, but also of Einstein's special theory of relativity. Relativistic effects become prominent mainly when the speeds involved are so high as to be comparable to the speed of light. We shall discuss basics of both relativity theory and quantum mechanics in this chapter.

5.1 Special theory of relativity

Mu mesons or muons (a type of unstable elementary particles) are produced when protons from the Sun are absorbed in the atmosphere. They travel typically at ~99% the speed of light: $v = 0.99c$. A muon has a certain lifetime, after which it disintegrates into something else. One may think that the lifetime should be something fixed and unique to the particle. Indeed it is, provided the observer measuring the lifetime is moving with the muon. Otherwise it is found to depend on its velocity relative to that of the observer. Observers at rest w.r.t. the moving muons would measure the average lifetime to be ~2.2 ms. But observers on Earth measure the lifetime of the moving muon to be ~15.6 ms. Strange but true. Einstein's special theory of relativity explains phenomena.

Newton laid the foundations of classical mechanics in the 17th century. His monumental *Principia* enunciated simple laws such as 'force equals mass multiplied by acceleration'. Newton's laws of mechanics, when augmented by his law of gravitation ('gravitational attraction between two masses is directly proportional to the product of the masses, and inversely proportional to the square of the distance between them'), dominated physics for the next 200 years. Newton postulated the existence of *absolute space*; space that is immovable and similar everywhere and at all times. He also presumed that gravitational interaction between any two bodies is *instantaneous*; that is, its speed of travel is infinite.

Maxwell's theory of classical electrodynamics, formulated in 1864, extended Newtonian classical mechanics to account for the motion of charged particles in electric and magnetic fields. Modern communication technology and information technology etc. are governed by the celebrated Maxwell equations.

The beginning of the 20th century saw the sharpening of challenges to the existing edifice of theoretical physics based on Newton's laws and Maxwell's equations. The first challenge was to the concept of *ether*, a hypothetical medium assumed by Huygens to be necessary for the propagation of electromagnetic waves. From such a presumption it followed that, observed from the Earth, light from different extra-terrestrial sources must travel at different speeds. But Michelson's famous interferometric experiment showed in 1887 that the observed speed of light does not depend on its direction with respect to the direction in which the Earth is moving. Thus the postulation of ether as a medium at rest in the universe did not serve any sensible purpose, and the idea was abandoned. Electromagnetic waves travel in empty space; there is no ether anywhere.

According to R. C. Tolman, Einstein's theory of relativity may be regarded as based on the fundamental idea of *relativity of all motion*. That all motion is relative is something we intuitively take for granted, but in the hands of Einstein this fundamental notion led to developments which changed the face of physics and cosmology.

There are two types of relative motion: uniform (i.e., constant-velocity) relative motion and non-uniform relative motion. They define, respectively, the content of Einstein's 'special theory' and 'general theory' of relativity.

Einstein's special theory of relativity is based on two postulates:

1. The laws of physics are the same for all observers in uniform (i.e., non-accelerating) motion relative to one another. This is the *principle of relativity*.

2. The speed of light in vacuum is the same for all observers, regardless of their relative motion or of the motion of the source of the light. [This was a rather bold assumption by Einstein, for which he was roundly criticized by many. But he turned out to be right.]

Laws of physics are represented by mathematical equations. Under the usual '*Galilean transformation*' of coordinates (which takes a reference frame S to another reference frame S'), Newton's laws of motion (e.g. Eq. 5.2 below) remain the same in S and S' if the two reference frames are moving w.r.t. each other with a *constant* velocity.

Offhand one would expect that Maxwell's equations would also remain invariant under the Galilean transformation, but this was not found to be the case. So either Newton's equations of motion or Maxwell's equations for electromagnetism needed to be modified. Einstein took the view that the Galilean transformation of coordinates should be replaced by some other transformation (called *Lorentz transformation*) so that the Maxwell equations remain invariant under it, irrespective of what happened to the Newton equations in the process (see, e.g., Krori (2010) for some details). The result was Einstein's special theory of relativity, of which Newtonian mechanics is only a special (limiting) case.

Maxwell had shown that light is an electromagnetic wave which travels with a speed of 3×10^8 m/s through ether, but the perceived existence of ether came as a hurdle in the formulation of an alternative form of coordinate transformation such that Maxwell's equations would remain invariant under it. 'Off with ether', said Einstein. He postulated that electromagnetic waves propagate through empty space devoid of any ether, and with a constant speed, no matter what the frame of reference is for measuring this speed. *The speed of light is a universal constant*, denoted by the symbol c.

The special theory of relativity has some remarkable consequences and predictions, which have been confirmed by experiment:

Relativity of simultaneity: Two events, simultaneous for one observer, may not be simultaneous for another observer if the observers are in relative motion.

Time dilation: Moving clocks tick more slowly than an observer's 'stationary' clock, as confirmed by the muon-lifetime experiment mentioned above. [For the same reason, GPS technology involves having to make substantial time corrections for coordinating the various clocks around the world.]

Length contraction: Objects get shortened in the direction they are moving w.r.t. the observer.

[Speed-of-light constancy, time dilatation, and length contraction are all interrelated. One can turn the whole narrative around, and say that constancy of speed of light follows as a consequence of time dilatation and length contraction (see Krauss 2017).]

Mass-energy equivalence:

$$E = mc^2 \tag{5.1}$$

Thus energy (E) and mass (m) are interconvertible; they are related by the square of the speed of light in vacuum (c).

Finiteness of maximum possible speed: No physical object, message or field can travel faster than the speed of light.

However, predictions of the theory become significant only for speeds comparable to the speed of light, which is very large indeed, being ~3 x 10^8 m/s.

Unlike the Newtonian concept of absolute time which does not change from one frame of reference to another, space and time cannot be distinguished in this theory, and must be treated as a unified *spacetime*. Two observers moving relative to each other at a *constant* velocity would observe the same laws of Nature in action. One of these observers, however, might record two events on distant stars as having occurred simultaneously, while the other observer would find that one had occurred before the other. Simultaneity does not exist for distant events. In other words, it is not possible to specify uniquely the time when an event occurs, without a frame of reference. The 'distance' or 'interval' between any two events can be described only by means of a combination of space and time, and not by either of them separately. The spacetime of four dimensions (three for space and one for time) in which all events occur is called the *spacetime continuum*.

[It is important to mention here that the idea of spacetime was introduced, not by Einstein, but by the mathematician Minkowski. He gave an improved interpretation of the early work of Einstein on special relativity. He argued that two observers moving relative to each other might be actually observing *different* 3-D projections of a 4-D reality, the latter being one in which space and time are on an almost equal footing (the so-called *Minkowski space*). I say 'almost', because in the 4-D spacetime, a length element s is defined as $s^2 = x^2 + y^2 + z^2 - t^2$. That negative sign before the time term makes all the difference. In the 4-D reality, there is no length contraction or time dilatation. These things occur only in the projections in the 3-D space we live in, there being different projections for differently situated observers. Einstein made the spacetime notion a central aspect of his general theory of relativity.]

Relativity became the premier guiding force in twentieth-century thought and art also: No independent absolute value exists, but rather truth is meaningful or significant only in a given context and time. Ditto for moral decisions. The relativity idea had a profound effect on artists, authors and musicians, and many new styles of literature, art, and music emerged in the early twentieth century.

Einstein's theories of relativity, both 'special' and 'general' were completely classical, i.e., non-quantum-mechanical. And Maxwell's equations of electrodynamics are consistent with the special theory of relativity.

5.2 General theory of relativity

Einstein's special theory of relativity was followed by his *general* theory of relativity (1907-

1915), which addressed the issue of gravity and acceleration.

Gravity is a force field, and any force causes acceleration. Consider two frames of reference, one (S_1) moving at a constant horizontal velocity and the other (S_2) having an accelerated motion in the same horizontal direction. If a ball is dropped in S_1, it falls vertically downwards. But from the perspective of S_2 the ball would follow a curved trajectory because S_2 is moving (w.r.t. S_1). For a scientist in S_1, the laws of physics are equations in flat space, whereas in S_2 they are equations in curved space. But the fact is that the laws are the same in S_1 and S_2. To achieve an invariance of the equations expressing the laws of physics, Einstein expressed them as *tensor equations*. [A law of physics expressed as a tensor equation in one laboratory or frame of reference has the same form in any other laboratory in any kind of relative motion w.r.t. the first laboratory. This is called *the principle of general relativity*.]

Einstein deduced from his theory that, contrary to Newton's assumption, gravitational waves, which mediate gravitational attraction between any two bodies, must travel at a *finite* speed, namely the speed of light. The theory interpreted gravitational interaction in terms of a *distortion of spacetime* by every object. The larger the mass of an object, the more is this distortion. When one object distorts spacetime, the effect is felt by every other object; this is how gravitational attraction between any two objects occurs.

Newton's second law of motion says that the acceleration (a) produced by a force F acting on an object of mass m is equal to F/m; i.e.

$$F = ma \tag{5.2}$$

The mass m that enters this equation is called the *inertial mass*. To understand why, let us recall Newton's first law of motion, which says that a body continues to be in a state of rest or uniform motion, unless acted upon by a force. Thus, the body exhibits *inertia* to any change of its state of rest or uniform motion; it resists a change in its state, and the resistance or inertia is proportional to its mass, the inertial mass.

Next, let us take Newton's law of gravitation. Consider any two bodies or objects which are having masses m_1 and m_2, and which are a distance r apart. The law says that the gravitational force of attraction between them is

$$F = Gm_1m_2/r^2 \tag{5.3}$$

Here G is the *gravitational constant*. [Some other such fundamental parameters are the speed of light c, and the Planck constant h.] The mass that enters this gravitation-law equation is *gravitational mass*.

The big question was: Is inertial mass the same as gravitational mass? Newton (and also Galileo) asserted that it is so, and that it is a happy coincidence that it is so. Einstein agreed that the two kinds of mass are the same, but he created a trail-blazing theory of gravity out of this 'happy coincidence', namely the general theory of relativity.

The sameness or equivalence of the two kinds of mass goes by the name of *equivalence principle*, an essential ingredient of Einstein's theory.

Einstein argued that the experience of being pulled downwards by the gravitational force on the surface of the Earth is equivalent to being inside a spaceship (far from any sources of gravity) that is being accelerated by its engines. From this, Einstein deduced that free-fall is actually *inertial motion* (i.e., constant-speed motion). Objects in free-fall do not accelerate, but rather the closer they get to an object such as the Earth, the more the time scale becomes stretched due to spacetime distortion around the planetary object. An object in free-fall is actually inertial, but as it approaches the planetary object the time scale stretches at an accelerated rate, giving the *appearance* that it is accelerating towards the planetary object. *An accelerometer in free-fall does not register any acceleration.* Falling under gravity is actually equivalent to floating in distorted space.

When an observer feels or detects the local presence of a force that acts on all objects in proportion to the inertial mass of the object, the observer can be taken to be experiencing an accelerated frame of reference. Imagine two reference frames, K and K'. K has a uniform gravitational field, whereas K' has no gravitational field but is uniformly accelerated by such an amount that objects in the two frames experience identical forces. To quote Einstein (1911):

We arrive at a very satisfactory interpretation of this law of experience, if we assume that the systems K and K' are physically exactly equivalent; that is, if we assume that we may just as well regard the system K as being in a space free from gravitational fields, if we then regard K as uniformly accelerated. This assumption of exact physical equivalence makes it impossible for us to speak of the absolute acceleration of the system of reference, just as the usual (i.e. special) theory of relativity forbids us to talk of the absolute velocity of a system; and it makes the equal falling of all bodies in a gravitational field seem a matter of course.

Einstein (1911) went further than that:

As long as we restrict ourselves to purely mechanical processes in the realm where Newton's mechanics holds sway, we are certain of the equivalence of the systems K and K'. But this view of ours will not have any deeper significance unless the systems K and K' are equivalent with respect to all physical processes, that is, unless the laws of Nature with respect to K are in entire agreement with those with respect to K'. By assuming this to be so, we arrive at a principle which, if it is really true, has great heuristic importance. For by theoretical consideration of processes which take place relatively to a system of reference with uniform acceleration, we obtain information as to the career of processes in a homogeneous gravitational field.

He combined the equivalence principle with his special relativity theory to predict, among other things, that clocks run at different rates in different gravitational fields, and that light rays bend in a gravitational field (*gravitational lensing effect*). This and many other predictions of the theory have been confirmed by experiment.

The central idea of Einstein's theory is that the presence of matter distorts or curves spacetime. Thus gravity is interpreted as not a force, but rather a curvature in the fabric of spacetime, and objects respond to gravity by following (or free-falling along) the curvature of spacetime existing in the vicinity of an object. This idea has shaped modern cosmology. It also marked a major inroad of geometry into physics.

The Cosmological principle

Einstein also enunciated the very fundamental *cosmological principle*. According to it, *on a large-enough scale, the universe is isotropic and homogeneous on average*. This means that the laws of physics are the same everywhere, and in all directions; there are no preferred directions or places.

Talking about the fundamental principles, we also have the so-called Copernican principle: *the universe is homogeneous.* [See Rudnicki (2014) for a detailed discussion of the so-called 'generalized Copernican cosmological principle'.]

5.3 Quantum mechanics

Classical mechanics is an approximation of quantum mechanics under certain conditions. Newton's three laws of motion are examples of laws of classical mechanics.

When does the approximation become invalid, making it necessary to invoke laws of quantum mechanics directly for understanding a natural phenomenon? One situation is when the spatial dimensions are extremely small. A tennis ball is so big that classical mechanics is adequate for describing its trajectory etc. An electron is so small that only quantum mechanics can explain its behaviour properly.

Is an electron a particle or a wave? Experimental evidence says that it has features of both. An electron has a certain mass and electric charge. If we apply an electric field, we can accelerate the motion of the electron, which we can correctly calculate by assuming that it is a particle having the known mass and charge. We may conclude that it is a particle. But now consider the following famous double-slit experiment first performed by Davisson and Germer in 1927:

They shot a beam of electrons through two parallel slits, and recorded the positions of the electrons on a flat screen on the other side (Fig. 5.1). What they found was that the electrons behaved, not as particles moving in straight lines, but as waves, forming a *diffraction pattern* like the one you would expect from a beam of light.

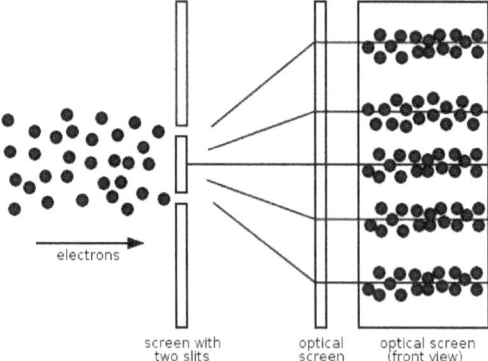

Fig. 5.1 The famous double-slit experiment for electrons. Image credit:
https://chem.libretexts.org/Textbook_Maps/Physical_and_Theoretical_Chemistry_Textbook_Maps/Ma
p%3A_Physical_Chemistry_(McQuarrie_and_Simon)/01%3A_The_Dawn_of_the_Quantum_Theory/1.
7%3A_de_Broglie_Waves_can_be_Experimentally_Observed

This established the *wave-particle duality* of elementary particles like electrons: They behave as both waves and particles, and one or the other of these aspects can be invoked, depending on the context.

There are serious consequences of this conclusion. A particle, by its nature, can be assigned a certain position or 'coordinates' in space. But we cannot do that for a wave. Consider the familiar sound waves in air. As a sound wave travels, there is compression and rarefaction in air. Some of this vibration of air reaches your ears, and you sense the sound. But can you say that the sound wave is here, and not there? No. It is everywhere; with different intensities, of course. So, if an electron has wave properties, it means that it is everywhere at the same time! We say that it is *delocalized*. This is one of the shocks that quantum theory inflicts on us. There are many others. And yet, we accept it as it is the most successful and *the most thoroughly tested theory*, or model of reality, ever.

The wave nature of electrons is indeed a reality. Otherwise we would not have been able to build the very important and much used *electron microscopes*. In these devices, electrons do what is done by light in an optical microscope.

Another challenge to classical physics came from the observed properties of radiation emitted by a hot body, the so-called *blackbody radiation*, on which Max Planck did some pioneering work in 1900. It became necessary to accept that radiation must exist in *discrete* packets or *quanta* called *photons*. Just as electrons have wave properties, light can also behave as if it is a collection of particles called photons. This was further established in 1905 by Einstein by an experiment involving the so-called *photoelectric effect*, and he was awarded the Nobel Prize for this work in 1921.

Thus in quantum mechanics, photons, as also 'material' particles like electrons, have both a wave aspect and a particle aspect. There is a wave associated with every photon. Let us consider a photon of energy E. For the electromagnetic wave associated with it, the velocity c, the wavelength λ, and the frequency v are related by the equation $c = v\lambda$. And the energy of the photon is given by

$$E = hv = hc/\lambda .$$ (5.4)

Here h is a fundamental constant of Nature, called *the Planck constant* ($h = 6.6260755 \times 10^{-34}$ Joule second).

Similarly, there is a wavelength associated with material particles like electrons. It is called the *de Broglie wavelength*, and is given by

$$\lambda = h/(mv),$$ (5.5)

where m is the mass of the particle, and v its velocity.

Energy is quantized in quantum mechanics. Postulation of the discrete nature of energy also explains an important feature of atomic physics. Consider an atom of hydrogen. There is a nucleus, namely a proton, and there is an electron which revolves around the nucleus. The proton is ~2000 times heavier than the electron. As the electron moves in an orbit around the proton, it may be expected to radiate energy because of the acceleration involved in any orbital motion. If this negatively charged electron indeed loses energy, it may be expected to gradually spiral into a decreasing-radius orbit, and then just merge with the positively charged proton. But this does not happen, and was explained in quantum mechanics in terms of the discrete nature of energy: All energy exists in the form of discrete quanta. For the hydrogen atom this means that the radius of the orbit of the electron cannot be smaller than a certain minimum value. And only a discrete set of orbits are allowed for the electron because only discrete values of energy are allowed, and the radiation emitted by the electron can possibly have only those energies which are equal to the differences of the discrete energy levels. When the electron has fallen to the lowest such level (called the *ground state*), it cannot radiate any more energy, and therefore it cannot get any closer to the nucleus.

Another pillar of quantum mechanics is Heisenberg's *uncertainty principle* (see Appendix A6 on quantum theory for some details). According to it, there are pairs of properties or variables (called *conjugate variables*) such that it is impossible to know with arbitrarily high precision the magnitudes of both members of the pair. The position x and the momentum component p_x along the x-axis are one such pair. If Δx is the uncertainty in the knowledge of x, and Δp_x the uncertainty in the knowledge of p_x, then the product $\Delta x \Delta p_x$ cannot be arbitrarily small; its value is always higher than a quantity of the order of the Planck constant h. To be precise, it has to be such that

$$\Delta x \Delta p_x \geq h/(4\pi) \tag{5.6}$$

Energy and time are another such pair of conjugate variables. Therefore

$$\Delta E \Delta t \geq h/(4\pi) \tag{5.7}$$

For the ground state of the hydrogen atom mentioned above, Δt can be arbitrarily large; after all, the electron can stay indefinitely in the ground state. Therefore it follows from the uncertainty principle that $\Delta E = 0$, implying that the ground state is extremely sharp. By contrast, the higher allowed energy states or levels are not sharp; they have a certain nonzero width in energy, because of the uncertainty in energy coming from the not-too-large value of Δt.

An electron has a position and a momentum, and we can only assign a probability distribution, or a *wave function*, to the values of these parameters. This implies a certain degree of quantum fuzziness, or *indeterminism*.

Quantum fluctuations and quantum field theory

The quantum uncertainty principle has a remarkable fallout. Suppose Δt is very small in a certain situation. Then the principle permits a correspondingly large value of ΔE. This means that, within the limits prescribed by the uncertainty principle, violations of the principle of energy conservation *can* occur via *quantum fluctuations* of energy.

The subject which combines quantum mechanics and the *special* theory of relativity is called *quantum field theory*. In this theory even vacuum is no longer something empty. It is a *quantum field*, alive with a spontaneous and perpetual creation and annihilation of *particle-antiparticle pairs* which are able to live within the time and energy uncertainty permitted by the uncertainty principle. Such quantum fluctuations have been occurring in vacuum all the time, right from the moment of the 'Big Bang'.

The Big Bang

It is believed by most cosmologists at present that our universe, as we know it today, emerged out of 'nothing' at a 'Big Bang moment', ~14.2 billion years ago (Krauss 2012, 2017). This was a quantum event because the spatial dimension of the system was extremely small, almost zero ($\Delta x = 0$). And this, in turn, means that Δp_x, and therefore ΔE, could become arbitrarily large at the moment of the Big Bang (see Eqs. 5.6 and 5.7). Our universe could have possibly emerged out of such a *quantum fluctuation*. This energy fluctuation got sustenance from the fact that the gravitational interaction was born at the same instant. This issue will be revisited later in this book. Meanwhile, it will be helpful to quote Lloyd (2006) here:

Quantum mechanics describes energy in terms of quantum fields, a kind of underlying fabric of the universe, whose weave makes up the elementary particles – photons, electrons, quarks. The energy we see around us, then – in the form of Earth, stars, light, heat – was drawn out of the underlying quantum fields by the expansion of our universe. Gravity is an attractive force that pulls things together. . . As the universe expands (which it continues to do), gravity sucks energy out of the quantum fields. The energy in the quantum fields is almost always positive, and this positive energy is exactly balanced by the negative energy of gravitational attraction. As the expansion proceeds, more and more positive energy becomes available, in the form of matter and light – compensated for by the negative energy in the attractive force of the gravitational field.

5.4 Summing over multiple histories

There are quite a few formulations of quantum mechanics. The one used by Hawking and Mlodinow (2010) in their book for explaining the existence of our present universe is Feynman's *sum-over-histories* formulation. Feynman was intrigued by some further variations carried out on the famous double-slit experiment described earlier in this chapter (Fig. 5.1). Suppose you close one slit and carry out the experiment. You record a pattern reminiscent of the single-slit experiment carried out earlier by Young for a beam of light: There is a diffraction pattern comprising of a central maximum and a number of secondary maxima. Now close this slit and open the other one, and repeat the experiment. You again get a similar diffraction pattern, a little displaced from the first one. But if you superimpose these two patterns, what you get is *not* the same as when the two slits are open *at the same time*. This is a most remarkable experimental result, which nobody can dispute.

Even if you reduce the intensity of the electron beam so much that the electrons come one at a time, you still get the same result. This is intriguing. How does a single electron 'know' which slit is open, or whether one or two slits are open?

During the 1940s, Feynman had formulated a new version of quantum mechanics, in which when a quantum particle goes from point A to point B, it has available to it *all* possible trajectories or 'histories'. This approach was in line with the Heisenberg uncertainty principle, according to which there is an essential uncertainty in the location of the electron, which means that the electron can be everywhere in the universe at the same time, and not necessarily on any particular trajectory. Of course, there is a stronger probability that it would be in the vicinity of the slits in the above experiment, but the probability cloud characterizing its position extends over all space.

So the electron in the above experiment actually samples all paths simultaneously, including the one in which it finds that one of the slits is closed. *All* possible histories or trajectories are equally real, each with its own probability value. Moreover, since the electron cloud extends over all space, all the alternative histories get enacted *simultaneously* (Fig. 5.2).

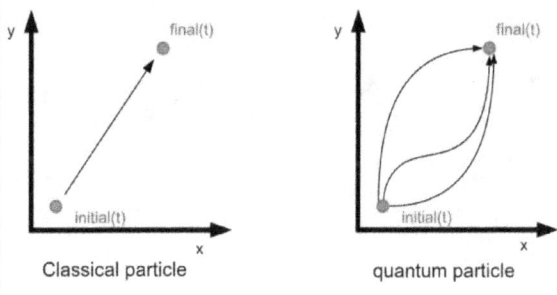

Fig. 5.2 Unlike a classical particle, a quantum particle can take all possible trajectories for going from an initial state to a final state. Image credit:

It may appear as if nothing much has been gained, because we can obtain the same results by treating the electrons as waves and solving the differential equation (namely the famous Schrödinger equation) for the 'wave function' describing the electrons. But there was a *conceptual breakthrough* here: We must carry out this sum over alternative or multiple histories for *any* quantum system, even for the birth and evolution of the universe, and not just for electrons. This is what was done by Hawking and Mlodinow (2010) in their model of the universe.

In Newtonian theory (or classical mechanics), the evolution from the past to the present is visualized as a definite succession of events, and there is only one history or trajectory. Not so in quantum theory. No matter how thoroughly and carefully we observe the present, the unobserved past, as also the future, is indeterminate, and exists only as a *spectrum of possibilities.*

This means that our universe does not have just a *single* past or history. We know the present, but the present could have evolved from any of a large number of possible histories, with various probabilities. Therefore we must sum over all those histories which can lead to the present we are in.

A remarkable fallout of this line of reasoning is that ours is only one of many possible universes. So there are *multiple universes*, and our *uni*verse is just one of them. It is a different matter whether we can ever interact with the other universes.

6. The Nature of Information

How did life emerge from nonlife? It did so through a very long succession of processes and events in which more complex structures evolved from simpler ones. Take any living entity; say the human body, or even a single-celled organism. The amount of *information* needed for describing the structure of a single biological cell is far more than the information needed to describe, say, a simple molecule like CO_2. The technical term one uses for this is '*degree of complexity*'. We say that a biological cell has a much larger degree of complexity than an atom or a molecule.

Let us tentatively (and very simplistically) define the degree of complexity of any object or system as the amount of information needed for describing its structure and function. It is necessary to become familiar with some of the jargon and basic concepts used in complexity science for dealing with questions such as the origin of life or the evolution of species. In particular, we should have a good idea about the nature of information.

Constituents of a complex system interact with one another, as also with the surroundings. Such interaction is basically *communication of information*.

Just what is information? Weiner (1948) developed a statistical theory of 'amount of information', in which *the unit amount of information* was defined as that transmitted by a single decision between equally probable alternatives.

Claude Shannon (1948) is regarded as the father of modern information theory. He provided a quantitative or numerical measure of information, and introduced the term 'bit', as the short form for 'binary digit'. Shannon's formulation was originally meant for designing better communication channels. An interesting aspect of modern communication theory is that the merit of a particular communication-channel design lies not just in how well the actual message is sent, but also in how well the channel *could* have sent all the other messages it might have been asked to convey (Tribus and McIrvine 1971).

I describe some formal aspects of information theory in Appendices A3 and A4. Here the basics are introduced in a non-mathematical language, not only of information theory, but also of its modern descendent, namely algorithmic information theory. It is instructive to adopt a historical approach and trace the development of ideas regarding the nature of information, and also information-processing (or what is now better known as *computation*).

6.1 Russell's paradox

At the beginning of the 20th century the mathematician David Hilbert asked a question in the theory of infinite sets: Starting from 1, 2, 3, . . . , what is the largest integer one can think of? Whatever that integer is, let ω be the first integer after all the finite integers. So we get 1, 2, 3, ... , ω. We need not stop there, and can go on adding the next higher integers to the set: 1, 2, 3, ... , ω, $\omega+1$, $\omega+2$, ... Let 2ω be the next integer after all the integers in this set: 1, 2, 3, ... , ω, $\omega+1$, $\omega+2$, ... 2ω. We can go on and on, and end up with:

$$1,2,3,..., \omega, \omega+1, \omega+2,...2\omega, 3\omega, 4\omega,..., \omega^2, \omega^3,..., \omega^\omega, \omega^{\omega^\omega}$$

This can go on. The next integer is ω to the power ω to the power ω . . . . forever.

Cantor proved a theorem which said that *for any infinite set there is a larger infinite set which is the set of all its subsets.* This is sometimes referred to as *Cantor's diagonal argument*.

Now suppose you apply this theorem to the '*universal infinite set*', which by definition is the set of *everything*. Application of Cantor's theorem here leads to a paradox because the theorem says that there is a set larger than the universal infinite set. This paradox was noticed by Russell. The seminal work *Principia Mathematica* by Whitehead and Russell (1925-1927), the first edition of which was published during 1910-1913, includes a new formulation of set theory by Russell, who was led to it during his sustained efforts to solve the following problem posed by him about sets:

Consider a set A, defined as a set containing all sets that are not members of themselves. Does A contain itself?

The two possible answers are Yes or No. If the answer is Yes, then there is a contradiction because set *A* is defined as a set containing sets which are *not* members of themselves.

If the answer is No, again there is a contradiction. Since *A* is defined as a set comprising of all sets which do not belong to themselves, it should contain itself. But according to the second answer, *A* does not contain itself.

Thus we have *incompatible propositions that imply one another*.

This famous paradoxical situation goes by the name of *Russell's paradox*. Apparently, Yes implies No, and No implies Yes. Russell's work for resolving this paradox led to a reformulation of mathematics in terms of his new theory of sets. His resolution of the problem amounted to imagining what we now call a *theoretical computer*, which was essentially a sequential logic machine. This imaginary machine carries out one logical operation at a time, i.e., it operates in discrete time. The answers about the set *A* are dealt with sequentially, *one at a time*. Thus, at any given point of time the answer may be, say, Yes. But the theoretical computer keeps running, and a few time steps later the answer becomes No. The program runs randomly in an infinite loop, alternating between Yes and No. Thus there is claimed to be no paradox because the answer is never Yes and No *at the same time*! Are you convinced by this explanation?

6.2 Hilbert's formal axiomatic approach to mathematics

Hilbert applied an approach different from that of Russell for tackling the problems posed by Cantor in set theory. He tried to use, and improve upon, Euclid's axiomatic method. The idea was to take the apparatus of symbolic logic to its extreme. He argued that one reason we get into contradictions in set theory is that words often have a vague meaning. So why not come up with a finite set of formal axioms and an *artificial* language for doing mathematics? This artificial language must have strictly precise grammatical and other rules.

Even more importantly, Hilbert wanted the rules of the symbolic logic to be so precise and 'artificial' that even a *mechanical proof checker* can be used to see if a proof based on the formal axioms is correct or not. This was Hilbert's important contribution, this way of making mathematics strictly black and white, with no shades of grey. The idea was to make mathematics completely certain and objective, with no room for human interpretations or subjectivity. He asserted that it should be possible to check the proof of a proposed theorem in a purely mechanical manner to see if there are any mistakes or not in the application of the agreed set of axioms.

Hilbert's goal was to put all pure mathematics on a formal and strictly objective footing. His idea was to formalize mathematics in terms of axioms and artificial language, so that everybody could agree on whether a proof is correct or incorrect. He hoped that mathematics would then comprise of *absolute truths*, and there would be a *theory of everything* in mathematics. He and his collaborators like John von Neumann did achieve considerable success in this direction.

But there was trouble in store for mathematics, as established by Gödel (1931). He, and also Turing (1936), showed that *it is impossible to formalize all of mathematics*. Nevertheless, Hilbert's (and also Russell's) work was the forerunner of Turing's (1936) work on what we now call the Turing machine (see Kurzweil 1998).

6.3 Gödel's incompleteness theorem

Laws and theories in science have been traditionally enunciated under three items of principle (Chaitin 2006):

1. *The Ockham's razor*: This is an approach that favours the smallest necessary number of assumptions, the smallest possible number of variables and parameters, and the smallest needed number of equations for formulating hypotheses. In other words, if there are two competing theories that explain the same set of experimental data, the simpler theory is better. In terms of computer algorithms (see below), the best theory is that which requires the smallest computer program for calculating (and hence explaining) the observations.

2. *Comprehension is compression*: If the observations are pointing to an underlying law, it should be possible to explain them in terms of an algorithm that requires a *smaller* number of bits than the number of bits required to represent the observed data. Even the most random (i.e., 'lawless') set of data can be 'explained' in terms of an algorithm of the same size. But we intuitively feel that we understand something in a better way if it is explained by a 'simple' theory, i.e., by an algorithm requiring a *smaller* number of bits than that needed for describing the entire set of data.

3. *Leibniz's principle of sufficient reason*: 'Everything happens for a reason'. If something is true, it must be true for a reason. This is why mathematicians try to discover the proof for the general case (i.e., prove theorems), rather than accepting something to be true just because a large amount of data seem to indicate that.

It was against such a background that Kurt Gödel (1931) created quite a commotion when he proved his *incompleteness theorem*, which demonstrates that mathematics contains statements that, if proved false, would render it inconsistent, but which cannot be proved to be true (see Dyson 1997). In other words, *some mathematical statements are true for no reason*.

Gödel's work went against Hilbert's belief that there is a theory of everything for mathematics. Hilbert (as also Leibniz) had visualized a world in which *all* truths can be calculated and proved by a 'universal coding'. This coding comprised of a few dozen axioms about sets, and it was sought to systematize all mathematics by deriving all mathematical truths from these axioms in a 'mechanical' way. Gödel proved that this is not possible. His theorem says:

No formal system encompassing elementary arithmetic can be at the same time both consistent and complete: Within any sufficiently powerful and noncontradictory system of language, logic, or arithmetic, it is possible to construct true statements that cannot be proved within the boundaries of the system itself.

Gödel established a procedure for encoding arithmetic statements as (large) numbers. Theorems in arithmetic had hitherto been speaking only *about* numbers. Gödel made them *self-*

referential and speak about themselves as well. With this interpretation of arithmetic theorems, he then constructed the now famous *Gödel sentence* which was a precise arithmetic statement, meaning

'*This statement is unprovable*',

or:

'This statement cannot be deduced from Hilbert's formal axiomatic system'.

If this statement is indeed provable, then it is a false statement. And if is not false but true, then it is unprovable, and we have incompleteness in our formal axiomatic system: We have a true statement that our formal system of logic and axioms is not able to capture. *Some truths are unprovable.*

The well-known *Goldbach conjecture* illustrates the point (see Gell-Mann 1994). The conjecture states that every even number greater than 2 is the sum of two prime numbers. Computer calculations have verified the truth of this conjecture even up to extremely large even numbers. But we can never be sure that the conjecture will not fail for some still larger even number. It has not been possible to *prove* the conjecture by rigorous mathematical reasoning. Suppose the Goldbach conjecture is *undecidable*. It would then be true without being provable because there could be *no* exception to it. The existence of any exception to the conjecture would disprove the conjecture, and that would contradict its undecidability.

It should be noted that Gödel put in some very clever thinking when he introduced a procedure (using prime numbers) for converting all arithmetic statements into numbers. His procedure was able to deal with pronouns like 'this', and even *self-referencing* pronouns like 'I'. This is something usually not found in mathematical formulas. Gödel converted the 'This statement is unprovable' entity into a large numerical statement in number theory, and although this converted entity looked like a statement in real mathematics, it was indirectly referring to itself, and saying that it is unprovable.

Gödel proved another incompleteness theorem, according to which: *No formal system can prove its own consistency.*

According to Chaitin (2006), Gödel's proof 'does not show that mathematics is incomplete. More precisely, it shows that individual formal axiomatic mathematical theories fail to prove the true statement "This statement is unprovable". These theories therefore cannot be "theories of everything" for mathematics. The key question left unanswered by Gödel was: Is this an isolated phenomenon, or are there many important mathematical truths that are unprovable?'

According to Chaitin's algorithmic information theory, which I shall outline shortly, some mathematical facts cannot be compressed into a theory because they are too complicated (in terms of algorithmic complexity). His work suggests that 'what Gödel discovered was just the tip of the iceberg'.

Before I discuss Chaitin's work on algorithmic information theory, we must take due note of the pioneering work of Alan Turing, who demonstrated that incompleteness or *undecidability* is quite pervasive in Nature.

6.4 Turing's halting problem

Hilbert had identified the so-called *decision problem* for defining a closed mathematical universe as follows: Is there a decision procedure that, given any statement expressed in a given language, will always produce either a *finite* proof of that statement, or else a definite finite construction that refutes it, but never both? In other words, can a precise *mechanical procedure* distinguish between provable and disprovable statements within a given system?

The answer to this question required a way to define 'mechanical procedure'. Turing (1936) provided an answer. He proved, among other things, that the notions of 'mechanical procedure', 'effectively calculable', and 'recursive functions' are actually one and the same thing.

I make a digression here to describe recursive functions.

Recursive functions

These are functions that can be defined by the accumulation of elementary component parts; they can be deconstructed into a finite number of elemental steps. A recursive *procedure* is one that calls on itself. Recursion is a process of defining or expressing a function or procedure in terms of itself. Each iteration of the recursive procedure is designed to produce a simpler version of the problem. This process continues until the algorithm reaches a subproblem for which the answer is already known or simple to work out in a nonrecursive manner (Kurzweil 1998).

As an illustration, consider the computation of the factorial function $n!$. The recursive rule here is: $n! = n(n-1)!$. The procedure calls on itself again and again, till it reaches the subproblem $1!$, for which the answer is a given: $1! = 1$.

Recursive functions are used extensively in game-playing algorithms. For chess-playing, for example, which involves two players, Kurzweil (1998) defines the following recursive formula:

PICK MY BEST MOVE: *Pick my best move, assuming my opponent will do the same. If I've won, I'm done.*

At each step of the recursion, consequences of each possible move are worked out. It is assumed that the opponent is equally intelligent, and will do the same. Since the number of possibilities to consider may blow up exponentially, the so-called *minimax procedure* is adopted to arrive at a strategy in reasonable time (this is explained in some detail in Appendix 10 on game theory). An expanding 'tree' of possible moves and countermoves is constructed (for a preassigned time), and an assessment of the freshest 'leaves' (extremities) of the tree that minimizes the chances of the opponent to win, and maximizes the chances of the game-playing program to win, is carried out. This information is then fed back down the branches of the tree for determining the best move of the program.

Turing's universal computer, and its halting problem

Returning to the work of Turing, his most important departure from the approach of Gödel was to bring in the idea of a (conceptual) computer for number crunching. He interpreted Hilbert's philosophy to mean that there should be a computer and a computer program for implementing Hilbert's idea of a mechanical procedure for deciding whether a given proof is correct or not.

The most famous and important notion he introduced was that of the *universal computer* (Turing 1936). He demonstrated that all digital computers are fundamentally equivalent, regardless of what is there in their innards. It is all 1s and 0s underneath. He used this approach to prove that *unsolvable mathematical problems do exist*: He proved the so-called *halting problem*:

There can be no general procedure to decide in advance whether a self-contained computer program will eventually halt (i.e., solve a posed problem and then halt). Often the only way to answer this question is by actually running the program to see whether it halts or not.

In other words, there is no computer program that can take another computer program and determine with certainty whether the first program halts or not. Thus there is no mechanical procedure that can decide in advance whether a program will eventually halt or not. And this means that there is no set of Hilbertian mathematical axioms that can enable us to prove whether a program will halt or not.

Turing invented an imaginary device now called *the Turing machine*. It consisted of a black box (it could be a typewriter or a human being) able to read and write a finite alphabet of symbols to and from an unbounded length of paper tape, and capable of changing its own '*m-configuration*' or 'state of mind'. In the context of computational science, Turing introduced the all-important notions of *discrete time* and *discrete state of mind. This made logic a sequence of cause-and-effect steps, and mathematical proof a sequence of discrete logical steps.*

The Turing machine embodies a relationship between a sequence of symbols in space and a sequence of events in time. Each step in the relationship between the tape and the Turing machine is governed by what we call a program. The program provides instructions to the machine for every conceivable situation. Even a simple program can generate a complicated output: *Complicated-looking behaviour need not require a complicated state of mind.*

Turing argued that all symbols, all information, all meaning and all intelligence that can be described in words or numbers can be encoded (and transmitted) as binary sequences of *finite* length (see Dyson 1997; Moravec 1999). In contrast to this, Gödel's theorem on undecidability in mathematics states that in all but the simplest mathematical systems there may be propositions that cannot be proved or disproved by any *finite* mathematical or logical process; the proof of a given proposition may possibly call for an infinitely large number of logical steps.

However, Turing did prove, like Gödel, that there were self-referential questions a universal Turing machine could not answer.

But perhaps outsiders could?

Turing's proof for the halting problem

Turing proved that the halting problem in unsolvable. The proof is as follows (Chaitin 2006): Suppose a general procedure H does exist that can decide whether any given computer program will halt. Knowing H, we can construct a program P that uses H as a subroutine. Suppose the size of program P is N bits, and P knows its own size. Clearly, P is large enough to contain the number N. Using the procedure H, the program P can take a look at all programs which are up to, say, $100N$ bits in size, and find out which of them halt and which do not. After this, P runs all the programs which had halted to see what outputs they had generated. Naturally, this will be a set of all digital objects with a degree of complexity (i.e., the number of bits required for their description) up to $100N$. At the end, P outputs the smallest integer not in this set, and then halts.

This means that we have shown that P has halted and it has given an output integer that cannot be produced by a program of size $\leq 100N$. This is logically inconsistent because the size of P in only N. Therefore H does not exist and thence P cannot be constructed; i.e., a general procedure H does not exist that can decide whether or not any given computer program will halt. QED. *Turing's halting problem is unsolvable*.

6.5 Elementary information theory

A bit has two states: 0 or 1. Shannon (1948) took the bit as the unit of information. One bit is the quantity of information needed (it is the 'missing' or 'not-yet-available' information) for deciding between two equally likely possibilities (for example, whether the toss of a coin will be 'heads' or 'tails').

The information content of a system is the minimum number of bits needed for a description of the system.

The term 'missing information' is assigned a numerical measure by defining it as the uncertainty in the outcome of an experiment yet to be carried out. The uncertainty may be high either because only one of a large number (N_s) of outcomes is possible, or, what is the same thing, the probability of a particular outcome out of many is inherently low.

Suppose we have a special coin with heads on *both* sides. What is the probability that the result of a spin of the coin will be 'heads'? The answer is 1, which is the maximum possible probability, i.e., certainty. Thus the carrying out of this experiment gives us zero information. We are certain of the outcome, and we get that outcome.

Next we repeat the experiment with a normal, unbiased, two-sided coin. There are two possible outcomes ($N_s = 2$). In this case the actual outcome gives us information which we did not have before the experiment was carried out (i.e., we get the missing information).

Suppose we toss two coins instead of one. Now there are four possible outcomes ($N_s = 4$). Therefore, any particular experiment here gives us even more information than in the last two situations.

To assign a numerical measure to information, we would like the following criteria to be met (Williams 1997):

1. Since information I depends on N_s, the definition of information should be such that, when we are dealing with a combination of two or more systems (e.g., two coins, or two dice), N_s for the composite system is correctly accounted for in the definition of information. For example, for the case of two dice tossed together or successively, $N_s = 6 \times 6 = 36$, and not $6 + 6 = 12$.

2. Information I for a composite or multivariate system should be a *sum* (and not, say, a multiplication) of the information for the components comprising the system.

The following relationship meets these two requirements:

$$N_s \sim base^I \qquad\qquad (6.1)$$

Suppose system X has N_x states and the outcome of an experiment gives information I_x. Let N_y and I_y be the corresponding quantities for system Y. For the composite system comprising of X

and Y,

$$N_x N_y \sim base^{I_x} base^{I_y} \tag{6.2}$$

Since $N_s = N_x N_y$, we get

$$N_s \sim base^{(I_x + I_y)} \tag{6.3}$$

Taking logarithms of both sides, and writing $I_x + I_y = I$, we get

$$I \sim \log N_s / \log(base) \tag{6.4}$$

What logarithm we take, and what proportionality constant we select, is a matter of context and choice. All such choices differ only by some scale factor, and the units. The important thing is that this approach for the definition of information has given us a correct accounting of the number of states or '*bins*' or classes (i.e., by a *multiplication* of the numbers of individual states), and a correct accounting of the individual measures of information (i.e., by *addition*).

All the cases considered above are *equiprobability* cases: The probability P_1 for getting 'heads' is 0.5, and so is the probability P_2 for getting 'tails'. Similarly, when the die is thrown, the probability P_1 that the face with '1' will show up is 1/6, as are the probabilities P_2, P_3, .. P_6 that '2', '3', . . '6' will show up. For such examples, the constant probability P is simply the reciprocal of the number of possible outcomes, or classes, or bins, i.e. N_s:

$$P = 1 / N_s \tag{6.5}$$

Eq. 6.4 can then be written in terms of P as follows:

$$I \sim \log(1 / P) / \log(base) \tag{6.6}$$

Introducing a suitable proportionality constant c, we get

$$I = c \log(1 / P) = -c \log P \tag{6.7}$$

This is close to the famous *Shannon formula* for missing information. Further mathematical details are given in Appendix A3.

Eq. 6.7 is similar to the well-known Boltzmann equation for entropy:

$$S = k \log W \tag{6.8}$$

Shannon information (I) is closely related to entropy (S) in statistical thermodynamics. Entropy is a measure of the number of possible microstates in which a system can exist, corresponding to a given macrostate. Similarly I is the amount of missing information before we perform an experiment the result of which we cannot predict with certainty; naturally, I is large if the possible number of outcomes of the experiment is large, and *vice versa*. Similarly S is large if the possible number of microstates is large and we have no idea about which microstate is actually occupied.

Suppose a system can have 8 (i.e., 2^3) possible microstates. Then the maximum possible value for entropy is $S_{max} = \log_2 8 = 3$.

In decimal notation we can label the microstates as 0, 1, 2, 3, 4, 5, 6, 7, or we can use binary notation and label them as 000, 001, 010, 011, 100, 101, 111. Suppose we know for sure that the state occupied is, say, 101. In this case, *available* information (J) has the value 3 because three digits are needed to specify the state 101. And the entropy S is zero in this case. So $J = 3$ and $S = 0$.

Next suppose there is a situation in which we have no knowledge about which of the 8 states is actually occupied. And all states have equal probability of being occupied. For such a case the available information J is zero. And entropy S is $\log_2 8$, or 3. Therefore $J = 0$ and $S = 3$.

It can be shown that *the sum of entropy and available information follows a conservation law:* $J + S = \text{constt}$.

In the above example, $S_{max} = 3$ and $J_{max} = 3$. In fact, the general conservation law can be stated as follows (Shannon and Weaver 1949; Layzer 1975; Bennett 1987a, b; also see Lin 2001, 2008):

$$J + S = J_{max} = S_{max} \; .$$

For real systems the possible number of microstates can be very large indeed. But it is not infinite, because it is limited by the uncertainty principle of quantum mechanics (see Appendix A6 on quantum theory). This principle says that there is an uncertainty (i.e., a lower limit on the precision) with which we can determine both the position and the momentum of a particle. Therefore any finite physical system can be completely specified in terms of a *finite* amount of information.

6.6 Entropy means unavailable or missing information

In Chapter 3 we considered the example of free expansion of a gas. We saw how, left to themselves (i.e., with no inputs from the outside), things are more likely to tend towards a state of greater disorder. In the double-sized chamber (Fig. 3.1), the gas has 2^n more accessible states, or microstates. This is a measure of the increase in entropy, as also of the higher degree of unavailable information. The symbol W was used for the number of microstates accessible to a system under consideration. This number doubled when the gas expanded to twice the volume.

We saw that one way of quantifying the degree of disorder is to say that entropy S, a measure of disorder, is proportional to W; i.e., $S \sim W$. But, as discussed in Chapter 3, Boltzmann did something even better than that for quantifying disorder by defining entropy S as proportional to the *logarithm* of W; i.e., $S \sim \log W$. This introduction of log W, instead of W, in the definition of entropy was done for the same reasons as those cited above for defining missing information I. In fact,

$$I \sim S \tag{6.9}$$

Also, the original (thermodynamic) and the later (statistical mechanical) formulations of entropy are equivalent. In the former, entropy of an isolated system increases because a system can increase its stability by obliterating thermal gradients. In the latter, entropy increases (and information is lost) because spontaneous obliteration of concentration gradients (and the ensuing more stable state) is the most *likely* thing to happen. *Concentration gradients get obliterated spontaneously because that takes the system towards a state of equilibrium and stability.*

For an isolated system, maximum stability, maximum entropy, and maximum probability all go together.

6.7 Algorithmic information theory

> *With Gödel it looks surprising that you have incompleteness, that no finite set of axioms can contain all of mathematical truth. With Turing incompleteness seems much more natural. But with my approach, when you look at program size, I would say that it looks inevitable. Wherever you turn, you smash up against a stone wall and incompleteness hits you in the face!*
>
> Chaitin (2001)

> *An infinite number of true mathematical theorems exist that cannot be proved from any finite system of axioms.*
>
> Chaitin (2006)

Algorithmic information theory (AIT) was founded by Kolmogorov (1965) and Chaitin (1987a, b, 2001, 2006).

How much information (in terms of number of bits) is needed for specifying the set of all positive integers: 0, 1, 2, 3, . . . ? The sequence runs all the way to infinity, so does it have an infinite information content? Something is wrong with that possibility. We can see that we can generate the entire sequence by starting from 0, and adding 1 to get the next member of the set, and then obtain the next member by adding 1 again, and so on. So, because of the *order* or *structure* in this sequence of numbers, an *algorithm* can be set up for generating the entire set of numbers. And the number of bits needed to write the corresponding computer program is small, and not at all infinite.

The number of bits needed to write the computer program for generating a given set of numbers or data is called the '*algorithmic information content*' (AIC) of that set of data. Algorithmic information theory (AIT) is the modern discipline which is a great improvement over classical information theory. But such ideas about *compression of information* have a long history.

Leibniz (1675) was amongst the earliest known investigators of compressibility (or otherwise) of information. He argued that a worthwhile algorithm or *theory* of anything has to be 'simpler than' the data it explains. Otherwise, either the theory is useless, or the data are 'lawless'. The idea of 'simpler than' is best expressed in terms of AIC defined above.

The information in a set of data can be compressed into an algorithm only if there is something nonrandom or ordered about the data. There must be some structure or regularity, and we must be able to recognize that regularity or 'rule' or 'law'. Then only can we construct the algorithm that generates the entire set of data. In fact, this is how we discover and formulate the laws of Nature, and understand natural phenomena. And the statements of the laws are nothing but a case of compression of information, using a smaller number of bits than the number of bits needed for describing an entire set of observations about Nature.

Consider two numbers, both requiring, say, a million bits for specifying them to the desired accuracy. Let one of them be an arbitrary random number, which means that there is *no defining pattern* or order or structure for specifying it. Let the other number be the familiar π (= 3.14159.....). The second number has very small AIC because a small computer program can be written for outputting it to a desired level of precision (π is the ratio of the perimeter of a circle to the diameter of the circle). By contrast, a random number (say 1.47373..59) has a much higher AIC: The shortest program for outputting it has information content (in terms of number of bits) not very different from that of the number itself, and the computer program for generating it can be only this:

Begin
Print "1.47373..59"
End

No significantly smaller program can generate this sequence of digits. The digit stream in this case has no redundancy or regularity, and is said to be *incompressible*. Such digit streams are called *irreducible* or *algorithmically random*.

Such considerations have led to the conclusion that there are limits to the powers of logic and reason. Gregory Chaitin has shown that certain facts are not just computationally irreducible; they are *logically* irreducible as well. The 'proof' of their truth must be in the form of additional axioms, without any reasoning. So there are severe limits to the powers of logic and reason.

Computation is characterized by three things:

1. The length of the computer program.

2. The time it takes to do the computation.

3. The computer memory needed for the job.

AIT largely ignores the second and the third aspect, and recognizes mainly the first for defining the information content of a given set of numbers or a given piece of information. In other words, it focuses on *program-size complexity*.

Chaitin makes the point that a theory can be likened to a computer program. The program calculates and explains a certain set of observations, and the smaller this program is (in terms of compression of information), the better is the theory.

When a set of observations or data cannot be described compactly in terms of axioms and/or theorems, there is no structure or order, or pattern in the data. Such a set of data is *logically random*. Something is random if the smallest program that calculates or generates it is the same size as it is, so there is no compression.

Chaitin's halting probability Ω

Chaitin introduced a number omega (Ω) to quantify the degree of logical randomness of any system, and to show the limited powers of reason. He demonstrated the existence of *an infinite stream of unprovable mathematical facts*.

Let the term 'program' imply 'the concatenation of the computer program and the data to be read in by the program'. Consider an ensemble of all such possible programs. What is the probability that a program chosen at random from this set will ever halt? The number Ω denotes that probability.

How do we choose a program at random for testing this? A program is a succession of bits (0s and 1s). Since we are considering all possible programs, any succession of bits is a possible program for testing its halting behaviour. We can flip a coin repeatedly to get a random sequence of bits. We go on adding random bits, one at a time, till the sequence of bits is a program that halts, if at all it can halt. The number Ω is the probability that the halting will occur (if at all) for the tested sequence of randomly generated bits.

These operations, of course, assume the presence of a computing machine for doing the job of testing. We also assume the use of a programming language. But it turns out that the crucial conclusions about halting or otherwise do not depend on these things: the actual values of Ω may depend on them, but not the general conclusions drawn. Our arguments can proceed by assuming a particular computer and a particular language for all computing.

Since the number Ω is a probability, it lies between 0 and 1. In binary notation it may look something like 0.110010101... The central point made by Chaitin is that the bits after the decimal point form an *irreducible* stream. Every 0 or 1 after the decimal point represents a fact, and the totality of these bits represents irreducible mathematical facts.

The number Ω can be regarded as an infinite sum. Each N-bit program that halts contributes $1/2^N$ to this sum. Each such program adds a 1 to the Nth bit. One may think that a precise value of Ω can be computed by adding all the bits for the programs that halt. This is not so. Although Ω is a perfectly well-defined specific number, *it is impossible to compute it in its entirety*. It is possible to compute only a few digits of Ω. For example, if we know that computer programs 0, 10, and 110 all halt, then $\Omega = 0.111$ up to three digits. But the first N digits of Ω cannot be calculated by a program of length significantly shorter than N. Knowing the first N digits of Ω will enable us to tell whether or not each program up to N digits in size ever halts. This means that at least an N-bit program is needed to calculate N bits of Ω.

Chaitin's Ω cannot be computed to arbitrarily high precision because if we know Ω exactly, we can solve Turing's halting problem, which is actually unsolvable.

Given any finite program, no matter how long, we have an infinite number of bits that the program cannot compute. This implies that, given any finite set of axioms, there are an infinite number of truths that are unprovable in that system. Ω is irreducible.

Thus a theory of everything for all of mathematics cannot exist. The number Ω has an infinite number of bits or mathematical facts that cannot be derived from any principles simpler than the string of bits itself.

This means that mathematics has an infinite degree of complexity, whereas any particular theory of everything can only have a finite degree of complexity.

Gödel's work had shown that individual formal axiomatic mathematical theories cannot prove the true numerical statement 'This statement is unprovable'. According to Chaitin (2006), Gödel left unanswered the key question: 'Is this an isolated phenomenon, or are there many important mathematical truths that are unprovable?' It now turns out that the number Ω provides an infinite number of true theorems that cannot be proved by any finite system of axioms.

Leibniz had stipulated that if something (a theorem) is true in mathematics, it is true for a reason, the reason being the proof of the theorem. But the bits of Ω are totally random, and therefore these mathematical truths are truths for no reason. They are true by accident, and are therefore unknowable.

6.8 Algorithmic probability and Ockham's razor

We are to admit no more causes of natural things than such as are both true and sufficient to explain their appearances. Therefore, to the same natural effects we must, so far as possible, assign the same causes.

Isaac Newton

Given the idea that neo-Darwinism-derived complexities might be superposed on those more conventionally generated by energy alone, perhaps we can agree that biology is physics with added features. That may be why, in physics, the simplest model for any particular phenomenon is usually the right one, whereas in biology historical accidents and accumulating complexity often invalidate the influence of Ockham's razor. That added degree of complexity might also explain why biology has no natural laws per se, only a set of coarse guidelines.

Chaisson, Cosmic Evolution

The famous Ockham's razor was mentioned above in Section 6.3. It makes sense to first choose the simplest of the two or more alternative theories for explaining a set of observations about a phenomenon. The proverbial Ockham's razor shaves away the unnecessary stuff, and only the simplest or the most parsimonious theory, which makes the smallest number of assumptions, usually survives in our criterion for a good theory.

But it is conceivable that the simplest theory may be wrong or inadequate. The idea of Ockham's razor (it is only an idea, after all) is that one should proceed to simpler theories until simplicity can be traded for greater explanatory power. Confronted with a multiplicity of candidate theories, we have to bring in likelihood or probability considerations ('Which theory is more *likely* to be right?'). It turns out that algorithmic information theory comes to our help here, and provides a certain degree of legitimacy to the philosophical-looking Ockham-razor approach. In AIT we define a parameter called *algorithmic probability* (AP). It is the probability that a random program of a given length fed into a computer will give a desired output; say the first million digits of π. Following Bennett and Chaitin's pioneering work done in the 1970s, let us assume that the random program has been produced by an unintelligent monkey. The AP in this case is the same as the probability that the monkey would type out the same bit string (a sequence of 0s and 1s), i.e., the same computer program as, say, a Java program suitable for generating the first million digits of π. The probability that the monkey presses the first key on the keyboard correctly is 1/2 or 0.5. The probability that the first two keys are pressed correctly is $(0.5)^2$ or 0.25. And so on. Thus the probability gets smaller and smaller for typing correctly a larger and larger sequence of bits. Therefore the longer the program, the less likely it is that the monkey would crank it out correctly. This means that the AP is the highest for the shortest programs, and *vice versa*.

Now suppose we are having a bit-string representing a set of data, and we want to understand the *mechanism* responsible for the creation of that set of data. In other words, we want to discover the law, or *the* computer program, among many we could generate randomly, which produced that set of data. According to the above AIT rationalization for Ockham's philosophy, the shortest such program is the most plausible guess, *because it has the highest AP*. The simplest explanation is *usually* (but not always) the right one.

The Ockham-razor idea has two parts: The principle of plurality; and the principle of parsimony, economy or succinctness. The former says that plurality should not be posited without necessity. And the latter says that it is pointless to do with more what can be done with less.

The celebrated scientific method (Wadhawan 2014) is implicitly based on three axioms:

- The existence of objective reality.
- The existence of natural laws.
- The constancy of natural laws.

In science we assume that theories or models of natural law must be consistent with repeatable experimental observations. This assumption is based on the above axioms, and Ockham's razor is often invoked in scientific debate:

'We could still imagine that there is a set of laws that determines events completely for some supernatural being, who could observe the present state of the universe without disturbing it. However, such models of the universe are not of much interest to us mortals. It seems better to employ the principle known as Occam's razor and cut out all the features of the theory that cannot be observed' (Hawking 1988).

Albert Einstein is famous for many one-liners, including the following: *Everything should be made as simple as possible, but not simpler*.

6.9 Algorithmic information content and effective complexity

I have discussed the various ways of defining the degree of complexity of a system in an earlier book (Wadhawan 2010). Algorithmic information content (AIC) is one such measure of complexity. It is the minimum number of bits needed to store the algorithm needed for computing the data or information for describing the structure and function of a system (and then stop computing). The larger this number is, the higher is the degree of complexity of that system. Thus AIC is a measure of how *hard* it is to represent a text or a bit stream using a computer program.

The crux of a highly complex system is actually in its *non-random* aspects. Gell-Mann (1994) defined *effective complexity* as roughly the length of a concise description of the *regularities* (as contrasted to the degree of randomness) of that system or bit string. By contrast, AIC refers to the length of the concise description of the *whole* system or string, rather than referring to the regularities alone. Gell-Mann's approach is clearly better for describing complexity.

We humans are *complex adaptive systems* (CASs). CASs have a learning feature, and such learning requires, among other things, the evolution of an ability to distinguish between the random and the regular. Effective complexity is related to the description of the regularities of a system by a CAS that is observing it.

A CAS separates regularities from randomness. Therefore a CAS provides the possibility of defining its complexity in terms of the length of the *schema* used by it for describing and predicting an incoming data stream. Gell-Mann defined the effective complexity of a system, *relative to a CAS that is observing it*, as the length of the schema used to describe its regularities.

6.10 Classification of problems in terms of computational complexity

AIT deals with the *information content* of a given set of numbers, or a given piece of information, or a given algorithm. By contrast, *computational complexity theory* (CCT) addresses the question of the computational *resources* needed, particularly the time required, for solving a problem. It classifies combinatorial problems in terms of the time taken, in principle, to solve them. It largely ignores the nitty-gritty about the computer available for solving the problem, or the algorithm used for the purpose.

Estimation of the time required for the computational solution of a problem is basically a counting problem, or a *combinatorial problem*. Therefore this subject overlaps substantially with statistical mechanics. The time taken to solve a problem on a computer depends on the computer and the algorithm. CCT however ignores these details and focuses instead on the inherent complexity of the problem, as reflected in the number of steps needed to solve the problem. A classification is therefore introduced as follows.

P-class problem. For a problem (say of size n) belonging to the P-class of complexity, i.e., for a *polynomial-class problem*, there exists a *deterministic algorithm* that solves it in polynomial time n^j, where j is an integer. A deterministic algorithm is one the operations of which are known in advance.

NP-class problem. NP stands for *nondeterministic polynomial time*. NP-class problems are those for which a nondeterministic algorithm exists that can verify a given solution of the problem in time that increases in a *polynomial* manner (rather than, say, an exponential manner) with the increasing size of the problem. A problem is of complexity class NP if one can *guess* a solution of the problem and then verify its correctness or otherwise in polynomial time. For an NP problem, a proposed solution can be *verified* 'quickly'. By contrast, a P problem is that which can be *solved* 'quickly' even without being given a solution. It is believed that P ≠ NP.

NP-complete-class problem. The term *NP-complete* is used for the hardest subset of NP-class problems such that, for a particular NP-complete problem, every instance of every problem in NP can be converted to an instance of this particular problem, and this conversion can be effected in polynomial time. No effective computer algorithms have been developed for *solving* such problems. The only known approach for possibly solving them requires an amount of computational time proportional to an exponential function of the size of the problem; or to a polynomial function on a nondeterministic computer that, in effect, guesses the correct answer. A fast solution of any NP-complete problem (if such a solution exists) can be translated into a fast solution of any other NP-complete problem.

A large number of NP-complete problems have been identified. The fact that no efficient algorithm has been devised for solving any of them encourages the conjecture that P ≠ NP.

6.11 'Irreducible complexity' deconstructed

Humans can be funny. Given two situations of comparable complexity, they have no difficulty in accepting one as having arisen spontaneously (with no designer or controller or creator involved), but they often insist on invoking a designer or creator ('God') for explaining the other. Consider the world economy. We all agree that it is extremely complex, and also that there is no central authority which designed what we see today. It has just evolved in complexity over time, and is self-regulating. There are sovereign nations, each with an economy of its own. There is trade among the nations, and within every nation. There are independent stock markets, various currencies, and exchange rates which vary on a daily basis, with no central authority controlling the rates at will. Things look so well regulated that Adam Smith made his famous (but figurative) statement about there being an '*invisible hand*' behind all the order and complexity of modern economy.

But come to the spontaneous evolution of biological complexity, and suddenly many people stop being rational, and postulate a God who must have created the complex life forms. This teleological argument was advanced, among others, by the British theologian William Paley: Suppose you are on a beach or an uncultivated field, and you come across a piece of rock. You find nothing striking about that (i.e., do not think of a creator of the rock), and move on. Next you see a watch. Chances are, you are quite clear in your head that the watch must have been made by a watchmaker: 'Since there is a watch there must be a watchmaker, because there is evidence of *design* in the watch'. Paley argued that, similarly, all the biological complexity we see around us points to the existence of a creator and designer, namely God.

People who subscribe to this line of reasoning have *hijacked* the phrase 'irreducible complexity' (IC) for making the point that, for example, 'a DNA molecule has complexity which cannot be explained or reduced to evolutionary causes involving evolution from simpler molecules. DNA controls the synthesis of proteins and yet, proteins must have pre-existed for the creation of DNA'. Therefore, so the argument goes, God must be invoked for explaining such IC.

[I say 'hijacked' because the scientific meaning of IC actually pertains to *logical randomness*: An irreducibly complex system is one for which no compression of information is possible, meaning that its apparent information content is not substantially more than that of the algorithm or theory or mathematical formula needed for expressing or explaining it; the apparent complexity cannot be reduced by discovering a rule or law that generated it, because none exists.]

Richard Dawkins demolished the if-there-is-a-watch-there-must-be-a-watchmaker argument at length in his 1986 book *The Blind Watchmaker: Why the Evidence of Evolution Reveals a Universe without Design*. If God created the irreducibly complex life forms, he must have been endowed with a still greater amount of complexity, information, and intelligence. But who or what created that? How can you explain an observed complexity by invoking an even greater first-cause complexity? I think it makes far better sense to assert (and prove) that the degree of terrestrial complexity rises incrementally, starting from less complex entities, aided by Darwinian natural selection and the availability of low-entropy or 'high-quality' energy from the Sun, all along remembering that we are dealing with thermodynamically *open* systems here.

As we shall see in the next chapter, Darwinian natural selection is *not* a totally random process. It acts on the genetic variation produced by random mutations and genetic drift, and helps the emergence of those individuals who have more of the adaptive traits useful for survival and reproduction. The probability for the spontaneous evolution of the existing life forms is indeed very low. *But a huge huge number of alternative life forms* could *have also evolved*. There are an enormous number of evolutionary pathways that *could* have been taken by organisms, and *any* of them could have been taken. This huge number should be multiplied with the very low probability of evolution of the existing life forms; then you get a much higher overall probability.

To understand this better, consider an analogy. You go out to the market place and see many people. Let us focus on any one of them. Nothing miraculous that you saw that person, right? Now wind back in time a little bit, so that you have not yet gone out to the market and, *before* going out, calculate the probability of meeting that particular person. Very low probability indeed. And yet you saw the person. It looks 'miraculous' only if you calculate the probability *after* the event, i.e., after singling out a particular person! There is just no case for thinking that the cosmic forces conspired or somebody pre-ordained that you would see that particular stranger today.

Feynman used to make fun of such tendencies. Here is what Bill Bryson (2003) wrote about him: 'The physicist Richard Feynman used to make a joke about *a posteriori* conclusions - reasoning from known facts back to possible causes. "You know, the most amazing thing happened to me tonight", he would say. "I saw a car with the license plate ARW 357. Can you imagine? Of all the millions of license plates in the state, what was the chance that I would see that particular one tonight? Amazing!" His point, of course, was that it is easy to make any banal situation seem extraordinary if you treat it as fateful.'

Another analogy would be helpful. Consider a deck of 52 playing cards. Any of the cards can be at the bottom. For each such possibility, there are 51 ways of choosing the second card for placing on top of the first. There are 50 possibilities for what the third card can be, and so on. The total number of possibilities is 52 x 51 x 50 x . . .3 x 2 x 1. This works out to $\sim 10^{68}$ ways in which the cards can be stacked one above the other, an enormous number indeed. There is only a 1 in 10^{68} chance that any particular stacking sequence will occur. And yet it would be highly illogical to infer that a miracle has occurred because we have observed a particular configuration and not another.

A flower is a work of art, but there is no artist involved. The flower evolved from lesser things, which in turn evolved from still lesser things, and so on, down the line, all the way.

7. Darwinian Evolution, Complex Adaptive Systems, Sociobiology

Amid all the recent interest in complexity, many point out that the future of science belongs more to biology, the study of complex systems, than to physics.

Margulis and Sagan

The theory of complex systems is perhaps going to look more like biology, with its myriad of species and of proteins, than physics, with its overreaching generalizations.

Herbert Simon

7.1 Darwinian evolution

In 1859, Charles Darwin announced one of the greatest ideas ever to occur to a human mind: cumulative evolution by natural selection.

Richard Dawkins

Natural selection is indeed an amazing engineer, but it works strictly without a drawing board. It does not think ahead. It only knows a good thing when it sees it in operation.

A. G. Cairns-Smith

The greatest single contribution to the subject of evolution of biological complexity was made by Charles Darwin. As Theodosius Dobzhansky said, 'nothing in biology makes sense except in the light of evolution'.

The year 2009 marked the second birth centenary of Darwin (see Sugden *et al.* 2009), as also 150 years of the publication of his celebrated book *On the Origin of Species by Means of Natural Selection*. The basic idea of biological evolution by natural selection is remarkably simple, yet of fundamental importance. Consider the mother-child relationship as an example. Mothers go through very substantial pain, hardship, risk to health and life, deprivation, sacrifices etc., and yet most of them are happy to bear a child and rear it. Why is that so?

In any population of females, there would be some variation regarding attitude to motherhood. There are likely to be some who avoid undergoing all the hardships and sacrifices involved. As a result, they do not get pregnant and bear children. Also, there are females in the population who are happy with the very thought of motherhood, pain and sacrifices notwithstanding. Such females would not only contribute their progeny to the population, the progeny, in turn, is likely to be like them, favourably disposed to the idea of motherhood. Over many generations, the result would be that the percentage of females not inclined to become mothers decreases, because no progeny means no representation of the tendency against motherhood in the population. This is *natural selection*: Nature 'selects' in favour of those females who are happy being mothers, and tends to gradually weed out those who are not. There is a gradual *evolution* of this trait in the population, so much so that, in due course, there are not many females left who do not wish to bear children.

Such reasoning can be generalized to other forms of biological evolution. Living organisms are thermodynamically open systems, i.e., they are constantly exchanging matter and energy with the environment. There is a fair amount of dynamic equilibrium between a living organism and its surroundings. The organism cannot survive if this equilibrium is disturbed too much, or for too long. The fact that an organism survives implies that, in its present form, it has been able to *adapt* itself to the environment. Also, if the environment changes slowly enough, organisms can evolve (over a long-enough time period) a new set of capabilities or features which enable them to survive even under the changed conditions. Over long periods of such evolutionary change, creatures may even develop into new species. This was the message of Darwin's (1859) bold theory of *evolution through cumulative natural selection.*

He demonstrated that adaptation to the environment was a necessary outcome of the exchange processes going on between organisms and their surroundings. A truly trailblazing consequence of his theory was that *all living organisms are the descendants of just one or a few simple ancestral forms* (see Dennett 1995).

Darwin started with the observation that, given enough time, food, space, and safety from predators and disease etc., the size of the population of any species can increase in each generation. But this indefinite (and exponential) increase does not actually occur; i.e., the so-called '*biotic potential*' of a species in the wild is actually never realized in full. In fact, usually only a very tiny fraction of the biotic potential is realized, meaning that, at least in the wild, only a small minority of the offspring reach maturity to produce the next generation of offspring; the rest die. Thus, there must be limiting factors in operation. Influenced by Malthusian ideas, Darwin imagined that if, for example, the available food is limited, only a fraction of the population can survive and propagate itself. But what decides who will survive and who will not?

Darwin's answer was that, since not all individuals in a species are exactly alike (i.e., since there is *variation* in the population), those better suited (or fitter) to cope with the prevailing conditions stand a better chance of survival ('*survival of the fittest*'). Moreover, the fittest individuals not only have a better chance of survival, they are also more likely to procreate. Thus, attributes conducive to survival and propagation have a better chance of getting *naturally selected* at the expense of less conducive attributes. And the effects of this natural selection accumulate over time, i.e., over generations. This is the process of *cumulative natural selection* recognized by Darwin.

It was also observed that children tend to resemble their parents to a substantial extent. The reason is that the progeny of better-adapted individuals in each generation, which survive and leave behind more offspring than others, acquire more and more of those features which are conducive for good adaptation to the existing or moderately-changing environment. A species perfects itself, or adjusts itself, for the environment in which it must survive, through the processes of both cumulative natural selection and *inheritance*.

Thus there are four basic features of Darwinian evolution:

1.   *Variability and variety* in members of a population in the matter of coping with a given environment.

2.   Likely *inheritance* of this variation by the next generation, with a few random modifications (mutations, replication errors, etc.).

3.   *Differential survival and reproductive success* of individual members of this new generation in the given environment.

4. *Establishment of a new population* more adapted to the environment, possessing new variations for passing on to the next generation.

Darwin's original notion of biological evolution by cumulative natural selection has been generalized for application to nonliving entities also. In fact, any population (for example that of bit-strings inside a computer) which comprises of individuals that (i) are similar but nonidentical, (ii) compete for resources and survival, (iii) replicate themselves from one generation to the next, and (iv) have a fair probability of passing on at least some of their winning characteristics to the next generation, is candidate for evolving into a population that is better and better adapted to the environment. If the environment changes, and if the changes are not too fast or drastic, the population evolves further, so as to survive and propagate itself even in the changed environment.

In any evolutionary process, what evolves is complexity. Darwinian evolution is about the evolution of biological complexity. Ironically, the word 'evolution' was not used by Darwin in his initial publications (Peters 1999). Instead, he talked about '*descent with modification.*' Nevertheless, his work was all really about the evolution of biological complexity in thermodynamically open systems. The concept of evolution has entrenched itself in the human psyche in a very profound and basic way, with far-reaching consequences (Dennett 1995).

Darwin changed the way we humans perceive ourselves. And the basic idea of evolution by natural selection has gone far beyond the precincts of biology. Apart from biological Darwinism, we speak of chemical Darwinism, quantum Darwinism, neural Darwinism, and what not. Deep down under, what evolves in any open system of interacting entities is complexity.

Biological entities embody enormous amounts of order, organization, and accumulated information. Can Darwinian evolution alone explain it? No. As Stuart Kauffman (1995b, 2000) emphasized, there is, in fact, an underlying complexity and order, on which Darwinian evolution operates further. Evolution of biological complexity is determined by two factors: *self-organization*, and *natural selection*. Self-organization or spontaneous ordering can occur in *any* thermodynamically open system, Darwinism or no Darwinism. Darwinian natural selection acts on this *existing* order and hones it further.

Complex systems evolve to states which have a coexistence of order and disorder, and the order and organization arise without any deliberate external cause. It is 'order for free'.

7.2 Complex adaptive systems

Although evolution of complexity occurs in every open dynamical system, it is important to take note of a special class of them called *complex adaptive systems* (CASs) (Gell-Mann 1994). CASs are dynamical complex systems that not only evolve, but also *learn* by making use of the information they have acquired.

Learning by a CAS requires, among other things, the evolution of an ability to distinguish between the *random* and the *regular*. CASs are those which can undergo processes like biological evolution or biological-like evolution. They do not just operate in an environment existing for them initially, but have the capability to even change the environment. For example, species, ant colonies, corporations, industries evolve to improve their chances of survival in a changing environment. Similarly, the marketplace adapts to factors like immigration, technological developments, prices, extent of availability of raw materials, and changes in tastes and lifestyles. Some more examples of CASs are: A baby learning to walk; a strain of bacteria evolving resistance to an antibiotic; a beehive adjusting to the decimation of a part of it; etc.

By contrast, a complex *material* is an example of a *nonadaptive* complex system (Wadhawan 2007). Galaxies and stars and other such complex objects are some more examples of nonadaptive complex systems. They are inanimate systems which evolve with time, but within the unchanging constraints provided by the initial conditions and by the environment. I shall discuss such systems in a later chapter.

I list here the characteristic features of CASs (Gell-Mann 1994; Holland 1975, 1995), also called *vivisystems* sometimes (Kelly 1994). I specifically consider systems that are large in terms of numbers of individuals or agents comprising the group (e.g., bees in a beehive).

1.   There is a *network* of the large number of individuals in the group, *acting in parallel*.

2.   Each individual *acquires information* about the surroundings and about itself.

3.   Each individual constantly reacts to what the others are doing. Therefore, from the vantage point of any individual in the network, *the environment is changing all the time*.

4.   A CAS identifies *regularities* in the information acquired by it, and condenses those regularities into a *schema* or conceptual model. It acts in the real world on the basis of that schema. CASs are *pattern seekers*.

5.   There can be many *competing schemata*, and the most suitable ones survive and evolve, based on the *feedbacks* received from the interactions with the environment.

6.   The control of a CAS is highly dispersed. *No one is really in command.*

7.   *Coherent behaviour* or *order* in a CAS arises from both competition and cooperation among the individuals themselves.

8.   *Emergent behaviour* (i.e., new, unpredictable behaviour) results from competition as well as cooperation among the individuals.

9.   A CAS may have *many levels of self-organization*. Individuals at one level may serve as the building blocks for individuals at the next higher level of hierarchy. In the human brain, for example, one block of neurons forms the functional regions for speech, another for vision, and still another for motor action. These functional areas link up at the next higher level, namely that of cognition and generalization.

10. In the light of new experience (obtained by feedback), CASs are constantly adjusting and rearranging their building blocks. This forms the basis of all learning, evolution, or adaptation in CASs. CASs are thus characterized by *perpetual novelty*. The processes of learning, evolution, and adaptation are basically the same. One of their fundamental mechanisms is the revision and recombination of the building blocks.

11. The CASs are constantly making *predictions*, thus anticipating the future. The predictions are based on various *internal models* of the world, and the models are constantly revised on the basis of new inputs; they are not static blueprints. Sheer large numbers and mutual exchange of information result in intelligence, '*swarm intelligence*' (Kennedy 2006). Beehives, ant colonies, and neural assemblies in a brain are examples of such swarm intelligence.

12. CASs have a certain *dynamism* not present in nonadaptive complex systems. And yet this dynamism is far from being total randomness. CASs have the ability to establish *a balance between order and chaos* (Fig. 7.1). This balance point is referred to as *the edge of chaos*. This point (or rather a membrane in phase space) represents the coexistence of order and chaos. Life signifies both stability and creativity, something that becomes possible in the vicinity of the edge of chaos.

13. The CASs have many *niches*, each of which can be exploited by an agent which has adapted itself to fill that niche. Filling up of a niche opens up new niches. The system is just too large and open to be ever in equilibrium. There is perpetual novelty, the stuff biological evolution is made of.

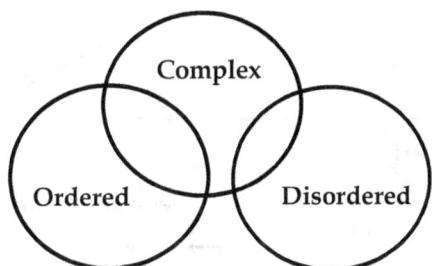

Fig. 7.1 Order (or structure) and disorder (or randomness) coexist in a complex system.

As pointed out by Gell-Mann (1994), the crux of a highly complex system is in its *non-random* aspects. He introduced the notion of *effective complexity*: It is roughly the length of a concise description of the *regularities* of that system or bit string. By contrast, algorithmic information content (AIC) (see Section 6.3) refers to the length of the concise description of the *whole* system or string, rather than referring to the regularities alone.

Suppose a bit stream is totally random. Then its AIC is infinite. But its effective complexity is zero, because a CAS observing the bit stream will not find any regularity in it, and therefore the length of the schema describing the regularities will be zero. At the other extreme, if the bit stream is totally regular, the AIC is very small (nearly zero), and so is the effective degree of complexity.

For intermediate situations, the effective complexity is substantial (i.e., different from zero). Thus, for effective complexity to be substantial, the AIC must not be too high or too low (Fig. 7.2). That is, the system should be neither too orderly nor too disorderly. For such situations, the AIC is substantial, but not maximal (for a given length of bit stream), and it has two contributions: The apparently regular portion (corresponding to the effective complexity), and the apparently random or stochastic portion. There is a critical balance between order and chaos, and this is the crux of what complexity is all about.

7.3 The inevitability of emergence of life on Earth

Darwin's published work hardly touches on the question of how life emerged out of nonlife; it was more about the evolution of life once it had appeared, and also about the emergence of new species.

The question of emergence of life out of nonlife is more in the realm of physics and chemistry, particularly the thermodynamics of open systems. As mentioned earlier in Chapter 4,

Darwinian evolution is only a special case of the more general evolution of all (thermodynamically open) dynamical systems. Naturally, the emergence of life out of nonlife is a physics problem. The recent work of Jeremy England (2013) should be viewed in that context. Let us recall that, as discussed in Section 1.3, emergence of a convection-based, and therefore more efficient, heat-dissipating mechanism is what creates the Bénard pattern. Such examples abound. One can generalize and say that *efficient dissipation of gradients can be taken as an organizing principle in physics*.

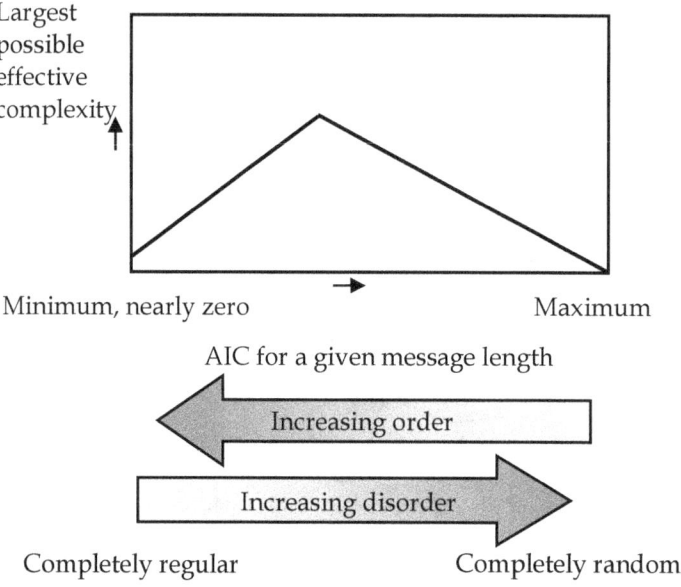

Fig. 7.2 Variation of the largest possible effective complexity with the algorithmic information content (AIC) of a given message length. (After Gell-Mann 1994.)

A sustained input of energy of any kind (e.g., electromagnetic energy form the Sun) results in the creation of ever stronger gradients. And as our statement of the second law of thermodynamics for open systems says, Nature abhors gradients, and organizing processes can occur if their occurrence helps obliterate gradients. In the context of living systems, a key fact is that living systems are much better than nonliving matter at capturing energy from the environment and dissipating it as heat. England (2013) has shown that when a group of atoms is driven by an external source of energy like the Sun or a chemical fuel, and is in contact with a thermal bath like the ocean, it would often structure itself gradually into a configuration such that it can dissipate energy *more and more efficiently*. Emergence of life is just one more step in that direction. The term '*dissipation-driven adaptation*' or '*dissipation-driven adaptive organization*' is an apt description of what happens in Nature very often.

Building on earlier work (but surprisingly making no mention of the extensive contributions of Stuart Kauffman), England (2013) has shown that all dynamical evolution is more likely to lead to structures and systems which get better and better at absorbing and dissipating more energy from the environment. Matter is adaptive to increasing amounts of energy inputs. Energy puts evolutionary pressure on matter.

England also makes the point that self-replication (the stuff biological reproduction is made of) is a good way of dissipating energy efficiently: 'A great way of dissipating more is to make more copies of yourself'.

7.4 Sociobiology, altruism, morality, group selection

A group comprising of many individual agents working together can be viewed as *a problem-solving system*. Each agent has some degree of autonomy, but may get overruled by the majority, in the larger interests of the group. This sort of teamwork is seen in many situations; e.g., in human sports and at the workplace, as also in social-insect colonies (see Chapter 18).

One of the great minds to have studied insect behaviour is E. O. Wilson, the Harvard naturalist. Wilson spent many decades decoding the biochemical communication mechanisms that ants use in order to function as a well-synched group. His discoveries led him to explore behaviour in all social organisms. He coined the word 'sociobiology', which is the study of the evolutionary basis of the behaviour of organisms. Wilson, in his 1975 book *Sociobiology: The New Synthesis*, took the scientific community by surprise with the assertion that biology also played a key role in human behaviour. At that time it was widely believed that human behaviour was purely culturally determined.

The triumph of Wilson's ideas was in presenting the self-organization view of group behaviour as a common aspect of both bee swarms and humans groups. The concept of the 'superorganism' was developed to express how a large group of non-intelligent agents can function as one highly intelligent entity. These entities can handle and synthesize large quantities of sensory data, and use the data to perform complex computations that are commonly associated with intelligence

Altruism and natural selection

Altruism can emerge in a species in spite of the fact that each individual is hard-wired to be selfish. It may appear at first sight that selfish individuals are more likely to survive and propagate their selfish-tendency genes. But if a group or a population as a whole has better survival chances if altruism prevails, than if rank selfishness prevails, altruism can emerge: Even though an altruist individual may not survive because it chooses to make sacrifices for the group, its altruistic genes will still survive in the population because the latter comprises of its brothers and sisters and other relatives. In such a situation, natural selection works in favour of promoting altruism in the gene pool. *In the human context, it should be clear to us that only a kind of collective altruism can ensure our survival as a species.*

At what level does natural selection drive biological evolution? Is it all about selfish genes and fertile individuals, or can '*group selection*' also occur? The group-selection idea involves altruistic behaviour conducive to the survival and propagation of a group as a whole, even at the cost of elimination of some individuals making the sacrifice for the sake of the group. Group selection is still a matter of debate, although it has been debunked by some experts.

Historically speaking, Darwin supported the idea of group selection (Mirsky 2009). He argued that, although moral men may not do better than immoral men at the level of the individual, tribes of moral men would 'have an immense advantage' compared to the survival and propagation rate of tribes with no moral scruples. But later opinion in the evolution community did not favour this postulate. The argument advanced was that at the genetic level it has to be 'every man for himself'. I quote Steven Pinker (https://www.edge.org/conversation/the-false-allure-of-group-selection#.T-fv_63C-jo.facebook):

'*I am often asked whether I agree with the new group selectionists, and the questioners are always surprised when I say I do not. After all, group selection sounds like a reasonable extension of evolutionary theory and a plausible explanation of the social nature of humans. Also, the group selectionists tend to declare victory, and write as if their theory has already superseded a narrow, reductionist dogma that selection acts only at the level of genes. . . . The more carefully you think about group selection, the less sense it makes, and the more poorly it fits the facts of human psychology and history. . . . . Group selection has become a scientific dust bunny, a hairy blob in which anything having to do with "groups" clings to anything having to do with "selection." The problem with scientific dust bunnies is not just that they sow confusion; ... the apparent plausibility of one restricted version of "group selection" often bleeds outwards to a motley collection of other, long-discredited versions. The problem is that it also obfuscates evolutionary theory by blurring genes, individuals, and groups as equivalent levels in a hierarchy of selectional units; ... this is not how natural selection, analysed as a mechanistic process, really works. Most importantly, it has placed blinkers on psychological understanding by seducing many people into simply equating morality and culture with group selection, oblivious to alternatives that are theoretically deeper and empirically more realistic'.*

Pinker summarizes his essay as follows:

'The idea of Group Selection has a superficial appeal because humans are indisputably adapted to group living and because some groups are indisputably larger, longer-lived, and more influential than others. This makes it easy to conclude that properties of human groups, or properties of the human mind, have been shaped by a process that is akin to natural selection acting on genes. Despite this allure, I have argued that the concept of Group Selection has no useful role to play in psychology or social science. It refers to too many things, most of which are not alternatives to the theory of gene-level selection but loose allusions to the importance of groups in human evolution. And when the concept is made more precise, it is torn by a dilemma. If it is meant to explain the cultural traits of successful groups, it adds nothing to conventional history and makes no precise use of the actual mechanism of natural selection. But if it is meant to explain the psychology of individuals, particularly an inclination for unconditional self-sacrifice to benefit a group of nonrelatives, it is dubious both in theory (since it is hard to see how it could evolve given the built-in advantage of protecting the self and one's kin) and in practice (since there is no evidence that humans have such a trait).

'None of this prevents us from seeking to understand the evolution of social and moral intuitions, nor the dynamics of populations and networks which turn individual psychology into large-scale societal and historical phenomena. It's just that the notion of "group selection" is far more likely to confuse than to enlighten — especially as we try to understand the ideas and institutions that human cognition has devised to make up for the shortcomings of our evolved adaptations to group living'.

Kerry Koyen (2012) (https://whyevolutionistrue.wordpress.com/2012/06/24/the-demise-of-group-selection/) has also argued strongly against group selection. And here is more from Pinker on morality (https://openparachute.wordpress.com/2009/02/): *'Nor is morality any mystery. Abstract, universal morality (e.g., a Kantian categorical imperative) never evolved in the first place, but took millennia of debate and cultural experience, and doesn't characterize the vast majority of humanity. More rudimentary moral sentiments that may have evolved – sympathy, trust, retribution, gratitude, guilt – are stable strategies in cooperation games, and emerge in computer simulations'.* ['Stable strategies' and 'cooperation games' are discussed in Chapter 30 and Appendix A10.]

8. Symmetry is Supreme

Long before any science, man was fascinated by symmetry.

Klaus Mainzer

Symmetry, as wide or as narrow as you may define its meaning, is one idea by which man through the ages has tried to comprehend and create order, beauty and perfection.

Hermann Weyl

Fundamental symmetry principles dictate the basic laws of physics, control the structure of matter, and define the fundamental forces of Nature.

Lederman and Hill

As far as I see, all a priori *statements in physics have their origin in symmetry.*

Hermann Weyl

In an earlier book I have discussed the subject of symmetry in some detail (Wadhawan 2011). Therefore this chapter touches only on its salient features briefly. The central statement is that the breaking of symmetry at a phase transition or a bifurcation point in phase space is a crucial factor responsible for self-organization, pattern formation, and the concomitant evolution of complexity.

8.1 Of socks and shoes

We are surrounded by symmetry and broken symmetry. Take the socks and shoes example. The two socks you wear are identical. We say that the pair possesses *permutation symmetry*: we can interchange (or permute) the two socks and the end result is as if we did not perform any permutation operation. The permutation here is a *symmetry operation*.

What about the two shoes? They are not identical. But you intuitively feel that there is something symmetrical about the pair. They are *mirror images* of each other. The act of reflecting across a mirror is the symmetry operation in this case. The reflection of the left shoe across a mirror looks identical to the right shoe, and vice versa. We say that the pair of shoes possesses *mirror symmetry*.

8.2 Connection between symmetry and conservation laws

The deep and unexpected relationship between conservation laws and symmetries of nature has been the single most important guiding principle in physics in the past century.

Lawrence Krauss (2017)

We speak of symmetry in any situation in which we can rearrange things (positions in space, values of quantities in time or direction, etc.) and still get the same answer for any physical question we may ask about the rearranged system. In the picture below, the flower has (approximate) '5-fold rotational symmetry': If we rotate it by $2\pi/5$ or $72°$, we cannot tell whether the rotation was indeed performed or not. The flower is 'invariant' under this *symmetry operation*.

Fig. 8.1 Flowers with an approximate 5-fold axis of symmetry.

Take the law of conservation of energy (also called the first law of thermodynamics). It says that the total energy cannot be enhanced or diminished. What that means is that if you measure the total energy today, and then again measure it tomorrow, you get the same value. The total energy is conserved, or remains *invariant*, as time progresses. So this is *time-symmetry* of total energy.

In fact, there is a deep connection between conservation laws and symmetry. According to Noether's theorem, *for every continuous symmetry of the laws of physics, there must exist a conservation law; for every conservation law, there must exist a continuous symmetry* (see Lederman and Hill 2008). For example, the time invariance of the laws of physics is linked to the law of conservation of energy. There are many more examples. Isotropy of space results in conservation of total angular momentum. Homogeneity of space gives us conservation of linear momentum. And since the laws of Nature do not change if the signs of all the electric charges are reversed, the total amount of electric charge of either sign cannot change.

8.3 Why so much symmetry?

We see so much symmetry in Nature, so the question arises: Why is that so? In my earlier book (Wadhawan 2011) I have argued at length that the basic underlying cause of this is simply the second law of thermodynamics for open systems ('minimization of free energy'). The second law is the mother of all organizing principles. An example of this is the symmetry principle, which says that *the effect is at least as symmetric as the cause*. Why should this be so? I quote from my book:

'Why is it that, during the dynamical evolution of an isolated system, symmetry cannot decrease? The answer lies in the second law of thermodynamics. The entropy of such a system cannot decrease, and symmetry and entropy both increase as a system evolves with time.

'Let us consider the example of ice and liquid water to understand the relationship between symmetry and disorder (or entropy).

'Which is more symmetric, ice or water? Water is more symmetric. Ice is a crystal, so its invariance under rotations is confined to only certain specific ('crystallographically self-consistent') rotations. Crystals are anisotropic objects, in general. By contrast, water looks the same from all angles. It is invariant under any and every rotation (an example of continuous symmetry). Water is *disordered*. By comparison, an ice crystal is an *ordered* object.

'*A disordered system is more symmetric compared to its ordered version, if any.*

'And entropy is a measure of disorder. The entropy or the degree of disorder of an isolated system cannot decrease with time. Therefore the symmetry of an isolated system cannot decrease with time. It can either increase or stay the same. This is what the symmetry principle also says.

'Nonequilibrium systems always have gradients of some kind or another. Natural phenomena are driven by the tendency to annul gradients, as demanded by the second law. Consider osmosis as an example. Suppose there is a membrane, with salty water on one side, and pure water on the other. The second law tells us that the difference in salt concentration will be reduced as time passes. On the whole, clean water will flow to the salty side and dilute it. This will happen spontaneously, not requiring the expenditure of any energy, as the system tries to reach a state of equilibrium.

'But *reverse* osmosis will require that energy be spent. This is how seawater desalination is done. We have to spend energy to force salt to go to the more salty side across the membrane, to make seawater potable' (Wadhawan 211).

To drive home further the central role played by the second law in the spontaneous emergence of symmetry in Nature, I consider in the next section (reproduced from Wadhawan 2011) the process of how a crystal grows.

8.4 Growth of a crystal as an ordering process

Consider a gas of noninteracting molecules. For it the most probable state of existence is one in which there is complete disorder in the mutual positions and velocities of the molecules. The system tends to attain this state of equilibrium, and then stays there. And the entropy has its maximum value when equilibrium has been reached.

Next, let us bring in short-ranged attractive and repulsive interactions among the molecules. The situation changes drastically. At very high temperatures, the disordering thermal tendency dominates and we have complete disorder and maximum entropy, and equilibrium. But as temperature is reduced, condensation to a liquid state takes place at some temperature, and there is some semblance of order now.

On further cooling, the liquid may form a crystal. A crystal is a highly ordered state of matter. When a crystal grows from a fluid, matter goes from a less ordered state to a more ordered state, so there is a local *lowering* of entropy. It should be remembered that we are dealing with an *open* system here, for which the second law of thermodynamics must be stated in terms of free energy $F (= E - TS)$, rather than entropy S. The law for open systems says that a process can occur if it minimizes F. But F depends not only on the entropy S, but also on the internal energy E. So a tighter binding of molecules in a crystal compared to the fluid state can possibly result in a larger drop in E compared to the drop in the magnitude of the term TS, thus giving a net decrease in F. The state with minimum F is the state of equilibrium.

If we have to compare two competing processes for which the difference in the entropy term is not significant, we can say that that process is favoured which minimizes the internal energy E. In particular, a crystalline material *self-organizes* into that space-group symmetry which results in the least internal energy, or the maximum binding energy.

The building blocks in a crystal are all identical. What could be the thermodynamic reason for that? It is the same as that for the fact that the molecules of a gaseous species are all identical. The crystalline material is like a particular chemical species, involving a large number of molecules and the interactions among them. The asymmetric unit of the crystal can comprise of one molecule, half a molecule, or more than one molecules of this chemical species. Say it is one molecule. A particular molecule got formed and is stable because it corresponds to the largest binding energy, or the lowest internal energy. It is highly unlikely that different portions of the chemical species will settle for *different* molecular shapes or sizes, because it is most probable that only one particular molecular shape and size has the largest binding energy, and any other configuration therefore has a smaller binding energy. So the asymmetric units or molecules are identical or equal because free-energy considerations demand that.

The symmetry of a crystal is synonymous with *identical or equal placement of equal parts* (Sheftal 1966a). When these equal parts self-assemble into a crystal, it is highly unlikely that the nature of the assembly will be different in different portions of the crystal. If one portion of the crystal finds for itself (through a process of trial and error) a least-energy configuration of neighbouring asymmetric units, it is most likely that other asymmetric units will also zero-in on the same mutual configuration, with the same binding energy per asymmetric unit. Only such an arrangement can ensure that the crystal as a whole has the lowest internal energy. If different parts of the crystal were to have different arrangements of the asymmetric units, then either the asymmetric units are not identical (not possible), or the interactions among the identical asymmetric units are not the same everywhere (again not possible).

Thus the symmetry of a crystal arises from the least-free-energy requirement imposed by the second law. The first law of thermodynamics (conservation of energy) is also involved: The tightest-binding state attained at equilibrium is the most likely state because energy is conserved. The bound molecules of the crystal cannot come apart spontaneously because the energy required to break the bonding cannot come out of nowhere.

8.5 Broken symmetry

> *What is the common link between symmetry and complexity? It is symmetry breaking as the origin of dynamics and variety of forms and systems in the world. Thus, symmetry and complexity are the spirit of nonlinear science.*
> Klaus Mainzer, *Symmetry and Complexity*

We see not only symmetry, but also so much *broken symmetry* (or reduced symmetry) around us. It is the second law again. If, under altered conditions, a system can lower its free energy further by making a phase transition to a different-symmetry phase or state, it would do so. The familiar example is that of liquid water freezing to crystals of ice on cooling to 0°C. Liquid water is highly symmetric: it looks the same from any direction, meaning that it is invariant to any rotation around any axis of rotation; it is isotropic. By contrast, a crystal of ice has a far lower degree of rotational symmetry; it is anisotropic.

Fig. 8.2 (called the 'Mexican-hat potential') provides a rationalization of spontaneous breaking of symmetry. The ball at the top could have rolled down along *any* direction, but once a random choice has been made spontaneously, the final state has a lower ('broken') symmetry. This figure also illustrates 'unstable equilibrium'. The top position has zero slope, and therefore corresponds to equilibrium. But this is *unstable* equilibrium because it is not robust against even a minor displacement away from equilibrium. Unstable equilibrium can lead to spontaneous breaking of symmetry as the system seeks a new state of stable equilibrium.

The moment of the birth of the universe was the moment of highest symmetry: No structure, no preferred direction, no inhomogeneity; just radiation and gravity. Details will be given in latter chapters, but here is what happened after that: The neonatal universe was hot and dense. A certain 'Higgs field' arose that soon condensed in space (rather like the more familiar Bose-Einstein condensation). It got momentarily stuck in a metastable high-energy state, before relaxing to its stable state. This relaxation released an enormous amount of energy (somewhat like what happens when water is heated under pressure to a superheated state at a temperature slightly above 100°C; on relaxation to the lowest-energy state stable at that temperature there is a violent release of trapped energy in the form of steam, and the concomitant expansion or 'inflation'). This very brief interlude of *cosmic inflation* occurred very soon after the birth of our universe (and was the real 'Big Bang'), during which even the tiniest of quantum fluctuations got amplified hugely, thus creating gradients. The emergence of matter from the earlier radiation field was a *symmetry-breaking phase transition* (mediated by the postulated Higgs field). The original radiation field was translation-invariant, and the appearance of matter broke, among other things, the translational symmetry. The Higgs field gives the particles their mass.

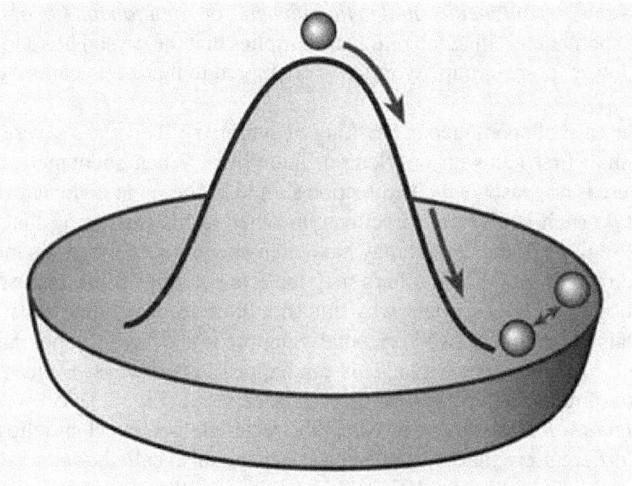

Fig. 8.2 The 'Mexican-hat potential'.
http://cerncourier.com/cws/article/cern/32522/1/CChig5_01_08

Nature annuls gradients because they embody states away from equilibrium, meaning that there is scope for lowering of free energy. Phase transitions occur to lower the free energy, and there is usually a concomitant lowering or 'breaking' of symmetry. *The entire history of the cosmos is one big saga of how symmetry-breaking transitions have occurred again and again.*

[Sometimes it is not very appropriate to use the term 'phase transition'; a better description is 'bifurcation in phase space', but that is only a technicality.]

At present, when objects interact with one another, only four types of interaction can be possibly involved: the gravitational interaction; the electromagnetic interaction; the weak nuclear interaction; and the strong nuclear interaction. But at the moment of the Big Bang there was only one 'unified' interaction. It is only because of the cooling of the universe that the four interactions known to us today emerged one by one, engendered by symmetry-breaking phase transitions.

Broken-symmetry considerations occupy centre-stage in theoretical physics. Particularly in particle physics, progress often means postulating, and then discovering, a new broken symmetry. Why only a broken (or 'hidden') new symmetry? Because an unbroken symmetry would have been manifest already, and would thus be *old* symmetry rather than new symmetry. Broken symmetries in Nature are not always very obvious; one has to go looking for them (Wadhawan 2011).

8.6 Symmetry aspects of phase transitions

Consider a crystal of iron. At high enough temperatures, it is in what is called a *paramagnetic* phase. Application of a magnetic field to this phase induces a response in the form of a small magnetic moment, and the magnitude of the magnetic moment induced by the applied field is proportional to the magnitude of the field applied.

When a crystal of iron is cooled, there comes a temperature at which it makes a phase transition to what is called a *ferromagnetic* phase. In this phase the crystal has a nonzero magnetic moment even when no external magnetic field is applied to it; we call this magnetic moment, for obvious reasons, *spontaneous magnetic moment* or *spontaneous magnetization*. The existence of this spontaneous magnetization also implies that the crystal has a lower directional symmetry (or a lower 'point-group' symmetry) compared to that of its paramagnetic phase.

So this is another case of spontaneous breaking of symmetry. There are several consequences of this, but we shall first focus on only one of them here. When spontaneous magnetization arises in iron, there is no reason why its direction should be the same in the entire crystal. There can be at least two such equivalent directions. In other words (assuming that this number is only two), some regions of the crystal may have their spontaneous magnetic moment pointing in a particular direction, and some others may have it pointing in the opposite direction. A geometrical analogy will help explain why this should be so. Imagine a square, two sides of which are vertical, and the other two horizontal. Suppose it undergoes a spontaneous reduction of symmetry, and becomes a rectangle. This can happen in two ways: Either the vertical pair of sides elongates and the other pair contracts, or vice versa (Fig. 8.3). So we end up getting two possible *orientation states*: those in which the rectangle is vertical and those in which it is horizontal. The different orientation states in a crystal are also called *domain states.*

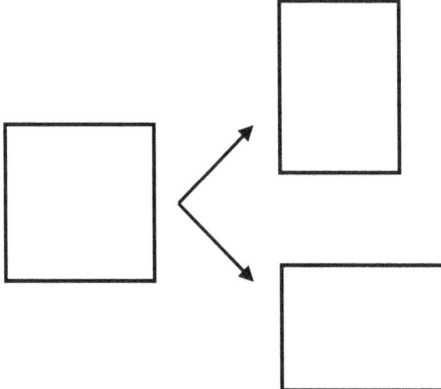

Fig. 8.3 When the symmetry of a square (left) lowers to that of a rectangle, it can happen in two ways (right).

A crystal of iron, on undergoing a symmetry-breaking transition to the ferromagnetic phase, splits into domains. All parts of a particular domain have the same direction for the spontaneous magnetization, and different domains can have different directions of spontaneous magnetization. The total number of such distinct directions is at least two, and can be higher. In fact there is a general theorem, from which it follows that *the larger the reduction of symmetry at a phase transition, the larger is the number of distinct orientation states permissible in the lower-symmetry phase* (Wadhawan 2000).

Thus a crystal of iron in its ferromagnetic phase can have domain structure, with magnetic moments of domains pointing in different directions.

Ferromagnetic phase transitions, characterized by the emergence of spontaneous magnetization (which is a macroscopic property), are an example of a more general class of phase transitions called *ferroic phase transitions* (FPTs). Another example of FPTs are *ferroelectric phase transitions*, in which there emerges a spontaneous polarization (rather than a spontaneous magnetization) when the crystal is cooled to a temperature below a critical temperature T_c.

Similarly, *ferroelastic* phase transitions are characterized by the emergence of spontaneous *strain* in the lower-symmetry phase (Wadhawan 2000). In fact, Fig. 8.3 is a schematic depiction of a ferroelastic phase transition; the square and the rectangle in this figure can be taken to represent the unit cells of the parent phase and the ferroelastic phase.

8.7 Latent symmetry

The symmetry of any entity is a manifestation of *equivalence* among different parts of the entity (Lederman and Hill 2004/2008). Because of this equivalence, when certain transformations (called *symmetry transformations*) are applied, the entity transforms back into itself, as if no transformations has been applied.

Very often, a symmetric object can be thought of as a *composite system* made up from equal or equivalent subparts. An object is said to possess *latent* symmetry (LS) if, when two or more copies of it are brought together or superimposed in a certain special way, the symmetry of the composite object is higher than what can be expected from the symmetry of the original object and the symmetry of the configuration chosen for constructing the composite object from its equal parts (Wadhawan 1987, 2000, 2011; Litvin and Wadhawan 2001, 2002; Litvin, Wadhawan and Hatch 2003).

The original object (or subpart) can be considered as the *building block* (BB), from which the composite object gets constructed by an assembly or juxtaposition or superposition of a number of equal objects, namely copies of the BB. Fig. 8.4 illustrates this by taking an isosceles triangle as the BB.

Two such isosceles triangles are shown in Fig. 8.4(a). The one on the left of the vertical line is the original, and the one on its right is a copy, obtained by reflection across the vertical line. The composite figure is a rhombus. It has two mirror planes, or rather mirror lines, of symmetry, one perpendicular to the horizontal diagonal (let us denote it by m_y), and the other perpendicular to the vertical diagonal (m_x). (The x-axis is taken as horizontal, and the y-axis as vertical.) The symmetry element m_y is a property of the original isosceles triangle, as also of its copy. And m_x is a consequence of the fact that we have *chosen* to juxtapose or superimpose the two triangles in a special way depicted in Fig. 8.4(a). So, there are no surprises so far.

The surprise comes when we take an isosceles triangle for which the apex angle θ has the special value 90°. We then get Fig. 8.4(b), which is a square. The square still has the m_y and m_x symmetries, just as the rhombus on the left has. But it also has the additional symmetry denoted by the symbol 4; i.e., if we rotate the square by an angle $2\pi/4$ (or 90°), we get back the same square. This is a symmetry element not present in the rhombus.

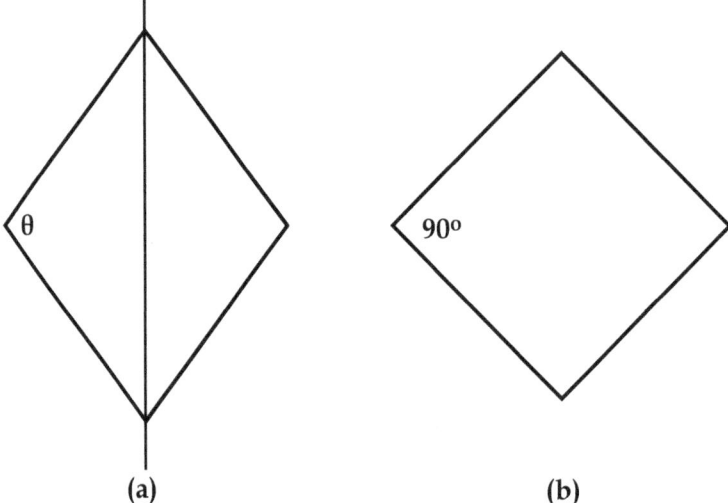

(a)   **(b)**

Fig. 8.4 (a) A composite object (rhombus) made up by joining two equal (mirror-reflected) isosceles triangles having an apex angle θ different from 90°. (b) Same as (a), except that the apex angle is now 90°. It has the symmetry of a square, which is higher than that of a rhombus.

We say that an isosceles triangle having an apex angle 90° has a *latent* symmetry (namely a 4-fold axis of rotation symmetry) which becomes *manifest* symmetry when two such equal triangles are juxtaposed in a special way shown in Fig. 8.4(b).

Latent symmetry may be significant from the point of view of complex systems, because many of them consist of interacting equal or identical objects. The bees in a beehive can be regarded as equal objects (assuming that they do not have 'free will', and that their behaviour comes entirely from instinct). Similarly, the neurons in a neural network can be treated as equal objects. At a more fundamental level, elementary particles like electrons lack individuality: Any two electrons with the same spin orientation are interchangeable; i.e., they possess permutation symmetry. *Is it the case that some types of emergence are the direct result of manifestation of latent symmetry?*

8.8 Latent symmetry and the phenomenon of emergence in complex systems

A complex system has emergent properties, i.e., it has properties that we cannot always expect or predict from the properties of its constituents or agents. From our discussion of latent symmetry above, we can say that if a complex system exhibits an emergent *increase* of symmetry, the increase may imply the presence of equivalent or equal subunits or agents which possibly possess latent symmetry; this latent symmetry leads to emergent symmetry when the subunits come together or interact is a special way necessary for the manifestation of the emergent symmetry.

Going back to crystallography, every crystal has at least the translational symmetry. The growth of a crystal from a fluid is an example of an open system in which there is a local increase of order (translational symmetry at least), and therefore a local decrease of entropy. This is reminiscent of local decrease of entropy or increase of order in an open system far away from equilibrium. The example of the growth of a crystal provides useful clues about how complexity evolves in systems comprising of equal parts or agents. The equal parts in the case of a crystal are the asymmetric units or the building blocks (BBs). And since crystals usually have directional symmetry also (and not just translational symmetry), it follows that these BBs (the agents) may have latent directional symmetry.

At least at an abstract level, there is an analogy between a crystal and a complex system that consists of equal parts or agents. *The manifest symmetry (if any) of such a complex system is an emergent property, arising from the interactions among its equal components. The components may have only latent symmetry, and not the manifest emergent symmetry exhibited by the complex system as a whole.*

It should be possible to extend the notion of latent symmetry from crystals to other systems consisting of equal, identical or equivalent objects or agents. Such systems at least have the exchange symmetry or *permutation symmetry*. They may even have symmetry described by the so-called *braid groups*.

To summarize, there are two important features of latent symmetry and its manifestation, namely emergent symmetry, which make it relevant to an understanding of a large class of complex systems, namely those involving equal or identical or equivalent agents:

1.   *Only certain special mutual configurations and linkages in a complex system of equal agents can make manifest the full inherent symmetry of the agents.*

2.   *When the conditions are conducive for the full manifestation of latent symmetry, i.e., when a system has hit upon or discovered the special mutual configurations mentioned above, there is an <u>unexpected</u> increase of symmetry (and decrease of order). This unexpected decrease of order is in line with the crux of complexity, namely the unexpected emergence of new features or properties.*

Thus, latent symmetry and its manifestations may be playing important roles in the self-assembly and self-organization of complex networks, leading, for example, to possible evolutionary or other advantages or disadvantages (Wadhawan 2011).

8.9 Broken symmetry and complexity

Broken symmetry is the opposite of emergent symmetry described above. Anderson (1972) emphasized the crucial role played by broken symmetry in the emergence of increased degree of complexity. Symmetry is broken at transitions from one hierarchical level of complexity to the next (higher) level of complexity. The domains in a ferromagnetic crystal are examples of broken-symmetry states. A ferromagnetic domain is a broken-symmetry (lower-symmetry) state with reference to the higher symmetry of the paramagnetic phase.

Broken symmetries are ubiquitous in Nature. And they always result in a higher degree of complexity. A crystal growing from a fluid is an example. A normal fluid like water has very high symmetry because its properties are isotropic. By contrast, an ice crystal resulting from the freezing of water has much lower symmetry. Practically all the physical properties of a crystal of ice are anisotropic, i.e., they are not the same in all directions, unlike those of liquid water. The symmetry of liquid water is a *continuous* symmetry, while that of an ice crystal is a *discrete* symmetry. The breaking of the continuous symmetry involved in the emergence of a crystalline solid from a fluid (like ice from water) has many consequences. The *emergence* of anisotropy of macroscopic properties is an example of this. Another is the emergence of *rigidity* (Anderson 1972; Chaikin and Lubensky 1996). Certain elastic moduli are zero for liquid water and nonzero for ice. Ice has a lower symmetry and more order compared to liquid water. Because of this, certain motions (e.g., shearing displacements) are resisted by ice but not by water. This is how rigidity emerges when the continuous symmetry of water is broken at the phase transition.

The lower symmetry of ice also means that its information content, or degree of complexity, is higher than that of water. Broken symmetry leads to a higher degree of complexity.

It is not a law that the degree of complexity of an open system must always increase as it evolves with time, although that is generally the case. In particular, the presence of latent symmetry can result in an *increase* in symmetry, and a corresponding *decrease* of the degree of complexity.

8.10 Symmetry of complex networks

Formal network theory is outlined in Appendix A9. Its relevance to complex systems arises from the fact that the units or subsystems of a complex system interact with one another, and we can view this as a *network* of interactions. A network is defined by a set of 'nodes' and a set of 'edges'. For example, the investors in a stock market are the nodes, and an interaction between any two of them is an edge (Barabási 2003).

Symmetry in complex networks has not been a much investigated subject, at least till recently. I have highlighted some recent important work in this field in my book on symmetry (Wadhawan 2011/2014). Some salient features are outlined here.

Complex networks often exhibit *permutation symmetry* (see Fig. A9.2 in Appendix A9). Trying to understand why such symmetry arises helps us understand the mechanisms behind the evolution of complexity in networks. Investigations have indicated that *similar linkage patterns* are responsible for the emergence of symmetry in certain complex networks (Xiao *et al.* 2008a, b). To get a feel for what this means, I cite an analogy: In the context of social networks, people with similar interests, educational background, specialization, age, etc. are more likely to have common friends. That is, people with similar properties are likely to have similar linkage patterns.

Another analogy, which can help understand why similar linkage patterns in many networks are responsible for their symmetry, can be drawn from the fields of crystal growth and crystallography. In Section 8.4 above, I have discussed how the growth of a crystal from an aqueous solution is really an *ordering* or symmetrizing process. In a dilute solution, the molecules can randomly collide and come apart. Sometimes they may stick together for substantial lengths of time (i.e., form clusters) before being split apart by thermal fluctuations. When the concentration of the solution is increased so as to be close to the saturation level, the chances that the clusters are bigger, and/or last longer, increase. For growing a crystal from the solution, the solution is maintained at a slightly supersaturated level (either by controlled evaporation of the solvent, or by controlled cooling of the solution). Under such conditions the average size of the clusters is quite large. As we saw in Section 3.4, there is a spread of cluster sizes, and those that are larger than a certain critical size are the *nuclei* which go on growing in size because their growth leads to a lowering of free energy. The growing crystal chooses a particular symmetry from among the 230 crystallographic space-group symmetries. How is this choice made?

Through a process of trial and error, involving the making and breaking of clusters, the system in some part of the solution may find some particular pattern of bonding which results in the maximum lowering of free energy. This is therefore *a stable linkage pattern* for the solute molecules. Since the molecules are the same everywhere, it is natural that the molecules elsewhere would also 'discover' the same minimum-free-energy linkage pattern and the same resultant crystallographic symmetry. Thus the unique symmetry that the grown crystal possesses is nothing but a consequence of similar linkage patterns occurring everywhere in the crystal. The chosen linkage pattern is that which is the most stable under the circumstances.

Thus the similar linkage patterns in complex networks are nothing but what I have called *building blocks* (BBs) in this chapter, and it is their latent symmetry which becomes manifest as the network grows (just like in crystal growth).

Why does emergent order arise just because an open system is far from equilibrium. One reason, as we shall see later in this book, is that the system has been pushed to the far-from-equilibrium situation by the sustained influx of lower-entropy energy, which gets assimilated as an ordered configuration. This role of negative entropy or available information has been duly emphasized by several authors for explaining the build-up of the degree of complexity of open systems. But the possible role of the latent symmetry of the agents (if any) in the reduction of the degree of complexity should also be investigated. It appears that this has not been done so far, except possibly in the context of complex networks mentioned here.

The second law of thermodynamics requires a minimization of free energy $F (= E - TS)$. The free energy has an enthalpy contribution E and an entropy contribution $-TS$. The latent symmetry of the BB affects the enthalpy term in the equation for the free energy. Crystal growth becomes possible because the conditions are such that there is a local lowering of enthalpy (because of the cohesive energy) which overrides the local lowering of entropy. The cohesive forces among the BBs win locally, leaving the problem of the local decrease of entropy to the heat bath in which the crystal grows: The slight local cooling of the growing crystal as more and more BBs attach to it is made up by a convection and conduction of heat from the heat bath to the crystal. The driving force for this is provided by the overall lowering of free energy consequent to the growth of the crystal from the fluid.

In crystals, the manifestation of latent symmetry of the BBs leads to the emergence of the final symmetry of the crystal, which is higher than the manifest symmetry of a single BB. The higher symmetry of the crystal becomes possible because the BBs are all identical. Otherwise we would not get the ordered object, namely the crystal. *Similarly, systems more complex than crystals but comprising of identical components or agents, may partly owe their occasional decrease of order (or increase of symmetry) to the fact that the components are all identical.* Since physical distances among the components are not of much relevance in a network, permutation symmetry has been correctly identified by Xiao *et al.* (2008a, b) as the manifest symmetry of the network underlying the complex system.

9. The Standard Model of Particle Physics

Let us begin by becoming familiar with the various types of forces or interactions underlying all phenomena in our universe, and also with the distinction made in particle physics between bosons and fermions.

9.1 The four fundamental interactions

There are four types of forces or interactions in our universe. The first is the *gravitational interaction*, or the gravitational force field. It is very weak, but is always present between any two particles or bodies. It is proportional to the product of the masses of the objects interacting. Since most of the celestial bodies are very massive, the gravitational force becomes very significant for them. Your weight is the gravitational force with which the rather massive Earth attracts you towards its centre.

Like charges repel, and unlike charges attract. Similarly, like magnetic poles (north-north or south-south) repel, and unlike magnetic poles (north-south) attract. Research showed that the electric interaction and the magnetic interaction are really two aspects of the same underlying phenomenon, so the term *electromagnetic interaction* was coined. This is the second of the four interactions.

The third is the *weak nuclear interaction*. It is operative, for example, inside the nuclei of radioactive materials, and is responsible for the emission of alpha-particles, beta-particles, etc. from inside such nuclei.

Lastly we have the *strong nuclear interaction*, which is very strong but very short-ranged, and is responsible for the large binding energies of nuclei: A rather large amount of energy is required for extracting a proton or a neutron from inside the nucleus of an atom.

Maxwell's theory of the electromagnetic interaction was a *classical* theory. So also was Einstein's theory of the gravitational interaction, namely the general theory of relativity. Quantum effects become very significant at sub-atomic length scales. Moreover, the early universe (at and immediately after the Big Bang) was also of very small dimensions. We therefore need a quantum formulation for all the four interactions.

In quantum theory, not only are the elementary particles quanta of mass/energy, even the force fields or interactions among the particles are mediated by quanta. For example, when two electrons interact, the fundamental particle which mediates the interaction is the photon. One electron emits a ('virtual') photon and recoils in the process. The other electron absorbs the photon and also recoils. This back and forth exchange of photons constitutes the electromagnetic interaction. *Quantum field theories* have to be formulated for all the four interactions, and only partial success has been attained so far.

Historically, the electromagnetic interaction was the first to be cast in a quantum-mechanical form, and this subject goes by the name of *quantum electrodynamics* (QED). Richard Feynman, who played a major role in the development of QED, had also formulated a *sum-over-histories* version of quantum mechanics (see Section 5.4). He used his famous *path-integrals* for working out the details of QED. But the QED theory ran into a conceptual problem. The summation over the infinitely many possible histories resulted in an infinite mass and charge for the electron, which was an absurd result. Feynman got over this problem by a procedure called *renormalization*, but I shall not go into its technical details here. Appendix A6 gives some additional information.

Why are there four different fundamental interactions, and not just one? The fact is that there was indeed only one interaction to start with, but as the very hot neonatal universe expanded and cooled, symmetry-breaking transitions occurred, resulting in the successive appearance of the various (less symmetric) interactions. Attempts continue to be made for 'unifying' the four interactions to see what kind of a theory emerges when this has been achieved. Such efforts run parallel to those for obtaining quantum-field-theoretic formulations for the four interactions.

The first to be unified were the electromagnetic interaction and the weak-nuclear interaction, resulting in what is called the *electroweak interaction*. A bonus point of this unification was that the renormalization procedure could be successfully carried out for the unified interaction for obtaining its quantum field theory (without encountering the 'infinities' problem mentioned above), whereas it was not achievable for the weak interaction separately.

The quantum field theory which successfully achieved renormalization for the quantum version of the strong nuclear interaction is called *quantum chromodynamics* (QCD). In this theory the proton and the neutron, as also some other particles, are envisioned as made up of a more fundamental set of particles called *quarks*. Quarks come in three *colours* (nothing to do with the usual meaning of colour): red, green, and blue, along with the respective *anticolours*. The quarks cannot exist as free, stable particles. Only those combinations of them can exist as free particles which do not have a net colour. For example, a colour and its anticolour cancel, giving a neutral net colour. Composite particles in which this occurs are the particles called *mesons*. Another possibility is that all three colours (one each), or all three anticolours, occur together in a composite particle. The name for such a composite particle is *baryon*. Protons and neutrons are examples of baryons.

In addition to colour, quarks have quantum parameters like 'up' (u), 'down' (d); 'charm', 'strangeness'; and 'top', 'bottom' (do not pay attention to the literal meanings of such words; they are just labels, with no literal meaning). Two up quarks and a down quark make a proton, and two down quarks and one up quark make a neutron.

The information given above is part of the so-called *standard model of particle physics*. In it, the electromagnetic interaction and the weak-nuclear interaction have been unified into the electroweak interaction, and a quantum version for it has been established. In addition, there is a quantum-field-theoretic formulation for the strong nuclear interaction (QCD), but no entirely satisfactory unification with any of the other interactions. The gravitational interaction has been neither quantized, nor properly unified with other interactions, although there is some progress in that direction.

9.2 Bosons and fermions

The Earth is a massive object, so it exerts an enormous gravitational pull on you. Have you ever wondered why is it that you do not get pulled all the way down to the centre of the Earth?! It is because of a principle in quantum mechanics, called the *Pauli exclusion principle*, which says that no two electrons (or other 'fermions') can have the same set of *quantum numbers*, or be in the same *quantum state*.

Any atom has a positively charged nucleus, and one or more electrons around the nucleus. Positive and negative charges attract, so why do not all the electrons get sucked into the nucleus, and thus eliminate the distance between the positive charges and the negative charges?

This is because of another principle of quantum mechanics, namely the *Heisenberg uncertainty principle* (see Chapter 5). The diameter of the nucleus is ~10^{-15} meter. The nucleus comprises of protons and neutrons, and each of them is ~2000 times heavier than an electron. If an electron were to get sucked into the nucleus because of the force of electric attraction (also called *Coulomb attraction*), it would be confined to a length of the order of 10^{-15} meter. This would be the uncertainty Δx in its position. We can plug this into the inequality embodying the uncertainty principle, namely Eq. 5.6, to get the corresponding uncertainty in its momentum p_x. From this momentum if we work out the kinetic energy ($= p_x^2/(2m)$), it turns out to be so large that the electron just cannot remain bound inside the nucleus. By contrast, a proton or a neutron *can* remain bound inside the nucleus because for it the kinetic energy is ~2000 times lower, because of the larger mass m.

So the electrons are outside the nucleus. And each atomic species (hydrogen, helium, carbon, etc.) has a *distinctive* distribution of electrons in specific orbits around the nucleus, and a distinctive 'valence' or proclivity for chemical bonding. The distinctive structure of atoms of any type (say carbon) is because electrons belong to a class of particles called *fermions*, and for fermions the Pauli principle says that no two of them can be in the same quantum state; so they separate out into more than one orbits. [The term 'fermion' is in honour of Enrico Fermi.]

All fundamental particles are either fermions or bosons ['bosons' in honour of S. N. Bose]. These two classes differ in a quantum parameter called *spin*. The spin of any fundamental particle can be either an integer (in a certain system of units), or a multiple of 1/2. That is, it is either integral (0, 1, 2, 3, ...) or half-integral (1/2, 3/2, 5/2, ...).

Although classical analogies can be very inadequate for explaining quantum mechanics, imagine a spinning top, nevertheless. We can associate an angular momentum with it: The faster the top spins, the higher is its angular momentum. And the angular momentum of a classical top can have any arbitrary value in a continuous range. Not so in quantum mechanics. Only discrete (or quantized) values are possible here, although quantum spin indeed has the identity or 'dimensions' of angular momentum. What is more, the spin value for an electron can be either 1/2 or -1/2, and nothing else.

Whereas the Pauli principle prevents electrons and other fermions from occupying the same quantum state, there is no such restriction for bosons. Unlike fermions, bosons *can* have the same set of quantum numbers. [A quantum state is specified by a set of quantum numbers. Spin is an example of such a quantum number; there are many others.]

How do electrons get distributed in various specific orbits (or *energy states*) around the nucleus if they cannot all occupy the same orbit? The answer lies in the wave nature of electrons. An electron would tend to be as close to the nucleus as possible, but the smallest (closed) orbit can be that which allows a wave to close back on itself smoothly and repeatedly.

The Pauli principle allows the smallest orbit to be occupied by two electrons, one with spin 1/2, and the other with spin -1/2. If there is a third electron, it must occupy the next permissible larger orbit, and so on. Only orbits which are compatible with the formation of 'standing waves' are possible. This is what defines their specificity or uniqueness.

An atom of sodium has 11 electrons around the nucleus. The first orbit (or *shell*) can take only two. The next shell can take 8, each with a distinct set of quantum numbers, and the 11th electron must go to the third shell.

Protons and neutrons are also fermions, with spin 1/2. By contrast, photons are bosons (spin =

1).

So, you and I exist because electrons and other fermions obey the Pauli principle, leading to the existence of distinct types of atoms and molecules. And it is the Pauli principle again which prevents the Earth from pulling us all the way down to its centre; electrons and other fermions (unlike bosons) just cannot come too close together in phase space.

Why are all fundamental particles either bosons or fermions? According to quantum field theory (to be discussed later), when 'matter' particles interact with one another, they do so by the emission and absorption of 'field' particles. For example, the electromagnetic interaction between two electrons is via an exchange of photons. And all matter particles are fermions, all field particles are bosons.

9.3 The standard model and the Higgs mechanism

As stated above, in the standard model the electromagnetic interaction and the weak nuclear interaction have been unified into a single *electroweak interaction*, and a quantum field theory of this unified interaction has been successfully formulated. Further, a quantum field theory of the strong nuclear interaction has also been established, but it has not been satisfactorily unified with other interactions (see below). A quantum theory of the gravitational interaction (*quantum gravity*), as also its unification with other interactions, is still work in progress.

During the 1970s, physicists formulated a number of *grand unification theories* (GUTs) which aimed at unifying the electroweak interaction and the strong interaction. Most of them predicted that the proton is not an eternally stable particle, and should decay with a 'half-life' of $\sim 10^{32}$ years. Experimental verification of proton decay has not been effected unambiguously, so the GUTs have not been an unqualified success. But there has been an unexpected fallout. Ideas from the GUTs have found applications in cosmology. A certain *scalar field* (the *Higgs field*, or rather a Higgs-like field) in the GUTs has provided a strong basis for supporting the idea of *cosmic inflation* postulated for the neonatal universe: The gravitational interaction arose 10^{-43} seconds (the *Planck time*) after the Big Bang. The other three interactions were still in a 'grand-unified' state. The spontaneous emergence (on further cooling of the universe) of a GUT scalar field 10^{-35} s after the Big Bang broke the *GUT symmetry*, and new interactions appeared one by one, the first to appear being the strong interaction.

Prior to the emergence of new interactions, the GUT scalar field had also caused a major and very rapid *inflation*, doubling the size of the baby universe every 10^{-34} s. This means that 100 doublings occurred within 10^{-32} s after the Big Bang. The inflation was so rapid that the quantum fluctuation just 10^{-20} the size of a proton could grow to a size of 10 cm in just 15 x 10^{-33} s. This is how the GUT scalar field kick-started the growth of the universe, *preventing a gravitational-pull mediated collapse of the quantum fluctuation that had initiated the universe.*

In the standard model there are 12 elementary particles of spin 1/2: six quarks and six *leptons*. Further, each such fermion has a corresponding *antiparticle*. For example, the antiparticle of an electron is a particle called the *positron* (of positron-emission tomography (PET) fame). Thus there are 24 fermions in all.

The six quarks are: up, down, charm, strange, top, and bottom quark. And the six leptons are: electron, electron-neutrino, muon, muon-neutrino, tau, and tau-neutrino.

The quarks interact via the strong nuclear interaction. They form 'colour-neutral' composite particles called *hadrons*. The hadrons are either baryons, or mesons. The baryons (protons and neutrons) contain three quarks, and the mesons comprise of a quark and an antiquark.

There are three types of 'gauge bosons': (i) photons; (ii) W^+, W-, and Z particles; and (iii) gluons. Out of these, photons and gluons are massless (zero rest mass) particles. Photons mediate the electromagnetic interaction. There are eight types of gluons, and they mediate the interaction among quarks. The W^+, W-, and Z gauge bosons have nonzero rest mass, and they mediate the (short-ranged) weak-nuclear interaction among quarks and leptons.

[A fourth type of gauge boson should also be mentioned here, namely the *graviton*. It is expected to be massless, and have spin = 2.]

In the standard model the Higgs field permeates all space. The *Higgs particle* (or *Higgs boson*) is the quantum of this field, and it has nonzero rest mass. The Higgs field is a scalar field (i.e., it has zero spin).

The Higgs field interacts, with different strengths, with the various particles. Particles that interact more strongly with it experience more resistance or drag to their motion, and thus are more massive. Some particles, such as photons, do not interact with the Higgs field at all, and are therefore massless (zero rest mass). In this way, the mass of everything is determined by the existence of the Higgs field.

Very high temperatures or energies (comparable to those that prevailed soon after the Big Bang) have to be created in the laboratory for creating and detecting the Higgs particle. Recently, the experiments conducted at the LHC have given conclusive proof that the Higgs particle exists (see Krauss 2017). This constitutes a major accomplishment of the rational human mind, which could predict the existence of the Higgs particle by adopting the scientific method for understanding natural phenomena.

You might be wondering: Why so much fuss about a Higgs mechanism for imparting mass to elementary particles? After all, what is so surprising about a particle having a mass? Why do we need a *mechanism* (the Higgs mechanism) for 'imparting' mass to the particles? We have to remember that in the beginning there was only a radiation field, as also the gravitational interaction. Particles with nonzero rest mass appeared via the Higgs mechanism. The Higgs field is a part of the theory of the electroweak interaction (see Krauss 2017). It has a role in breaking the symmetry between the electromagnetic interaction and the weak interaction. This symmetry existed at very high temperatures, and was broken spontaneously as the universe cooled a little and the Higgs field appeared. Particles with nonzero rest mass arose when this symmetry was broken by the emergence of the Higgs field.

The Higgs field pervades all space, and the physical laws governing it have high symmetry. Yet the present actual state of the Higgs field has a spontaneously-broken or lower symmetry. The field takes a particular value at low energies, and pervades all space, rather like the energy associated with vacuum in quantum field theory. Particles acquire their masses by interacting with the pervading Higgs field.

One has to go beyond the present standard model of particle physics for creating a grand theory of how our universe came into existence. Are there particles even more elementary than 'the veritable zoo' of particles envisaged in the standard model? Yes, there are. We shall discuss that later in the book

.

10. Cosmology Basics

If there were no life other than on the Earth the universe would seem to be an absurdly large object if life were its primary goal.

Paul Frampton (2010)

10.1 The ultimate causes of all cosmic order and structure

The presently dominant (though often challenged) theory of the cosmos says that our universe was born from a quantum fluctuation of energy, which was immediately followed by a very rapid and huge inflation of space, and has been expanding and cooling since then. This implies *a continuing creation of gradients*. As discussed in Chapter 3, the essence of the second law of thermodynamics for open systems is that Nature tends to find ways to annul gradients, and that this results in the spontaneous creation of local order and complexity in certain pockets. So the incessant expanding and cooling of our universe is the ultimate cause of all the cosmic self-organization and structure at various scales.

In an expanding and cooling universe there are two opposite tendencies at work. On the one hand there is a tendency to maximize the overall entropy, as demanded by the second law for isolated systems. On the other hand there is a perennial creation of spatial and thermal gradients, which make an effectively negative contribution to the overall entropy. Consequently, although the universe does tend to attain a state of equilibrium, it never quite reaches it (Chaisson 2001). It is this ever persistent state of departure from equilibrium (and substantial departure at that) which is the root cause of all evolution of complexity. [At a more local level, so far as our Earth is concerned another major contributor to the rising levels of complexity is the sustained influx of energy from our Sun.]

Another key factor has been postulated by cosmologists, namely the occurrence of ultra-minute quantum fluctuations that occurred *during* the inflation period, within 10^{-35}s after the birth of the universe (Ade *et al*. 2014; Krauss 2014). These fluctuations got amplified very quickly and are the primary source of all the structure we see today, including stars, galaxies, and clusters of galaxies.

Thus there are three key factors responsible for the evolution of *cosmic* complexity:

1. The second law of thermodynamics for open systems.
2. The continual expansion and cooling of our universe.
3. The occurrence of quantum fluctuations during the inflation period, very soon after the birth of the universe.

In view of the above, it is instructive to take a good look at the birth and the expanding nature of our universe. At the time of this writing there are already voices pointing out that perhaps cosmology has got it all wrong, and that there was really no Big Bang (Frampton 2010). I shall discuss this issue near the end of this chapter. It appears that it will take some time before the experts agree on an alternative theory of cosmology, if at all there is a need to revise the present theory drastically or to abandon it altogether. In any case, any new theory will also have to give a convincing explanation of so much of existing experimental data on which the Big Bang theory has been based.

10.2 The Big Bang and its aftermath

> *In the beginning there was light. But more than this, there was gravity. After that, all hell broke loose . . . .*
>
> Lawrence Krauss (2017)

If we look at the clear sky at night, we see stars which *seem* to be at the same location every night. It was therefore natural for humans in ancient times to postulate a *static universe*. The invention of the telescope and the development of spectroscopic techniques made us change our views on the nature of the cosmos. In 1929 Edwin Hubble, after sustained observations of the stars and the galaxies, announced his finding that the universe is not static, but expanding. We now call it *Hubble expansion*. In fact, Hubble also reported that, for practically all the galaxies (the exceptions being some of the galaxies in our immediate neighbourhood), the more distant a galaxy is from us, the faster it is moving away from us. If the distance of a galaxy is D and its velocity away from us is v, then its magnitude is

$$v = HD ,\tag{10.1}$$

where H is the so-called *Hubble parameter*.

Combining the general theory of relativity and Einstein's cosmological principle gives the so-called *Friedmann equation*, which parameterizes the expansion of the universe in terms of a *scale factor*. This scale factor depends on time, and characterizes the typical distance between galaxies. We can combine the known value of the Hubble parameter and the presently known composition of the universe to calculate the scale factor for any point of time in the past. When we do this we come to the conclusion that, 13.7 ($\pm 2\%$) billion years ago, all the observed matter must have been confined to just one point in spacetime. This point is taken as the birth moment for the universe, and we say that the universe started at that point of time (and was soon followed by the inflationary episode, the real Big Bang).

There is an unsavoury aspect of this exercise. The Friedmann equation becomes *singular* at the birth of the universe; i.e., the density and the temperature become infinite, and therefore the general theory of relativity becomes inapplicable there. Many attempts have been made to bypass this problem of the *initial singularity* (see, e.g., Frampton 2010; Hawking and Mlodinow 2010).

The Planck time

To deal with time scales very near the Big Bang episode, the so-called Planck time is used. It is worked out from the fundamental constants c (the speed of light in vacuum), G (Newton's gravitational constant), and h (the Planck constant), and has a value of $\sim 10^{-44}$ second.

For a duration of the order of the Planck time after the origin of the universe, the spatial dimensions involved were extremely small. Therefore quantum effects must have been very important, and densities involved very high, so gravitational effects must have been very important. So what we need here is a marriage of quantum mechanics and the general theory of relativity. But such a theory of *quantum gravity* still eludes us to a large extent, although there has been progress (see, e.g., Hawking 2001; Hawking and Hertog 2002, 2006).

Quantum cosmology

One may argue that quantum mechanics can save us the embarrassment of the initial singularity. The subject of *quantum cosmology* has been built around this assertion. We invoke the Heisenberg uncertainty principle: $\Delta x \Delta p_x \sim h$. This means that Δx need not be zero, meaning that the size of the universe at its birth was not zero, though it was very small. This is one way of getting around the initial-singularity problem.

The uncertainty relation $\Delta E \Delta t \sim h$ is also relevant here. Suppose Δt is vanishingly small. That means that ΔE can be arbitrarily large, and the principle of energy conservation *can* be violated within a time duration Δt. This quantum fluctuation of energy can explain how the universe could emerge *out of nothing* (Hawking and Mlodinow 2010; Krauss 2012, 2017).

We can also explain the emergence of positive energy (radiation or/and matter) out of nothing if there is a *simultaneous* emergence of a balancing amount of negative energy. This negative energy arose because the Big Bang was accompanied by the emergence of the gravitational interaction, which is an attractive interaction. Any attractive interaction engenders a negative contribution to the total energy because one needs to expend (positive) energy to break free from the binding force of the attractive interaction. (By contrast, a repulsive interaction (like the one between two positive charges or two negative charges) makes a positive contribution to the overall energy.)

So the birth of our universe need not be associated with a singularity.

It is also possible that the birth was associated with an instability, rather like the instability that results in a phase transition or a 'bifurcation' in phase space.

Cosmic microwave background

The Big Bang theory was proposed in 1930 by Georges Lemaître, and developed by other physicists, notably George Gamow. Fred Hoyle, Hermann Bondi, Thomas Gold, and Jayant Narlikar formulated an alternative *steady-state theory* of the universe in 1948. This latter theory implies an infinitely old universe, with no 'beginning'. The Bell Labs scientists Arno Penzias and Robert Wilson discovered in 1965 *the cosmic microwave background* (CMB) radiation that Gamow had predicted to be a consequence of the Big Bang. The observation of this radiation, a relic of the early universe, apparently delivered a body-blow to the steady-state theory. However, the last word has not been said yet about what is the correct model for the universe. The steady-state-universe model has much to commend itself. It has even been suggested that, although something spectacular did happen at the so-called 'Big-Bang moment', that point in spacetime was perhaps not the moment of birth of our universe. There is perhaps an unending series of Big Bangs and Big Crunches.

The present universe is pervaded by the CMB having an average temperature of ~3 K. The corresponding wavelength (~10 cm) falls in the microwave part of the electromagnetic spectrum. The CMB provides a glimpse of what the universe was like when it was just ~380,000 years old. As we shall see later, that was when atoms formed by the combination of ions.

It is a very cold universe indeed. This temperature varies inversely as the scale factor that characterizes the size of the universe. As the universe expands the CMB temperature falls. Conversely, if we move backwards in time the temperature must have been much higher.

In 1992 it was discovered by NASA's Cosmic Background Explorer satellite that the CMB actually has a small variation of ~0.001% as we look in different directions (Turner 2009). This variation has been attributed to a very slight lumpiness in the distribution of matter when the

CMB photons set out on their journey into space. This lumpiness got amplified by a huge factor with the passage of time, and was the originator of the structure we see today in the form of galaxies and clusters of galaxies.

The surface of last scatter

In the history of the cosmos, something very important happened ~380,000 years after the Big Bang. At that time the visible universe was ~1000 times smaller than it is now. So its temperature was 1000 times higher, i.e. ~3000 K. This is about the highest temperature at which hydrogen atoms are stable; at higher temperatures the electron separates from the proton, so we only have ions (an ionized plasma of protons and electrons) and not atoms.

As the universe cooled below this temperature, the ions 'recombined' (or rather *combined*) to form neutral atoms of hydrogen. It so happens that neutral atoms are not good scatterers of photons, but ions in general, and electrons in particular, are. So when the ions combined to form atoms, the universe changed from opaque to transparent. At higher temperatures the photons of the CMB were strongly scattered by the ions, and not many of them escaped into space, so the big blob was rather opaque. When the scattering of photons became much less, they could escape into space, and the universe became transparent.

We can thus imagine a boundary of the universe, which was a *surface of last scatter* for the photons of the CMB: This surface was no longer 'visible' after the universe became transparent, 380,000 years after the Big Bang. The photons from this epoch of the universe had a wavelength of ~0.1 mm.

What was there before the Big Bang?

So our universe has been expanding ever since the Big Bang. But what was there *before* the Big Bang? This is not an easy question to answer, so the debate continues. Here are some possible scenarios (Turner 2009; Ali and Das 2015):

•   There was nothing before the Big Bang: No energy, no matter, no time.

•   There was a 'primeval state' described by the (still unformulated) theory of quantum gravity, and then 'quantum emergence' of spacetime occurred (Ali and Das 2015).

•   Ours is only *one* of the many possible universes (parallel universes, or *multiple universes*; see Appendix A6 on quantum theory for more on this). They keep emerging from eternal space.

•   The Big Bang was only the latest event in the cyclic and ever-repeating sequence of events: Big Bang, expansion, contraction, Big Crunch; and then Big Bang and expansion again.

Into what is the universe expanding?

What lies 'outside' the expanding universe? For answering this question one has to understand the nature of spacetime. The space of our ordinary experience has three dimensions, so spacetime has four dimensions. *The expansion of the universe is in 4-dimensaional spacetime, and not in 3-dimensional space.*

Four dimensions are hard to visualize, but three are easier. Imagine 3-dimensional spacetime: only two dimensions for space, and one for time. This spacetime can be likened to a balloon, on which there are spots representing galaxies. Expansion of the universe in spacetime can be likened to the inflation or expansion of this balloon. As the balloon expands, the spots representing galaxies move away from one another (Fig. 10.1).

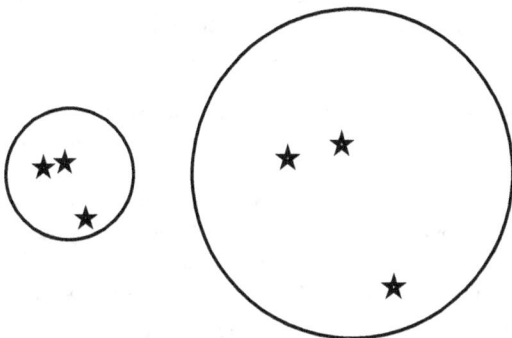

Fig. 10.1 Here a 2-dimensional balloon (left) represents a universe having two spatial dimensions, the third dimension being that of time. As spacetime stretches like an inflating balloon (right), distances between stars and galaxies increase. The farther two galaxies are from each other, the faster their distance increases with time. Not only matter, even the wavelength of radiation also stretches as space expands.

In this hypothetical situation, all creatures are 2-dimensional, meaning that they exist only on the surface of the balloon, and there is no third space dimension available for going into. Everything happens on the surface of the balloon; there is no 'outside' into which the balloon is expanding. And since the balloon is a closed surface, there are no boundaries. Of course, the real situation is that the expanding balloon is in a 4-dimensional hyperspace, in which three dimensions are for what we ordinarily perceive as space, and one is for time. There is no boundary and there is no 'outside'. Our universe is expanding in spacetime, and not in space or time separately. There is no question of expanding *into* something, because there is nothing outside the spacetime in which the universe is expanding.

10.3 Dark matter and dark energy

A major upheaval occurred in cosmology during the late 1990s. Till then it was believed that the expansion of the universe should gradually slow down because of the attractive gravitational potential building up among the cosmic entities. That is, gravitational attraction should result in a *deceleration* of the rate of expansion. But observational data told us in 1998 that just the opposite is happening. The cosmic expansion appears to be *accelerating*, rather than decelerating.

The existence of something weird, namely *dark energy*, was postulated for explaining this observation (Frampton 2010). It is called dark because we cannot see it directly; its presence is felt only through its effects on visible matter.

Another important fact of modern cosmology is that there was, in fact, a *transition* from decelerated expansion to accelerated expansion some time more recent than when the visible universe was half its present size.

Visible matter derives most of its mass from baryons. Protons and neutrons are examples of baryons, and they are much heavier than, say, electrons. Modern cosmology tells us that only 4% of the total energy in the universe is in the form of baryonic matter. [As indicated by the famous Einstein equation, mass m and energy E are related: $E = mc^2$.]

It has been postulated that a full 24% of the total energy content of the universe is in the form of (nonbaryonic) matter called *dark matter*. Dark matter is believed to be made up of what are called *weakly interacting massive particles* (WIMPs) (Frampton 2010). The existence of a fundamental particle called the *neutralino* has been postulated in this context. Another candidate particle is the so-called *axion*, which is much lighter than the neutralino.

The remaining 72% is dark energy. It is an unknown form of energy that is believed to have made the expansion of our universe to *accelerate*, ~10 billion years ago.

What is the difference between dark energy and dark matter? To answer that, let us begin by recalling the well-known Boyle's law we learnt in school. It says that, for a given mass m of a gas, its volume V multiplied by the pressure p it exerts on the container is a constant which depends on the temperature of the gas ($pV = \text{constt}$). At any particular temperature the molecules of the gas have a certain average kinetic energy, which determines p. That is why p increases if the temperature is increased. We can make the Boyle's-law statement independent of the mass m by talking in terms of density ρ ($\rho = m/V$), rather than volume V. Then the Boyle's law says that $p/\rho = \text{constt}$.. This equation is an example of what is called an *equation of state*.

The particles in baryonic matter and dark are essentially at rest, so the pressure they exert is zero.

Photons travel at the speed of light and they exert positive pressure. For them the equation of state can be derived to be $p/\rho = 1/3$.

Dark energy is weird in that it exerts *negative* pressure. For it, $p/\rho = -1$ in one of the imagined cosmic scenarios. Thus pressure is equal to negative of density (in a certain system of units) in the simplest of its manifestations. *Dark energy exerts a repulsive force or negative pressure*, tending to blow the universe apart, and this is countered to a certain extent by the gravitational force of attraction.

There is a viewpoint that there is perhaps no need to postulate the existence of the still-unobserved and weird dark energy and dark matter, and that the acceleration of expansion of the universe can be explained as being due to the fact that our galaxy perhaps lies near the centre of *a giant cosmic void* (Clifton and Ferreira 2009). Admittedly, this idea is just about as far-fetched as the postulation of dark matter and dark energy.

An idea has been floated that dark matter could be comprised of baryonic matter (protons, neutrons) if it is all tied up in 'brown dwarf stars', or in dense chunks of heavy elements. The technical term for this possibility is *massive compact halo objects* (MACHOs).

A more likely explanation is that, rather than the normal baryons, their 'supersymmetric' partners ('sbaryons') (see below) are involved. This makes sense because it explains why such dark matter does not interact with ordinary mater, except through gravity.

But another possibility is that dark matter is not baryonic at all, and is rather made up of particles like axions or WIMPs. In the supersymmetry modification of the standard model, there is something called *R symmetry*, which means that the number of superparticles must be conserved in every process. It follows from this that there must exist a *lightest supersymmetric particle* (LSP), which is stable because there is no lighter superparticle into which it can decay under the R-symmetry constraint. Typically, the LSP is a 'linear combination' of superpartners of: the Higgs boson (*Higgsino*), the photon (*photino*), and the B gauge boson (*bino*).The term *neutralino* has been coined for such an LSP; it is an example of a cosmological WIMP.

An important feature of modern cosmology is that, when one calculates the number of neutralinos that survived annihilation in the early universe, the agreement with the known amount of dark matter turns out to be extremely good; in fact, too good to be true!

When it was observed in 1998 by the Hubble Space Telescope that the expansion of the universe is accelerating, three explanations were offered for this:

1. Perhaps the acceleration can be explained in terms of a long-discarded version of Einstein's theory of gravitation, the one that contained a *cosmological constant*. In Einstein's theory, it is possible for more space to come into existence. In one version of his theory there was a cosmological constant, 'put in by hand' (and later withdrawn). This version makes a prediction that 'empty space' can possess energy of its own. Since this energy is a property of space itself, the energy density would not be diluted as the universe expands. As space expands, more energy comes into existence, and the universe expands faster and faster.

2. Perhaps there is some unknown *energy-fluid* that fills all space.

3. Perhaps Einstein's theory is wrong, and a new theory is needed that would include a new field that can explain the acceleration.

A whopping ~74% of the universe is dark energy. Dark matter makes up ~22%. The rest (including normal matter) adds up to less than 4% of the universe.

Both dark matter and normal matter pull the universe together. But dark energy does the opposite: It pushes the universe apart. Further observations and their interpretations have conjured up the following scenario: Till ~5 billion years ago, the universe was not having an accelerated rate of expansion. Dark matter dominated the early universe, but dark energy overtook the influence of dark matter ~5 billion years ago. As the universe expands, the domination of dark energy over the effect of dark matter is getting stronger and stronger. Why should that be so? As stated above, one explanation can come from Einstein's general-relativity theory, with cosmological constant included.

Another explanation for how space possesses energy comes from quantum field theory. In quantum physics when we speak of vacuum, we really mean a space which has a certain minimum-energy state.

We may not understand dark matter much, but it is just as well that we have already discovered or postulated it. The curvature of spacetime in the universe depends on the overall mass/energy density of the universe. Since the curvature, i.e., the geometry, is observed to be flat or Euclidean, rather than curved (either spherical or hyperboloid), the mass/energy density must have a certain critical value. This is indeed found to be the case, provided we include the contribution from dark matter.

Why has the geometry of our universe been flat, right after the Big Bang? An answer to this *Flatness Problem* comes from the inflation postulate I describe in the next section in some detail. The exponentially rapid expansion ('inflation') of our universe from a size much smaller than that of a proton to the size of a tennis ball (or more) smoothed out our spacetime to make it very flat. And dark matter has contributed to this flat geometry.

I must mention here that the very existence of dark matter and dark energy is being questioned by more and more scientists, but any alternate model would require a total overhaul of modern cosmology.

10.4 Cosmic inflation

Alan Guth was the originator of the 'cosmic inflation' postulate, which solves, among other things, the Horizon Problem and the Flatness Problem.

The present size (diameter) of the visible part of the universe is ~30 billion light years. Therefore its size at the time of the emergence of the surface of last scatter must have been ~30 million light years (i.e., 1000 times smaller). That was when the CMB photons reaching us today (in the form of cosmic microwave background radiation) started their journey into space. But the age of the universe at that time was just 0.3 million years. Therefore information or light signals could have travelled a distance of only 0.3 million light years. This is 100 times less than the diameter of the surface of last scatter. Therefore there was apparently no reason for the distant parts of the surface of last scatter to be causally connected and be as smooth as we find them to be.

The inflation postulate attends to this problem. For postulating inflation, Guth was inspired by the analogy of a first-order phase transition. As an example, consider the freezing phenomenon in water. As liquid water is cooled, its free energy changes. There comes a temperature (the freezing point, 0°C) below which a different phase, namely ice, has a lower free energy than liquid water. Therefore, as demanded by the second law of thermodynamics, a phase transition should occur from liquid water to ice. Yes, but that is not the full story.

Suppose you take extremely pure water, and also ensure that there are no disturbances like vibrations etc. Then you find that you can 'supercool' it to a metastable state; i.e., it continues to be a liquid even below 0°C. This happens because each phase of H_2O is stable in a certain range of temperatures and, since it is a first-order phase transition, there is no reason that 0°C be the temperature where one stability range ends and the other begins. There is a small range of temperatures around 0°C in which *both* phases are stable, meaning that liquid water and ice can coexist. But below 0°C ice does have a lower free energy than liquid water. Therefore even the slightest disturbance to supercooled water can make it undergo *rapidly* the arrested phase transition to ice. When this happens, the trapped excess free energy gets released. We call it 'latent heat'.

Something similar was postulated by Guth to have happened to the nascent universe. The very hot universe was cooling rapidly. The original Higgs-like scalar field (the high symmetry of which had not yet been broken by the symmetry-breaking transition) got supercooled into a metastable ground state, even as the universe cooled past the point where a Bose-Einstein type condensation of the scalar field should have occurred, accompanied by the transition. An enormous amount of energy got trapped in this metastable state. Guth argued that this energy would be gravitationally repulsive, and therefore, when the transition did occur at a certain point in the supercooling process, the trapped energy caused the universe to expand (*inflate*) by a huge amount in a very short time. The volume of the tiny universe increased by a factor of at least 10^{78} during the inflation, which occurred during the early part of the electroweak epoch: It started ~10^{-36} seconds after time-zero, and ended ~5×10^{-33} seconds later.

Let us put in some numbers. Even for the entire inflation period, we have $\Delta t \sim 10^{-33}$ sec only. The uncertainty principle says that $\Delta E \Delta t \geq h/(4\pi)$. Since $h/(4\pi) \sim 10^{-27}$ erg sec, we get ΔE ~10^6 erg, or ~10^9 GeV. Any value of ΔE larger than this is also possible as a quantum fluctuation (resulting in the appearance and disappearance of 'virtual particles') provided Δt is appropriately smaller than 10^{-33} sec. Normally this would be of no serious significance if the virtual particles could disappear within the time Δt. But space was doubling in length every 10^{-37} seconds during inflation, so *the momentary inhomogeneities created by quantum fluctuations got yanked apart rapidly, and were frozen in space when inflation ended*; there was no going back.

The 'false vacuum' energy was of essentially the same nature as that coming from Einstein's cosmological constant, or dark energy.

Predictions of this cosmic-inflation model are in conformity with the characteristic density inhomogeneities recorded in 2006 in the CMB map (for a flat-geometry universe at the end of the inflation epoch).

The expansion of space during the inflation interlude was much faster than the speed of light. This is possible. According to the special theory of relativity, nothing can travel through space at a speed faster than that of light. But there is no limit on the speed with which *space itself* can expand.

Let us note two more things here: (i) There was and is gravitational interaction among the frozen fluctuations, and the gravitational potential makes a negative contribution to the total energy, so the total energy content of the universe was and is close to zero. (ii) 380,000 years after the Big Bang the evolution of the structure frozen at the end of the inflation episode led to what we see in the CMB map today. In due course, this structure evolved into the present configuration of galaxies, clusters of galaxies, life, people. You can appreciate why the inflation postulate is such an integral part of modern cosmology (see Turner, Baumann and Steinhardt 2014, Krauss 2017).

There is more to it, still. *Inflation also explains or predicts why there should be a multiverse* (i.e., many universes). To understand that, let us hark back to the phase transitions in water, but this time let us consider boiling instead of freezing. Above 100°C, vapour or steam is a more stable phase of H_2O than the liquid phase. And it is again a first-order phase transition, meaning that there is a range of temperatures in which liquid water and steam can coexist. Because of thermal fluctuations and other local variations, bubbles of steam start appearing even below 100°C, and they do so with increasing frequency as the temperature is increased, till the whole system starts boiling. Something similar happened during cosmic inflation, and this phenomenon is called *chaotic inflation*. There was a time interval during which inflation occurred, and at any instant during this interval a different universe could inflate and go its separate way, like a bubble of steam in the analogy given here.

So, not just our universe, but a whole lot of universes, could have emerged before the end of the inflation era, each with its own value for the cosmological constant and for other fundamental constants, and its own laws of physics. There is a *multiverse*, rather than just one (i.e., our) *uni*verse.

10.5 Supersymmetry, string theories, M-theory

Particle physics and cosmology have moved beyond the standard model, and concepts like supersymmetry have been formulated. At and soon after the Big Bang the temperatures and energies were so high that there was only one fundamental interaction, and not four. As our universe cooled a little, symmetry-breaking transitions occurred and different interactions appeared one by one. New fields and matter arose as a result of these transitions. Since there was only radiation and no matter to start with, the present distinction between matter particles (fermions) and field particles (bosons) is also a result of broken symmetry. We call it *supersymmetry* (SUSY). If supersymmetry could be restored by going to high enough temperatures or energies, the distinction between fermions and bosons would vanish.

As we go up the symmetry ladder, more and more 'dimensions' come into existence. Supersymmetry involves symmetry operations in a certain n-dimensional superspace, four of these dimensions being the spacetime coordinates we perceive in our world. The most important new symmetry emerging from the supersymmetry description of Nature is that for every particle with spin J (a boson), there must be another particle with spin $J \pm 1/2$ (a fermion). We would see a 'degeneracy' (sameness) if this supersymmetry could be realized by going to high-enough temperatures, and the masses of the two partner particles would then become equal. But since the supersymmetry got broken at a certain temperature, the masses are now different.

The standard model does not include quantization of the gravitational interaction, and its unification with the other three interactions. This means that additional broken symmetries need to be postulated and verified, with an attendant increase in the number of dimensions of the hyperspace in which the symmetry transformations operate. *String theories* attempt to do that. They involve an extension of the conceptual framework of quantum field theory.

The uncertainty principle is one reason why it is so hard to formulate a quantum theory of gravity (Einstein's general theory of relativity is a wholly classical theory). The uncertainty principle applies to pairs of 'conjugate parameters'. For example, the position of a particle along the x-axis and its momentum component along the same direction are one such pair of conjugate parameters. A second such pair is the value of a field and its rate of change: The more accurately one is determined, the more uncertain the value of the other becomes. This means that there is *no such thing as empty space*: An empty space would mean that both the value of a field and its rate of change are exactly zero; and this is not allowed by the uncertainty principle. Thus when we speak of vacuum in quantum physics, we really mean a space which has a certain minimum-energy state. This state is subject to *quantum fluctuations*, which means that virtual pairs of particle-antiparticle can make momentary appearances (within the limits prescribed by the uncertainty principle), and then disappear by annihilating each other. There are infinitely many such virtual pairs possible, each having energy, implying that the vacuum state should have infinite energy. But an infinite-energy vacuum state would curve the universe to an infinitely small size, according to the general theory of relativity. This is not what actually happens, so our theory is plagued by an 'infinity problem' again.

In 1976 the idea of supersymmetry was put forward in this context. In supersymmetry theory, force particles (bosons) and matter particles (fermions) are symmetry-related, or rather *super*symmetry-related. This scenario has the potential to solve the above infinity problem: It turns out that the infinities from matter-related virtual particles are all negative, while they are all positive for force-related virtual particles, so they can cancel each other out.

The notion of *supergravity*, which emerges when we invoke supersymmetry, has the potential to unify gravity with the other three interactions.

The idea of supersymmetry had actually originated earlier when string theories were being formulated. In string theories the elementary particles are envisaged, not as points, but rather as patterns of vibration that have length but no width ('strings'). The various string theories are consistent only if spacetime has 10 dimensions, rather than 4. We *see* only four dimensions because the other six have 'curled up' into a space of very small size. An analogy will help understand this. Consider a straw you use for drinking lemonade. Its surface is 2-dimensional: We need two numbers or coordinates for specifying the location of any point on it. But if the straw is extremely thin (say a million-million-million-million-millionth of a centimetre), it is practically 1-dimensional; the other dimension has just curled up into near-nothingness in terms of visibility.

Introduction of supersymmetry into a string theory leads to the idea of *superstrings*. Earlier, there appeared to be at least five different string theories (or rather superstring theories), and millions of ways in which the extra dimensions could be curled up. But many experts are now convinced that the five superstring theories, as also supergravity, are merely different approximations to a more fundamental theory called the *M-theory*, each superstring theory being valid in different (but overlapping) situations.

M-theory involves 11 dimensions instead of 10. It is this extra dimension which unifies the five string theories. Moreover, M-theory allows for not just strings (which are 1-dimensional objects), but also point particles, 2-dimensional membranes, etc., all the way up to 9-dimensional entities (called '*p-branes*', with *p* running from 0 to 9). M-theory is the unique supersymmetric theory in 11 dimensions.

A crucial feature of M-theory is that its mathematics restricts the ways in which the dimensions of the internal space can be curled-up. Thus the theory comes up with *unique* (rather than arbitrary) values for the fundamental constants and the 'apparent' laws of physics corresponding to any particular mode of curling.

M-theory needs to be verified adequately. If confirmed, it may well be the long-coveted 'theory of everything' (TOE).

10.6 Has modern cosmology got it all wrong?

Is our universe really expanding? Was there a Big Bang at all? Is the universe really homogeneous and isotropic (rather than lumpy and anisotropic)? Is 'dark energy' a valid concept at all? Already there is a growing tribe of cosmologists whose answer to each of these questions is 'no' (see, e.g., Wolchover 2011; Hartnett 2011; Mitra 2013; Crothers 2014; LPPhysics 2014; Grant 2014).

Here are some of the objections to the currently accepted model of cosmology, particularly the Big Bang part of it. A case is being made that *the observations are, in fact, consistent with a non-expanding universe with no Big Bang* (Wolchover 2011; Mitra 2013; LPPhysics 2014):

1. Any superhot explosion, like the Big Bang, would have generated a certain small amount of lithium. Yet, as astronomers have observed older and older stars, the amount of lithium observed has been found to be less and less and, in the oldest stars, is *less than one-tenth of the predicted level.* This, however, accords with non-Big-Bang predictions that explain the production of light elements by stars and cosmic rays within the galaxies themselves.

2. The Big Bang theory requires the existence of dark matter. Yet, multiple lines of evidence, especially observations of the motions of galaxies, show that dark matter does not exist. Similarly, the existence of dark energy was postulated to explain the recent acceleration in the rate of expansion of the universe. But this latter observation may well be due to the way our particular region of the cosmos is drifting through the rest of space. Our relative motion makes it look like the universe is expanding faster and faster, while in actuality its expansion is slowing down.

3. In the Big Bang theory, the universe is supposed to start off completely smooth and homogenous (after the inflationary interlude). But as telescopes have peered farther and farther

into space, larger and larger structures of galaxies have been discovered which are too large to have been formed in the time since the Big Bang.

4. The inflation that is supposed to have occurred immediately after the birth of our universe should have smoothed out any large-scale asymmetries in the universe. The cosmic background radiation (CBR) should be perfectly symmetrical. But the CBR in fact shows *strong evidence of asymmetries* from one side of the sky to the other that, although small, could not have been produced by the ultra-symmetric inflation that hypothetically occurred in the neonatal universe.

5. Apparently, there was no spacetime before the Big Bang. But if there was no spacetime, the hypothesis of a quantum fluctuation in spacetime, leading to the origin of the universe, does not make sense. [But this objection seems to have been taken care of by the work of Ali and Das (2015).]

6. Current cosmology is based on two basic principles: The Copernican Principle (*the universe is homogeneous*), and the Cosmological Principle (*the universe is both homogeneous and isotropic*). Starting with early studies of the cosmic microwave background (CMB), and culminating with results from the COBE and the WMAP satellites, scientists were faced with a signal at the largest scales of the universe, a signal that indicated that we are in a special place in the universe. The Copernican and cosmological principles require that any variation in the radiation from the CMB be more or less randomly distributed throughout the universe, especially on large scales. Results from the WMAP satellite indicated that when looking at large scales of the universe, the noise can be partitioned into 'hot' and 'cold' sections, and this partitioning is aligned with our ecliptic plane and equinoxes. This partitioning and alignment results in an axis through the universe, which scientists dubbed '*the axis of evil*', because of the damage it does to their theories. This axis passes right through our tiny portion of the universe.

It is now being asserted by some that *the evidence is consistent with an evolving but non-expanding universe, which had no beginning in time and no Big Bang.* Here is how Mitra (2013) attacks the existing theory: 'Newtonian Cosmology was apparently plagued with the problem of infinite gravitational force, and Einstein's General Relativity apparently ushered in the revolutionary concept of a closed finite non-singular static universe. Later, Big Bang model (BBM) essentially incorporated non-static versions of similar relativistic model. Simultaneously the concept of a Cosmological Constant or a repulsive vacuum energy density got incorporated either for Inflation or for Dark Energy. We dismantle this nearly century old edifice by presenting several exact proofs showing that Cosmological Constant or Dark Energy is non-existent, and Einstein's Static universe is just the Minkowski vacuum. By using the just found Schwarzschild form of the FRW metric we show that FRW metric too is actually the Minkowski vacuum! It is suggested that physical universe is quasi-Newtonian where, for any given galaxy, finite gravitational potential is due to interaction of nearest neighbours while the infinite background forces cancel one another due to symmetry. Such a universe is likely to have a fractal structure as suggested by observations. The cosmic redshift might arise due to asymmetric spread of wave packets associated with line emissions from distant galaxies. The cosmic background radiation might be due to thermalization of star lights in an eternal universe as suggested by Hoyle, or it might be superposition of gravitationally redshifted quiescent "Eternally Collapsing Objects'', the supposed "Black Hole Candidates". The atmosphere of hot ECOs may synthesize not only light elements but infuse fresh hydrogen from flares of ECO plasma'.

[A word about 'black holes': Black holes (BHs) were earlier believed to be regions of spacetime from which nothing, not even light, can escape. It is a prediction of the general theory of relativity that if a region has matter of extremely high density, then the prevailing gravitation field can be so strong that there can occur so much curving of spacetime that everything that comes too near to the so-called *event horizon* of this object gets lost forever, because even photons cannot escape from it to convey any information to the outside world. All the information that went into the BH should just stay there forever. This leads to a certain *information paradox*, which has turned out to be quite tough to resolve. In quantum mechanics we can speak only in terms of probabilities for the occurrence of anything. But probabilities can be meaningful only if, on adding up the probabilities of all possible occurrences, the result is equal to unity. Thus, an important tenet of quantum theory is that *information is never truly lost, nor is it truly copied*. The general relativity theory is a *classical* theory. The BH information paradox arose when Stephen Hawking showed, around 1974-75, that if a BH is surrounded by *quantum fields*, it would radiate quantum particles ('Hawking radiation') and shrink in size gradually, *all the way to zero size*. The question is: Does the information inside the BH also get lost? It cannot be, if quantum theory is not to be violated. So, either the gravitational theory or the quantum theory must be modified if they are to be mutually compatible.]

Mitra (2014) continues the attack in a later paper: 'Even if one would assume the astrophysical massive compact objects (MCOs) to be Black Holes (BHs), no energy can be extracted from them because neutral vacuum BHs cannot acquire any (induced) electromagnetic property, neither can any current emerge from the central singularity. This is so despite wishful models claiming the contrary by attributing the Event Horizon (EH) or an imaginary "membrane" with wishful electromagnetic properties. Similarly various Quantum Gravity (QG) theories too attribute various imaginary and mysterious properties like "Brick Wall", "Fire Wall" with the EH even after claiming that the *vacuum* EH is a perfectly regular spacetime without any special property! The *vacuum* EH is also associated with imaginary material structures and entropy in a completely self-contradictory manner. To legitimize such contradictions & fudge, the "Holography" principle is invoked by which the information contained within the 3-D BH interior is hypothesized to be encoded on the 2-D EH. Further, some QG theories try to explain gravity & BH entropy (S_{BH}) in terms of random motion of "atoms of vacuum" of dimension ~ ℓ_p (Planck Length). But since $\ell_p \to 0$ as $\hbar \to 0$, a classical vacuum would possess infinite entropy by such a hypothesis and so spacetime may not be granular ever. It is asserted that though BHs correspond to exact General Relativistic solutions, the relevant integration constants are zero, i.e., a Schwarzschild BH has $M = 0$ (Mitra, JMP 2009), and Kerr BHs too correspond to $M = a = 0$, implying $S_{BH} = 0$ & BHs are asymptotic ground states of preceding collapse which radiates away entire mass-energy, angular momentum & entropy. Thus the finite mass BH Candidates must be Quasi-BHs. It has been shown that the most natural case for Quasi BHs are ultra-magnetized hot quasi-static balls of plasma, Magnetospheric Eternally Collapsing Objects (MECOs) radiating at their Eddington Luminosity. Spinning MECOs behave like ultramagnetic GR pulsars and may naturally explain high energy astrophysical phenomena. Since there is no true BH, there (is) no quantum "Information Paradox", no need for "Holography", no need to bid farewell to physical reality. As spactime membrane gets infinitely stretched and no singularity is formed, there is probably no need for any fictitious QG. Gravity may always remain classical and separated from other interactions like oil & water.'

The assertion by some cosmologists that the universe is not expanding is based, among other things, on a reinterpretation of the experimental data ('red shift') on which the edifice of the expanding-universe model had been based so far. Here is what Lerner, Falomo & Scarpa (2014) write: 'The Tolman test for surface brightness (SB) dimming was originally proposed as a test for the expansion of the universe. The test, which is independent of the details of the assumed cosmology, is based on comparisons of the SB of identical objects at different cosmological distances. Claims have been made that the Tolman test provides compelling evidence against a static model for the universe. In this paper we reconsider this subject by adopting a static

Euclidean universe (SEU) with a linear Hubble relation at all z (which is not the standard Einstein–de Sitter model), resulting in a relation between flux and luminosity that is virtually indistinguishable from the one used for ΛCDM models. Based on the analysis of the UV SB of luminous disk galaxies from HUDF and GALEX datasets, reaching from the local universe to z ~ 5, we show that the SB remains constant as expected in a static universe.

'A re-analysis of previously published data used for the Tolman test at lower redshift, when treated within the same framework, confirms the results of the present analysis by extending our claim to elliptical galaxies. We conclude that available observations of galactic SB are consistent with a SEU model.

'We do not claim that the consistency of the adopted model with SB data is sufficient by itself to confirm what would be a radical transformation in our understanding of the cosmos. However, we believe this result is more than sufficient reason to examine this combination of hypotheses further.'

There are also doubts about the interpretation given to some recent experimental data about the early universe. See https://www.nextbigfuture.com/2014/05/problems-with-big-bang-expanding.html.

Of course, the establishment is fighting back. This is what Krauss and Wilczek (2014) wrote: 'It is commonly anticipated that gravity is subject to the standard principles of quantum mechanics. Yet some (including Einstein) have questioned that presumption, whose empirical basis is weak. Indeed, recently Freeman Dyson has emphasized that no conventional experiment is capable of detecting individual gravitons. However, as we describe, if inflation occurred, the Universe, by acting as an ideal graviton amplifier, affords such access. It produces a classical signal, in the form of macroscopic gravitational waves, in response to spontaneous (not induced) emission of gravitons. Thus recent BICEP2 observations of polarization in the cosmic microwave background will, if confirmed, provide empirical evidence for the quantization of gravity. Their details also support quantitative ideas concerning the unification of strong, electromagnetic, and weak forces, and of all these with gravity.'

[Gravitation waves are said to have been observed in the recent LIGO experiments, though not from the Big bang.]

I must mention here that explanations and rationalizations are available for quite a few of the objections listed above. For example, it has been pointed out that there are limits to how much of the universe we can observe. We are not looking at 'the whole universe'. Therefore, it could be that the apparent violations of the Copernican principle and the cosmological principle are just 'local' anomalies, not present in the universe as a whole. Random chance dictates that some pockets of the whole universe will have larger or smaller fluctuations than others, and those fluctuations might even be aligned entirely by coincidence. In other words, the axis of evil could well be an illusion, a pattern that would not seem amiss if we could see more of the universe. Or it could be that the fact that the axis of evil appears to be roughly aligned with the ecliptic means that there is something close by that is interfering with the measurements.

In any case, from our vantage point the observable universe, by definition, is that which has us at the centre.

More recently, a renewed debate occurred regarding the validity of the Big Bang postulate in general, and the cosmic inflation idea in particular. Ijjas, Steinhardt and Loeb (2017) pointed out that the various inflation theories have too many adjustable parameters, so this is not good science because it makes the theories untestable. They also asserted that, contrary to the 'official' interpretation, the very precise data obtained from the Planck satellite 'disfavoured the simplest inflation models and exacerbated the long-standing foundational problems with the theory, providing new reasons to consider competing ideas about the origin and evolution of the universe' (also see Ijjas *et al.* 2013). These authors further advocated an alternative scenario, in which our universe began, not with a bang but with a bounce (the *Big Bounce*; see Moskowitz (2016)) from a previously contracting cosmos (an idea which goes well with the theory that our universe has no beginning or end).

The establishment was quick to repudiate many of these assertions. In a letter signed by a group of 33 physicists (Guth *et al.* 2017), it was pointed out that the inflation theory made several predictions which have proved to be correct: 'The standard inflationary models predict that the universe should have a critical mass density (that is, it should be geometrically flat), and they also predict the statistical properties of the faint ripples that we detect in the cosmic microwave background (CMB). First, the ripples should be nearly "scale-invariant," meaning that they have nearly the same intensity at all angular scales. Second, the ripples should be "adiabatic," meaning that the perturbations are the same in all components: the ordinary matter, radiation and dark matter all fluctuate together. Third, they should be "Gaussian," which is a statement about the statistical patterns of relatively bright and dark regions. Fourth and finally, the models also make predictions for the patterns of polarization in the CMB, which can be divided into two classes, called E-modes and B-modes. The predictions for the E-modes are very similar for all standard inflationary models, whereas the levels of B-modes, which are a measure of gravitational radiation in the early universe, vary significantly within the class of standard models.

'The remarkable fact is that, starting with the results of the Cosmic Background Explorer (COBE) satellite in 1992, numerous experiments have confirmed that these predictions (along with several others too technical to discuss here) accurately describe our universe. The average mass density of the universe has now been measured to an accuracy of about half of a percent, and it agrees perfectly with the prediction of inflation'.

They go on: 'Like any scientific theory, inflation need not address all conceivable questions. Inflationary models, like all scientific theories, rest on a set of assumptions, and to understand those assumptions we might need to appeal to some deeper theory. This, however, does not undermine the success of inflationary models. The situation is similar to the standard hot big bang cosmology: the fact that it left several questions unresolved, such as the near-critical mass density and the origin of structure (which are solved elegantly by inflation), does not undermine its many successful predictions, including its prediction of the relative abundances of light chemical elements. The fact that our knowledge of the universe is still incomplete is absolutely no reason to ignore the impressive empirical success of the standard inflationary models'.

I end this chapter by quoting Lawrence Krauss (2017) (http://bigthink.com/videos/big-bang-evidence-higgs-gravity-waves-with-lawrence-krauss):

'Our picture of the earliest moments of the universe has been evolving, and I'm happy to say, in some sense has more empirical support than it did before. The discovery of the Higgs field implies that you can get fields that freeze in empty space. And that's a central part of what we think happened in the very early universe.

'And if we can detect gravitational waves from the Big Bang we'd have a window on the universe back to a time when it was a billionth of a billionth of a billionth of a billionth of a second old, answering questions about the origin of the universe as we know it — ideas that I speculated upon in my last book (Krauss 2012), for example — for which we have new evidence that I've described in my new book (Krauss 2017).

'But because the temperature of the universe and the energies and particles were so extreme at that early time — when the entire visible universe was contained in a region that was smaller than the size of an atom — there's a wonderful symbiosis between large scales and small scales.

'And if we can probe the early universe back to a time that I described we'll actually be probing physics on scales that are much smaller than we can see at the Large Hadron Collider, 12 orders of magnitude smaller in scale (or higher in energy) than we can probe with our highest-energy accelerator now. . . .

'When the universe was a billionth of a billionth of a billionth of a billionth of a second old our current picture suggests: A field very similar to the Higgs field froze in space, but it was in what is called a metastable state. Sort of like… if you have a beer party and you put beer in the freezer because you forgot to until the few minutes before the party, and then during the party you forget that it's in the freezer, and you take it out later. And it's there — liquid — and you open it up, and suddenly it turns to ice, and the bottle cracks: The beer is in a metastable state. At that temperature it would rather be frozen except it's under a high pressure. The minute you release the pressure it freezes instantaneously, releasing a lot of energy. As our universe cooled we think the same thing happened; basically a field got frozen but in the wrong configuration, and as the universe cooled, suddenly — boom! — like those beer bottles, it changed its state, releasing a huge amount of energy, creating the hot Big Bang.

'Now the interesting thing is, while it was in that metastable state and storing energy, general relativity tells us that if you have a field in empty space that's storing energy it produces a gravitational effect that's repulsive, not attractive. So during that brief time gravity is repulsive, and the expansion of our universe started speeding up faster and faster and faster, and the size of our universe (we think) increased by a factor of 10 to the 30th in scale, or at least 10 to the 90th in volume, in a time interval of a billionth of a billionth of a billionth of a second. That means it went from the size of an atom to the size of a basketball in a short time, and that rapid expansion produced characteristics which pervaded the universe today: The fact that our observed universe looks flat, the fluctuations, and the cosmic microwave background radiation all came from quantum fluctuations that happened during inflation.

'Inflation is the only First Principles idea that in principle explains why our universe looks the way it does. And what's wonderful about it is it doesn't require any exotic ideas of quantum gravity or theories we don't have, it's based on ideas that are central to our current understanding of the standard model of particle physics, just extrapolating them somewhat. So it's very well-motivated; even though it is hard to believe that it could have happened, we think it did'.

11. Uncertainty, Complexity, and the Arrow of Time

Because of time, beings move on;
Because of time, they grow up;
In time they reach their end;
Time, though unshaped, possesses shapes.

Maitri Upanishad

It was the incorporation of time into the conceptual scheme of Galilean physics that marked the origins of modern science.

Ilya Prigogine, *The End of Certainty*

11.1 Irreversible processes, and not entropy, determine the arrow of time

Dynamical systems evolve in complexity as a function of time. Thus understanding the nature of time is relevant for understanding certain aspects of complexity. The second law of thermodynamics (through which the concept of entropy arises) is of central importance in the theory of complexity. This is the law which also determines *the arrow of time*. We often come across the statement: 'Increase of entropy occurs in the direction of increasing time'. This perception of reality has been questioned from time to time (Layzer 1975; Prigogine 1980, 1996; Carroll 2008; Callender 2010). A proper statement would be:

Irreversible processes occur in the direction of increasing time.

As we shall see in the next section, irreversible processes *can* result in a decrease of entropy and increase or creation of order (as against disorder and chaos) in certain pockets of the universe. But irreversible processes always occur in the direction of increasing time.

11.2 Irreversible processes *can* lead to order

Neither the historical nor the thermodynamic arrow of time can be observed at the microscopic level. The motion of a single sugar or tea molecule generates neither information nor entropy. 'Order' is a macroscopic concept, a property of systems made up of many particles; it has no meaning when it is applied to individual atoms or molecules. In the physics of elementary particles the world changes but does not evolve.

David Layzer (1975)

According to Prigogine (1996), the traditional formulation of the laws of physics describes only an idealized stable world, not quite like the unstable and evolving real world. He emphasized that the irreversibility of a real process cannot be always equated with creation of disorder only. This is true not only for systems driven far away from equilibrium, but even for steady-state processes. *The arrow of time can certainly be a source of order also.* Even an experiment like that involving simple thermal diffusion is enough to make the point (Prigogine 1996):

Consider a container (Fig. 11.1) in which there is a mixture of two gases, say hydrogen and nitrogen at a certain temperature T_1. Let the number of molecules of the two types be the same. At equilibrium the mean velocity of the lighter atoms (hydrogen) is higher because the mean kinetic energy for the molecules of the two gases is the same. And since all parts are at the same temperature the two types of atoms are distributed uniformly everywhere (Fig. 11.1(a)).

Now raise the temperature on the left to a higher value T_2. The equilibrium has been disturbed, and both types of molecules on the left have a higher mean velocity. But among them the increase in the mean velocity of the lighter (hydrogen) molecules is more than that of the heavier molecules of nitrogen. Therefore at steady state there will be a larger number of hydrogen molecules in the right part of the container. Correspondingly, a larger number of nitrogen molecules will diffuse to the left.

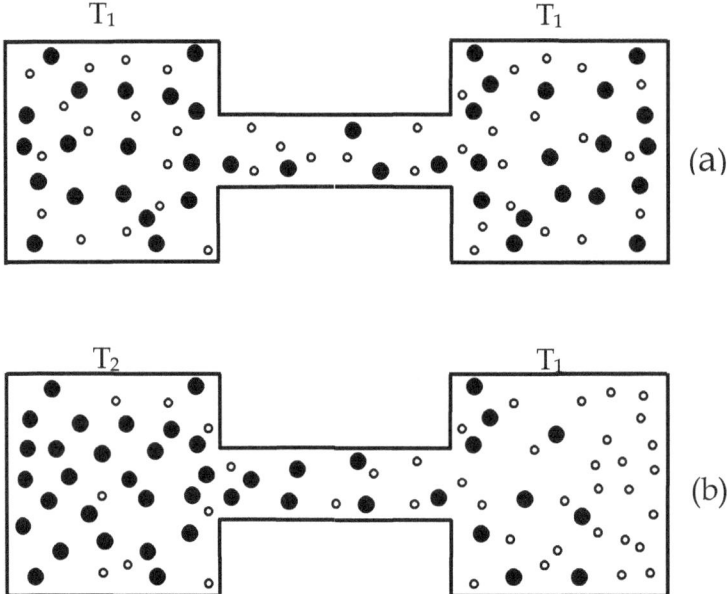

Fig. 11.1 (a) Mixture of two gases in equilibrium at a temperature T_1. (b) The temperature of the left side of the container is raised to a higher value T_2. Now both types of molecules on the left have a higher mean velocity, but the increase in mean velocity is higher for the lighter-gas molecules (small unfilled circles). Therefore more of them diffuse to the right than in the opposite direction. At steady state the number of lighter molecules is more on the right-hand side.

This is an irreversible process, and yet disorder has been reduced, and therefore order created: The probability that a molecule of nitrogen will be found in the left part of the chamber is higher for Fig. 11.1(b) than for Fig. 11.1(a), which is measure of a higher degree of order.

It should be realized that it is the *flow* of heat that has made this ordering possible. There has to be a heat bath which supplies heat to the left part so that its temperature is maintained at the higher value T_2. Similarly, heat must be continuously extracted from the right-hand part for keeping its temperature at a value less than T_2. There is a flow of heat all the time. 'The system evolves to a steady state in which one component is enriched in the hot part and the other in the cold part. The entropy produced by the irreversible heat flow leads to an ordering process, which would be impossible if taken independently from the heat flow' (Prigogine 1996). Contrary to popular perception, the arrow of time is clearly associated with *ordering* here.

11.3 The arrow of time and the early universe

> *The cosmic steps of symmetry breaking initiate the expansion of the universe. Thus, they determine the cosmic arrow of time breaking the symmetry of time.*

Klaus Mainzer, *Symmetry and Complexity*

> *Exactly as biological evolution cannot be defined at the level of individuals,*
> *the flow of time is also a global property.*
>
> Ilya Prigogine, *The End of Certainty*

In physics we have laws of Nature; and we have constraints, initial conditions, boundary conditions, or symmetry conditions. By and large the laws have time-reversal symmetry at the *microscopic* or elementary-particle level. Naturally, one wonders whether the existence of the arrow of time is because of the initial conditions at the birth of the universe. There are physicists who believe this to be the case (Layzer 1975).

As an analogy, consider the motion of the planets around our Sun. The law of gravitation decides the orbits, but there is nothing in this law that can explain why the planets should revolve around the Sun in the *same* direction, or why most of their orbits are nearly circular rather than being strongly elliptical. These other features (*regularities*) can be explained only in terms of a theory of how the planetary system got formed. There must have been initial conditions which made these special features possible.

Such a theory would hark back to certain statistical regularities in the primordial system from which the various stars and their planetary systems originated. But one cannot just stop at this primordial system as the origin of things. One must speculate about the chain of events backwards in time all the way, till the scenario in the very early universe has been theorized about and checked against available evidence. Layzer (1975) carried out such an analysis using a thermodynamic cum information-theoretic approach. His conclusions were as follows: 'We have now traced the thermodynamic arrow and the historical arrow to their common source: the initial state of the universe. In that state microscopic information is absent and macroscopic information is either absent or minimal. The expansion from that state has generated entropy as well as macroscopic structure. Microscopic information, on the other hand, is absent from newly formed astronomical systems, and that is why they and their subsystems exhibit the thermodynamic arrow'.

Layzer went on to say that 'If the theories that I have presented here are correct, however, not even the ultimate computer - the universe itself - ever contains enough information to completely specify its own future states. The present moment always contains an element of genuine novelty and the future is never wholly predictable'.

11.4 When did time begin?

Did time have a beginning? The present belief among many scientists is that our universe began at the 'moment' of the Big Bang (a Planck-time interval after the all-important quantum fluctuation). That event, which occurred ~14.7 billion years ago, is taken as defining the zero or the origin of the spacetime system of coordinates. What was there before the Big Bang? If something was indeed there, it must have evolved as a function of time to the state existing then.

A proper answer to this question will have to await the formulation of a satisfactory theory of quantum gravity. But what we know already is that the wholly classical general theory of relativity (which deals with the nature of the gravitational interaction) breaks down at the origin of the spacetime coordinate axes. It is applicable only for *later* spacetime points than the origin. Even if it is applicable for points earlier than the 'singularity' that the Big Bang was, it cannot be depended upon to give sensible physics right through the singularity on to the other side in which we live.

Hawking and Mlodinow (2010) have suggested a good way out of this perplexing situation. I mentioned their notion of *model-dependent realism* in Chapter 2. Suppose we work with a model of reality in which there was nothing before the Big Bang. Then the question 'what was there before the Big Bang?' does not bother us anymore because it gets answered by our model. But is it a good model of reality? Yes. In fact this is the best model we can formulate which is in agreement with all the observational data we have at our disposal *at present*.

Time began with the Big Bang.

If, however, new data become available which are better modelled by, say, a cyclical origin and death of the universe, then the new model would be the new reality. There is no reality beyond what our most successful model tells us.

11.5 Uncertainty and complex adaptive systems

> *Uncertainty is not necessarily bad or synonymous with risk. Complex systems use uncertainty to their advantage as they adapt to changes in their environment and learn to be resilient to unexpected shocks. Uncertainty then, rather than being the source of so many problems, becomes a necessary element if a market and a society are to remain free.*
>
> Edgar Peters, *Patterns in the Dark*

Some details about the nature of information, as also of uncertainty, are given in Appendix A3. There are two main types of uncertainty: vagueness, and ambiguity (Peters 1999). *Vagueness* implies an inability to define something precisely by establishing a definite cutoff for classification purposes. *Ambiguity* means lack of sufficient information about the possible outcomes of our actions, thus making unclear the choice between two or more objects or actions.

Uncertainty, complexity, and time are interrelated. Defining uncertainty rigorously is a nontrivial exercise. If we say that uncertainty means unknown future, it amounts to presuming that the future is fixed, and merely needs to be discovered and accurately predicted. This is not correct (Peters 1999). Everything happens in real time, and the flow of time occurs in only one direction. It is not possible to step out of the flow of time and see what lies ahead.

As argued by Peters (1999), complexity and the arrow of time generate uncertainty. If complexity gets reduced, uncertainty stands a chance of getting eliminated. Too much complexity means too much freedom, and therefore high uncertainty.

Real time entails change. The succession of events changes the universe, and time progresses. The concept of probability (see Appendix A2) has a meaning only if there is prior knowledge of *all* the possibilities of the consequences of an action. This is seldom the case. The unpredictable nature of complexity implies that the future state of a complex system cannot be from among a fixed and known number of possibilities. There is always an element of uncertainty, and also *novelty* and *surprise*.

Complex adaptive systems (CASs) are able to both generate novelty and also assimilate novelty during the process of adaptation and learning. In CASs, order and randomness coexist. Thus uncertainty is a component of complexity.

How can complexity imply both order and uncertainty? The order is global (it is on a large scale), and the uncertainty is in the details. For example, no two oak trees are identical, and yet there is a global sameness in oak trees (determined by the DNA). We sow a seed, and we can be sure of the *type* of tree that would germinate and grow. This is the order part of the complexity involved. The uncertainty part is that we have no way of predicting the exact details of any particular tree.

The purpose or goal of a CAS is not a final state of static equilibrium. Instead, it may be survival, efficiency, etc. This requires resilience against uncertain or unexpected situations. The CAS may use uncertainty to generate order, and may produce more uncertainty in the process.

Random systems are clearly different from complex systems. In a random system, passage of time is of no significance. But is a CAS, there is evolution as a function of passing time. Although there is unpredictability, a CAS does depend on the past to interpret the present. CASs are path-dependent without being predetermined. And there is no central planner involved.

12. The Cosmic Evolution of Complexity

12.1 Our cosmic history

The chronology of cosmic events after the origin of our universe has been pieced together by modern cosmology to be as follows.

10^{-43} second. The quantum gravity era. As the very hot plasma expanded after the cosmic explosion, it also cooled. The temperature 10^{-43} seconds after the birth of the universe was $\sim 10^{32}$ K.

10^{-36} second. Cosmic inflation started. It lasted till $\sim 5 \times 10^{-33}$ second after the beginning of the universe. It created a large patch of space filled with quarks. The temperature was $\sim 10^{27}$ K, and other forms of matter (leptons, gauge bosons, and several other elementary particles) also appeared, as also antimatter.

10^{-30} s. One potential type of dark matter called *axions* gets synthesized. Matter and antimatter are on equal footing at this stage.

10^{-11} s. Matter wins over antimatter.

10^{-10} s. A second potential type of dark matter called *neutralinos* gets synthesized. The electro-weak interaction splits into the electromagnetic interaction and the weak interaction.

10^{-5} s. The temperature has fallen to $\sim 10^{12}$ K. This is when the quarks formed the protons and the neutrons, and the antiquarks formed antiprotons. The collisions between protons and antiprotons left behind mostly protons, as well as photons.

0.01 - 1 s. Collisions among electrons and positrons occur, leaving behind mostly electrons.

1 - 300 s. Enough cooling of the universe has occurred (to $\sim 10^9$ K), so that *nuclei* (not atoms yet) of deuterium, tritium, helium, and lithium get formed by the coming together of protons and neutrons.

380,000 years. More cooling results in the possibility of capture of electrons by the nuclei, so that *atoms* get formed. This is a major event, because there is a concomitant release of the till-then-trapped *cosmic microwave background* (CMB), which was to be detected by our instruments in the present era, and which constitutes the strongest validation of the Big Bang theory. The wavelength of this radiation increases as the universe expands and cools, and is currently in the microwave regime; hence the name 'cosmic microwave background'. The photons of the CMB we observe today started their journey 380,000 years after the Big Bang from the so-called *surface of last scatter*.

380,000 years - 300 million years. Gravity continues to amplify the density differences in the gas that fills space.

300 million years. Stars and galaxies appear.

1 billion years. One billion years after the Big Bang is the present limit of how far into the past our instruments can give us data about the conditions at that time.

3 billion years. This is when the formation of stars peaked. This is also when clusters of galaxies formed.

9 billion years. Our solar system is formed.

10 billion years. Dark energy takes hold and the expansion of the universe begins to *accelerate*.

13.7 billion years. The present.

12.2 We are star stuff

> *Matter is much older than life. Billions of years before the sun and earth even formed, atoms were being synthesized in the insides of hot stars and then returned to space when the stars blew themselves up. Newly formed planets were made of this stellar debris, the earth and every living thing are made of star stuff.*
>
> Carl Sagan

About ten million years after the Big Bang, enough cooling and expansion had occurred to fill the universe with a mist of particles, mostly hydrogen and some helium, as also some types of elementary particles (including neutrinos), some electromagnetic radiation, and perhaps some other, unknown, particles. The universe was just cold, dark, and formless at that stage.

Then, when enough further cooling had occurred, some quantum-mechanical primordial fluctuations in the densities of the particles resulted in a clumping of some of the particles, rather like the cluster formation and nucleation that precedes the growth of a crystal from a fluid. The presence of such clumped particles brought the gravitational forces into prominence, resulting in a cascading effect. Portions of the mist began collapsing into large swirling clouds. Over a period of a few hundred million years, huge galaxies, each containing billions of young stars of various sizes, formed and began to shine. The formless darkness of the initial period was gone.

The large superstars among these were strongly bright spheres, the brightness coming from the nuclear fusion of hydrogen and helium in their interiors, made possible by the prevailing extreme temperatures and pressures. This is how many of the heavier elements got formed in the interiors of these large stars.

The emergence of heavier elements by the process of nuclear fusion continued steadily until the element iron (Fe) started forming. The iron nucleus is *the most stable* of them all, having the largest binding energy per nucleon. [Protons and neutrons inside the nucleus are jointly called nucleons; and 'binding energy' is defined as the amount of energy required to extract a nucleon from inside the nucleus and take it far away from it]. Therefore iron cannot fuse with one or more nucleons and release radiative energy of the nuclear process; such a process would not lower the potential energy and the free energy. Consequently, the presence of iron acts as a '*poison*' for the nuclear fusion process. Thus the appearance of iron marked the beginning of the end of the available nuclear fuel, and therefore the end of the life of the star. In due course, the smaller among such stars simply ceased to shine, shrinking into cold and dead entities.

But a very different fate awaited the larger stars. No longer able to sustain their size because of the progressively decreasing processes of nuclear fusion of elements, they began to *collapse* under their own immense gravitational pull. A rapid change occurred in their interiors. Under the immense squeezing generated by the collapse, the iron-element core imploded. This resulted in a new state of matter as the electrons and the protons in the atoms were squeezed together. The dominant process of interaction was the electroweak interaction:

$$p^+ + e^- \rightarrow n^0 + v_e \tag{12.1}$$

(i.e., protons and electrons combined to produce neutrons and electron-neutrinos).

Thus, this collapse led to a compression of the star to an extremely dense ball of pure neutron matter, with the neutrino cloud bursting outwards, resulting in an explosion of the outer shell of the star (the *supernova explosion*). This is how the synthesized elements (up to the atomic number (Z) for iron), residing in the outer shell of the star, were scattered into the universe, accompanied by a brilliant flash of light.

A consequence of such supernova explosions (which still occur from time to time, and illuminate the galaxies with brilliant flashes of light) was the emergence of clouds of dust and gas and the debris containing heavy elements. These clouds encircled the galaxies in spiralling arms. *The intensity of the supernova explosions and the temperatures involved were so high that elements heavier than iron were also synthesized and scattered into space.*

The chemical elements in our bodies have come from that star stuff: In the outer portions of the spirals occurred a condensation of the dust, the clouds and the debris, resulting in the formation of the second generation of (smaller) stars, including our Sun, as also planets, moons, comets, asteroids, etc.

Our solar system was formed when the universe was ~9 billion years old. In the initial period, our Earth underwent several violent upheavals (bombardment by comets and meteors, as also huge earthquakes and volcanic eruptions). By the time the Earth was ~2.5 billion years old, its continents had formed. Life appeared in due course, which further influenced the ecosphere in a major way. In particular, free oxygen (as opposed to oxygen chemically bound with other elements) was liberated as a 'waste product' by the algae that consumed carbon dioxide present in the atmosphere and in the oceans.

Two billion years ago, our Earth was extremely radioactive as well. The heavier-than-iron elements produced in the outer shell of the exploding stars during the supernova explosion were/are radioactive, as their binding energy per nucleon was lower than that of iron: Such elements can increase their binding energy per nucleon (and thus attain a more stable state) by undergoing nuclear fission, either spontaneously or with the assistance of free neutrons. Uranium was among the heaviest elements produced during the last few seconds of the supernova explosion. Thus this element was a part of the Earth right from the beginning.

The above narrative indicates that the degree of complexity (i.e., the information content) of the universe has been, by and large, increasing since the moment of the Big Bang. It has been argued by Chaisson (2001) that this cosmic evolution, which will go on deep into the future, if not forever, is a consequence of the fact that the universe has been expanding ever since its creation. The increasing complexity that we see around us can be ultimately traced to the expanding and cooling nature of the universe.

13. Why Are the Laws of Nature What They Are?

Why is there something rather than nothing?
Why do we exist?
Why this particular set of laws, and not some other?

Stephen Hawking

13.1 The laws of Nature in our universe

Why are the laws of Nature what they are? This used to be a very profound question till recently. Developments in physics during the last few decades have now made it rather trivial and trite.

Such questions used to be in the domain of philosophy, and human history can boast of a truly dazzling succession of great philosophers. But the question now is: What is the true worth of an otherwise great philosopher who was/is innocent about the finer points of quantum mechanics? Modern physics has come up with plausible answers to fundamental questions over which philosophers fretted for centuries. No wonder, Hawking and Mlodinow (2010) wrote this, somewhat facetiously perhaps:

Traditionally these are questions for philosophy, but philosophy is dead. Philosophy has not kept up with developments in science, particularly physics. Scientists have become the bearers of the torch of discovery in our quest for knowledge.

An answer to the question 'Why are the laws of Nature what they are?' comes from M-theory. According to it, there are actually 11 dimensions. We see only four because the rest of them have got 'curled up' so much that they are not visible to us. There are $\sim 10^{500}$ different modes of curling up, meaning that that many different universes are possible. One of them is the universe we inhabit. The *apparent laws* of a universe depend on how the extra dimensions in this universe got curled up. [We say 'apparent laws', because the more fundamental laws are those of the M-theory.]

Thus, there are multiple universes, each with its own set of apparent laws. We just happen to be living in a universe with a certain set of laws and a certain set of values for the fundamental constants. If the laws of a universe are not conducive to emergence and evolution of life, living beings cannot possibly exist in that universe, discussing such questions. It is as simple as that.

In Newtonian physics, the past was visualized as a definite series of events. Not so in quantum physics. No matter how thoroughly and accurately we observe the present, the unobserved past, as also the future, is indeterminate, and exists only as a 'spectrum of possibilities'. This means that our universe does not have just a single past, or history. Since the origin of the universe was a quantum event, Feynman's sum-over-histories formulation for going from spacetime point A to spacetime point B occupies centre-stage. But we have knowledge only about the present state of the universe (point B), and we know nothing about the initial state A. Therefore, as emphasized by Hawking, we can only adopt a '*top down*' approach to cosmology, wherein every alternative history of the universe exists simultaneously, and the histories relevant to us are those which, when summed up, have a high probability of giving us our present universe (point B).

The picture that emerges is that many universes emerged spontaneously, simultaneously or otherwise. Most of these universes are not relevant to us because their apparent laws are not conducive to *our* emergence and survival. The M-theory offers ~10^{500} possibilities of start-up universes. We have to single out those which correspond to the curling up of exactly those dimensions which we find to be the case for the universe we inhabit. Further, we have to select those histories which reproduce, for example, the observed mass and charge of the electron, and other such observed fundamental parameters.

[Something about 11 dimensions cropped up recently from an unexpected quarter. I shall describe in Chapter 42 the work of Markram, who has been working on the *Blue Brain Project*, aimed at creating a digital reconstruction of the brain by reverse-engineering the mammalian brain circuitry. He was quoted as saying something very intriguing: 'We found a world that we had never imagined . . . there are tens of millions of these objects even in a small speck of the brain, up through seven dimensions. In some networks, we even found structures with up to eleven dimensions' (see http://www.zmescience.com/medicine/neurons-high-dimensional-structures/).

What if the M-theory does not get due validation? The redeeming fact is that the multiverse idea would be still intact; via the cosmic-inflation theory.

Euclidean geometry holds true in our universe; i.e., ours is a *flat-geometry* universe. Which is just as well. As explained by Krauss (2012), only a flat-geometry universe can satisfy the requirement that the sum total of positive and negative contributions to the overall energy of the universe add up to zero. The energy-conservation law was not violated when our universe emerged out of 'nothing'. The total energy is still zero.

13.2 The anthropic principle

Even slightly different values for some of the fundamental constants of Nature would have led to entirely different histories of the cosmos, making our emergence and existence very different, if not impossible. Why do these parameters have the values they have? According to a 'weak' version of the so-called anthropic principle:

The parameters and the laws of physics can be taken as fixed; it is simply that we humans have appeared in the universe to ask such questions at a time when the conditions are just right for our life.

Life as we know it exists only on planet Earth. Here is a partial list of necessary conditions for its existence:

1. Availability of liquid water is one of the preconditions for our kind of life. Around a typical star like our Sun, there is an optimum zone (popularly called the *Goldilocks zone*), neither so hot that water would evaporate, nor so cold that water would freeze, such that planets orbiting in that zone can sustain liquid water. Our Earth is one such planet.

2. This optimum orbital zone should be circular or nearly circular. Once again, our Earth fulfils that requirement. A highly elliptical orbit would take the planet sometimes too close to the Sun, and sometimes too far, during its cycle. That would result in periods when water either evaporates or freezes. Our kind of life needs liquid water all the time.

3. The location of the planet Jupiter in our Solar system is such that it acts like a *massive gravitational vacuum cleaner*, intercepting asteroids that would have been otherwise lethal to our survival.

4. Planet Earth has a single relatively large Moon, which serves to stabilize its axis of rotation.

5. Our Sun is not a binary star. Binary stars *can* have planets, but their orbits can get messed up in all sorts of ways, entailing unstable or varying conditions, inimical for life to survive and evolve.

It is not only that the planet we live on is conducive to our existence; even the universe we live in (with its operative set of laws of physics) is so. The 'cosmological' or 'strong' version of the anthropic principle says just that:

Our universe has the fundamental constants and the laws of physics that are compatible with our existence; had they been different (i.e., inimical to our existence), we would not be here, discussing the principle.

The chemical elements needed for life were forged in certain stars, and then flung far into space through supernova explosions. This required a certain amount of time. Therefore the universe cannot be younger than the lifetime of the stars. The universe cannot be too old either, because then all the stars would be 'dead'. Thus, according to the cosmological anthropic principle, life exists only when the universe has the age that we humans have measured it to be, and has the physical constants that we measure them to be.

Rees (1999), in his book *Just Six Numbers*, listed six fundamental constants that together determine the universe we see. Their mutual values are such that even a slightly different set of these six numbers would have been inimical to our emergence and existence. Consideration of just one of these, namely the strength of the strong nuclear interaction (which determines the binding energies of nuclei), is enough to make the point. It can be roughly defined as that fraction of the mass of an atom of hydrogen which is released as energy when two hydrogen atoms fuse to form an atom of helium. Its value is 0.007 (in certain units), which is just right (give or take a small acceptable range) for any known chemistry to exist. And no chemistry means no life. Why?

Our chemistry is based on reactions among the 90-odd elements. Hydrogen is the simplest among them. Many of the other elements in our universe got synthesised by fusion of hydrogen atoms. This nuclear fusion depends on the strength of the strong nuclear interaction, and also on the ability of a system to overcome the intense Coulomb repulsion between the fusing nuclei. Existence of intense temperatures is one way of overcoming the Coulomb repulsion. A small star like our Sun has a temperature high enough for the production of only helium from hydrogen. The other elements in the periodic table must have been made in the much hotter interiors of stars larger than our Sun. The value 0.007 for the strong interaction determined the upper limit on the mass number of the elements we have here on Earth and elsewhere in our universe. A value of, say, 0.006, would mean that our universe would contain nothing but hydrogen, making impossible any chemistry whatsoever. And if it were too large, say 0.008, all the hydrogen would have disappeared by fusing into heavier elements. No hydrogen would mean no life as we know it; in particular there would be no water without hydrogen.

Similarly for the other fundamental constants of our universe.

But why? Why does the universe have these values for the fundamental constants, and not some other set of values? A fallout of Hawking's model for our universe is that even the strong anthropic principle acquires validity, provided it is stated properly and in the context provided by the M-theory. The new version of the strong anthropic principle goes something like this:

Out of the various possible universes, our universe just happens to have the fundamental constants and physical laws it has; other universes (which we cannot observe) have different laws of physics and different values for the fundamental constants. Our existence in our universe has been possible because we have evolved to be compatible with our apparent laws of physics and our set of fundamental constants; other universes may or may not be conducive to life of any kind.

14. The Universe is a Quantum Computer

The natural dynamics of a physical system can be thought of as a computation in which a bit not only registers a 0 or 1 but acts as an instruction: 0 means 'do this' and 1 means 'do that'. The significance of a bit depends not just on its value but on how that value affects other bits over time, as part of the continued information processing that makes up the dynamical evolution of the universe.

Seth Lloyd (2006)

Natural phenomena can be viewed as just computations. The laws of Nature are quantum-mechanical, and the universe is one big quantum computer. The constituent particles of the universe (electrons, quarks, photons, etc.) and their mutual interactions are, at their most fundamental level, *bit-streams of information*, the time-evolution of which is governed by the laws of physics. Each elementary particle registers one bit of information. When the particles interact with one another, they transform and process the information, bit by bit. Each collision between elementary particles is like a logical operation in a computer. Thus the laws of quantum mechanics and of elementary-particle physics are the algorithms which determine how the universe evolves through a succession of computations, *quantum computations*. Since an elementary particle can be everywhere (subject to a certain probability distribution; see Appendix A6 on quantum theory), every atom, electron or photon participates in the registering and processing of information. And as the universe quantum-computes, its degree of complexity evolves (usually increases).

14.1 Quantum computation

The amount of time the quantum computer takes to perform the simulation is proportional to the time over which the simulated system evolves, and the amount of memory space required for the simulation is proportional to the number of subsystems or subvolumes of the simulated system. The simulation proceeds by a direct mapping of the dynamics of the system onto the dynamics of the quantum computer.

Seth Lloyd (2006)

Till now we have had only limited success in the construction of quantum computers. If we can understand better how our quantum computers register and process information, we can understand better how the interrelated physical systems in our universe treat the information in and around them and become more and more complex (Lloyd 1996).

In quantum mechanics we have the wave-particle duality. Moreover, we can only speak in terms of *probabilities* regarding the location and momentum of, say, an electron. An electron can be in two or more places at the same time. Therefore, a quantum bit, or *qubit*, can register both 0 and 1 at the same time, implying that a quantum computer can perform millions of operations *simultaneously*. Naturally, such a computer is far more efficient than a classical computer.

The nuclear spin is a good example of a qubit. If 'spin up' is taken as having a bit value 0, then 'spin down' has a bit value 1. In the wavemechanical description of spins, we can identify clockwise spin as 'spin up', and write the wave function as $|0\rangle$. Similarly, counterclockwise spin is 'spin down', and the wave function is $|1\rangle$.

Quantum mechanical wave functions can be combined or 'superposed'. In the case of spin waves, it can be shown that the sum $|0\rangle+|1\rangle$ corresponds to a spin along an axis perpendicular to the one used for spin-up or spin-down. Thus spin-up plus spin-down is *spin-sideways*. Similarly, $|0\rangle-|1\rangle$ corresponds to a spin opposite to that of $|0\rangle+|1\rangle$

Any computation involves bit-flips. The nuclear spins or qubits can be flipped by using a magnetic field. When a magnetic field is applied, the nuclear spins precess around the field direction (the so-called *Larmor precession*). The angular frequency of the precession is proportional to the strength of the field applied. We can control the spin direction by applying the field for a time which is a suitable fraction of the time period of the Larmor precession. For example, one can start with a spin $|0\rangle$ and apply the field for a time that is one-fourth of the time period of precession; this gives a spin state $|0\rangle+|1\rangle$. Similarly, if the field is applied for three-quarters of the time period, we get the spin state $|0\rangle-|1\rangle$. In fact, a variety of spin orientations can be obtained by adjusting the time of application of the field. Like in classical physics, the spin rotations preserve the amount of information.

In classical computation, the four logic operations AND, OR, NOT, and COPY constitute the *universal set of operations*, meaning that any computation can be done in terms of them. They all carry out one-to-one mappings, meaning that if we take the output as the new input, we get back the original input. Moreover, all these operations conserve information; they do not destroy it.

In quantum computation, apart from spin rotation, one more operation is needed for performing universal computations, namely the *controlled-NOT* operation (Lloyd 2006). It takes two bits as inputs, the first bit serving as the control bit. If the first bit is 0, it does nothing to the second bit. If the first bit is 1, the second bit is flipped from 0 to 1, or from 1 to 0. Thus 00 changes to 00; 01 changes to 01; 10 to 11; and 11 to 10. It is a reversible operation; applying it twice restores the initial pair of bits. Rotations of qubits and the controlled-NOT logical operation together constitute a universal set of computations in quantum computation. The total number of possible logical operations here is much more than in classical computation because a variety of rotations of the spins or qubits are possible. And they all conserve information (Lloyd 2006).

14.2 Quantum entanglement

In quantum computation a state like, say, $|01\rangle$ represents two simultaneous waves corresponding to two qubits, the first in state $|0\rangle$ and the second in state $|1\rangle$. When a controlled-NOT operation is in action, information from the first qubit spreads to and 'infects' the second qubit. The operation creates *mutual information* between the two qubits, because the second qubit now knows whether the first qubit is $|0\rangle$ or $|1\rangle$. This creation of shared information, apparently out of nothing, is a characteristic feature of the quantum-mechanical version of the controlled-NOT operation, not present in classical computation. If a classical system is in a definite microstate, implying zero entropy, then each of its constituents is also in a definite state, with zero entropy. For example, if two classical bits are in a state 01, then it means that the first bit is in state 0 and the second bit is in state 1. By contrast, consider a quantum bit in a definite state, comprising of the correlated states of two qubits. Its entropy is zero. But this does not imply that the component qubits also are necessarily in definite states. This property is known as *quantum entanglement* (see Gilder 2008).

In an entangled state we may have information about the definite state of the system *as a whole*, and yet not know with certainty the states the component pieces are in. Consider a qubit in a definite state $|0\rangle+|1\rangle$. Its entropy is zero. But it has contributions from both $|0\rangle$ and $|1\rangle$. The qubit $|0\rangle+|1\rangle$ registers both $|0\rangle$ and $|1\rangle$ simultaneously, and we can speak only about the *probabilities* of the states in which $|0\rangle$ and $|1\rangle$ are. Entanglement enables quantum computers to perform much more efficiently than classical computers (Farías *et al.* 2009).

A qubit can register a 0 and 1 at the same time. If 0 corresponds to 'do this', and 1 to 'do that', then a quantum computer does both 'this' and 'that' at the same time. This *quantum parallelism* is what makes quantum computers so efficient: A single processor can perform several tasks at the same time.

14.3 The universe regarded as a quantum computer

> *Paradigms are highly useful. They allow us to think about the world in a new way, and thinking about the world as a machine has allowed virtually all advances in science, including physics, chemistry, and biology. The primary quantity of interest in the mechanistic paradigm is energy. This book advocates a new paradigm, an extension of the powerful mechanistic paradigm: I suggest thinking about the world not simply as a machine, but as* a machine that processes information. *In this paradigm, there are two primary quantities, energy and information, standing on an equal footing and playing off each other. . . . The conventional mechanistic paradigm gives no simple answer to the question of why the universe in general, and life on Earth in particular, is so complex. In the computational universe, by contrast, the innate information-processing power of the universe systematically gives rise to all possible types of order, simple and complex.*
>
> Seth Lloyd (2006)

Lloyd (2006) has argued at length in support of his contention that the universe is a quantum computer: 'The universe is the biggest thing there is and the bit is the smallest possible chunk of information. The universe is made of bits. Every molecule, atom, and elementary particle registers bits of information. Every interaction between those pieces of the universe processes that information by altering those bits. That is, the universe computes, and because the universe is governed by the laws of quantum mechanics, it computes in an intrinsically quantum-mechanical fashion; its bits are quantum bits. The history of the universe is, in effect, a huge and ongoing quantum computation. The universe is a quantum computer'.

And what is the universe computing? *It computes its own behaviour.* What are the algorithms for this computation? The laws of quantum mechanics.

The universe has been expanding ever since the Big Bang, and it is conceivable that the space into which it is expanding is infinite (this is really a simplistic statement, because we have only spacetime, and not separate space and separate time). But the speed of light is not infinite. Therefore the causally connected part of the universe has a finite size, limited by what has been called the '*horizon*' (Lloyd 2006). The quantum computation being carried out by the universe is confined to this part. As this universe expands, the size of the causally connected region increases, which, in turn, means that the number of bits of information within the horizon increases, as does the number of computational operations.

Thus the expanding quantum-computing universe is a reason for the continuing increase in the degree of complexity of the universe.

The expanding universe creates gradients and, as Nature finds ways to annul the gradients, entropy increases on the whole, although the possibility of self-organization and order in certain pockets is always there. This is how the degree of complexity increases in certain pockets of the universe. In the present chapter we have seen that the quantum-computer picture of the universe is an alternative but very instructive way of understanding the relentless increase of complexity.

There are limits to how far we can carry out our reasoning processes for understanding or describing natural phenomena. Beyond a limit, it is like in *Alice in Wonderland*: *'Do cats eat bats? Do cats eat bats?' Sometimes she asked, 'Do bats eat cats?' For you see, as she couldn't answer either question, it didn't much matter which way she put it.*

15. Chaos, Fractals, and Complexity

We believe that it is impossible to discuss complex systems without mentioning the concept of chaos. The emergence of the concept of chaos is one of the biggest scientific achievements this century, and has inspired interest in complex systems by radically changing many fundamental beliefs in science.

Kaneko and Tsuda (2000)

15.1 Nonlinear dynamics

Dynamics is perhaps the oldest branch of physics. It is the mathematical study of how systems change with time.

The dynamics of a system may be linear or nonlinear. Linear dynamics involves linear mathematical operators. The latter have the property that their action on a sum of two functions is equal to the sum of the action of the operator on each function.

Nonlinear dynamics means that the output is not linearly proportional to the input. Consider a system described by the equation $y(t) = cx(t)^2$. If you double the input x, the output y does not double; it becomes four times. But if $y(x) = cx(t)$, we are dealing with linear dynamics. Linearity also respects *the principle of superposition*: If $x_1(t)$ and $x_2(t)$ are two solutions of an equation, then $c_1 x_1(t) + c_2 x_2(t)$ is also a solution. This is not so for nonlinear dynamics. Nonlinear operators are such that, e.g., $(x_1(t) + x_2(t))^2 \neq x_1(t)^2 + x_2(t)^2$.

Chaos is about *highly* nonlinear dynamics, although not all nonlinear systems are chaotic.

In normal conversation we use the term 'chaos' to describe a condition of utter confusion, totally lacking in order or organization. In scientific terms, chaos is defined as sustained and disorderly-looking long-term evolution of a deterministic nonlinear dynamical system that satisfies certain mathematical criteria. To the extent that noise can be ignored, all chaos is deterministic chaos.

According to Kaneko and Tsuda (2000): 'In fact, chaos exists everywhere in our lives. We are tempted to imagine that if chaos is such an ordinary phenomenon, perhaps humans discovered chaos and defined the concept in ancient times. Actually, in many mythical stories chaos is described as a state in which heaven and earth are not divided (a state in which everything is mixed). Chaos is also described as an "energy body" responsible for the creation of heaven and earth'.

In ancient Indian philosophy, the concept of *Brahmman* is propounded for this 'energy body'; and the 'state in which everything is mixed' is referred to as *Pralaya*.

Apart from nonlinearity (or rather, *because* of the presence of strong nonlinearity), a chaotic system is characterized by *unpredictable* evolution in space and time, in spite of the fact that the differential equations or difference equations describing it are deterministic.

The concept of a 'mode' is well-known in dynamics. Any periodic motion can be expressed as a superposition or summation of elementary sinusoidal motions (*Fourier components*), such motions or modes being a fundamental sinusoidal motion and its harmonics. This is possible even for nonperiodic linear motion, except that the discrete summation of the modes is replaced by an integral over a continuum of modes.

But when chaotic motion is decomposed into Fourier components or modes, an uncountably infinite number of modes appear as a continuous spectrum: Chaos contains infinitely many but countable periodic motions, and uncountably many nonperiodic motions (Kaneko and Tsuda 2000). The motions are unstable, and this instability leads to *a sensitive dependence on initial conditions*.

Chaos theory (Appendix A8) is about finding the underlying order in apparently random data. It is about a certain class of nonlinear dynamical systems which may be either *conservative* or *dissipative*. Conservative systems do not lose energy over time.

A dissipative system, by contrast, loses energy; e.g., via friction. As a consequence of this, it always approaches some limiting or asymptotic configuration (namely an *attractor* in phase space).

Dynamical phenomena may be either *discrete* or *continuous*. The former can be described by difference equations, which can be solved by iteration. The latter require differential equations for their description.

Iteration can be likened to feedback: The output of a discrete dynamical system at any particular time step serves as the input for determining the output of the next time step.

15.2 Extreme sensitivity to initial conditions

> *The flapping of a single butterfly's wing today produces a tiny change in the state of the atmosphere. Over a period of time, what the atmosphere actually does diverges from what it would have done. So, in a month's time, a tornado that would have devastated the Indonesian coast doesn't happen. Or maybe one that wasn't going to happen, does.*
> Ian Stewart, *Does God Play Dice? The Mathematics of Chaos*

The unpredictability feature of chaotic systems, mentioned in the above section, is because of extreme sensitivity to initial conditions. This is popularly referred to as *the Butterfly effect*. The above quotation from Ian Stewart explains the genesis of this phrase.

Poincaré, near the end of the 19th century, had recognized the existence of chaos. The mathematics of celestial motion yielded complex solutions when the number of bodies was greater than as little as 3. But his work was not noticed, as also the work of several other scientists on chaos in the early 20th century. It was the meteorologist Edward Lorenz who in 1961 established the field of chaos theory as we know it today. Even his work went unnoticed for nearly a decade.

Lorenz was working on the problem of weather prediction. He started with the standard (Navier-Stokes) equations of fluid dynamics and simplified them greatly for carrying out his computer-simulation studies regarding the dynamics of the atmosphere. The three equations he set up are described in Appendix A8. The remarkable discovery he made, by accident, was that the predictions his model made depended in a crucial way on the precision with which he specified the values of the three adjustable parameters (r, σ, and b) in his equations. Two calculations, identical except that they differed in the value of one of these control parameters by, say, 0.000001, made totally different long-term predictions. The dynamics was not just nonlinear; even the slightest variations in the initial conditions gave wildly different results after a certain number of time steps.

Further investigations led to the discovery of what is now known as the *Lorenz attractor*. It was found that, although the results of the calculations were very sensitive to the values of the three control parameters, in every case the output always stayed on a double spiral, shown in Fig. 15.1.

Fig. 15.1 A plot of the Lorenz attractor for $r = 28$, $\sigma = 10$, and $b = 8/3$.
http://en.wikipedia.org/wiki/File:Lorenz_attractor_yb.svg

This was new science. Till then only two types of dynamical systems were known: Those in steady state, and those in which the system undergoes periodic motion, repeating its configuration periodically. Fig. 15.1 depicts a third kind of ordered dynamics. The system does not settle to a single-point attractor or a closed-loop attractor in phase space. But it is not random dynamics either. There is order, except that the phase-space trajectory never comes to the same point again. First one spiral is traced, and then the other, and then again the first; so on. The Lorenz attractor shown in Fig. 15.1 belongs to a new family, called *strange attractors*.

15.3 Chaotic rhythms of population sizes

The Lorenz equations are a bit complicated for illustrating chaotic dynamics. A well-known classic model of population dynamics is much simpler for this purpose. It is an iterative or feedback model, and was formulated in 1845 to understand the long-term population dynamics of any species. How does the population of a species, confined to a certain geographic area, vary from year to year? Obviously, the population in a year $t+1$ will depend on the population in the previous year t: $x_{t+1} = k_1 x_t$ (incidentally, this equation is an example of a *difference equation*). The proportionality constant, k_1, is a measure of the growth rate of the population x. [We have 'normalized' this equation by dividing the actual population number by the maximum value achievable by it, so that $0 < x < 1$.]

But there are other factors to consider. For example, if the population in a year t becomes too large, there can be an increased decimation of the population, either from predators, or due to shortage of food, or due to the increased competition in the reproduction dynamics. Therefore the model should include a negative term, which should depend on x_t, but should make a smaller contribution to x_{t+1} than the leading term $k_1 x_t$. Thus a simple modelling equation can be

$$x_{t+1} = k_1 x_t - k_2 x_t^2 .$$

It turns out that the essential physics of the problem can be captured even in a '1-dimensioinal' version of this model. This is done by taking $k_1 = k_2 = k$, resulting in the following *logistic difference equation*: $x_{t+1} = k x_t (1 - x_t)$.

Details are given in Appendix A8 about how the population dynamics depends on the value of the parameter k. A fascinating variety of dynamics, including chaos, is observed as k takes various values. The results can be summarized as in Fig. 15.2.

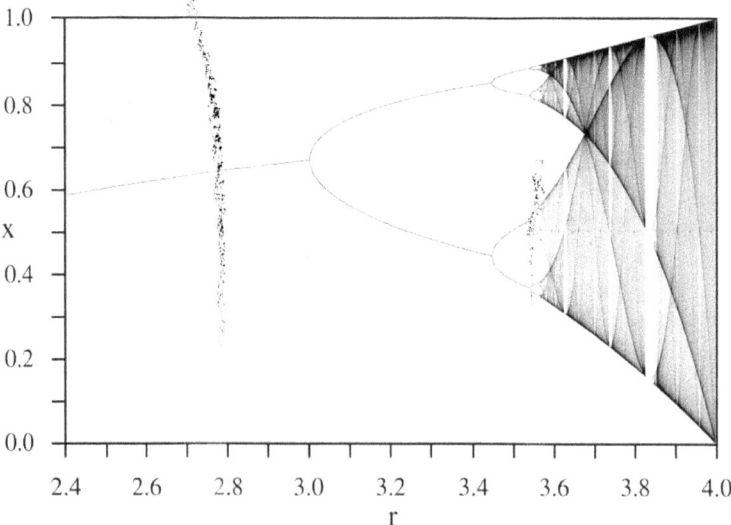

Fig. 15.2 Bifurcation diagram for the population-dynamics logistic equation (with k replaced by the symbol r.). Wikipedia commons, from which this image has been taken, carries the following description: 'The image was created by forming a 1601 x 1001 array representing increments of 0.001 in r and x. A starting value of x=0.25 was used, and the map was iterated 1000 times in order to stabilize the values of x. 100,000 x-values were then calculated for each value of r and for each x value, the corresponding (x,r) pixel in the image was incremented by one. All values in a column (corresponding to a particular value of r) were then multiplied by the number of non-zero pixels in that column, in order to even out the intensities. Values above 250,000 were set to 250,000, and then the entire image was normalized to 0-255. Finally, pixels for values of r below 3.57 were darkened to increase visibility.' Image credit: https://en.wikipedia.org/wiki/Bifurcation_diagram
http://upload.wikimedia.org/wikipedia/commons/7/7d/LogisticMap_BifurcationDiagram.png

For example, if we take $k < 1$ (or rather $r < 1$ in Fig. 15.2), and carry out the iteration by starting with some initial population value x_0, and calculate x_1 from it for the next year, and then calculate x_2 from x_1, and so on, we find that the population eventually becomes zero (not shown in Fig. 15.2 because results for $r < 2.4$ have not been plotted in it). This eventual or final value of x, denoted by x^*, is an *attractor*. There is a *basin of attraction* such that every starting value x_0 in this case is eventually drawn or 'attracted' towards the attractor $x^* = 0$. Thus for $k < 1$ we have a *fixed-point attractor* at the zero value of the population; the conditions are too inimical for the population to survive.

A different population dynamics is predicted by the logistic equation for $1 < k < 3$. Now $x*$ is not zero; rather it increases from zero to ~0.667 as k is increased from 1 to 3.

A fundamentally different dynamics emerges for $3 < k < 4$: The population trajectory no longer converges to a single fixed-point attractor. Further, the trajectory becomes increasingly sensitive to the value of k. For $k = 3.4$, the trajectory has not one but two fixed points: at $x_1* \approx 0.452$ and $x_2* \approx 0.842$. This means that, from year to year, the population oscillates between ~45% and ~84% of the maximum possible value. We now have a *two-point attractor*. Such an oscillating system is described as having a *period 2*.

Thus, Fig. 15.2 shows a bifurcation point in the curve near and beyond $r = 3$. For any particular value of r (or k), the population oscillates between two limiting values from year to year. This periodic attractor is actually just the beginning of still more complex dynamics as the value of k is increased further.

There is a critical value $k \approx 3.4495$ beyond which we get a *four-point attractor*. For example, for $k = 3.5$ we get $x_1* \approx 0.875$, $x_2* \approx 0.383$, $x_3* \approx 0.827$, $x_4* \approx 0.501$. Such successive bifurcations of each attractor into two, such that there are 4, 8, 16, 32, .. etc. fixed points, occur with smaller and smaller increases in k.

Finally, we move into the *chaotic regime* of complexity for values of k above ~3.57. The periods now double every time k is increased by even an infinitesimally small amount. The number of points that comprise the *chaotic attractor* is practically infinite, and the trajectory looks quite erratic, although there are some ranges of k values for which there is apparent stability.

15.4 Fractal nature of the strange attractor

> *Fractals in nature originate from self-organized critical dynamical processes.*
>
> Bak & Chen (1989)

> *Chaos, which is regarded as an information channel, has characteristics in which information mixes in the binary space of state variables and self-similar structures appear in the information. This characteristic establishes sufficient conditions correctly to transmit arbitrary external information and is common to chaos discovered so far in living systems.*
>
> Kaneko and Tsuda (2000)

An inspection of the chaotic regime in Fig. 15.2 shows some white strips. A closer look at these strips reveals little windows of order, where the equation goes through the bifurcations again (as it did for $3 < k < 3.57$) before returning to chaos. This is *self-similarity*: The graph has a copy of itself hidden deep inside it. This is an important aspect of chaos. The strange attractor associated with chaos has '*fractal*' character.

The idea that Nature is full of fractal configurations was first put forward and investigated by Mandelbrot (1977, 1982). A fractal structure has *scale invariance* and *self-similarity*: It looks the same (self-similar) at just about any level of magnification or change of scale (Fig. 15. 3).

The famous Koch's snowflake illustrates the point. The recipe (the local mechanism) for creating it is very simple: A fractal object has an *initiator* and a *generator*. In Fig. 15.4, the initiator is an equilateral triangle, shown in the top left corner. For obtaining the generator, we partition each side of the triangle into three equal parts, remove the middle one-third, and replace the gap so created by two segments of the same length as the other segments, in the form of a peak (i.e., we add an equilateral triangle at each gap). The resulting form is shown on the top right corner in Fig. 15.4.

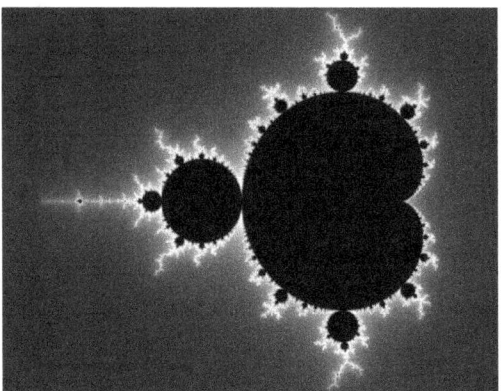

Fig. 15.3 The Mandelbrot set is a famous example of a fractal. The generating equation for it is a complex quadratic polynomial: $z_{n+1} = z_n^2 + c$. The complex number c is a part of the set if, on starting with $z_0 = 0$ and applying the iteration repeatedly, the absolute value of z_n never exceeds a certain number, no matter how large n becomes. Image credit: http://en.wikipedia.org/wiki/File:Mandel_zoom_00_mandelbrot_set.jpg

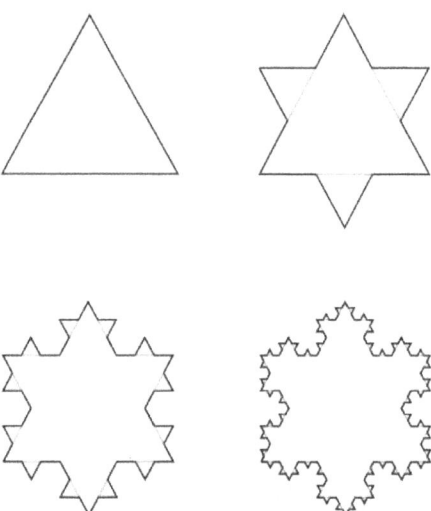

Fig. 15.4 The first four iterations of the Koch snowflake. Image credit: http://en.wikipedia.org/wiki/Koch_snowflake

This process is repeated indefinitely, applying the generator procedure to each straight segment. The next two iterations are shown in the lower half of Fig. 15.4.

Koch's snowflake looks complex, but has an underlying simplicity, once we have identified the generating mechanism. Similarly, the chaos generated by the logistic equation is highly complex, even though the underlying mechanism is controlled by just one parameter, the growth rate k (or r).

The self-similarity exhibited by chaotic dynamics and other fractal patterns points to *the sameness of the underlying causes at all length scales*. No wonder, an enormous number of natural entities have fractal shapes.

The notion of *fractal dimensions* is introduced to reflect the fact of self-sameness. The fractal dimension of the Koch snowflake is worked out to be ~1.26. The curve is coarser than a 1-dimensional smooth curve or line. Since it is more crinkly, it is better at taking up space. However, it is not as good at filling up space as a square (which has a dimension value 2), since it does not really have any area. So it makes sense that the dimension of the Koch curve is somewhere between 1 and 2.

15.5 Chaos and complexity

As we have seen in this chapter, chaos is not all randomness. Order and disorder coexist in a chaotic system. One can make a similar statement for any complex adaptive system also. There is overlap between chaos and complexity.

Fig. 15.2 in this chapter and Fig. 7.2 in Chapter 7 are depictions of this fact. The chaotic regime in Fig. 15.2 has pockets of order. Similarly, Fig. 7.2 emphasizes the fact that the degree of complexity is the highest when there is an optimum proportion of order and disorder.

In the language of algorithmic information theory (Chapter 6), chaos has the largest (but not infinite) degree of complexity. By contrast, random or noisy systems have an infinite degree of complexity. The crux of a complex system lies in its non-random parts or regularities. In Chapters 6 and 7, effective complexity was related to the description of the regularities of a system by a CAS that is observing it. For effective complexity to be substantial, the system should be neither too orderly nor too disorderly. For such situations, the algorithmic information content (AIC) is substantial, but not maximal (for a given length of bit stream), and it has two contributions: The apparently regular portion (corresponding to the effective complexity), and the apparently random portion. This critical balance between order and disorder is the crux of complexity. In a later chapter I shall link this 'optimum proportion' to the very important *edge-of-chaos* notion in complexity science.

Whereas chaos violates reductionism, it does not respect holism either. Chaos, though unpredictable, is deterministic, and there are *pockets of order* in what appears to be random dynamics. Equations with just three control variables can have chaotic solutions. One can say that, starting from the observed seemingly random behaviour, one can carry out 'reduction' to just three variables, even though the number of particles involved is very large. Chaos is neither reductionistic nor holistic. It is a case of islands of order in a sea of disorder.

According to Bak (1996), chaos theory cannot explain complexity. Chaos theory explains how simple, deterministic systems can sometimes exhibit unpredictable behaviour, but complex-looking behaviour occurs in such systems *only for a specific value of the control parameter*; i.e., the complexity is not *robust*. There is no *general* power-law behaviour, which is a signature of complex critical systems.

Thus, practically all really interesting complex phenomena have a lot in common with chaos. It follows that a good description of complexity should maintain its distance from both reductionism and holism.

Complexity has a domain of existence: On the one end of this domain there may be islands of order in a sea of disorder; on the other end there may be islands of disorder in a sea of order.

16. Cellular Automata as Models of Complex Systems

The sciences do not try to explain, they hardly even try to interpret; they mainly make models. By a model is meant a mathematical construct which, with the addition of certain verbal interpretations, describes observed phenomena. The justification of such a mathematical construct is solely and precisely that it is expected to work.

John von Neumann

16.1 Cellular automata

The notion of cellular automata was put forward in the 1940s by Stanislas Ulam, a colleague of John von Neumann (see Neumann 1963, 1966). What Ulam suggested to Neumann was to consider a digital programmable universe, in which 'time' is imagined as defined by the ticking of a cosmic clock, and 'space' is a discrete lattice of cells, each cell occupied by an abstractly defined simple computer called a *finite automaton*. Simple local rules determine the state of any cell at any discrete point of time. There are only a finite number of states available to a cell or an automaton. These states could be, say, a few colours, or a few integers, or just 'dead' or 'alive', etc. At each tick of the digital clock, every automaton changes over to a new state determined by its present state and the present states of the neighbouring cellular automata (CA). The rules by which the state of each automaton changes at a given instant of digital time are the equivalent of the physical laws of the universe. There is thus a *state transition table*, which determines how each automaton changes for each of the possible configurations of the states of the neighbouring cells.

The notion of cellular automata got a big boost through the work of Stephen Wolfram (1983, 1984a, 1984b, 2002). He established that CA have wide-ranging applicability and, in particular, have features of nonlinear dynamical systems. He emphasized that the complex phenomena we see in the universe can be usually thought of as the running of *simple programs*. The best and often the only way to understand these phenomena is by modelling them on a computer with the help of CA, rather than by working out the consequences of (necessarily) idealized and approximate mathematical models based on a set of differential equations. Approximations are not a good idea for modelling complex systems. Chaotic systems are a striking example of this. Even the slightest of errors in the specification of a control parameter can lead to a totally different and unpredictable phase-space trajectory. Wolfram's idea of a 'simple program' typically has the following ingredients:

- A set of transformation or operation rules (usually 'local' rules).
- Data to operate on.
- An engine that applies the rules to the data.

In a cellular automaton, the data enter only at the beginning of the computation ('initial conditions'), and the engine keeps applying the same deterministic rules to the outputs of its previous application of the rules. Extremely complex-looking patterns can be generated by any of a large number of simple programs investigated by Wolfram (2002). Fig. 16.1 is an example.

16.2 Conway's Game of Life

A particularly popular example of CA is the *Game of Life* invented (around 1970) by John Conway. It provides a graphic demonstration of *artificial evolution*, as the evolving structures can be seen on a computer screen.

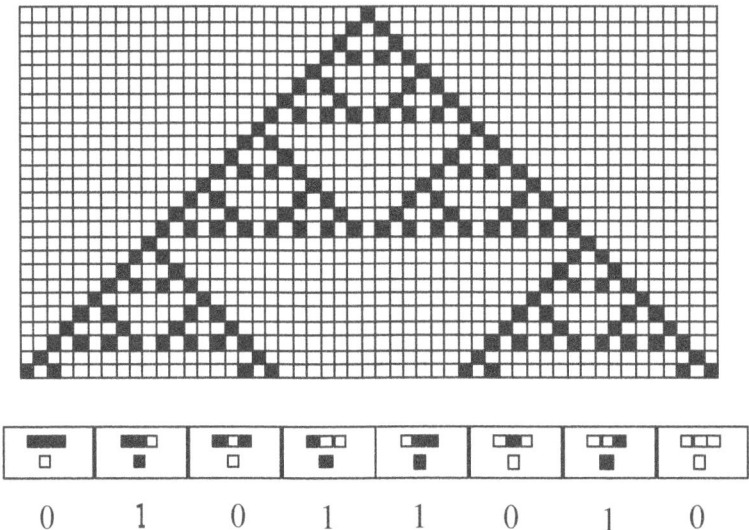

$$0 \quad 1 \quad 0 \quad 1 \quad 1 \quad 0 \quad 1 \quad 0$$

Fig. 16.1 . An example of a simple cellular automaton generating a complex-looking pattern. Shown here is a 1-dimensional cellular automaton. It consists of a row of squares, and each square can be either black or white. Starting from just one such row of squares, each time the system is updated, a new row of squares is created just below the previous row, following a simple rule. The simple transformation rule operative in this figure says that a square in the new row should be black only if one or the other, but not both, of its vertically-above predecessor's neighbours is black. Shown in a separate figure at the bottom is a graphical depiction of this rule, in which 8 block are drawn, corresponding to the 8 possible configurations of three neighbouring cells, each configuration determining the colour of the cell in the next row. Starting with a single black square in the top row of squares, this rule produces a complex pattern of nested triangles. Shown at the bottom, below the graphical depiction of the operating rule for this CA, are the numbers 0, 1, 0, 1, 1, 0, 1, 0, where 0 corresponds to a white cell, and 1 corresponds to a black cell. Treating this sequence of digits as a binary number, we get 01011010, which is the number 90 in the decimal system. Wolfram therefore calls this CA as generated by *Rule 90*. The sequence of the eight blocks for depicting this rule has been so chosen as to generate the smallest possible number for labelling or classifying the rule.

One starts with a 2-dimensional lattice of square cells, each cell randomly taken to be either black or white. Let black mean that the 'creature' denoted by a cell is 'alive', and let white mean that the corresponding creature is 'dead'. Very simple local rules are introduced for how the cells will change from one time step to the next. For example, if a cell has two or three neighbours which are alive (i.e. black), then the cell becomes alive if it was dead to start with, or remains alive if it was already so. If the number of neighbours is less than two, the cell dies of 'loneliness'. And if the number of neighbours is more than three, again the cell dies, this time due to 'overcrowding'. Remarkable patterns emerge on the computer screen when this program is run.

Every run is different, and it is not possible to exhaust all possibilities. The live cells organize themselves into coherent and ever-changing patterns, like real creatures in Nature. Fig. 16.2 shows the so-called 'Gosper glider gun'. A 'gun' is a configuration of the Game of Life which repeatedly and endlessly shoots out moving objects such as 'gliders'. For more details, see http://en.wikipedia.org/wiki/Conway%27s_Game_of_Life.

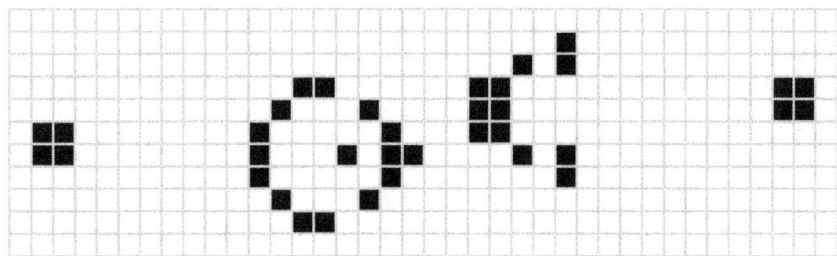

Fig. 16.2 The Gosper glider gun. Image credit: Bryan Burgers.
http://en.wikipedia.org/wiki/File:Game_of_life_glider_gun.svg

16.3 Self-reproducing automata

In the late 1940s, Neumann got interested in the question: Can a machine be programmed to make a copy of itself? Can there be self-reproducing machines? (see Neumann 1963, 1966). To bring out the essence of self-reproduction, Neumann imagined a thought experiment. Consider a machine moving around on the surface of a pond. The pond contains all sorts of machine parts. Our machine is a '*universal constructor*'; i.e., given a recipe for constructing any machine, it can search the pond for the right parts and construct the desired machine. In particular, it can construct a copy of itself if the requisite description is known to it.

But it is still not a *self-reproducing* machine, because the copy it has constructed of itself has no information about its own description for constructing another copy of itself. Neumann argued that for this to be possible, the original machine must have a '*description copier*'; i.e., a mechanism for duplicating the original description and for attaching this duplicate description to the new copy it is constructing of itself. The offspring will now have the wherewithal for a sustainable self-reproduction in the so-called *Neumann universe*. Thus any self-reproducing system must play two roles: It should serve as an algorithm that can be executed during the copying and constructing process; and it should serve as a data bank that can be duplicated and attached to the offspring.

These two predictions were confirmed for real-life systems in 1953 when Watson and Crick determined the structure of the DNA molecule. The DNA molecule not only serves as a data base for synthesizing various proteins etc., but also unwinds and makes a copy of itself when the biological cell divides into two (Fig. 16.3).

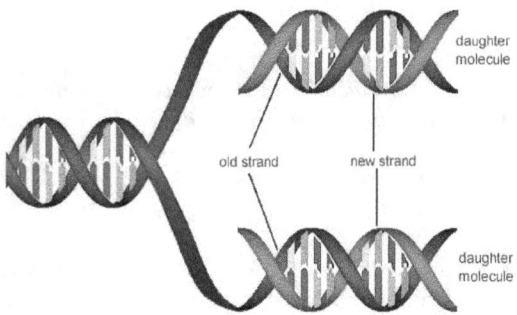

Fig. 16.3 Self-replication of DNA. The daughter DNA molecules produced consist of one progeny and one parental strand.
http://www.ied.edu.hk/biotech/eng/classrm/class_gene2.html

For testing his thought experiment, Neumann used the CA suggested by Ulam (described above). He proved that there existed at least one cellular-automaton pattern which could reproduce itself (Neumann 1966). The pattern he found involved a large lattice of cells, with 29 possible states for each cell. This was an important result because it meant that self-reproduction was possible in machines, and was not confined to living beings only.

Or, you may well say that *'living beings' are nothing more than self-replicating machines*; only their level of complexity is a bit too much for us to comprehend fully at present.

16.4 The four Wolfram classes of cellular automata

The cellular automaton shown in Fig. 16.1 is a special case of general 1-dimensional CA (Wolfram 1983, 1984a, b). For the general case, a site i on a row of cells at time t can carry any of the values a_i from the set 0, 1, ... , k-1. We can, for example, associate a different colour with each member of the set. The value a_i of the site for each i is updated in discrete time steps according to a deterministic rule which depends on the neighbourhood of the site:

$$a_i^{(t+1)} = \varphi[a_{i-r}^{(t)}, a_{i-r+1}^{(t)}, ..., a_{i+r}^{(t)}].$$

For Fig. 16.1, $k = 2$ and $r = 1$, i.e., there are just two colours (black and white), and only nearest neighbours on the previous time step decide the colour of a cell.

An extensive empirical analysis of all 1-dimensional CA by Wolfram (1984a) showed that the CA rules and the patterns generated by them (even when we start from random or disordered initial conditions) can be generally divided into just *four universality classes*:

Class 1 CA

In Class 1 CA, evolution from almost any initial state leads finally to a unique homogeneous state. This is like the occurrence of a limit point or a point attractor in the phase space of a nonlinear dynamical system. Class 1 patterns are *static* or *repetitive*, after the initial few time-steps.

Class 2 CA

In Class 2 CA, there is ultimately a sequence of simple stable or periodic structures. This corresponds to the occurrence of limit cycles or periodic attractors in phase space. Class 2 patterns are *nested*.

Both Class 1 and Class 2 CA are *predictable* after their static, repetitive or nested nature has been discerned, and are therefore *computationally reducible*. Fig. 16.1 is an example of a Class 2 CA.

Class 3 CA

Compared to Class 1 or Class 2 CA, Class 3 CA go to the opposite extreme. They generally exhibit *chaotic* or *aperiodic* long-time behaviour. Such CA grow indefinitely. Their patterns are often self-similar or scale-invariant. They are characterized by a fractal dimension, with $\log_2 3$ or ~1.59 as the most common value. They correspond to *strange attractors* in phase space.

Class 4 CA

Class 4 CA are the most relevant from the point of view of complex behaviour. For them the patterns grow and contract irregularly. The long-time behaviour is *neither fully predictable nor totally chaotic*. There are complicated localized structures, some of which propagate with time. There are regions of coherent structures that propagate, grow, split apart, disappear, or recombine in all sorts of ways. The pattern is changing all the time; it never settles down. The long-time behaviour is *undecidable* (typical of a complex system). The Game of Life mentioned above is an example of a Class 4 automaton.

The simplest possible rules for 1-dimensional CA are those with $k = 2$ and $r = 1$. Detailed analysis (Wolfram 2002) shows that there are 256 distinct types of them. Each type corresponds to a specific rule for generating a row of cells from the previous row. Thus there are 256 rules in all, and the caption to Fig. 16.1 explains why the pattern shown there is labelled as generated by *Rule 90*.

Fig. 16.4 is an interesting example of the possible operation of Rule 30 in Nature.

Fig. 16.4 A textile cone snail (*Conus textile*) shell, similar in appearance to Rule 30, a cellular automaton with chaotic behaviour. Location: Cod Hole, Great Barrier Reef, Australia.
Image credit: Richard Ling <richard@research.canon.com.au>
http://upload.wikimedia.org/wikipedia/commons/7/7d/Textile_cone.JPG

Rule 30 for going from one row of a cellular automaton to the next is as follows (https://en.wikipedia.org/wiki/Rule_30):

Current pattern	111	110	101	100	011	010	001	000
New state for centre cell	0	0	0	1	1	1	1	0

The binary number 00011110 is equal to 30. Hence the label 'Rule 30'.

16.5 Universal cellular automata

Any process that follows definite rules can be regarded as a computation. Thus the CA can carry out computation, as can Turing machines, and many other systems in Nature (Wolfram 2002). In fact, as discussed in Chapter 14, the universe is a quantum computer. In computations carried out by humans on computers, the computer programs define the rules of computation. In Nature, the rules of computation are nothing but the laws of Nature.

The notion of a *universal computer* emerged from the work of Alan Turing in the 1950s, and this launched the computer revolution. It was demonstrated that it is possible to build universal machines with a fixed underlying construction, but which can be made to perform different computations by being programmed in different ways. With suitable programming, any computer system or computer language can be ultimately made to perform the same set of tasks. Can at least some of the cellular automata (CA) be universal computers? 'Yes' according to Wolfram (2002). He has described the construction of CA that can be *universal cellular automata* (UCA). The rule for one such automaton is extremely simple. In fact, it is a somewhat complicated version of *Rule 90* illustrated in Fig. 16.1.

What is not simple in the universal-automaton version of Rule 90 is that, unlike in Fig. 16.1, each cell is represented by a block of 20 cells. Each block encodes the colour of the cell it represents, as also the rule for updating the colour at the next time step. Thus there are 19 possible colours for each cell. The new colour of each cell depends on the previous colours of a total of five cells. Thus there are, in principle, 2,476,099 cases to consider. Further, even the nearest-neighbour restriction in the formulation of the rule underlying the automaton is removed, and next-nearest-neighbour interactions are also recognized, so that there are 32 cases in each rule, compared to 8 before.

Wolfram (2002) demonstrated that such an automaton can ultimately emulate any CA with any set of rules, irrespective of how many neighbours are involved or how many colours are involved. Thus nothing fundamental can be gained by using rules more complicated than those used for defining this universal automaton. Given this automaton, more complicated rules can always be emulated just by setting up appropriate initial conditions.

The above universal automaton is quite complicated. The breakthrough achieved by Wolfram (2002), after years of investigations, was that even a simple rule, namely his *Rule 110*, with just two colours and only nearest-neighbour interactions, is one of the UCA. That is, *Rule 110, with various choices of initial conditions, can emulate any computation.* The crucial idea making this possible is to build up components from combinations of localized structures that the Rule 110 produces. This has been demonstrated to work, allowing us to get a large number of different kinds of components without having to increase the complexity of the underlying rules.

The typical behaviour of Rule 110, with random initial conditions, has been found to be that of a large number of localized structures that move around and interact with one another in complicated ways. In fact, this is a feature of all Class 4 CA. Perhaps most of them are UCA. Even Class 3 CA, which are characterized by chaotic or fractal behaviour, may turn out to include many which are UCA.

The UCA can emulate Turing machines, mobile automata, substitution systems, register machines, RAMs, logic circuits, etc.

In UCA the rules are simple but the initial conditions can be very complicated. There is also the question of efficiency.

17. Wolfram's 'New Kind of Science'

Indeed, I even have increasing evidence that thinking in terms of simple programs will make it possible to construct a single truly fundamental theory of physics, from which space, time, quantum mechanics and all the other known features of our universe will emerge.

S. Wolfram (2002)

17.1 Introduction

A new interpretation of complexity was given by Wolfram (2002) in his monumental book *A New Kind of Science* (NKS). The book continues to raise debate, even though there has also been very good progress in applying the NKS approach to a variety of problems in science.

I mentioned in Chapter 6 the computational or algorithmic irreducibility analysed by Chaitin. This is *information irreducibility*. It means that the observed digit stream is such that there is no theory or model more compact than simply giving the string of bits directly; there is no program for calculating the string of bits that is substantially smaller than the string of bits itself. A second kind of irreducibility, namely *time irreducibility* (or computational irreducibility), was analysed on a computer by Wolfram (2002). It pertains to a physical system for which there are no computational shortcuts, and for which the *quickest* way to see what the system does is just to run the computer program modelling the system.

Wolfram's basic approach to complexity and computational irreducibility is as follows. According to him, our conventional approach in science has been to steer clear of systems which are not computation-reducible. We have tended to model the phenomena we observe in terms of mathematical equations, which we then try to solve. Such a philosophy of science works well for systems that are simple, rather than complex. For achieving any modicum of success in understanding complex systems, we should investigate 'simple programs' and their interactions, rather than mathematical equations. Naturally, such an approach has become possible because of the availability of powerful computers. Wolfram's viewpoint is that the phenomena we see in the world around us should be thought of as the running of 'simple' computer programs. The best and perhaps the only way to understand these phenomena is by modelling them on a computer, rather than by working out the consequences of the inevitably idealized and approximate mathematical models based on a set of equations. I introduced this approach while describing cellular automata (CA) in the Chapter 16. Exploiting the immense power of modern computers, one can generate a huge repertoire of the consequences of all sorts of simple programs, as embodied in the corresponding CA. Then, for understanding the basics of a given complex system one tries to see if its observed behaviour pattern can be matched with any of the archived CA. If yes, then the simple program used for generating that particular CA pattern is a good *starting point* for, say, building models and finding a likely explanation of the time or space evolution of the complex behaviour observed in the actual physical system. Thus, rather than using CA to *mimic* or model the observed behaviour of complex systems, Wolfram advocates their use to reveal unknown aspects of the systems that they model.

The '*Principle of Computational Equivalence*' (PCE), enunciated in Wolfram's (2002) book (and described below in the next section), is a major component of Wolfram's NKS approach to understanding natural phenomena. He dares to go where no scientist would venture readily, namely attacking research problems of immense complexity. As explained above, one of the ways he does this is by constructing his '*computational universe*', which is a huge repertoire of '*patterns*' generated by running a large number of conceivable cellular automata, and then 'mining' this universe for possible solutions for the problem at hand.

According to Wolfram (2012a), 'There are typically three broad categories of NKS work to date: pure NKS; applied NKS; and the NKS way of thinking. . . Pure NKS is about studying the computational universe as basic science for its own sake – investigating simple programs like cellular automata, seeing what they do, and gradually abstracting general principles. Applied NKS is about taking what one finds in the computational universe, and using it as raw material to create models, technology and other things. And the NKS way of thinking is about taking ideas and principles from NKS – like computational irreducibility or the Principle of Computational Equivalence – and using them as a conceptual framework for thinking about things'.

May 14, 2009 marked the 7th anniversary of the great NKS book. The extremely powerful computational search engine *Wolfram|Alpha* was released at about the same time. In an email to people on his mailing list, Wolfram wrote as follows:

'The trickle of academic work aimed directly at pure NKS – the basic investigation of simple programs and the computational universe – has turned into a stream, though tremendous opportunity for growth remains. There is an immensely complex web of systematizable knowledge out there in the world. And before NKS, I would have assumed that to handle something of this complexity would have required building a system that is somehow correspondingly complex – and in practice completely out of reach. But from NKS we have learned that even highly complex things can have their origins in simple rules and simple programs. And this is what inspired me to believe that building Wolfram|Alpha might be possible. As a practical matter, many algorithms in Wolfram|Alpha were found by NKS methods – by searching the computational universe for programs that achieve particular purposes. And there is a curious sense in which the discoveries of NKS about computational irreducibility are what make Wolfram|Alpha possible'.

17.2 Wolfram's principle of computational equivalence (PCE)

The immense number of CA examined by Wolfram (2002) led him to the formulation of the *Principle of Computational Equivalence* (PCE). According to this principle:

*Almost all processes that are not 'obviously simple' (*i.e., those not described by the classes 1 and 2 of CA in the previous chapter*) correspond to computations that are of equivalent degree of complexity.*

In other words, irrespective of the simple or complicated nature of the rules or the initial conditions of a process, any such process will always correspond to a computation of equivalent difficulty or sophistication.

The genesis of the PCE, which is indeed a 'bold hypothesis', and needs thorough confirmation or proof, lies in the idea of computational universality (see the notion of 'universal cellular automata' (UCA) described in Chapter 16): It is possible to construct universal systems that can perform essentially any computation, and which must therefore all be capable of exhibiting the highest level of computational sophistication. This is reminiscent of the well-known *Church-Turing thesis*, according to which, any process or computation governed by specific rules that can be performed at all can be performed by a Turing machine or equivalent universal computer. The fact that even a simple rule like Rule 110 is a UCA implies that computational sophistication does not necessarily imply sophisticated or complex rules. It does not matter how simple or complicated either the rules are, or the initial conditions for a process are; so long as the process itself does not look obviously simple (e.g., purely repetitive or purely nested), it will almost always correspond to a computation of equivalent sophistication.

The PCE implies that simple CA no more elaborate than the Game of Life are computationally equivalent to powerful computer systems. Once you get beyond very simple systems, all systems attain the highest level of complexity possible, and are computationally equivalent to all other non-simple systems. The PCE, according to Wolfram (2002), 'tells us what kinds of computations can and cannot happen in our universe [and] summarizes purely abstract deductions about possible computations, and provides foundations for more general definitions of the very concept of computation'. No system can ever carry out explicit computations that are more sophisticated than those carried out by systems like cellular automata and Turing machines.

The PCE, though still under debate, may have far-reaching consequences for science, and for much else. Wolfram has applied it to a wide-ranging set of situations. He has demonstrated that simple programs can be used to explain space and time, perception, and also for rationalizing a variety of fundamental facts in physics, biology, and many other sciences.

Here is what an article in Wikipedia about the NKS says about the PCE http://en.wikipedia.org/wiki/A_New_Kind_of_Science: 'The PCE has been criticized for being vague, unmathematical, and for not making directly verifiable predictions; however, Wolfram's group has described the principle as such, not a law, theorem or formula. It has also been criticized for being contrary to the spirit of research in mathematical logic and computational complexity theory, which seek to make fine-grained distinctions between levels of computational sophistication. Others suggest it is little more than a rechristening of the Church-Turing thesis. However, the Church-Turing thesis imposes an upper limit while Wolfram's PCE suggests the nonexistence of intermediate degrees of computation sending a computational system either to the upper level (universal) or to the lowest degree, explained by Klaus Sutner (http://portal.acm.org/citation.cfm?id=771538) in terms of physics-like computation as a zero-one law claiming that, in practice, constructing actual computers with intermediate degrees is highly artificial and hasn't ever been done, hence endorsing Wolfram's intuition captured in his PCE'.

17.3 The PCE and the rampant occurrence of complexity

How is complexity defined in Wolfram's scheme of things? According to him, the apparent complexity in Nature follows from *computational equivalence*: We tend to consider behaviour complex when we cannot readily reduce it to a simple summary. If all processes are viewed as computations, then doing such reduction in effect requires us as observers (complex adaptive systems) to be capable of computations that are more sophisticated than the ones going on in the systems we are observing. But the PCE implies that usually the computations will be of the same sophistication – providing a fundamental explanation of why the behaviour we observe must seem to us complex (see the website www.wolframscience.com).

Let us now see how the PCE rationalizes the rampant occurrence of computational irreducibility (or complexity) in Nature. For this we have to address the question of comparing the computational sophistication of the systems that we study with the computational sophistication of the systems that we use for studying them. The PCE says that, once a threshold has been crossed, any real system must exhibit essentially the same level of computational sophistication. *And this applies to our own perception and analysis capabilities also; according to the PCE they are not computationally superior to the complex systems we seek to observe and understand.* Beyond a certain threshold, all systems are computationally equivalent.

If predictions about the behaviour of a system are to be possible, it must be the case that the system making the predictions is able to *outrun* the system it is trying to make predictions about. But this is possible only if the predicting system is able to perform more sophisticated computations than the system under investigation. But the PCE does not allow that. Therefore, except for simple systems (mainly Class 1 and Class 2 CA), no systematic predictions can be made about their behaviour at a chosen time in the future. Thus there is no general way to shortcut their process of evolution. In other words, most such systems are computationally irreducible.

This line of reasoning also helps understand the illusion of free will. The human brain is clearly too complex not to be a computationally irreducible system. This means that, even if all the rules are known and all the initial conditions have been specified, we cannot predict the end results of the computations carried out by the brain. The only way to know an end result is to actually run the system to see what happens; whether the program 'halts' or not. There is also the added complication of extreme sensitivity to initial conditions for chaotic or quasi-chaotic systems. All told, this is how the illusion of unpredictability or 'free will' arises. There is the illusion that a human being (and perhaps other creatures) can 'decide' the future action arbitrarily; at will.

A real example of a computationally irreducible cellular automaton is given on page 740 of Wolfram (2002). This cellular automaton follows a 3-colour *totalistic rule*. In a totalistic rule, the new colour of a cell depends only on the average colour of cells in its neighbourhood, and not on their individual colours.

As Wolfram emphasizes, *the whole idea of doing science with mathematical formulas makes sense only for computationally reducible systems*. For others there are no computational shortcuts; practically the only way of knowing a future configuration is to actually run through all the evolutionary time steps. Wolfram (2002) argues that his NKS is ideally suited for that purpose. One can generate the consequences of all sorts of simple programs defined by the corresponding CA. For understanding the basics of a given complex system observed in Nature, one tries to see if the observed behaviour pattern can be matched with any of the archived complex patterns. If yes, then the simple program used for generating that particular CA pattern is a likely 'explanation' of (or at least a good starting point for model-building and understanding) the time or space evolution of the complex behaviour observed in the actual physical system under study.

The PCE apparently puts an *upper limit* on complexity by implying that no system can carry out explicit computations that are more sophisticated than those carried out by CA or Turing machines.

17.4 Why does the universe run the way it does?

Is there a fundamental deterministic rule from which all else follows? Conventional wisdom says that randomness is at the heart of quantum mechanics, and because of this randomness the universe has infinite complexity (as quantified by the '*K-S entropy*' for a random system; see Section A4.2). Wolfram suggests that this may or may not be so. According to him, there may be no real randomness; only *pseudo*-randomness, like the randomness produced by random-number generators in a computer. The computer generates these numbers by using mathematical equations, and what we get are actually deterministic sequences of numbers. Wolfram gives the analogy of $\pi = 3.1415926...$ Suppose you are given, not the whole equation, but only a string of digits coming from far inside the decimal expansion. It would look random, *in the absence of complete knowledge*. In reality it is only pseudo-random. Wolfram puts forward the viewpoint that, similarly, the randomness we see in the physical world may really be pseudo-randomness, and the physical world may actually be deterministic. It is simply that we do not know the underlying law, which may well be a simple cellular automaton. [But there is also a computational-irreducibility aspect to this scenario (Wolfram 2002), as discussed in Chapter 6.]

In addition, there is a human or anthropic angle to the meaning of complexity. Let us go back to the equation $\pi = 3.1415926 \ldots$ It has only a small information content or degree of complexity: A small algorithm, using the fact that π is given by the ratio of the circumference of a circle to its diameter, can generate the entire information contained in this equation. But if we humans do not have knowledge about the entire sequence of digits, but are given only a string of digits coming from far inside the decimal expansion, then the degree of complexity is just about as large as the length of the string, and can, in principle, be infinite. For us humans, the degree of complexity of a system depends on our knowledge about the system. As more knowledge is acquired by us, the degree of complexity may keep collapsing. Of course, this happens only for systems which are not irreducibly complex. If the complexity of a system is irreducibly or intrinsically large, our increasing knowledge about the system can have little effect on its degree of complexity.

Wolfram argues that the universe is fundamentally rule-based. It is one big ongoing computation, of which everything is a part. He discusses how the concept of '*causal networks*' can be used to explain how space, time, elementary particles, motion, gravity, relativity, and quantum phenomena all arise. According to him, causal networks can represent everything that can be observed. This presumes that Nature is fundamentally discrete. Wolfram goes on to suggest that any instance of complex behaviour we observe is produced by a universal system: 'I suspect that in almost any case where we have seen complex behaviour . . . it will eventually be possible to show that there is universality.'

17.5 Criticism of Wolfram's NKS

The impact of Wolfram's book has been truly wide-ranging, with applied NKS emerging as the largest group of applications. I quote Wolfram (2012a) again:

'*Let's start with the largest group: applied NKS. And among these, a striking feature is the development of models for a dizzying array of systems and phenomena. In traditional science, new models are fairly rare. But in just a decade of applied NKS academic literature, there are already hundreds of new models: Hair patterns in mice. Shapes of human molars. Collective butterfly motion. Evolution of soil thicknesses. Interactions of trading strategies. Clustering of red blood cells in capillaries. Patterns of worm appendages. Shapes of galaxies. Effects of fires on ecosystems. Structure of stromatolites. Patterns of leaf stomata operation. Spatial spread of influenza in hospitals. Pedestrian traffic flow. Skin cancer development. Size distributions of companies. Microscopic origins of friction. And many, many more*'.

While there are many enthusiasts, there are also many critics of NKS. A criticism levelled against Wolfram's NKS is that his cellular automata lack the predictive power of existing theories developed around conventional, i.e., calculus-based, mathematics. While this may be a somewhat valid point in a general sense, it is not very relevant in the context of complexity. Complex systems *are* unpredictable, except possibly that one can sometimes explain/predict to some extent a level of complexity in terms of the previous lower level of complexity.

Kurzweil (2005) has remarked that even the most complex CA discussed by Wolfram do not have the evolution feature so crucial to the question of complexity. I think this is because the CA discussed by Wolfram are not open systems. There is no influx of energy or negative entropy or information into the CA that run simple programs. The NKS should be extended to overcome this deficiency. In fact, as we shall see later in this book, this is what Langton (1989) did to some extent in his pioneering work on *adaptive computation*.

Another interesting comment about the efficacy or otherwise of the NKS as providing a theory of the evolution of the universe is that of Lloyd (2006): 'The idea of using cellular automata as a basis for the theory of the universe is an appealing one. The problem with this argument is that classical computers are bad at reproducing quantum features, such as entanglement. Moreover, as has been noted, it would take a classical computer the size of the whole universe just to simulate a very tiny quantum-mechanical piece of it. It is thus hard to see how the universe could be a classical computer such as a cellular automaton. If it is, then the vast majority of its computational apparatus is inaccessible to observation'.

Wolfram (2012b) has reviewed the various responses to his work. Personally, I find the attitude of several conventional scientists very intriguing, even disappointing. There are any number of extremely complex problems challenging us for a solution. The traditional approach in science has been to model the system under investigation in terms of a few differential equations, and try to solve them under suitable 'boundary conditions'. We feel elated if our model embodies the 'essential physics' of the problem, and even makes some verifiable predictions. And we feel absolutely thrilled if the predictions also turn out to be true. But the wicked thing about most of the real-life complex systems is that *any simplifying assumption for modelling them can kill the very essence of the problem*. You can do two things when faced with such a situation. Either stay away from working on such research problems, or do what NKS suggests. Staying away is not necessarily a good idea. For how long can you go on working only on *simple* or *simplifiable* research problems? Complexity requires a radically new approach to how science has to be done. NKS is one such approach.

Some critics of NKS tend to snigger at what has been achieved by it. I would take them seriously if they had some better alternatives to offer. They have none.

As mentioned above, a criticism levelled against Wolfram's NKS is that his CA lack the predictive power of theories developed around conventional, i.e., calculus-based, mathematics. So far as complex systems are concerned, they *are* unpredictable anyway, except possibly that one can sometimes explain/predict the level of complexity in terms of the previous *lower* level of complexity. In any case, is this criticism really valid? Suppose you have succeeded in identifying some archived simple program from Wolfram's computational universe as providing a reasonably good match with the complexity pattern observed in Nature. Such a simple program is clearly giving you a good hint about the basic interactions and physics involved. You can even create 'predictions' by tinkering with the simple program and generating the modified patterns, and checking them against experiment. If such a prediction gets confirmed reasonably well, you are perhaps on the right track so far as gaining an insight into the *basics* of the complex phenomenon is concerned. What more can you ask for? Getting onto the right track is half the battle won. Just build on that great start, by *any* means.

I quote from Wolfram (2012b) for more on this issue: 'Another theme in some reviews is that the ideas in the book "do not lead to testable predictions". Of course, just as with an area like pure mathematics, the abstract study of the computational universe that forms the core of the book is not something which in and of itself would be expected to have testable predictions. Rather, it is when the methods derived from this are applied to systems in nature and elsewhere that predictions can be made. And indeed there are quite a few of these in the book (for example about repeatability of apparent randomness) — and many more have emerged and successfully been tested in work that's been done since the book appeared.

'Interestingly enough, the book actually also makes abstract predictions — particularly based on the Principle of Computational Equivalence. And one very important such prediction — that a particular simple Turing machine would be computation universal — was verified in 2007.'

What about the future of NKS? Wolfram (2012c) gushes with optimism and expectation. And the tribe of NKS enthusiasts continues to grow in size and diversity. There is a reason for this, namely the universal applicability of NKS for just about anything of interest to us humans. To make my point, I can do no better than reproduce the last few paragraphs of Wolfram's (2002) book:

'It would be most satisfying if science were to prove that we as humans are in some fundamental way special, and above everything else in the universe. But if one looks at the history of science many of its greatest advances have come precisely from identifying ways in which we are not special — for this is what allows science to make ever more general statements about the universe and the things in it.

'Four centuries ago we learned for example that our planet does not lie at a special position in the universe. A century and a half ago we learned that there was nothing very special about the origin of our species. And over the past century we have learned that there is nothing special about our various physical, chemical and other constituents.

'Yet in Western thought there is still a strong belief that there must be something fundamentally special about us. And nowadays the most common assumption is that it must have to do with the level of intelligence or complexity that we exhibit. But building on what I have discovered in this book, the Principle of Computational Equivalence now makes the fairly dramatic statement that even in these ways there is nothing fundamentally special about us.

'For if one thinks in computational terms the issue is essentially whether we somehow show a specially high level of computational sophistication. Yet the Principle of Computational Equivalence asserts that almost any system whose behaviour is not obviously simple will tend to be exactly equivalent in its computational sophistication.

'So this means that there is in the end no difference between the level of computational sophistication that is achieved by humans and by all sorts of other systems in nature and elsewhere. For my discoveries imply that whether the underlying system is a human brain, a turbulent fluid, or a cellular automaton, the behaviour it exhibits will correspond to a computation of equivalent sophistication.

'And while from the point of view of modern intellectual thinking this may come as quite a shock, it is perhaps not so surprising at the level of everyday experience. For there are certainly many systems in nature whose behaviour is complex enough that we often describe it in human terms. And indeed in early human thinking it is very common to encounter the idea of animism: that systems with complex behaviour in nature must be driven by the same kind of essential spirit as humans.

'But for thousands of years this has been seen as naive and counter to progress in science. Yet now essentially this idea — viewed in computational terms through the discoveries in this book — emerges as crucial. For as I discussed earlier in this chapter, it is the computational equivalence of us as observers to the systems in nature that we observe that makes these systems seem to us so complex and unpredictable.

'And while in the past it was often assumed that such complexity must somehow be special to systems in nature, what my discoveries and the Principle of Computational Equivalence now show is that in fact it is vastly more general. For what we have seen in this book is that even when their underlying rules are almost as simple as possible, abstract systems like cellular automata can achieve exactly the same level of computational sophistication as anything else.

'It is perhaps a little humbling to discover that we as humans are in effect computationally no more capable than cellular automata with very simple rules. But the Principle of Computational Equivalence also implies that the same is ultimately true of our whole universe.

'So while science has often made it seem that we as humans are somehow insignificant compared to the universe, the Principle of Computational Equivalence now shows that in a certain sense we are at the same level as it is. For the principle implies that what goes on inside us can ultimately achieve just the same level of computational sophistication as our whole universe.

'But while science has in the past shown that in many ways there is nothing special about us as humans, the very success of science has tended to give us the idea that with our intelligence we are in some way above the universe. Yet now the Principle of Computational Equivalence implies that the computational sophistication of our intelligence should in a sense be shared by many parts of our universe — an idea that perhaps seems more familiar from religion than science.

'Particularly with all the successes of science, there has been a great desire to capture the essence of the human condition in abstract scientific terms. And this has become all the more relevant as its replication with technology begins to seem realistic. But what the Principle of Computational Equivalence suggests is that abstract descriptions will never ultimately distinguish us from all sorts of other systems in nature and elsewhere. And what this means is that in a sense there can be no abstract basic science of the human condition — only something that involves all sorts of specific details of humans and their history.

'So while we might have imagined that science would eventually show us how to rise above all our human details what we now see is that in fact these details are in effect the only important thing about us.

'And indeed at some level it is the Principle of Computational Equivalence that allows these details to be significant. For this is what leads to the phenomenon of computational irreducibility. And this in turn is in effect what allows history to be significant—and what implies that something irreducible can be achieved by the evolution of a system.

'Looking at the progress of science over the course of history one might assume that it would only be a matter of time before everything would somehow be predicted by science. But the Principle of Computational Equivalence — and the phenomenon of computational irreducibility —now shows that this will never happen.

'There will always be details that can be reduced further — and that will allow science to continue to show progress. But we now know that there are some fundamental boundaries to science and knowledge.

'And indeed in the end the Principle of Computational Equivalence encapsulates both the ultimate power and the ultimate weakness of science. For it implies that all the wonders of our universe can in effect be captured by simple rules, yet it shows that there can be no way to know all the consequences of these rules, except in effect just to watch and see how they unfold.'

18. Swarm Intelligence

How intelligence, or something approaching intelligence, can arise in a complex system comprising of non-intelligent individuals? This happens only in *complex adaptive systems* (CASs), which I introduced in Section 7.2. In such systems the individual members are able to respond collectively (to the advantage of the system) to changes in the environment. And the interesting thing is that the underlying rules of response and interaction are generally very simple.

Network theory is very relevant in this context. I have elaborated on some formal aspects of network theory in Appendix A9. The basic idea is simple. The individual members of a CAS can be regarded as the 'nodes' of a network. And those of the nodes are connected by an 'edge' for which there is some interaction between the individuals represented by the nodes. Once this correspondence is established, the entire power of formal network theory can be brought to bear on an analysis of the behaviour of the CAS. Beehives and ant colonies are good and well-investigated examples of how such CASs operate as a single, intelligent, superorganism. It turns out that we humans can draw some practical lessons from studies on such CASs. No wonder, CASs and swarm intelligence have attracted the attention of managers from a variety of institutions and enterprises (see Chapter 44).

18.1 Emergence of swarm intelligence in a beehive

The beehive is an example of a *self-organized superorganism*. Each honeybee has hardly any intelligence to speak of, but the hive as a whole possesses '*swarm intelligence*'.

The queen bee emits a pheromone, namely *trans*-9-keto-2-decenoic acid. [Pheromones are chemicals that play the role of signals among members of the same species.] This 'queen substance' is secreted by the mandibular glands of the queen bee. Worker bees lick the queen's body. They move around and regurgitate the chemical, so that it spreads in the hive. The pheromone has several effects on the bees:

(i) The ovaries of the worker bees do not develop.

(ii) They raise larvae in such a way that the young bees cannot become queens, so that the queen has no rivals so long as she is secreting the pheromone. [All female honeybees, including the queen bees, develop from larvae which are identical genetically. Those fed on a certain 'royal jelly' become fertile queens, while the rest remain sterile workers.]

(iii) The pheromone guides a husband to the queen on her nuptial flight.

(iv) The pheromone promotes the consummation of the marriage.

When the secretion of the pheromone by the queen bee stops, the above processes are annulled. The worker bees become fertile again, and daughter queens can be raised.

Each hive has a distinctive scent, common to all its members. This enables the bees to recognize members of the same hive, and repel foreigner bees.

How does a large group like this take decisions? Evolutionary processes have made the beehive an effective decision-making unit, *even though nobody is in command*. The queen bee is *not* in command. Let us see how, for example, the beehive collectively decides on a new site for setting up another hive.

In late summer or early spring, when the sources of honey are aplenty, a large colony of bees (typically with ~10,000 bees) splits into two. A daughter queen and about half the population stays back in the old hive, and the rest, including the queen bee, leave so that they can start a new hive at a carefully selected site. How is the new site chosen?

Typically, a few hundred worker bees go scouting for possible sites. The rest stay bivouacked on a nearby tree branch, conserving energy, till an acceptable new site has been selected. An acceptable site for nesting is typically a cavity in a tree with a volume greater than 20 litres, and an entrance hole smaller than 30 cm^2. The hole should be several meters above the ground, facing south, and located at the bottom of the cavity.

The scout bees come back and report to the swarm about possible nesting sites by dancing a *waggle dance* in particular ways. Typically there are about a dozen sites competing for attention. During the report, the more vigorously a scout dances, the better must be the site being championed.

Deputy bees then go to check out the competing sites according to the intensity of the dances. They concur with the scouts whose sites are good by joining the dances of those scouts. That induces more followers to check out the lead prospects. They return and join the show by leaping into the performance of *their* choice.

By compounding emphasis (*positive feedback*), the favourite site gets more visitors, thus increasing further the number of visitors. Finally, the swarm flies in the direction indicated by mob vote.

Evolutionary processes have been operative, not only in the development of the waggle dance as one of the signalling mechanisms in the beehive (other signalling mechanisms being via the specific pheromone mentioned above), but also for some other survival instincts of the bees. Seeley *et al.* (2006) discovered a good example of this in the speed-versus-accuracy feature of the final decision for the new nesting site. Typically, there are about a dozen nesting sites competing for attention. Do all but one get eliminated by compounding emphasis and consensus? No, not necessarily.

Eliminating all but one site will adversely affect the speed of the decision-making process, which may be detrimental to the overall welfare of the hive (e.g., higher energy costs). Rather than going by consensus (agreement among practically all the scout bees), the hive goes by *quorum* (sufficient number of scout bees visiting any site). The swarm flies to occupy a site which is seen to be visited by ~150 bees. This *may* turn out to be an erroneous decision, but not very likely to be so. A compromise is reached between accuracy in the selection of the nesting site, and speed with which the final decision is taken. Through a process of Darwinian natural selection and evolution, the species has fine-tuned itself for what is best for its survival and propagation. *Emergent behaviour and biological evolution go hand in hand.*

Let us now review the salient features of what goes on in this CAS.

1. The system comprises of a large number of *distributed* members or agents, namely the honeybees, acting in parallel.

2. In the network-theory parlance, each bee is a 'node' of the network, and a possible line ('edge') joining any two nodes represents the interaction (communication) between those two

bees. The interactions between nodes or agents are through *signals* (waggle dance, pheromone secretion and ingestion, etc.).

3. An edge in the network represents *exchange of information*.

4. The nodes (bees) have sensors for receiving the information (visual, chemical, tactile).

5. Processing of information occurs in the brains of the bees, aided by 'instinct' or in-born tendencies (*internal rules*). That the bees survive and flourish is proof that evolutionary processes have led to the development of adequately appropriate internal rules.

6. The beehive is a complex *adaptive* system. For example, if a fraction of the population is decimated, the remainder would quickly readjust, and carry on as before.

7. There is no central command.

8. The individuals are *autonomous*, but what they do is influenced strongly by what they 'see' others doing.

9. Emergent behaviour arises through sheer large numbers and effective communication and interaction. In the present case, swarm intelligence emerges, of which no single member is capable alone.

10. The network or web of bees possesses *nonlinear causality* of peers influencing peers. 'Nonlinear' means that even small causes may induce disproportionately large responses. And different initial conditions can lead to dramatically different end results. The reason for the nonlinearity can be sought and found in the *positive feedback* feature, or in *the law of increasing returns*.

11. The evolutionary (internal-rules) part of the behaviour of the bees can also be explained in terms of emergent behaviour. The networked swarm is adaptable and resilient, and it nurtures small failures so that large failures do not happen frequently. This not only helps survival and propagation, but also favours *novelty*. The large number of combinations and permutations possible among the interacting agents has the potential for new possibilities. And if heritability is brought in, individual behaviour and experimentation leads to *perpetual novelty*, the hallmark of evolution.

Investigation of one type of complex system can provide insights into what may be happening in other complex systems. An obvious case in point is: How to understand human intelligence as a kind of swarm intelligence. Human intelligence emerges from the interactions among neurons, in spite of the fact that any particular neuron is as dumb as can be.

18.2 Ant logic

> *Ants are the history of social organization and the future of computers.*
> Kelly (1994)

Ants are social insects. Around 20% of all land animals are ants. There are more than a million ants for every human. An ant is a small dumb creature, not able to see far (only a few feet). Considering its size, the landscape in which it moves around must appear very rugged to it. Then how is it that ants in an ant colony are able to find food rather rapidly and generally by the shortest route, in spite of the fact that there is no one in command of operations?

It is a case of swarm intelligence again. In such a swarm, each individual has little or no intelligence, and it only follows some simple local rules, and yet the swarm as a whole ends up possessing intelligence. It is instructive to understand the basic processes involved in an ant colony, the more so because *ant logic* has already found several applications in the field of artificial evolution and in computational science (Bonabeau, Dorigo and Theraulaz 2000).

An ant colony is a remarkable parallel-processing superorganism. The ants function independently and simultaneously, and communicate with one another, unknowingly perhaps, via pheromones. A number of scout ants set out in search of food, going in different directions independently and randomly. They emit the pheromone all the time, both while going away from the nest and while returning to it. It follows that a trail used by many ants will have a strong pheromone odour. The pheromone evaporates slowly with time, i.e., its strength on a trail is a decaying function of time.

Suppose one of the many scout ants has accidentally discovered the shortest usable path to food (unknown to other ants, and even to itself, to start with). Then it will naturally be able to travel to the food, and come back by the same path (guided by the pheromone trail), in the shortest time, compared to other ants which did not happen to take this route. The to and fro journey along this trail will result in twice as much pheromone along it, compared to a trail which is twice as long. Different ants traverse different trails, and the trails may intersect. At trail-crossings the ants divert to the trail with the strongest odour, thus further strengthening its odour (law of increasing returns, or positive feedback). Ultimately, all ants follow the shortest route to food, in spite of the fact that no design work, or planning, or supervision, has gone into this. Swarm intelligence indeed.

This self-reinforced emergence of the optimum trail is also an example of '*autocatalysis*', and is an instance of *pattern-formation processes* in complex systems.

18.3 Positive and negative feedback in complex systems

Positive and negative feedback mechanisms play a central role in the emergence of all sorts of complexity. Examples include population variations from year to year, vagaries of the stock market, as also stable patterns of behaviour in a group or swarm.

Positive feedback is also referred to as the *law of increasing returns*, pointed out above in the context of the emergence of swarm intelligence in a beehive and in an ant colony. In the case of bees, for example, the potential nesting site favoured by deputy bees gets more visitors, thus increasing further the number of visitors: '*. . them that has get more votes, the have-nots get less*' (Kelly 1994).

Examples of negative feedback abound in our day-to-day life. You are steering a car on a straight road, and you sense that it is headed a little to the right. You take (opposite) corrective action by steering it a little to the left. If you find that it has swung more to the left than you intended, you bring it back to the straight path by taking opposite corrective action again by turning it a little to the right. The opposite action here is a case of negative feedback.

The interplay between positive and negative feedback is well illustrated by the logistic difference equation I discussed in Section 15.3 in the context of population dynamics: $x_{t+1} = kx_t(1-x_t)$. We can rewrite it as: $x_{t+1} = kx_t - kx_t^2$. The first term in this equation represents positive feedback, and the second (with a negative sign in front of it) negative feedback.

Suppose $k > 1$. The first term ($x_{t+1} = kx_t$) then will lead to an ever-increasing population: Whatever the population x_t is at a given time step t, it would be greater at the next time step $t+1$, and so on. A case of the law of increasing returns.

The second term, which represents negative feedback, provides a counter effect to the runaway population increase which can result from the first term.

As we saw in Section 15.3, the parameter k determines which of the two terms dominates. For $k < 1$ the population eventually becomes zero.

For $1 < k < 3$, the population eventually settles at the value $x* \sim 0.667$, rather than $x* = 0$.

For certain higher values of k, the system exhibits crazy behaviour, as outlined in Section 15.3. The interplay between positive and negative feedback can result in fascinating complex behaviour in a CAS.

19. Nonadaptive Complex Systems

Galaxies and stars and other such complex objects are examples of nonadaptive complex systems. They are inanimate systems which do evolve with time, but do not have the learning ability; there is no goal to be achieved by adaptation or learning (Schneider and Dettenwanger 2007). To the extent that everything interacts with everything else, such systems do affect their environment also, but they do not have the kind of evolution features one associates with complex adaptive systems (CASs).

Even certain *materials* (to be described below) are nonadaptive complex systems. They are invariably *composite* materials, although sometimes their composite nature is unintentional, rather than by human design.

19.1 Composite materials

Composite materials are made up of two or more components or phases, which are strongly bonded together in some desired connectivity pattern (Wadhawan 2007). They are inhomogeneous solids. Artificial composites are designed to perform specific jobs. Natural composites may be either biological or nonbiological. The former have evolved as a consequence of some specific evolutionary processes.

Composites can be classified as being either *structural* or *nonstructural*; the latter category is better known as *functional* composites. A structural composite typically has a matrix, a reinforcement, and a filler. RCC (reinforced cement concrete) is a structural composite, as are wood and particle-board. Ferrofluids (or ferromagnetic fluids, or magnetofluids) are an example of functional composites, in which ~10 nm-sized crystals of a ferromagnetic material (e.g. Fe_3O_4) are dispersed in a fluid such as kerosene or transformer oil (Wadhawan 2007). Since the crystal size is so small, application of a magnetic field induces a very large response in the colloidal suspension. The result is a strong anisotropy of field-induced properties (e.g., optical properties and viscosity), which can serve a functional purpose in a device.

A *nanocomposite* is defined as one in which at least one of the constituents or phases has at least one of the dimensions below 100 nm.

Complex materials are a subset of functional composites, with the defining feature that not all their properties can be explained in a reductionistic manner from those of the component phases comprising the composite. Of particular interest in this context are the so-called *metamaterials*, which are artificially fabricated nanocomposites; novel properties can be created in them by tailoring the shape, size, composition, symmetry, and connectivity etc. of the component materials.

Interesting dilemmas are faced regarding the definition of a composite material as we go to sub-nanometre length scales. Crystals are normally regarded as single-phase materials, and not composites. But they can also be viewed as 'Nature's own composites', with molecules serving as the component 'phases' comprising the crystal composite (Newnham and Trolier-McKinstry 1990).

19.2 Ferroic materials

I introduced phase transitions in Section 8.6. Additional information about them is given in

Appendix A7. Ferroic phase transitions (FPTs) are a special and very important subclass of phase transitions. Across such a transition there is change in the directional (or point-group) symmetry of the crystal. Concomitantly, at least one new macroscopic physical property arises in the lower-symmetry (or ferroic) phase. Ferromagnetic phase transitions are one type of examples of FPTs, and spontaneous magnetization is the new property which arises in the ferromagnetic phase. Ferroelectric and ferroelastic phase transitions are some other examples of FPTs. More details are given in Section A7.10.

Ferroic materials are those materials which can potentially undergo at least one FPT as a function of some control parameter like temperature (Wadhawan 2000).

19.3 Multiferroics

Multiferroics are the most important examples of nonbiological natural composites, as also of complex materials. A material in which two or more ferroic properties coexist is a multiferroic; with the further proviso that, in them, two or all three of the electric, magnetic and elastic interactions compete in a delicately balanced manner, meaning that even a very minor local factor can tilt the balance locally in favour of one or the other. Dominance of any one of these interactions would normally give a fairly definite ground-state configuration for the ferroic. But if there is close (*hairy edge*) competition between two or all three, there occur *competing ground states* (Dagotto 2005; Wadhawan 2007). A consequence of this is that, in the same crystal, different portions may order differently. Even the slightest of local perturbations (defects, inclusions, voids, composition variations, etc.) can tilt the balance in favour of ferroelastic, ferroelectric, or ferromagnetic ordering, which occurs over mesoscopic length scales.

Ferroelastic and ferroelectric phase transitions are *structural* phase transitions, meaning that there is a change of crystal structure caused by the movement of atomic nuclei to their new equilibrium sites. By contrast, purely ferromagnetic phase transitions are essentially *nonstructural* in nature: Magnetic ordering *can* occur without significant alteration of the positions of the atomic nuclei. In a multiferroic it can transpire that the tendency for structural, electronic, and magnetic ordering occurs at the same temperature, or over a narrow range of temperatures. This has the important consequence that there is no longer a unique ground state. Instead, there are competing ground states.

This means that, in the same crystal, different portions may order differently, no matter how competent a job the crystal grower has done for growing a 'perfect' single crystal. Even the slightest of local perturbations (defects, inclusions, voids, composition variations, etc.) can tilt the balance in favour of ferroelastic, ferroelectric, and/or ferromagnetic ordering over mesoscopic length scales. The single crystal is no longer 'single', in the sense that any randomly picked up unit cell may not necessarily be identical to another unit cell elsewhere.

The crystal structure of a multiferroic is intrinsically inhomogeneous.

Transition-metal oxides are particularly prone to exhibiting multiferroic behaviour involving a strong correlation of spin, charge, and elastic degrees of freedom (Dagotto 2005). Some examples of such systems are (Wadhawan 2007): manganites; superconducting cuprates; systems exhibiting martensitic phase transitions; relaxor ferroelectrics; etc. For them, local and global deviations from stoichiometry are the rule rather than the exception. This happens because any local deficiency in the oxygen content is readily accommodated by an appropriate change in the valency of the transition-metal cation. There are thus varying degrees of dynamic charge transfer between the transition metal and the oxygen ions, apart from the fact that the oxygen ions are highly polarizable.

Some characteristic features of multiferroics are summarized here (see, for example, Shenoy *et al.* (2005), and references therein for more details; also see Wadhawan (2007)):

- Multiferroics are *strongly correlated electronic systems*; i.e., there is a strong correlation among the structural and electronic degrees of freedom.

- This strong interaction leads to *nonlinear response* and *feedback*.

- Nonlinearity leads to *coherence* of structures over mesoscopic scales, which control macroscopic response.

- Nonlinearity also results in *spatio-temporal complexity*, including the occurrence of *landscapes* of metastable or coexisting phases (in the same single crystal), which can be easily disturbed by the slightest of internal or external fields.

- There is a competition between short-range and long-range ordering interactions, resulting in *a hierarchy of structures* (of different length scales) in the mesoscopic patterns.

- Multiferroics are *intrinsically multiscale*, not only spatially, but also temporally: There is usually a whole range of relaxation times.

- The term *glassy behaviour* (as opposed to the behaviour of crystalline materials) is used in the context of systems that exhibit noncrystallinity, 'nonergodicity', hysteresis, long-term memory, history-dependence of behaviour, and multiple relaxation rates (Angell 1995). Multiferroics usually display a variety of glassy properties.

- In most multiferroics, the same intra-unit-cell distortions are responsible for both *local* anisotropic chemistry and *long-ranged* anisotropic elastic behaviour. Any local atomic ordering occurring in a unit cell of the crystal has a push-pull effect on neighbouring unit cells; i.e., it creates a local displacement field, which then influences further neighbouring unit cells, thus having a knock-on effect which results in an elastic propagation of the disturbance or ordering to distant parts. This coexistence of short- and long-ranged interactions is responsible for the self-organization of patterns of hierarchies.

19.4 Spin glasses

Spin glasses are an example of a very special type of magnetic ordering in crystals in which the spins on embedded magnetic atoms are randomly oriented and located (Mydosh 1993; Guerra 2006). They are the archetypal example of nonadaptive complex systems.

One class of materials which exhibit spin-glass properties are alloys made from nonmagnetic metals and small amounts of 'good-magnetic-moment' atoms like Mn, Fe, Gd, Eu. Two simple examples are $Cu_{1-x}Mn_x$ and $Au_{1-x}Fe_x$. Let us consider a pure Cu crystal (at $T = 0$ K), in which we progressively increase the doping by Mn. The dopant Mn atoms substitute for some of the host Cu atoms at their official crystalline sites. When there are only a few Mn atoms, and they are distributed randomly throughout the host Cu lattice, their magnetic spins do not interact with one another significantly, and they just point along random directions. If the distribution is truly random, the sum total of all the magnetic moments is zero.

But as more and more Mn atoms are brought in, their degree of isolation decreases, and there are regions with Mn atoms on neighbouring lattice sites. Now, since each of the Mn atoms carries a spin, the question of interaction of spins on two neighbouring Mn atoms arises. The spins may align either parallel to each other (*ferromagnetic coupling*), or antiparallel to each other (*antiferromagnetic coupling*).

The latter case is particularly interesting in the context of spin glasses. Imagine a triangle ABC in the lattice, all corners of which are occupied by Mn atoms, and only nearest-neighbour interaction is significant (Fig. 19.1). Suppose that spin A is pointing up. Then both spins B and C should point down (because they both couple antiferromagnetically with spin A). Suppose spin B indeed points down (as shown in Fig. 19.1). Spin C should also point down, but this is inconsistent with the fact that the interaction between B and C is also antiferromagnetic. So spins B and C are *frustrated*, not knowing whether to point up or down! Spin A is also frustrated, because we could as well start from spin B and go around the triangle. Thus, there is frustration all around!

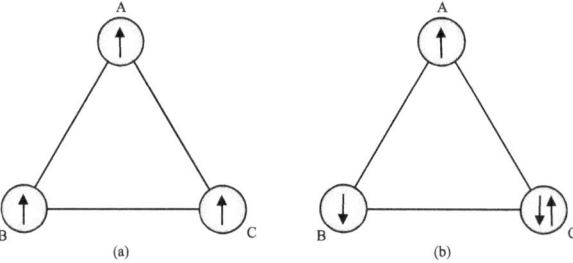

Fig. 19.1 Ferromagnetic interaction (a), and antiferromagnetic interaction (b). The triangle in (b) depicts frustration. There is no frustration in (a); all three spins point the same way, and this is compatible with pairwise ferromagnetic interaction. There is frustration in (b) because the spin at any of the sites has a consistency problem, no matter whether it points up or down. [Wadhawan 2007]

A consequence of the frustration is that the system has a large number of competing ground states, all of similar total energy, but with *activation barriers* separating them.

I have skipped details, but the fact is that there can be a competition between ferromagnetic and antiferromagnetic couplings, and the former do not lead to frustrating scenarios. Imagine a potentially ferromagnetic system, with the spins all pointing up, or all pointing down (Fig. 19.1(a)); everything is self-consistent, and there is no frustration. And, in the right temperature range, ferromagnetic coupling can lead to the formation of parallel-pointing clusters of spins in some parts of the crystal, while some other parts have antiferromagnetic couplings.

Next, let us bring in thermal fluctuations. Suppose a particular spin gets flipped because of some thermal effect. Naturally, there will be an effect on the neighbouring spins coupled to it. The end result is that the boundaries between the various clusters move in a near-random but correlated manner.

19.5 Relaxor ferroelectrics

Relaxor ferroelectrics exhibit a broad and frequency-dependent maximum in their permittivity *vs.* temperature response in the radio-frequency (RF) regime of frequencies, and this response function falls off rapidly at high frequencies. The permittivity values are high to very high (see Wadhawan 2007).

Lead-based relaxor ferroelectrics have the general formula $Pb(B_1B_2)O_3$, where $B_1 = Mg^{2+}$, Zn^{2+}, Ni^{2+}, Sc^{3+}, etc., and $B_2 = Nb^{5+}$, Ta^{5+}, W^{6+}, etc. The archetypal relaxor ferroelectric is lead magnesium niobate (PMN), $Pb^{2+}(Mg^{2+}_{1/3}Nb^{5+}_{2/3})O^{2-}_3$. The unit cell of this crystal is a cube. The Pb^{2+} ions occupy the corners of the cube, and the three O^{2-} ions sit at the centres of the faces of the cube. The body centre of the cube is occupied either by Mg^{2+} or by Nb^{5+}, with a relative occupation frequency of 1 to 2, thus ensuring that the average ('global') valence for the occupants of the body-centres is 4 (= (2 + 5x2)/3), the same as that of Ti^{4+} in $BaTiO_3$ ($Ba^{2+}Ti^{4+}O^{2-}_3$).

Any particular unit cell has either Mg^{2+} or Nb^{5+} at the body centre, and therefore there is a *charge-imbalance* at the unit-cell level, because the needed valence for charge balance is 4+. This unit-cell charge-imbalance is the basis of practically all the remarkable properties exhibited by PMN, and is the equivalent of the frustration of spins mentioned above.

How the structure gets over the local charge-imbalance problem has been a subject of considerable investigation and debate (see Wadhawan 2007). According to one of the models, there exist *1:1 ordered regions* in the crystal, separated by Nb-rich regions. In this kind of 'chemical' ordering, neighbouring unit cells are occupied by Mg and Nb ions, so that there are <111> planes occupied alternately by Mg- and Nb-based unit cells. But this cannot go on beyond a certain length scale (~2-3 nm) because it corresponds to an average charge of (2+5)/2 or 3.5 at the body centre, whereas the required value for charge balance is 4. Such negatively-charged 1:1 ordered clusters (called *chemical clusters*) are therefore separated by positively charged regions (called *polar clusters*) in which all unit cells have only the Nb ion at the body centre. The chemical clusters and the polar clusters extend over only a few nanometres each, and their size distribution is of a dynamically changing nature. At such small sizes they can be readily turned around by thermal fluctuations, and the boundaries between them are constrained to move in a correlated manner. Under the influence of an RF probing field, the polar clusters reorient as a whole, thus accounting for the very large permittivity values. At high probing frequencies, the fluctuations in the direction and size of the polar clusters are not able to follow the rapid variations of the probing field. Consequently the dielectric permittivity falls off rapidly with increasing frequency; hence the name *relaxor*.

PMN readily forms a solid solution with PT (lead titanate), giving us the material PMN-PT, which offers additional flexibility because one can change the proportion of PMN and PT (i.e., the stoichiometry or composition) to work with the best proportion for a given purpose. Like temperature or pressure, composition can also be used as a control parameter, and phase transitions can occur at specific values of the composition. Of special interest is the 65/35 proportion, which corresponds to the so-called *morphotropic phase boundary* (MPB). The MPB is a nearly vertical phase boundary in the phase diagram, meaning that the two solid phases coexist in a near-equilibrium state over a very wide temperature range (Wadhawan 2007).

19.6 Relaxor ferroelectrics as vivisystems

It has been suggested that the positively and negatively charged nanometre-sized clusters in PMN-PT can be likened to bees in a beehive: they behave as a swarm, with rudiments of swarm intelligence (Wadhawan *et al.* 2005; Wadhawan 2007). Such a material is therefore like what Kelly (1994) called a vivisystem, which is any strongly interacting, quasi-autonomous, large, distributed system. Here is why:

• Each cluster is autonomous, but is influenced by what happens to other clusters.

- PMN-PT exhibits the so-called *shape-memory effect* (SME). Suppose we start with some shape of a specimen of PMN-PT, and cool it through the relaxor-ferroelectric transition temperature under the action of a deforming force. When the force is removed, most of the deformation persists, as if the specimen has been deformed plastically (rather than elastically). But when the specimen is heated back to the parent-phase temperature, it recovers the shape we started with, unlike what you expect for plastic bending. It is as if it has a memory of the shape it had when the experiment was started. This is the so-called *one-way* SME, because only the high-temperature shape is remembered, and not the low-temperature shape also.

- It has been demonstrated that, after a certain minimal amount of 'training', PMN-PT also exhibits the so-called *two-way* SME, meaning that there is a memory of the shape during both heating and cooling (Wadhawan *et al.* 2005). There is a whole range of temperatures such that the specimen remembers its shape at every temperature, irrespective of whether it is being heated or cooled through that range of temperatures.

- The 'training' mentioned above amounts to a very simple thermomechanical treatment of the specimen across the phase transition. This treatment makes the specimen *learn* and *remember* its shape at different temperatures.

- The composition of PMN-PT chosen by Wadhawan *et al.* (2005) for these experiments was close to the MPB (65/35). This was deliberate, so that the system is *poised on an instability*, at the *hairy edge between chaos and order*.

- Like the spin glasses mentioned earlier, relaxors like PMN-PT have a glassy character, and glassiness is often linked to *long-term memory* (Stillinger 1995). The memory arises because of the 'nonergodic' nature of the system. The rough and rugged energy landscape of a nonergodic system leads to some retention of the memory of the thermomechanical training to which the specimen is subjected for making it remember its shape for a whole range of temperatures. The memory can be, for example, in the form of 'nucleation sites' favouring the formation of a particular variant or domain-type of the ferroelectric-ferroelastic phase of PMN-PT, thus contributing to the two-way reversibility of the shape memory.

Attempts by Anderson (1988, 1989) and others at understanding the physics of spin glasses and their electrical analogues (relaxor ferroelectrics), as also neural networks, protein folding, etc., have led to the development of a fairly common and general theoretical framework, applicable to all sorts of vivisystems. Although it is useful and convenient to make a distinction between CASs and nonadaptive complex systems, it is clear from the above discussion that this distinction is not a sharp one always.

20. Self-Organized Criticality, Power Laws

History, including that of evolution, is just "one damned thing after another." We can explain in hindsight what has happened, but we cannot predict what will happen in the future. The Danish philosopher Soren Kierkegaard expressed the same view in his famous phrase "Life is understood backwards, but must be lived forwards."

Per Bak (1996)

We have seen in Chapter 1 how the second law of thermodynamics for open systems is responsible for the self-organization so rampant in complex systems. It turns out that complex systems not only self-organize, they also tend to inch towards a state of what Bak, Tang and Wiesenfeld (1987, 1988) called *self-organized criticality* (SOC). This is a very powerful idea in complexity science. In fact, according to Bak (1996), SOC is so far the only known general mechanism to generate 'robust' complexity (see Section 20.3 below for the meaning of 'robust' here).

Bak defined a complex system as that which consists of a large number of components, has large variability, and the variability exists on a wide range of scales. According to him, large 'avalanches', not gradual change, are responsible for qualitative changes of behaviour, and may form the basis for emergent phenomena in complex systems. A surprisingly large number of phenomena, processes and behaviours can be understood in terms of this very basic feature of complexity.

20.1 The sandpile experiment

We can understand the essence of the SOC idea by considering the very simple 'sandpile experiment'. Imagine a tabletop on which grains of sand are drizzling down steadily. To start with, the flat sandpile just grows thicker with time, and the sand grains remain close to where they land. A stage comes when the sand starts cascading or sliding down the sides of the table. The pile gets steeper with time, and there are more and more *sandslides*. With time the sandslides (*avalanches* or *catastrophes*) become bigger and bigger, and eventually some of the sandslides may span all or most of the pile. The average slope now becomes constant with time, and we speak of a *stationary state* (Fig. 20.1).

This is a system far removed from equilibrium. Its behaviour has become *collective*. Falling of just one more grain on the pile may cause a huge avalanche (or it may not). The sandpile is then said to have reached a *self-organized critical state*, which can be reached from the other side also by starting with a very thick pile: the sides would just collapse and roll off until all the excess sand has fallen off.

The state is *self-organized* because it is a property of the system and no outside influence has been brought in. It is *critical* because the grains are critically poised: The edges and surfaces of the grains are interlocked in a very intricate pattern, and are just on the verge of giving way. Even the smallest perturbation can lead to a chain reaction (avalanche), which has no relationship to the smallness of the perturbation. The response is unpredictable, except in a statistical-average sense. The period between two avalanches is a period of tranquillity (*stasis*) or equilibrium; *punctuated equilibrium* is a more appropriate description of the entire sequence of events.

Fig. 20.1 A sandpile with an SOC shape.
http://www.huffingtonpost.com/mario-livio/from-sandpiles-to-bureauc_b_2005591.html

20.2 Power-law behaviour and complexity

*Power laws mathematically formulate the fact that in most real networks the
majority of the nodes have only a few links and that these numerous tiny
nodes coexist with a few big hubs, nodes with an anomalously high number
of links.*

A.-L. Barabási, *Linked*

*It should be fully appreciated that the concept of a power-law distribution is
counterintuitive, because it may lack any characteristic scale. The property
prevented the use of power-law distributions in the natural sciences until the
recent emergence of new paradigms (i) in probability theory, thanks to the
work of Lévy and thanks to the application of power-law distributions in
several problems pursued by Mandelbrot; and (ii) in the study of phase
transitions, which introduced the concept of scaling for thermodynamic
functions and correlation functions.*

Montegna and Stanley (2000)

In a system that is in an SOC state, big avalanches are rare, and small ones frequent. And all
sizes are possible. There is *power law behaviour*: the average frequency of occurrence, $N(s)$,
of any particular size, s, of an avalanche is inversely proportional to some power τ of its size
(Fig. 20.2): $N(s) = s^{-\tau}$.

A log-log plot of this power-law equation gives a straight line (because $\log N(s) = -\tau s$ is rather
like the equation $y = mx$ for a straight line), with a negative slope determined by the value of
the exponent τ. The system is *scale-invariant*: Mostly the same straight line holds for all values
of s. Large catastrophic events (corresponding to large values of s) are consequences of the
same dynamics which causes small events.

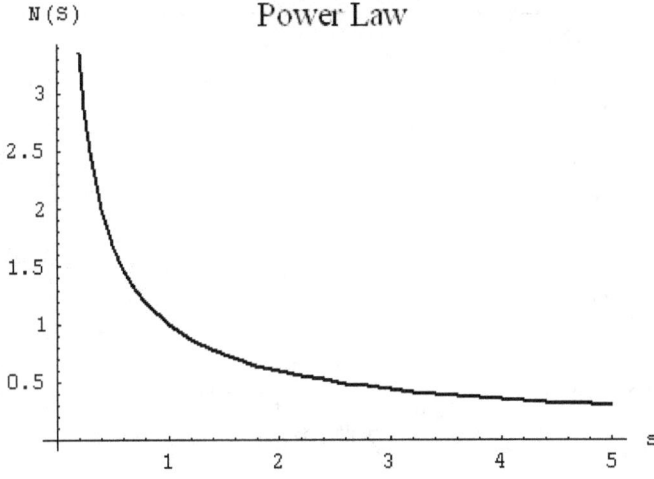

Fig. 20.2 Typical power-law behaviour.

Here are a couple of examples from real-life data.

Any country has a few cities with very large populations, and many cities with small populations. Fig. 20.3 presents data for Germany.

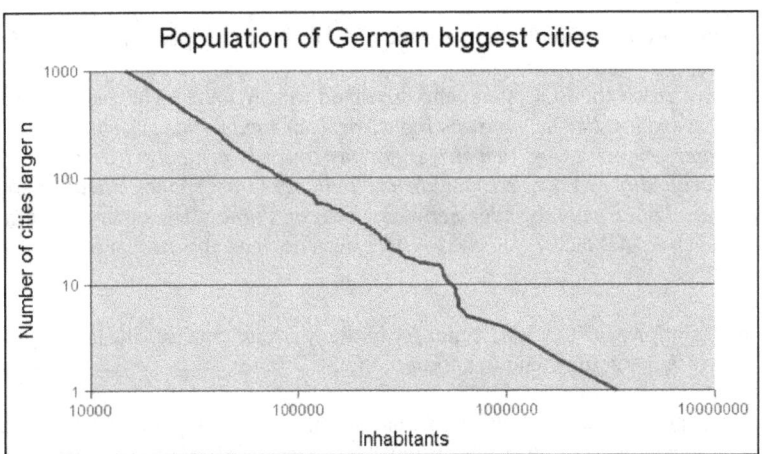

Fig. 20.3 Power-law distribution observed for the cities of Germany. Plotted here on a log-log scale is the cumulative distribution of population. The *y*-axis shows the number of cities having a population equal to or greater than a given population on the *x*-axis.
http://commons.wikimedia.org/wiki/File:Powercitieslnrp.png

Similarly the number of earthquakes of magnitude M occurring in, say, a year is found to be proportional to $10^{-\tau M}$ (*the Gutenberg-Richter law*). Fig. 20.4 shows a graph of all the earthquakes recorded in 1995. The straight line is a plot of the Gutenberg-Richter equation with $\tau = 1$. The value of τ varies a little from area to area, but worldwide it is $\tau \approx 1$. This is known as *Zipf's law*.

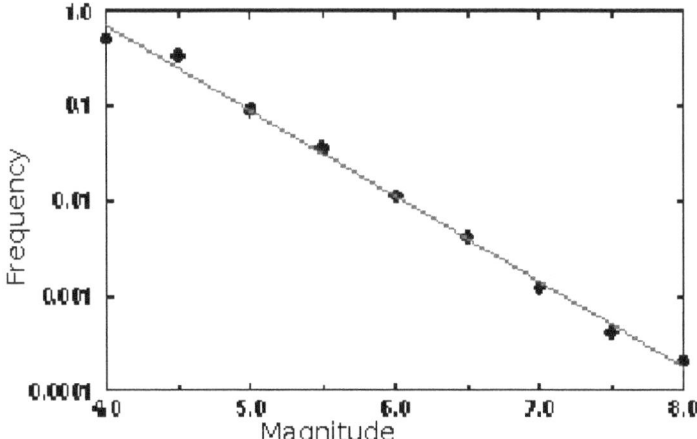

Fig. 20.4 Frequency of earthquakes (on a log scale) *vs.* their magnitude.
http://www.nld.ds.mpg.de/~soc/examples.php

Power-law behaviour is an observational reality, and it spans a very wide variety of phenomena in complex systems, each exhibiting a characteristic value of τ. Because of the complexity and unpredictability of the phenomena involved, it has not been possible to build realistic theoretical models which can predict the value of the critical exponent τ for a system.

Bak (1996) gave several examples to make the point that *Nature operates at the SOC state.*

How do systems reach the SOC state, and then tend to stay there? The sandpile experiment provides an answer. *Just like the constant input drizzle of sand in that system, a steady input of energy, or water, or electrons, can drive systems towards criticality, and then they self-organize into criticality by repeated spontaneous pullbacks from super-criticality, so that they are always poised at or near the edge between chaos and order.* The openness of the system (to the inputs) is a crucial factor. A closed system need not have this tendency to move towards an SOC state.

Nature operates at or near the SOC state. As in the sandpile experiment, large systems may start from either end, but they tend to approach the SOC state.

As we shall see again and again in this book, biological and other types of dynamical evolution are nothing but examples of *self-organization.* The constant influx of low-entropy energy (or rather Gibbs free energy) leads to the evolution of higher and higher levels of complexity and order.

Bak (1996) theorized that even life is a self-organized critical phenomenon. Similarly, in biological evolution, mass extinctions, as also 'punctuated equilibria', can be understood as SOC phenomena.

A comprehensive review of the present status of the SOC paradigm in some areas of research has been given by Watkins *et al.* (2016). I quote from the abstract of their paper: 'Introduced by the late Per Bak and his colleagues, self-organized criticality (SOC) has been one of the most stimulating concepts to come out of statistical mechanics and condensed matter theory in the last few decades, and has played a significant role in the development of complexity science. SOC, and more generally fractals and power laws, have attracted much comment, ranging from the very positive to the polemical.'

It is instructive to list in one place the types of natural phenomena wherein we observe power-law behaviour:

1. Thermodynamically small systems. They follow Tsallis thermodynamics (see Section A1.5). For them, the Boltzmann factor has a power-law form, rather than the exponential form characteristic of thermodynamically large systems.

2. Fractal systems. They have no characteristic length scale (see Chapter 15).

3. Systems in the SOC regime, as discussed in this chapter. 'The applications of the SOC paradigm range from geosciences (including earthquakes, forest fires, solar activity, and rainfall) to the financial markets, while the renewed emphasis on the heavy tails generated by SOC processes has had a direct influence on modern network science and its applications' (Sharma *et al.* 2012). ['Heavy tails' or 'fat tails' refers to the slow decay typical of power-law distributions, unlike the case of exponential decay for Gaussian or 'bell curve' distributions]

4. $1/f$ noise systems, or 'pink noise' systems (f is frequency). For random processes that are correlated in space or time, the autocorrelation function decays as a power law (rather than as an exponential law), implying long-range space or time dependence. Long-term memory processes occur in them ('long-term memory' is defined in terms of persistence of the observed autocorrelations). Systems with this characteristic occur widely in Nature. For them it is difficult to predict extreme events such as floods, earthquakes, business-market crashes, etc. Zipf's law mentioned above in this chapter is an example.

5. Systems in which 'critical phenomena' occur in the vicinity of a phase transition (see Section A7.13).

6. Social-systems data with *Pareto distribution*. The Pareto distribution is a power-law probability distribution that was originally used for modelling the distribution of incomes in a population. It has since found application in a variety of social, scientific, geophysical, and many other types of observable phenomena. It is a skewed distribution with a slowly decaying tail, meaning that much of the data are in the tail. It is also related to the *Pareto principle* or the 80-20 law, according to which 20% of all people receive 80% of all income. Barabási (2003) has listed many other situations in which this law holds: 80% of profits are produced by only 20% of the employees. 80% of customer service problems are created by only 20% of the consumers. 80% of decisions are made during 20% of meeting time. And so on.

7. 'Degree distribution' in scale-free networks and generalized random networks (see Sections A9.7 and A9.8).

20.3 Robust and nonrobust criticality

In general, complex behaviour is exhibited mainly by open systems far from equilibrium. For closed equilibrium systems, complex universal behaviour can arise only under some very specific conditions. Critical phenomena at continuous phase transitions in crystals are an example of this (see Appendix A7). At the critical point, the system passes from a disordered state to an ordered state. The important point is that the system *has to be brought* very close to the critical point to observe complex behaviour, namely scale-free fluctuations of the 'order parameter', giving transient ordered domains of all sizes. Such complex criticality is *not robust*; it occurs only at the critical point, and not at other temperatures (Bak 1996).

According to Bak (1996), chaos is another such example of nonrobust criticality, and therefore 'chaos theory cannot explain complexity'. As discussed in Chapter 15 and Appendix A8, chaos theory explains how simple, deterministic systems can sometimes exhibit unpredictable behaviour, but complex-looking behaviour occurs in such systems *only for some specific range of values of the control parameter(s)*; such complexity is not robust. There is no general power-law behaviour, which is a signature of complex critical systems.

A third example of criticality which is not self-organized, and therefore not robust, is that of a nuclear-fission reactor. The reactor is kept critical (neither subcritical nor supercritical) by the operator by the use of control rods, and cannot, on its own, tend towards (and be stable around) a state of self-organized criticality.

Robustness is an essential feature of SOC. Fragile criticality is not conducive to the adaptive evolution of a complex system.

21. Characteristics of Complex Systems

Complexity refers to the condition of the universe which is integrated and yet too rich and varied for us to understand in simple common mechanistic or linear ways. We can understand many parts of the universe in these ways but the larger more intricately related phenomena can only be understood by principles and patterns—not in detail. Complexity deals with the nature of emergence, innovation, learning and adaptation.

Arthur Battram (1998): *Navigating Complexity*

As difficult as it may be for us to engage in a trans-disciplinary approach to the whole system, and as little we may be used to dealing with complex occurrences—it will not pay off if we try to make our decision-making easier on ourselves by simply ignoring the complexity of the world we live in. It is just as impossible to escape from this complexity than it is to escape from the complexity of our own being.

Most importantly, we need to accept that we are much more entangled with the complex systems of our environment and the biosphere, than our conventional mode of linear cause and effect thinking with its method of dividing the world into categories tries to make us believe. It is simply not a case of here humanity and there nature. We, ourselves are Nature, our billions of biological cells are a part of her—and all the technology we have ever created is included in nature.

Frederick Vester

This chapter lists the main points made in the book so far.

1. Complex systems (CSs) are usually comprised of a large number of interacting or connected components. They are characterised by variability, rather than sameness or uniformity, and this variability extends over several length scales and/or time scales.

2. Linear dynamical systems cannot exhibit complex behaviour. All complex systems are nonlinear, although not all nonlinear systems may be complex. Some nonlinear systems may be simple or simplifiable, in the sense that a reductionistic approach to their investigation (a hallmark of traditional science) can be quite successful.

3. Unlike simple or simplifiable or non-complex systems, it is usually not possible for us to 'take the CS apart', and study its subparts separately. The CS has to be treated as a whole. Even the environment may have to be factored-in in the big picture. Of course, sometimes one can model some parts separately and then place them in the context of the CS as a whole.

4. In simple systems it is usually possible to ignore detail at small scales and assume uniformity and smoothness. By contrast, chaos is one example of complexity for which small details do matter, in the sense that they can blow up into unpredictable large effects. Fractal systems are another such example. They have self-similar structure at all length scales.

5. Scaling has a wider range of existence than in just fractals. In simple systems there is a

particular length scale at which a phenomenon or effect appears, and it fades away for larger scales, usually as $f(x) \sim \exp(-x/\lambda)$. Here λ is the characteristic length scale. But in a CS the variation does not often have a characteristic length scale, and *power-law behaviour is very common*: $f(x) \sim x^\alpha$ (Fig. 21.1B). For such a situation the scaling rule is $f(ax) = a^\alpha f(x)$. Here α is the *scaling exponent*. There is no characteristic length; instead there are scales within scales. For a scale a times larger, the phenomenon has the same appearance, but scales by a^α. In complex systems, power-law behaviour is the rule, rather than the exception.

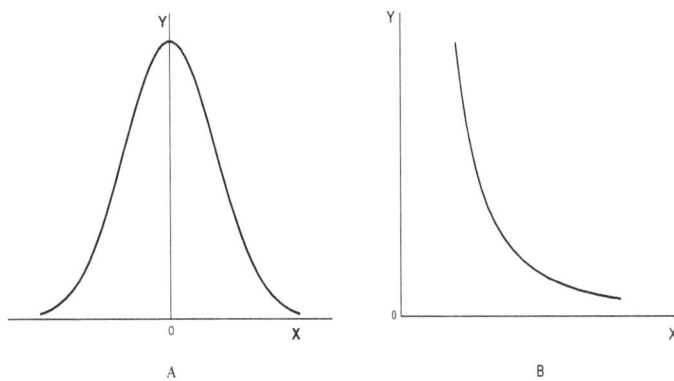

Fig. 21.1 Bell-curve or Gaussian behaviour (A), *vs.* power-law behaviour (B). Curve A has a characteristic length, but there is none for B.

6. The renormalization-group approach offers a way of dealing with the scaling properties of a system. Critical phenomena in the vicinity of a phase transition are an example of this. One starts with the governing equations of the system, and then scales up. This gives the same set of equations, but with different coefficients. This step amounts to averaging over finer details. This is done again and again till a so-called *fixed point* is reached in phase space.

7. Traditional reductionistic science usually uses differential equations for analytical modelling. Usually this amounts to assuming smoothness and *analyticity* at all levels of detail. Also, coarse-graining or smoothening is often done for dealing with simple systems. But for complex systems, all length scales or time scales may be important. That is why computer simulations (using *difference* equations) rather than analytical methods (using *differential* equations) fare better when dealing with CSs. This also explains why fractals (for static methods), iterative maps (like the logistic difference-equation in chaos theory), and cellular automata are often the only approaches available for modelling CSs.

8. Complex behaviour occurs only in *open systems*, meaning that energy and/or matter should be able to flow through them. This raises the possibility that, unlike the case of a closed system, entropy *can* decrease locally, without violating the second law of thermodynamics. Since entropy is a measure of disorder, decreased entropy means increased order or information content. And the information content of a complex system can build up with time.

9. As made further clear by the work of England (2013), if a sustained input of energy drives a system far from equilibrium, the system tends to develop a structure or tendencies which

enable it to dissipate energy more and more efficiently. This is called *dissipation-driven adaptive organization*. (Also see Horowitz and England 2017, and Kachman, Owen and England 2017.)

10. Complex systems can self-organize spontaneously. Consider liquid water, cooled slowly. A temperature comes at which it changes to ice. Ice has highly organized structure, namely its very regular crystal structure. By comparison, liquid water is disorganized or disordered. This is *self*-organization because all that we have done is to cool the system, and there was no input of information regarding the organized structure that ice should adopt. It is also an example of *increase in information content*. Compared to liquid water, a large amount of information got locked in when water changed to ice. The atomic arrangement in ice exhibits a pattern which is missing in the atomic arrangement of liquid water. The phrase '*order for free*' was used by Stuart Kauffman for such self-organization.

11. The phase transition from water to ice, mentioned above, is a special case of what is called a *bifurcation in phase space*. During the dynamical evolution of a complex system, such bifurcations may occur repeatedly, leading to more and more locking-in of information, *something that could not have been anticipated* (because of the random choice of alternative trajectories a CS may make at every bifurcation point in phase space).

12. Per Bak (1996) pointed out the proclivity of complex systems towards a special type of self-organization, namely *self-organized criticality* (SOC). This was a very important contribution to the field of complexity science.

13. Systems that are too ordered or too disordered (or chaotic) are not very interesting from the point of complexity (Fig. 21.2). Although various proportions of order and disorder are conceivable, complexity thrives best at the *edge of chaos*. The dynamics of complexity around the edge of chaos is ideally suited for evolution that does not destroy self-organization.

14. In a complex adaptive system (CAS), the randomness is local, and the order or structure is global. For example in a market, local randomness creates innovation and resilience. In free-market economies, competition is the source of local randomness, and regulation maintains global structure (Peters 1999).

15. Complex systems create order. Paradoxically, complexity also breeds uncertainty. How can complexity generate both order *and* uncertainty? It is all a matter of scale. The order is global. Uncertainty is in the details (Peters 1999).

16. Because of the random choice of bifurcation paths, unexpected properties *emerge* in a CS. *Emergence is the hallmark of complexity*.

17. The bifurcations in the phase space of CSs are often referred to as *complexity transitions*.

18. Since energy and matter are always flowing through an open CS, it is ever in a position to explore new possibilities (*perpetual novelty*).

19. *Positive feedback* is an important mechanism of how self-organization can occur in a CS. However, it is not the only possible mechanism for this. Often, *chain reactions* achieve

something similar. And *negative feedback* provides the necessary antidote for maintaining a state of optimal balance and perpetual novelty.

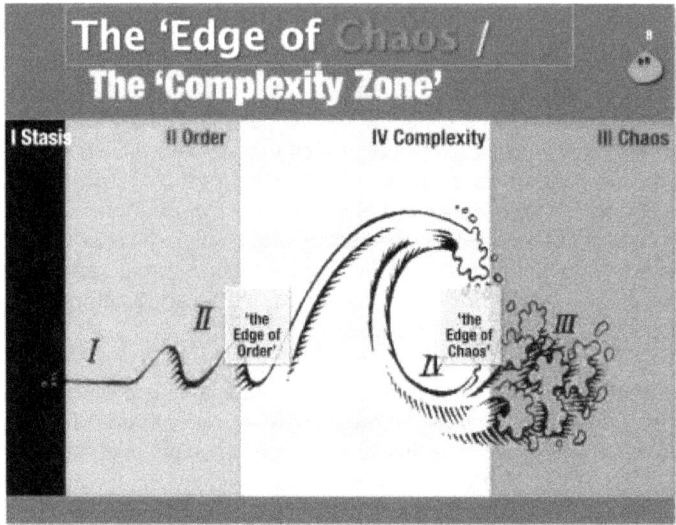

Fig. 21.2 Arthur Battram's graphic depiction of the 'complexity zone' between the 'edge of order' and the 'edge of chaos'. Image credit:
https://plexity.wordpress.com/zomes-at-the-edge-of-chaos-good-enough-systems-a-p-battram-independent-contributions-to-playwork-theory-1/

20. In a CS, simple networked interactions can result in extremely complex behaviour. For example, swarm intelligence emerges in a beehive or an ant colony through the operation of very simple local interactions among the bees / ants. This conclusion can be extended to all sorts of large complex systems, and lies at the heart of how we can answer, with a fair degree of success, some of the most difficult questions we ask about ourselves and about the universe we live in (Wadhawan 2010). For example, we can explain human intelligence as a kind of swarm intelligence, the swarm here being that of neurons.

21. *Local rules dominate the evolution of complexity*. They lie at the heart of pattern formation and self-organization in complex systems. As emphasized by Wolfram, what we have mostly in natural phenomena are *simple* local rules (or 'simple programs') in operation. He asked: *What secret it is that allows nature seemingly so effortlessly to produce so much that appears to us so complex (?)*. His answer was: Local rules in action. The formidable cellular automata (CA) formalism built up by him is a formalisation of the local-rule paradigm, and is a powerful tool for understanding a wide range of natural phenomena.

22. *Nature is creative via the complexity-evolution route*. Human creativity takes the same route. The goal of a CAS is not a static equilibrium. It is, instead, a dynamic and evolving state that is ever changing and ever creative. The details as to how it achieves this state are never constant. Because a complex system does not need a specific sequence of events to achieve its goal of a stable state, it is resilient to unexpected changes in its environment, and it creatively adapts itself to maintain its goal. It uses uncertainty to generate order, but the unique features of the order it generates are not predictable. Even as a complex system

turns uncertainty into order, it also generates more uncertainty (Peters 1999).

Some key points: For understanding a complex system, it is a good idea to look for the local rules in operation. The local-rules approach has been given a formal cloak in the subject of cellular automata (CA). The idea is that when you cannot work with differential equations, try working with difference equations. An additional powerful approach is to map a complex system to a mathematical network. This enables us, for example, to identify the type of formal network the CS belongs to, and to exploit the mathematical framework of network theory. Moreover, in a network, it is the presence of a direct link or edge between two nodes that defines proximity (and therefore the local rule in operation), and not necessarily physical proximity. If the networking individuals are 'thinking' creatures, in comes game theory with all its available insights. Biological evolution is a subset of dynamical evolution, and for understanding biological evolution the central idea is that of natural selection by Darwinian evolution. It is any day a good idea to get the hang of the finer points of Darwinian evolution, and for that the most important book in my opinion is Daniel Dennett's *Darwin's Dangerous Idea: Evolution and the Meanings of Life* (1995). Finally, the bottom line for everything is that practically all natural phenomena can be understood in terms of the second law of thermodynamics: Nature abhors gradients, and tends to annul them in efficient ways. The second law is the over-arching self-organization principle, from which all other self-organization principles follow.

Complexity refers to the condition of the universe which is integrated and yet too rich and varied for us to understand in simple common mechanistic or linear ways.

We can understand many parts of the universe in these ways but the larger and more intricately related phenomena can only be understood by principles and patterns – not in detail.

Complexity deals with the nature of emergence, innovation, learning and adaptation.

The Santa Fe Group (1996)

My thesis has been that one path to the construction of a nontrivial theory of complex systems is by way of a theory of hierarchy. Empirically a large proportion of the complex systems we observe in nature exhibit hierarchic structure. On theoretical grounds we could expect complex systems to be hierarchies in a world in which complexity had to evolve from simplicity. In their dynamics hierarchies have a property, near-decomposability, *that greatly simplifies their behaviour. Near-decomposability also simplifies the description of a complex system and makes it easier to understand how the information needed for the development or reproduction of the system can be stored in reasonable compass.*

Herbert Simon (1996)

Complex adaptive systems are constantly revising and rearranging their components in response to feedback from the environment. Examples are to be found in the evolution of organisms, the brain changing connections between neurons, firms reshuffling their departmental structure, countries realigning their alliances.

At some deep, fundamental level, all these processes of learning, evolution and adaptation are the same. And one of the fundamental mechanisms of adaptation in any given system is the revision and recombination of the building blocks. ... New opportunities are always being created by the system. It is therefore essentially meaningless to talk about a complex adaptive system being 'in equilibrium': the system can never achieve balance. It is always moving on. ...

Agents in the system can never 'optimise' their 'fitness' or their utility. The space of possibilities is too vast; they have no practical way of finding the optimum. The most they can ever do is to change and improve themselves relative to what the other agents are doing. In short, a complex adaptive system is characterized by perpetual novelty.

Arthur Battram (1998)

II. Pre-human Evolution of Complexity

Life emerged out of nonlife at a certain stage in the evolution of complexity in our universe, and in due course we humans emerged and evolved. One of our greatest achievements has been the adoption of the all-important *scientific method* for investigating and understanding natural phenomena and formalizing and recording our knowledge of Nature in a logical, rational, objective, and systematic manner. And practically all this ever-growing mass of scientific knowledge is available to everybody.

This part of the book deals with how complexity and order evolved from the interplay of the 'blind forces' of Nature, *before* the emergence of humans. Emergence of humans made a sea change in the evolutionary history of complexity. Humans as a species not only have an exceptionally high degree of intelligence (and the physical ability to make sophisticated tools and machines), they also have the ability and knowledge to *control* the course of evolution to a certain extent.

In this part I discuss self-organization and emergence in nonliving systems and in pre-human living systems. Evolution of complexity after we humans emerged, including 'artificial' evolution, will be discussed later in Part III of the book.

Complexity is considered by many to be the single most important scientific development since general relativity and it promises to make sense of no less than the very heart of the Universe. Using it, scientists can find order emerging from seemingly random interactions of all kinds, from something as simple as flipping coins through to more challenging problems such as the patterns in modern jazz, the growth of cancer tumours, and predicting shopping habits.

Neil Johnson (2009)
Simply Complexity: A Clear Guide to Complexity Theory

22. Evolution of Structure and Order in the Cosmos

22.1 The three eras in the cosmic evolution of complexity

In Chapter 12 I summarized our cosmic history, all the way from the Big Bang onwards. The degree of complexity (i.e., the information content) of our universe has been, by and large, increasing, mainly because Nature tends to annul gradients of all types.

Chaisson (2001) identified three eras in the cosmic evolution of complexity.

1. In the beginning there was only radiation, with such a high energy density that there was hardly any structure or information content in the universe; it was just pure energy.

2. As the universe cooled and thinned out, a veritable phase transition, or *bifurcation* in the phase-space trajectory, occurred, resulting in the emergence of matter coexisting with radiation. This marked the start of the second era, in which a high proportion of energy resided in matter, rather than entirely in radiation.

3. The third era was heralded by the onset of *'technologically manipulative beings'*.

22.2 Chaisson's parameter for quantifying the degree of complexity

At the birth of our universe, the information content of our universe was nil. There was just a single force field or radiation field, with no alternative states, so the information was nil (recall Eq. 6.7, the Shannon-information equation: $I = - c \log (1/P)$; when $P = 1$, we get $I = 0$.).

Very soon, structure appeared and the information content, or the degree of complexity, started increasing. A new way of quantifying the cosmic degree of complexity (a new *complexity metric*) was introduced by Chaisson (2001). He emphasized the importance of a central physical quantity for this purpose, namely *free-energy rate density*, or 'specific free energy rate', denoted by Φ. Chaisson emphasized the fact that *'energy flow is the principal means whereby all of Nature's diverse systems naturally generate complexity*, some of them evolving to impressive degrees of order characteristic of life and society'. The *flow* refers to a rate of input *and output* of free energy. If the input rate is zero, a system would sooner or later come to a state of equilibrium, marking an end to the evolution of complexity. If the output rate is zero, there would be disastrous consequences of the relentless energy inflow.

The measure introduced by Chaisson, namely the energy flow per unit time per unit mass, has the units of power. Other similar quantities in science are: luminosity-to-mass ratio in astronomy; power density in physics; specific radiation flux in geology; specific metabolic rate in biology; and power-to-mass ratio in engineering. Chaisson estimated the values of this parameter for a variety of systems. The results are amazing, and important. Here are some typical estimated values (in units of erg s^{-1} g^{-1}):

Galaxies (Milky Way)	:	0.5
Stars (Sun)	:	2
Planets (Earth)	:	75
Plants (biosphere)	:	900
Animals (human body)	:	20,000
Brains (human cranium)	:	150,000

Society (modern culture)   :   500,000

Thus the degree of complexity of our universe can be seen to be increasing very rapidly (Fig. 22.1).

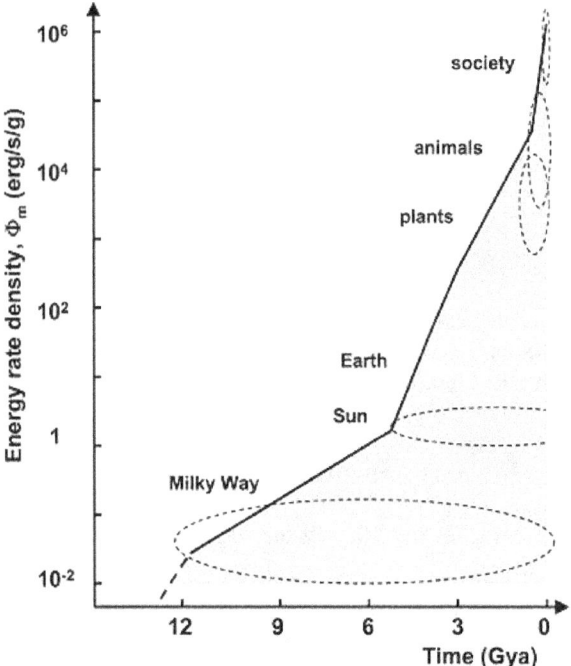

Fig. 22.1 The rapid (essentially exponential) increase of degree of complexity as a function of time, as estimated by Chaisson in terms of free-energy rate density. Image credit: http://www.informationphilosopher.com/solutions/scientists/chaisson/

We humans are responsible for much of the recent increase in complexity. When we emerged on the scene (through Darwinian evolution), we developed a relatively large brain and a high level of intelligence, *and* also the ability to develop spoken and written language. Moreover, our opposable thumb made it easier for us to build machines. Development of powerful computers followed in due course, as also immense telecommunication networks. Information build-up and its flow is the stuff we humans thrive on.

At present there are no indications of life anywhere else in our universe. Leave aside creatures with intelligence comparable to or surpassing that of humans, even the most primitive extra-terrestrial life has not been found yet. Therefore, *from the vantage point of increase of complexity, emergence of humans has turned out to be something of cosmic importance*. We are 'star stuff' and our emergence was of cosmic importance! Therefore it behoves us that we manage our ecosphere in a responsible and mature manner.

22.3 Cosmic Evolution of Information

When spacetime began (presumably at the Big Bang), the classical information content of our universe was nil. Thus, in the beginning, the effective complexity was zero. This is consistent with the viewpoint that the universe emerged out of 'nothing' (see Krauss (2012) for the meaning of 'nothing').

As the early universe expanded, it pulled in more and more energy out of the quantum fabric of spacetime. Under continuing expansion, a variety of elementary particles got created, and the energy drawn from the underlying quantum fields got converted into heat, meaning that the initial elementary particles were of very high kinetic energies, and increasing in number rapidly. Therefore the entropy of the universe increased rapidly. And high entropy means that the particles required a large amount of information to specify their coordinates and momenta. This is how the degree of complexity of the universe grew in the beginning.

Soon after that, quantum fluctuations, resulting in density fluctuations and clumping of matter, made gravitational effects more and more important.

But why does the degree of complexity go on increasing? To answer that, I have to refer to the concept of algorithmic probability (AP) introduced in Section 6.8 while discussing Ockham's razor. Ockham's razor ensures that short and simple programs or 'laws' are the most likely ones to explain natural phenomena, which in the present context means the explanation of the increase of complexity in the universe. I explained this by introducing the metaphor of an unintelligent monkey typing away randomly the digits 1 and 0, each such sequence of binary digits offering a possible 'simple program' for generating an output that may explain a set of observations.

The quantum-mechanical laws of physics are the simple (computer) programs, and the universe is the computer. But what is the equivalent of the monkey, or rather a large number of monkeys, injecting more and more information and complexity into the universe by programming it with a string of random bits? According to Seth Lloyd (2006), '*quantum fluctuations are the monkeys that program the universe*'.

Some experts are of the view that the universe will continue to expand, and that it is spatially infinite. But the speed of light is not infinite. Therefore, the *causally connected* portion of the universe has a finite size, limited by what has been called the '*horizon*' (Lloyd 2006). The quantum computation being carried out by the universe is confined to this part. Thus, for all practical purposes, the part of the universe within the horizon is what we can call 'the universe'. As this universe expands, the size of the causally connected region increases, which in turn means that the number of bits of information within the horizon increases, as does the number of computational operations. Thus the expanding universe is the reason for the continuing increase in the degree of complexity of the universe.

The expansion of the universe is a necessary cause (though perhaps not a sufficient cause) for all evolution of complexity, because it creates *gradients* of various kinds: '*Gradients forever having been enabled by the expanding cosmos, it was and is the resultant flow of energy among innumerable non-equilibrium environments that triggered, and in untold cases still maintains, ordered, complex systems on domains large and small, past and present*' (Chaisson 2202). The ever-present expansion of the universe gives rise to gradients on a variety of spatial and temporal scales. And, '*it is the contrasting temporal behaviour of various energy densities that has given rise to those environments needed for the emergence of galaxies, stars, planets, and life*' (Chaisson 2002).

In the grand cosmic scenario, there was only *physical* evolution in the beginning, and it prevailed for a very long time. While the physical evolution still continues, the emergence of life started the phenomenon of biological evolution. This led to a rapid increase in the rate of increase of degree of complexity, because living systems are very efficient at annulling gradients (Margulis and Sagan 2002; England 2013).

22.4 Why so much terrestrial complexity?

From cosmic complexity, let us now zero-in on terrestrial complexity. Why is the terrestrial complexity increasing all the time? Apart from the cosmic expansion, there is a locally identifiable reason for this: our Sun has been bombarding our ecosphere with low-entropy or 'high-grade' energy, which has been creating gradients. Why 'low-entropy'? Entropy is defined as $dS = dQ/T$, and the average value of the temperature T for the Sun is huge compared to that for the Earth. On entering our ecosphere, the energy of the photons coming from the Sun gets degraded to a large extent through the processes of 'thermalization', namely dissipation into states of much lower average temperature (and therefore a correspondingly high value for the entropy). A small fraction of this energy, however, gets trapped as free energy. It is stored in our ecosphere in the form of simple or complex molecules. Some of the energy-rich simple molecules in which the free energy gets stored or trapped are: H_2S, FeS, H_2, phosphate esters, HCN, pyrophosphates, and thioesters. In the history of chemical and biological evolution on Earth, such simple molecules contributed to the evolution of complex molecules characterising life. When food is consumed by a living organism, its processing by the organism builds up a high *information content* for the organism, even though there is always a net rise in the global entropy (as demanded by the second law of thermodynamics).

23. The Primary and Secondary Chemical Bonds

We saw in the Chapter 12 how atoms came into existence. From atoms to molecules was the next step in the evolution of complexity. Formation and evolution of molecules, i.e., *chemical evolution*, preceded the emergence and evolution of life. Under the influx of low-entropy energy from the Sun, and probably aided by the presence of certain rocks (which helped lower the activation barriers for chemical reactions), the various atoms and molecules underwent chemical reactions, resulting in the emergence of molecules of higher and higher information content or complexity.

Why do chemical reactions occur? Why do atoms form molecules, or one molecule changes to another? They do so to lower the free energy, or, what is the same thing, to acquire stability for the system as a whole.

The chemical symbol H is used for an atom of hydrogen, which is the first element in the periodic table of elements. It has a nucleus, which is just a proton in this case, and there is an electron orbiting around the nucleus. The electron has a negative charge, exactly equal in magnitude to the positive charge of the proton. Taking this quantity as the unit of charge, we say that an H atom has a *charge number* 1 ($Z = 1$). Taking the mass of the proton as the unit of mass, we say that H has a *mass number* 1 ($A = 1$). [The electron is ~2000 times lighter than the proton, so the atom and the nucleus have practically the same mass here.]

Element number 2 in the periodic table is helium (chemical symbol He). There are two protons in its nucleus, and two electrons orbiting around the nucleus. There are also two neutrons in the nucleus. Neutrons are so called because they are charge-neutral. The mass of a neutron is only slightly greater than that of a proton. So, for the He atom, $Z = 2$, and $A = 4$.

Life on Earth is based on *organic chemistry*, i.e., the chemistry of the carbon atom, denoted by the symbol C. For this atom, $Z = 6$, and $A = 12$.

23.1 The primary chemical bonds

Consider hydrogen gas. Does it exist as a collection of hydrogen atoms? No. If there are many such atoms around, they spontaneously form hydrogen *molecules*. A molecule of hydrogen is denoted by the symbol H_2. It consists of two nuclei of hydrogen, and there are two electrons orbiting around them. Why does hydrogen 'prefer' to exist as H_2, rather than as H? Because H_2 is more *stable* that H. Why? Consider the two electrons of H_2. Quantum mechanics tells that they have no individuality and are indistinguishable. Let us consider any of them. Since positive and negative charges attract one another, this electron stays close (*but not too close*) to the two nuclei. [But for the Heisenberg uncertainty principle of quantum mechanics, the electrons of all the atoms would have gone *right into their nuclei*, and you and I would not be here, discussing chemical evolution of complexity!] Naturally, the positive charges on the *two* nuclei of H_2 are better than only one positive charge in H, when it comes to exerting an attractive force on the electron (Fig. 23.1). Thus H_2 is more stable (it has a higher magnitude of internal energy) than H because the former is a more strongly bound entity. Thus H atoms form H_2 molecules spontaneously, because by doing so the overall free energy gets reduced (the second law of thermodynamics demands that the free energy be as small as possible).

Formation of H_2 from two atoms of H is an example of *spontaneous increase of chemical complexity*: An H_2 molecule has a higher degree of complexity than an H atom because more information is needed for describing the molecule than the atom.

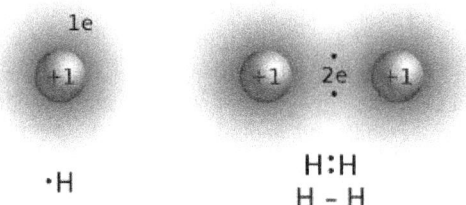

·H

$$H \!:\! H$$
$$H - H$$

Fig. 23.1 An atom (left) and a molecule (right) of hydrogen gas.
http://en.wikipedia.org/wiki/File:Covalent_bond_hydrogen.svg

What is the nature of the bonding between the two atoms of H_2 or H-H? It is described as *covalent bonding*. Each of the two H atoms contributes its electron to the chemical bond between them, and the two electrons in the bonding region belong to both the nuclei.

Another kind of chemical bonding is the so-called *electrovalent bonding*, or *ionic bonding* (Fig. 23.2).

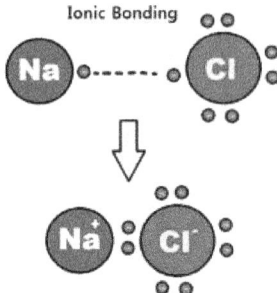

Fig. 23.2 Ionic bonding between sodium and chlorine. Image credit:
https://en.wikibooks.org/wiki/Structural_Biochemistry/Chemical_Bonding/Ionic_interaction

Ionic bonding occurs between oppositely charged ions. Take sodium chloride (NaCl). For the Na atom, Z = 11, and for the chlorine atom, Z = 17. The laws of quantum mechanics are such that an atom of Na is more stable if it is surrounded by only 10 electrons, instead of 11. Similarly, Cl is more stable if it has 18 electrons, rather than 17. They can solve the problem together by getting readily 'ionized'; i.e., an Na atom can become a positively charged ion Na^+ by losing an electron (called the *valence electron*), and a Cl atom can become a negatively charged ion Cl^- by gaining an electron. The two oppositely charged ions can lower the overall potential energy (and therefore the free energy) by coming close to each other, thus forming an ionic bond between them.

The third important and generally strong type of bonding is *metallic bonding*. It occurs in metals like aluminium (Al), copper (Cu), silver (Ag), gold (Au), etc. Take the case of Al. For it, Z = 13. But like an atom of Na considered above, it is more stable if it has just 10 electrons around the nucleus. So Al atoms, when in the close vicinity of many other Al atoms, lose their three valence electrons to a common pool, and these valence electrons become the common property of all the Al ions. A lump of Al metal is held together by this cloud of negatively charged electrons, attracted to, compensating for, and shielding, the positive charges on the Al ions (Fig. 23.3).

The covalent, electrovalent, and metallic bonds described above are the so-called *primary bonds*. They are strong bonds. Diamond, for example, consists entirely of covalently bonded carbon atoms, and is an extremely hard material. In metals also the atoms are quite strongly bonded to one another, as are the atoms in a crystal of sodium chloride in which the electrovalent interaction dominates.

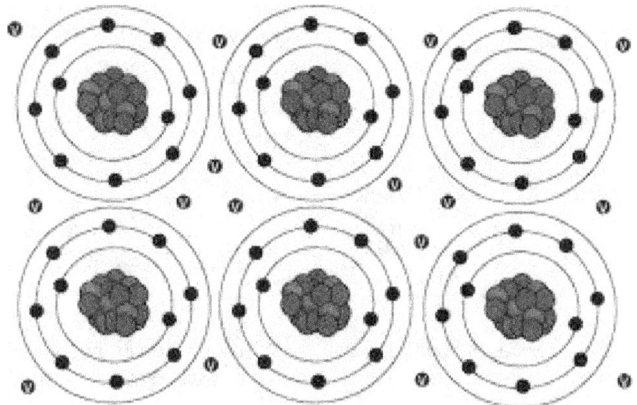

Fig. 23.3 Metallic bonding among aluminium atoms.
Image credit: http://www.ndt-ed.org/EducationResources/CommunityCollege/Materials/Structure/metallic.htm

23.2 The secondary chemical bonds

There are two main types of secondary chemical interactions: hydrogen bonding and the van der Waals interaction.

Take the example of a water molecule, H_2O or H-O-H. The oxygen atom forms covalent bonds with the two hydrogen atoms. Each such bond (O-H) has two electrons associated with it, one coming from hydrogen and one from oxygen (there are some simplistic statements in this chapter, but that is deliberate). The electron distribution around the hydrogen nucleus in such a bond is somewhat different from that in a *symmetrical* bond like C-C in the structure of diamond. The oxygen nucleus has a charge number (Z) equal to 8, which is much more than the charge number 1 of H, so the former hogs (attracts towards itself) a larger share of the electron charge cloud associated with the covalent bond; we say the oxygen atom is very *electronegative* compared to the hydrogen atom. This makes the nucleus of the hydrogen atom somewhat less shielded by its electron than when there was no bonding of any kind. For similar reasons, the oxygen nucleus and its charge cloud of electrons are together a little more negative than what they would be in an isolated atom of O. The end result is that the O-H bond in a water molecule is like a tiny electric *dipole*. There are two such bonds in the H-O-H molecule, so there are two positive ends and one negative end. The upshot is that, because of this asymmetric charge separation ('*asymmetric bond*'), the entire water molecule has a net '*dipole moment*'. Therefore we say that water is a *polar liquid*; it is comprised of a large number of little dipoles.

Any electric dipole, because of a separation of its net positive charge and net negative charge by a certain distance, can be made to reorient itself by applying an electric field. Even when no external electric field is applied, the water molecules, being dipoles, tend to orient themselves such that a positive end (a hydrogen end) of one molecule points towards (and is attracted by) the negative end (the oxygen end) of another molecule. So we speak of *hydrogen bonds*, denoted in this example by O-H...O (Fig. 23.4).

In a hydrogen bond, for obvious reasons, the electronegative atom not covalently bonded to the hydrogen atom (oxygen in the case of H_2O) is called the *proton acceptor*, whereas the one covalently bonded to this hydrogen is called the *proton donor* (oxygen again, but of a *different* H_2O molecule, in the above example).

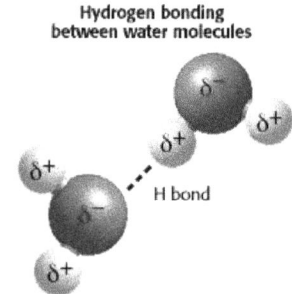

Fig. 23.4 A hydrogen bond between two molecules of water. Image credit:
http://www.biology.arizona.edu/biochemistry/tutorials/chemistry/page3.html

As we shall see below, *the most crucial aspect of the hydrogen bond in the evolution of chemical and biological complexity is that it is of intermediate strength, not as strong as a typical covalent bond, and yet not as weak as the van der Waals interaction (or the London dispersive interaction)*, which I describe next.

The van der Waals interaction is very weak, and it is *always* present between any two atoms or molecules. Quantum-mechanical fluctuations in the electronic charge cloud around an atom or molecule result in a transient charge separation, or the creation of a transient dipole or multipole moment. Concomitantly, the electric field emanating from this transient multipole moment induces a multipole moment on any neighbouring atom or molecule. This results in a very small, very short-ranged, attraction between the two atoms or molecules (Fig. 23.5).

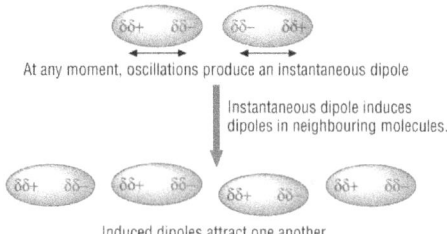

Fig. 23.5 The always-attractive van der Waals interaction between two atoms or molecules.
http://www.chemhume.co.uk/ASCHEM/Unit%201/Ch3IMF/chapter_3__chemical_bonding_andc.htm

23.3 The hydrogen bond and the hydrophobic interaction

Hydrogen bonding is particularly important for biological evolution. The reason is that it is not as strong as the three primary types of chemical bonding described above, and yet is not as weak as the ubiquitous van der Waals bonding. The typical strength of a hydrogen bond is such that, at the usual temperatures at which biological phenomena occur (in our highly conducive ecosphere), thermal fluctuations (and some activation-barrier-lowering local conditions) can readily (but not too readily) break the hydrogen bond. The easy making and breaking of hydrogen bonds provides the necessary flexibility for any thermodynamically open chemical system to try out the various candidate reaction-pathways and configurations. And then a kind of '*chemical natural selection*' does the rest.

The energy required to break a bond is a measure of its strength. The covalent bond is the strongest, with a typical bonding energy of ~400 kilocalories (kcal). For the ionic bond, it is typically ~200 kcal. For the metallic bond it varies widely, from a rather low value for mercury to a very high value for tungsten, but it is still strong bonding.

For van der Waals bonding, the strength is just ~1 kcal. *Hydrogen bonding has intermediate strength; typically ~14 kcal.*

Thus the attractive energy involved in hydrogen bonding is ~10 times larger than the energy of thermal fluctuations, at the temperatures of usual interest for biological systems. And yet it is much lower than the binding energy of a covalent bond. This leads to an important conclusion:

> *At most of the temperatures of interest for our usual biological systems, it is difficult for thermal fluctuations to break covalent bonds (unless 'catalysis' or other activation-barrier-lowering factors are involved), but not too difficult to break hydrogen bonds.*

Water is a polar liquid. This is so because the oxygen atom is highly electronegative. By contrast, the carbon atom is hardly so. In a typical C-H bond, the pair of electrons in the covalent bond is almost equally shared by C and H nuclei, resulting in a near-zero dipole moment. The Z number for C is 6, which is not too different from the value 8 for oxygen. Then why does the C atom in the C-H bond not hog a major share of the electron charge cloud, unlike in the O-H bond? The reasons have to do with the laws of quantum mechanics controlling the distribution of electrons around the C atom and the O atom. Incidentally, the O atom is not only highly electronegative (and the C atom is not), the O atom is also highly *polarizable* compared to the C atom.

Thus hydrocarbons (i.e., compounds comprising mainly of C and H atoms) tend to be nonpolar materials, unlike water. This is very important for the evolution of biochemical complexity on our planet. Suppose we bring such a material in contact with water. The two do not mix, because the bonding is polar in one case and nonpolar in the other. The nonpolar material disturbs the hydrogen bonding in water, so the water molecules re-establish the hydrogen bonding among themselves by going into regions *around* the nonpolar material. Two such regions of a nonpolar liquid are shown in Fig. 23.6. Over a period of time the two portions of the nonpolar material will encounter one another, combine, and form one larger nonpolar region that is excluded from the water matrix. This combined state is energetically more favourable than the one in which the nonpolar regions were separate, so the combined state will persist. The net result is that it appears as if the nonpolar material has a 'phobia' for water. So we speak of *hydrophobic interaction* (see Chandler 2005).

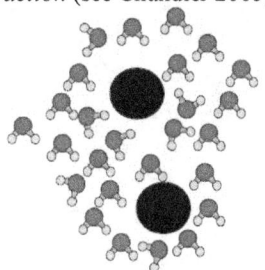

Fig. 23.6 Hydrophobic exclusion occurs when water and a nonpolar material are brought together. There are two such regions containing the hydrophobic material. Each of them is excluded from the water matrix. Credit: http://academic.brooklyn.cuny.edu/biology/bio4fv/page/hydropho.htm

All processes occur so as to minimize the free energy. The hydrophobic exclusion or segregation of the nonpolar fluid is no exception. The hydrogen bonding in water is something dynamic: The hydrogen bonds are formed and broken and formed, and so on. The donor and acceptor oxygen atoms interchange their roles continuously. The hydrophobic interaction is a so-called 'entropic effect' because in this case the lowering of the free energy involves a lowering of the entropy term as the hydrogen bonding around the intruder nonpolar component has a more ordered or quasi-crystalline character than in the original water matrix.

In the mixture, the hydrogen bonds in water are partially reconstructed by building a water 'cage' around the hydrocarbon molecules, and the water molecules that form the cage or 'solvation shell' have substantially restricted mobilities. This leads to significant losses in translational and rotational entropy of the water molecules and makes the process unfavourable in terms of free-energy lowering of the system. This unfavourable free energy is directly proportional to the number of carbon-hydrogen bonds in the hydrocarbon molecule. Therefore, by aggregating together or clumping the nonpolar molecules reduce the surface area exposed to water and their disruptive effect gets minimized.

Many organic molecules predominantly have the hydrocarbon structure, but often there are the so-called *functional polar groups* attached to them. The molecules of cholesterol, fatty acids, and phospholipids are some examples. They have a nonpolar or hydrophobic end, and a polar or *hydrophilic* end. Something interesting happens when you put such molecules in water. They *self-organize* such that the hydrophilic ends point towards water, and the hydrophobic ends get tucked away (or rather tucked in), thus avoiding interfacing with water. It is for such reasons that oil and water do not mix. On the other hand, alcohol and water mix readily (no stirring needed) because both are polar liquids.

24. Cell Biology Basics

All tissues in animals and plants are made up of biological cells, and all such cells normally come from other cells. Generally, a cell may be either a *prokaryote* or a *eukaryote* (Fig. 24.1). The former is an organism that has neither a distinct 'nucleus', nor other specialized subunits or *organelles*. Examples include bacteria and blue-green algae. Unicellular organisms like yeast are eukaryotes. Such cells are generally separated from the environment by a semi-permeable membrane. Inside the membrane there is a *nucleus* and the *cytoplasm* surrounding the nucleus. Multicellular organisms are all made up of eukaryote-type cells. In them the cells are highly *specialized* (we call it *cell differentiation*), and perform the function of the organ to which they belong.

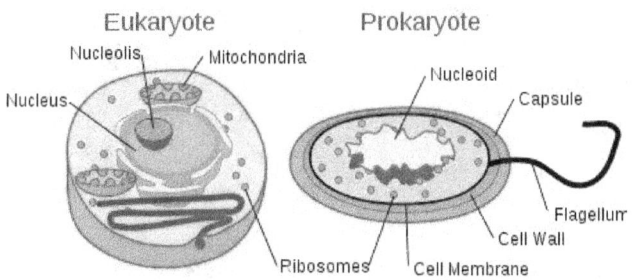

Fig. 24.1 Eukaryote and prokaryote cells.
http://en.wikipedia.org/wiki/Cell_%28biology%29

The nucleus contains *nucleic acids*, among other things. With the exception of viruses, two types of nucleic acids are found in all cells: RNA (ribonucleic acid) and DNA (deoxyribonucleic acid). Viruses have either RNA or DNA, but not both (but then viruses are not cells). Apart from having a nucleus, a eukaryotic cell has *mitochondria, ribosomes*, and *vacuoles*. Plant cells also have *chloroplasts*. Mitochondria make energy out of food. Ribosomes make proteins. Vacuoles are used for storage of water or food. Chloroplasts use sunlight to create food by photosynthesis.

DNA is a very long molecule that has the genetic information encoded in it as a sequence of four different molecules called *nucleotides*: adenine (A), thymine (T), guanine (G), and cytosine (C). There is a double backbone of phosphate and sugar molecules, each carrying a sequence of the 'bases' A, T, G, C. This backbone is coiled into a double helix, like a twisted ladder (Fig. 16.3). In this double-helix structure, the base nucleotide A bonds almost always to base nucleotide T (via a weak hydrogen bond), and G bonds to C (Fig. 24.2). The sequence of base pairs along the backbone defines the *primary structure* of a DNA molecule.

The chemical composition of RNA is quite similar to that of DNA, except that it has the base uracil (U) instead of thymine (T) (Fig. 24.3).

DNA contains the codes for manufacturing the various proteins. All the proteins in our body are chain-like structures made up from just 20-odd *amino acids*. Depending on the sequence of the amino acids (this sequence defines the *primary* structure of the protein), the chain of molecules 'folds' in a specific way which defines the *secondary* structure of the protein (Fig. 24.4).

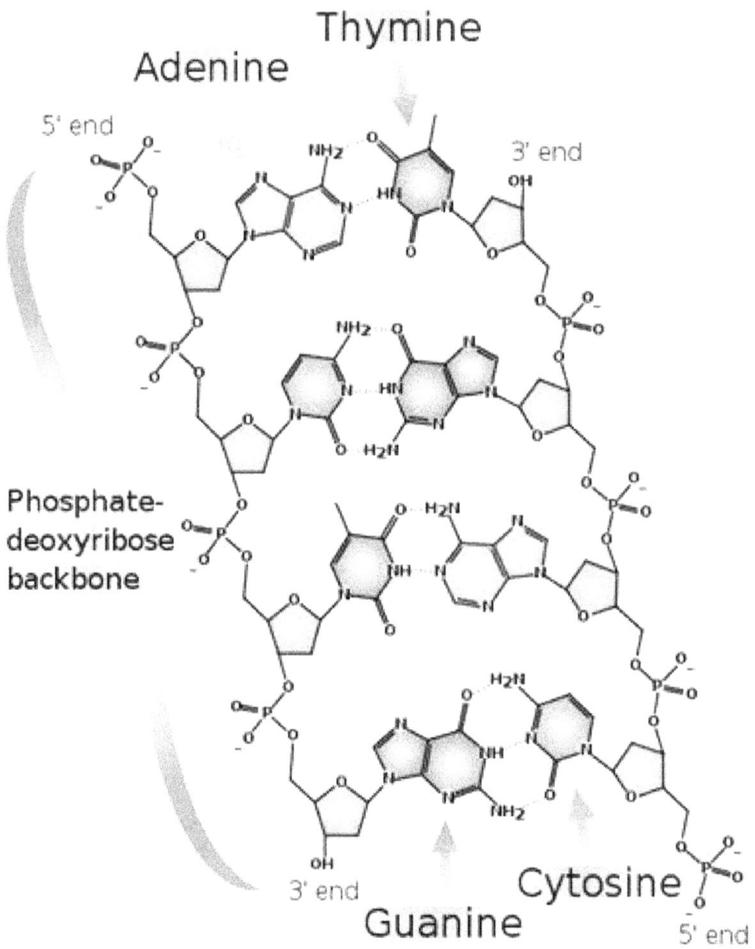

Thymine

Adenine

5' end

3' end

3' end

Cytosine

Guanine

5' end

Phosphate-
deoxyribose
backbone

Fig. 24.2 Some details of the DNA molecule.
http://en.wikipedia.org/wiki/DNA

Production of a protein in the cell nucleus involves *transcription* of a stretch of DNA (this stretch is called a *gene*) into a portable form, namely the *messenger RNA* (or mRNA). The mRNA is then *translated* into the corresponding protein: The mRNA molecule travels to the cytoplasm of the cell, where the information is conveyed to the ribosome. This is where the encoded instructions are used for the synthesis of the protein. The code is read, and the corresponding amino acid is brought into the ribosome. Each amino acid comes connected to a specific *transfer RNA* (tRNA) molecule; i.e., each tRNA carries a specific amino acid. There is a three-letter recognition site on the tRNA that is complementary to, and pairs with, the three-letter code sequence for that amino acid on the mRNA.

Such *one-way* flow of information from DNA to RNA to protein is the basis of all life on Earth. This is *the central dogma of molecular biology*: It says that information can flow only from DNA to RNA to protein, and cannot flow in the reverse direction, i.e., from protein to RNA to DNA, or from protein to DNA.

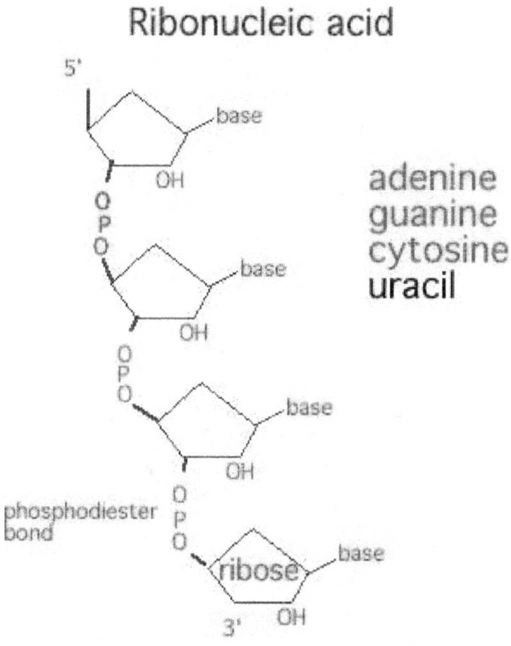

Ribonucleic acid

Fig. 24.3 The RNA molecule.
http://academic.brooklyn.cuny.edu/biology/bio4fv/page/molecular%20biology/rna.html

Three letters (out of the four, namely the bases A, T, C, G) are needed to code the synthesis of any particular protein. The term *codon* is used for the three consecutive letters on an mRNA. The possible number of codons is 64, and only 20 amino acids are processed by these codons. So there is redundancy: The linking of most of the amino-acid-triplets for synthesizing a protein can be coded by more than one codon.

[An apparent 'violation' of the central dogma occurs in certain cephalopods like octopuses, squid, and cuttlefish (Liscovitch-Brauer *et al.* 2017). They do *RNA editing* of over half of their transcribed genes for recoding the synthesis of certain proteins. For this they use enzymes to pluck out specific adenosine RNA bases, and replace them with a different base, namely inosine. This has seriously constrained the evolution of the cephalopod genome (https://phys.org/news/2017-04-smart-cephalopods-genome-evolution-prolific.html#jCp.]

There are ~60-100 trillion cells in the human body. In this multicellular organism (as also in any other multicellular organism), almost every cell (red blood 'cells' are an exception) has the same DNA, with exactly the same primary structure. The nucleus contains 95% of the DNA, and is the control centre of the cell. The DNA inside the nucleus is complexed with proteins to form a structure called *chromatin*.

The fertilized mother cell (the *zygote*) divides (*self-replicates*; see below) into two cells. Each of these again divides into two cells, and so on. Before this cell division (*mitosis*) begins, the chromatin condenses into elongated structures called *chromosomes*. A *gene* is a functional unit on a chromosome, which directs the synthesis of a particular protein. Humans have 23 pairs of chromosomes. Each pair has two non-identical chromosomes, derived one from each parent.

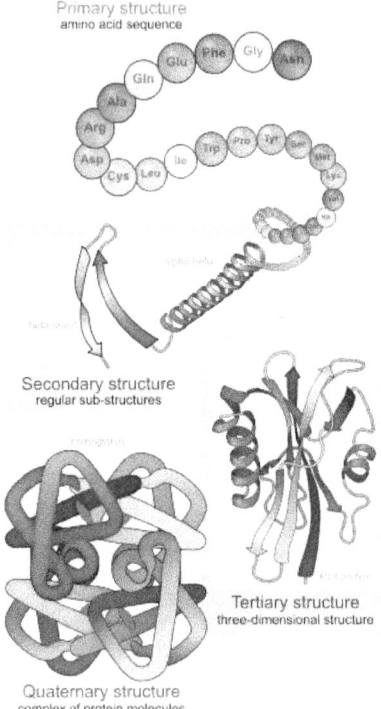

Fig. 24.4 The primary, secondary, and tertiary structure of a protein.
https://en.wikipedia.org/wiki/File:Main_protein_structure_levels_en.svg

During cell division, the double-stranded DNA splits into the two component strands, each of which acts as a *replication template* for the construction of the complementary strand (Fig. 16.3). 'Complementary strand' means that for every A on the original template these is a T on the new strand; similarly, there is a C for every G; A for T; and G for C. At every stage, the two daughter cells are of identical genetic composition (they have identical *genomes*), the same as that of the parent cell. In each of the 60 trillion cells in the human body, the genome consists of around three billion nucleotides.

25. Evolution of Chemical Complexity

As we have seen in Chapter 23, the lowering of free energy at the atomic scale occurs as follows: If two atoms are close to each other, they will bond together to form a molecule if the molecule has a lower free energy than that of the two separate atoms.

Atoms form molecules, molecules may self-organize into assemblies, and so on, all leading to an increase of chemical complexity. And in open systems the set of conditions can change with time. Naturally, every time this happens, a further round of self-organization must occur, again governed by the second law of thermodynamics. This is *chemical evolution*. As we shall see in due course in this book, chemical evolution has led to biological evolution. And evolution and emergence are the hallmarks of complexity.

We shall now discuss what can favour the spontaneous formation of molecular assemblies, i.e., coming together and binding of a number of molecules.

25.1 Of locks and keys in the world of molecular self-assembly

The important phenomenon of *molecular recognition* becomes operative when we consider the possibility of interactions among molecules to form *assemblies* (or '*supramolecular aggregates*'). Those types of molecules are more likely to form assemblies or aggregates which have a certain degree of mutual *complementarity*.

There are two types of complementarity to consider: That of *lock-and-key-like shapes*, and that of *complementary charge distributions* (remember, positive charge attracts negative charge). These complementarities, if present, enable portions of two molecules to fit snugly into each other, thus lowering the overall potential energy, and thence the free energy. This is naturally a more stable configuration because thermal fluctuations or other perturbations are less likely to knock the snugly-fitting (and therefore more tightly bonded) molecules apart. This is the essence of chemical self-assembly in Nature. Fig. 25.1 illustrates it with a schematic example, that of binding between an enzyme and a substrate.

Self-assembly of supramolecular aggregates is like crystal growth, except that the end product carries a lot more information; i.e., it is more complex.

The role of molecular complementarity was discovered by the Nobel Laureate Paul Ehrlich (see Avery 2003). As a student he was working on the newly discovered aniline dyes, which he was using for staining biological cells. He found that each dye stained only a particular type of tissue or a specific species of bacteria, and not others. What happens is that the dye molecule moves around in the solution till it finds a '*binding site*' nicely fitting (complementing) the shape of one of its side chains, like a key fitting its own lock. For better binding the complementarity of such a 'lock' and 'key' should be not only spatial, but also electrostatic; otherwise the '*specificity*' is not very strong: Not only should the two *shapes* be complementary, even the regions of positive excess charge on one molecule should be complementary to regions of negative excess charge on the other molecule.

Here are some examples of spatial and charge complementarity in Nature:

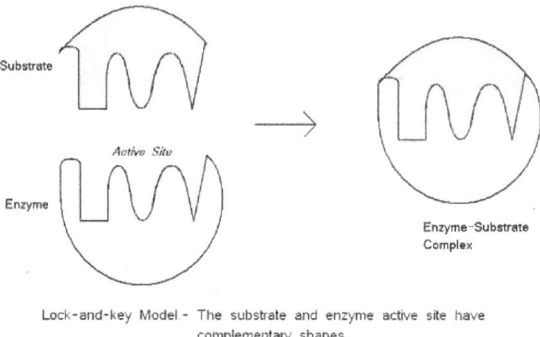

Lock-and-key Model - The substrate and enzyme active site have complementary shapes

Fig. 25.1 Illustration of the lock-and-key model for explaining why certain enzymes are more likely to bind to certain substrates.
http://en.wikibooks.org/wiki/Structural_Biochemistry/Enzyme/Active_Site

- The complementarity between the active site of an enzyme and the substrate for the enzyme.

- The well-known 'base-pair complementarity' between DNA strands.

- Self-assembly of viruses and subcellular organelles.

- Receptors located on the surface of cells binding only to a very limited number of substrates (often only one).

Let us remind ourselves that supramolecular aggregates, normally formed under near-ambient conditions, do not usually involve the strong covalent interaction. Instead, they are governed by weak, i.e., noncovalent or secondary, interactions (van der Waals; weak-Coulomb; hydrogen bond; hydrophobic; etc.). This means two things: (i) The aggregates are stable if the ambient conditions do not change too much. (ii) If the conditions change, the bonds in a supramolecular assembly at or near room temperature can get readily broken and re-formed, until the system has found its new stable configuration. *Near-reversibility of bonding is a very important feature of self-assembly through molecular recognition.*

Biological and other 'soft' materials can self-assemble into a variety of shapes, and over a whole range of length scales. There is usually some amount of water present, and the most important factors mediating molecular self-assembly are the hydrogen bond and the hydrophobic interaction. *We owe our lives to the hydrogen bond.*

Life and its evolution depend on the hydrogen bond. This bond is much weaker than the covalent bond, and yet strong enough to sustain self-assembled biological structures, enabling them to withstand reasonably well the disintegrating influences of thermal fluctuations and other perturbations. Hydrogen bonding, and the associated hydrophobic interaction, has the right kind of strength to enable superstructures to self-assemble without the need for irreversible chemical reactions. And yet, under appropriate conditions, there is a strong element of reversibility associated with these weak interactions, enabling the spontaneous making and breaking of assemblies until the lowest-free-energy configurations (mostly lock-and-key arrangements) have been attained.

Incidentally, self-assembly *per se* is a far more ubiquitous phenomenon than just molecular self-assembly. Some examples are: crystals; liquid crystals; bacterial colonies; beehives; ant colonies; schools of fish; weather patterns; galaxies.

Self-assembly may be either static or dynamic. The former occurs in systems which are in local or global equilibrium, and which do not dissipate energy (e.g., crystals). Dynamic self-assembly is more relevant from the point of view of evolution of complexity, and always involves dissipation of energy. Some examples are: oscillating and reaction-diffusion reactions; weather patterns; galaxies.

25.2 Self-organization of matter

> *The term* [self-organization] *and others like it (those with the prefix 'self-')* *are deceptive in that such ordering is actually occurring not by itself, as* *though by magic, but only with the introduction of energy.*
>
> Chaisson (2001)

> *In nature and in social systems, there are many processes that self-organize;* *that is, independent elements spontaneously begin cooperating and acting as* *one entity without an organizer. In the weather, we see these processes as* *hurricanes or tornadoes. In social systems, they have been variously* *described as bull markets, the 'invisible hand' of the free markets, or the* *madness of mobs.*
>
> Edgar Peters, *Patterns in the Dark*

In many ways we humans have arrogated to ourselves the notion of 'us *versus* Nature'. If *we* organize something, it is not self-organization. If order, structure, pattern, or organization arises without our intervention, we call it self-organization. In self-organization, globally coherent pattern-formation emerges from the local interactions of the constituents. There is 'parallel processing' and 'distributed action', there being no central coordinator.

Self-organization is rampant in Nature. It can occur because it is not forbidden by the second law of thermodynamics for open systems (Schrödinger 1944), and because our Sun, as also the expanding nature of our universe, ensures a persistent supply of free energy or negative entropy.

It takes information to describe an organized system or structure. It has been established by the work of stalwarts like Maxwell, Boltzmann, Gibbs, Szilard, and Shannon that free energy contains or engenders information. Our Earth, an open system, has been receiving a large input of information (as Gibbs free energy), mainly from the Sun. Most of this input into the biosphere of the earth dissipates as heat, or is re-emitted back into the cosmos. But a small fraction gets stored as *cybernetic information*. The term cybernetics was introduced by Weiner (1948) for 'the entire field of control and communication theory, whether in the machine or in the animal'.

The cybernetic information (also called *semiotic information*) is stored in our ecosphere in the form of simple or complex molecules. In the history of chemical evolution on Earth, these simple molecules contributed to the evolution of complex molecules characterizing and supporting life. When food is consumed by a living organism, its processing by the organism builds up a high information content for the organism, even though there is always a net increase of the global entropy. The nonbiological versions of this build-up of information or complexity are called *social progress* or *cultural evolution*.

To get a feel for some of the factors which influence self-organization, let us consider specifically the case of molecular self-organization or self-assembly. The undercurrent of self-organization in complex systems runs through the entire book.

Because of the hydrophobic interaction, polar and nonpolar liquids do not mix. By contrast, alcohol and water mix so readily that no stirring is needed; both are polar liquids. When polar and nonpolar liquids do not mix, what kind of patterns does the segregated system form? I have argued elsewhere (Wadhawan 2011) that in any self-aggregation process, a symmetrical conglomeration always corresponds to the least-free-energy situation. Fig. 25.2 shows some beautifully symmetric shapes adopted by phospholipids in water. Beautiful high-symmetry self-assemblies like *micelles, liposomes*, and *bilayer sheets* may ensue because of the hydrophobic interaction. Art without artist.

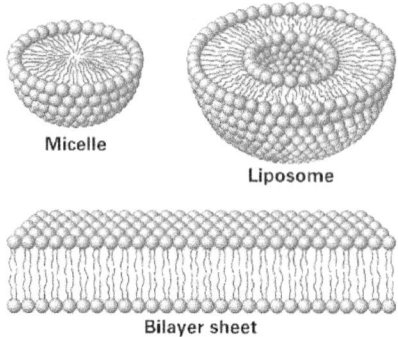

Fig. 25.2 Cross section of the different structures that phospholipids can take in an aqueous solution. The circles are the hydrophilic heads and the wavy lines are the fatty acyl side chains. Image credit: Mariana Ruiz Villarreal.
http://bioweb.wku.edu/courses/biol22000/2Bonds/Lecture.html

The second law of thermodynamics for open systems is the only self-organization principle there is (subject to the constraints of the first law of thermodynamics). Much of the symmetry we see in Nature is a consequence of this law (Wadhawan 2011).

Fig. 25.2 is also a good example of *chemical adaptation*. Given a set of environmental conditions, molecules in a system tend to self-organize so as to minimize the overall free energy. Even *internal* changes in the molecular system present a changed environment to every member of the set of molecules. And molecular configurations are changing all the time. Thus, chemical adaptation and evolution occur in an open system of molecules all the time.

Does 'natural selection' and 'survival of the fittest' also occur in chemical evolution? Indeed it does, because when the resources are limited, there is competition among the alternative molecular-reaction pathways. As a result, only the fittest pathways win, so far as consumption of precursor molecules and energy-rich molecules is concerned. Such considerations aroused special interest for explaining the origin of life-sustaining molecules. Some pioneering work in this direction was done by Melvin Calvin (1969), who introduced the idea of *autocatalysis* as a mechanism for molecular selection (see below).

A very important factor in the chemical evolution of complexity is the *reversibility* of the non-covalent bonding between molecules. It is the key to self-assembly. Through a process of molecular trial and error a snugly fitting arrangement of molecules gets found, and once that happens, there is no going back. The tightly bound self-assembly is unlikely to get broken again, even though only non-covalent interactions are involved. But there can still be a *going forward*. A change of environment induces the assembly to *adapt* again, by exploiting the reversible nature of the non-covalent interactions.

Lipid bilayers and micelles (Fig. 25.2) are striking examples of chemical self-organization. The amount of information contained in such assemblies is very high. This information is distributed among the shapes of the component molecules, *and* in the interaction patterns among them. The build-up of this information involves a succession of stages: molecular recognition; self-assembly; self-organization; and chemical adaptation and evolution.

Jean-Marie Lehn, a pioneer in the fields of supramolecular chemistry and evolution of chemical complexity, defined self-organization as the '*spontaneous but information-directed generation of organized functional structures in equilibrium conditions*'. The necessary information ('coding') for self-organization is contained in the molecular-recognition and self-assembly proclivities of the component molecule. This coding also determines how the self-assembled edifice further self-organizes into a functional structure in equilibrium.

Vesicles provide a good, even dramatic, example of self-organization in nonliving complex systems. Vesicles are spherical supramolecular assemblies separating an aqueous interior volume from an external solvent by means of one or more lipid bilayers (Fig. 25.3). Vesicles can form naturally, or they can be prepared artificially. The latter are known as *liposomes*. Given the right conditions, lipids can self-assemble into giant vesicles the size of biological cells. The basic driving force for their self-assembly is the hydrophobic interaction. As vividly described by Menger and Gabrielson (1995), vesicles can mimic the living cell in many ways, even though they are not living entities:

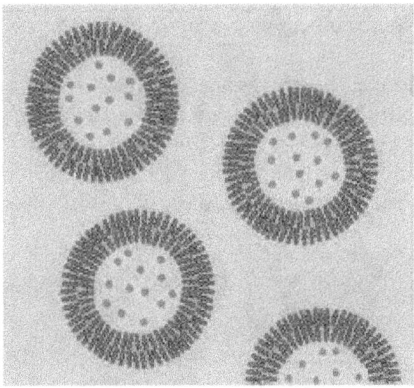

Fig. 25.3 Vesicles.
http://en.wikipedia.org/wiki/File:Lipid_vesicles.svg

When a giant vesicle, which happens to have a smaller vesicle inside it, is exposed to octyl glucoside, the smaller vesicle can pass through the outer membrane into the external medium ('birthing'). The resulting injury to the membrane of the host vesicle heals immediately. Addition of cholic acid, on the other hand, induces a feeding frenzy in which a vesicle grows rapidly as it consumes its smaller neighbours. After the food is gone, the giant vesicle then self-destructs (a case of 'birth, growth, and death'). Such lifelike morphological changes were obtained by using commercially available chemicals; thus these processes should be assigned to organic chemistry, and not to biology or even biochemistry.

As mentioned earlier, self-organization is a far more ubiquitous phenomenon than something at just the molecular level. Here are some more examples:

• A laser is a self-organized system. Under properly engineered conditions, photons spontaneously group themselves into a configuration in which they all move in phase, resulting in a powerful laser beam.

• A hurricane is a self-organized system. The steady influx of energy from the Sun draws water from the oceans, as well as drives the winds. And mild tropical winds may grow into an organized configuration of a hurricane when some critical threshold is crossed.

• A living cell is a self-organized system, which organizes itself all the time, depending on the environment.

• An economy is a self-organizing system. The demand for goods and services, as also the demand for labour, constantly reorganizes the economy in a spontaneous way, without any central controlling authority.

25.3. Emergence of autocatalytic sets of molecules

Life depends on molecules of RNA, DNA, proteins, polysaccharides, etc. How did such large molecules ('polymers') get synthesized spontaneously from their building blocks, namely nucleic acids, amino acids, sugars, etc.?

Adaptation and evolution occurred repeatedly in thermodynamically open chemical systems, and in due course, chemical evolution led to the emergence of life-sustaining *macromolecules*. Pioneering work in this direction was done by Calvin (1969), who introduced the idea of *autocatalysis* as a mechanism for emergence of large molecules.

A polymer is a long molecule, the backbone of which usually comprises of a chain of covalently-bonded repeat units (*monomers*) (Fig. 25.4). There can be variations, either in that the bonding is not covalent everywhere, or in that not all subunits are identical. Examples of polymers and polymer solutions include plastics such as polystyrene and polyethylene, glues, fibres, resins, proteins, and polysaccharides like starch.

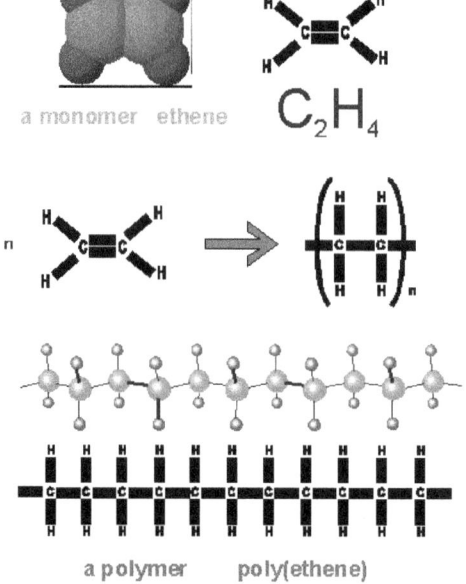

Fig. 25.4 Formation of the polymer polythene.
http://www.petervaldivia.com/technology/plastics/index.php

Short polymers can form spontaneously in Nature, aided by the lock-and-key mechanism discussed above. Suppose a monomer has a shape and charge distribution such that another monomer can fit snugly into some part of it. There are random collisions among the monomers in a fluid medium, and usually they do not stick together, and simply bounce off after a collision. But once in a while the collision may be such that the two monomers have just the right orientation for a lock-and-key fitting. Then the chances of the two sticking together and forming a *stable* dimer are much larger.

Dimers can lead to trimers, trimers to tetramers, and so on, resulting in polymers. But this can at best be a rather low-probability, and therefore very slow process, and only short polymers can possibly form spontaneously in reasonable time (Cairns-Smith 1985; Zimmer 2004).

Then how did very long polymers like RNA, DNA get formed? The answer has to do with the emergence of *autocatalytic sets of molecules*. Catalysis is a process that facilitates or speeds up a chemical reaction. Often a chemical process involves two or more intermediate reactions. A catalyst is a molecule that speeds up the production of an end product of the chemical process by participating in the intermediate reactions, but separating itself at the end of the chain of reactions, thus becoming available all over again for further catalysis. Often, a chemical reaction may almost never occur if no catalyst is present. Enzymes are examples of proteins that assist (catalyse) chemical reactions in biological systems.

Photosynthesis carried out by green plants in the presence of sunlight is another familiar example involving catalysis. Chlorophyll is the catalyst here. Through a number of intermediate reactions, the net reaction is:

$$CO_2 + H_2O \rightarrow C_6H_{12}O_6 + O_2$$

That is, carbon dioxide combines with water to produce glucose and oxygen. This is the oxygen we breathe.

Autocatalytic sets of molecules are those which can catalyse the synthesis of themselves. Autocatalysis requires that a given 'factor' (say A) should be able to convert a substrate or precursor B into a new factor of the same type: $A + B \rightarrow 2A + C$. The idea of autocatalysis was introduced by Calvin (1969) as a mechanism for 'molecular selection', with implications for how life emerged on Earth. Stuart Kauffman (1991, 1993, 1995, 2000) carried Calvin's idea much further for explaining how long polymers like RNA could form spontaneously during the course of chemical evolution on Earth (also see Farmer, Kaufmann and Packard 1986a). In the so-called *primordial soup*, namely a fluid in contact with rocks of various types, there existed small molecules of amino acids, nucleic acids, etc. Given enough time, some of them must have undergone random polymerization reactions of various types; i.e., *short* polymers got formed. It is entirely possible that at least some of these short polymers, with side chains and branches hanging around, acted as *catalysts* for facilitating the production of other molecules which may also be catalysts for another chemical reaction. Thus: A facilitates the production of B, and B does the same job for C, and so on. Given enough time, and a large enough pool containing all sorts of molecules, it is quite probable that, at some stage a molecule, say Z, will get formed (aided by catalytic reactions of various types), which is a catalyst for the formation of the catalyst molecule A the system started with.

Once such a loop closes on itself (Fig. 25.5), it would head towards what we now call '*self-organized criticality*' (SOC). There will be more production of A, which will lead to more production of B, and so on. The plausibility advantage of this scenario visualized by Kauffman is that there is no need to wait for random reactions. And once a threshold has been crossed, the system may inch towards the *edge of chaos*, and acquire robustness against destabilization.

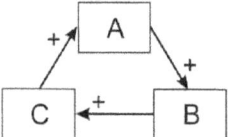

Fig. 25.5 A closed loop of autocatalytic reactions.
http://www.library.utoronto.ca/see/SEED/Vol2-1/Ulanowicz/Ulanowitz.htm

Kauffman argued that this order, emerging out of molecular chaos, was *akin to life*: The system could consume (metabolize) raw materials, and grow into more and more complex molecules. It progressed into a situation where the forebears of DNA started appearing. Once a set of autocatalytic reactions had established itself, it went on incrementally evolving into still more complex sets of molecules. Chance events and/or new external conditions resulted in the emergence of a slightly more complex version of, say, one of the molecules in the autocatalytic set. A further round of chemical Darwinism and evolution of a new set of autocatalytic set of molecules followed. And so on, till molecules as complex as RNA, DNA and proteins emerged on the scene, which have life-sustaining and life-propagating properties.

So, one version of what might have happened is something like this (Avery 2003): There was little or no molecular oxygen in the original atmosphere around the Earth. A variety of local energy sources were, of course, present (undersea hydrothermal vents; ultraviolet radiation; volcanic energy; radioactive nuclei; lightning; meteoric impacts). Under these conditions, amino acids, nucleotides, and other building blocks of the future living organisms got synthesized on the rocks submerged in the seas, and in the atmosphere around the Earth. Several energy-rich molecules, mentioned at the beginning of this chapter, were also produced. An era of *chemical Darwinism* or *molecular Darwinism* followed next, in which autocatalytic systems of molecules competed with one another for the limited supply of precursors and energy-rich molecules. These sets of molecules had at least three of the features of what constitutes life: They 'ate' the energy-rich molecules; they reproduced themselves; and they competed with other autocatalytic sets of molecules for survival.

25.4 Positive feedback, pattern formation, emergent phenomena

> *Nature seeks to accomplish the most "building" with the least energy, and with the most economic use of materials. Nature is always looking for the tightest fit, the shortest path, the least energy expended. Nature modifies and adapts its basic patterns as needed. Constantly shaping materials to the demands of an ever changing environment. The study of pattern formation is the quest to find the hidden unity that lays under all the modifications and adaptations that are constantly churning out of the natural world.*
>
> Ian Stewart

The above discussion about the emergence of autocatalysis (an important step towards the origins of life) reinforces the central theme of this book, namely that open systems tend to move towards states of increasing order and complexity (read information content) under the influx of free energy or negative entropy. The emergence of life is, of course, the most dramatic manifestation of emergent phenomena in open systems.

The question of the emergence of order and complexity in open systems was investigated in the 1970s by the Nobel Laureate Prigogine (see Nicolis and Prigogine 1989; Prigogine 1998). An important feature of his work on open systems was that self-reinforcement or *positive feedback* is also needed for self-organization or pattern formation to occur. Only then can small chance events become magnified, sometimes taking the system to new valleys in phase space (*basins of attraction*). This 'law of increasing returns' (Arthur 1990; Kelly 1994), or positive feedback, illustrated above by autocatalytic sets of reactions, can often lead to a multiplicity of qualitatively new outcomes. The actual outcome may depend on chance initial conditions, a feature characteristic of highly nonlinear systems. For such systems the sum total can be very different from the linear superposition of the parts.

Emergent behaviour in large nonlinear open systems is very common in Nature (Davies 2005). It has evolutionary significance. Chemical evolution, leading to the emergence of life, is a remarkable example of that.

25.5 Pattern formation: the BZ reaction

I discussed *dissipative structures* and *bifurcation points* in Chapter 4 in the context of systems driven far from equilibrium. An instability can arise at a certain sufficient departure from equilibrium, resulting in a new set of phenomena, like oscillating chemical reactions, non-equilibrium spatial structures, chemical waves, etc. Such spatiotemporal self-organizations in dissipative structures are not limited to just chemical reactions, but occur in all sorts of situations in Nature, including those in living systems. Spontaneous pattern formation is a consequence of this (Fig. 25.6).

Fig. 25.6 Examples of pattern formation in Nature. [Photographs by the author.]

Alan Turing is regarded as a pioneer in the field of pattern formation in Nature. From 1952 to the end of his life in 1954, he worked on mathematical biology, particularly on the chemical basis of the phenomenon of *morphogenesis* (Turing 1952). His reaction-diffusion equations are central to the field of pattern formation. Turing's theory predicted the now-famous oscillatory reactions called *Belousov-Zhabotinsky (BZ) reactions*. These were first observed in the 1960s.

BZ reactions are a vivid example of nonequilibrium-thermodynamic systems and nonlinear chemical oscillators (Cross and Hohenberg 1993). Here is what Turing (1952) wrote in his seminal paper: 'It is suggested that a system of chemical substances, called morphogens, reacting together and diffusing through a tissue, is adequate to account for the main phenomena of morphogenesis. Such a system, although it may originally be quite homogeneous, may later develop a pattern or structure due to an instability of the homogeneous equilibrium, which is triggered off by random disturbances. Such reaction-diffusion systems are considered in some detail in the case of an isolated ring of cells, a mathematically convenient, though biologically unusual system. The investigation is chiefly concerned with the onset of instability. It is found that there are six essentially different forms which this may take. In the most interesting form stationary waves appear on the ring. It is suggested that this might account, for instance, for the tentacle patterns on Hydra and for whorled leaves. A system of reactions and diffusion on a sphere is also considered. Such a system appears to account for gastrulation. Another reaction system in two dimensions gives rise to patterns reminiscent of dappling. It is also suggested that stationary waves in two dimensions could account for the phenomena of phyllotaxis. The purpose of this paper is to discuss a possible mechanism by which the genes of a zygote may determine the anatomical structure of the resulting organism.'

The BZ reaction illustrates some of these ideas in the chemistry realm. Typically, it involves the following:

- A one-electron redox catalyst;
- an organic substrate that can be readily brominated and oxidized;
- bromated ions in the form of $NaBrO_3$ or $KBrO_3$;
- catalysts such as Ce^{3+}/Ce^{4+} salts or ferroin; and
- an organic substrate (normally malonic acid, instead of citric acid used originally by Belousov).

Sulphuric acid or nitric acid is used as the solvent. The system shows oscillations between an oxidized state and a reduced state, with a concomitant change of colour (Fig. 25.7).

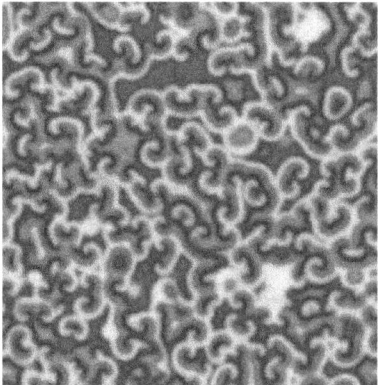

Fig. 25.7 Computer simulation of the BZ reaction occurring in a Petri dish.
http://en.wikipedia.org/wiki/File:The_Belousov-Zhabotinsky_Reaction.gif.

This is an example of an auto-catalytic reaction-diffusion process. The diffusion is of a nonlinear nature, and the process involves both excitatory and inhibitory molecules (i.e., both positive and negative feedback). Suppose we use the following notation: $X = HBrO_2$; $Y = Br^-$; $Z = Ce^{4+}$; $P = HOBr$; $A = BrO^{3-}$; $B = BrMS$. Then a simplified form of the reactions involved can be depicted as follows:

$A + X \rightarrow X + P$
$A + Y = 2P$
$A + X \rightarrow 2X + Z$
$2X \rightarrow A + P$
$B + Z \rightarrow fY$

The third equation is an example of autocatalysis, and the concomitant nonlinearity. There are bifurcations in phase space, and the dynamics enters the chaotic regime for a certain range of values for the control parameters.

26. What is Life?

Perhaps there is no clear dividing line between life and nonlife. Nevertheless, the emergence of what many of us intuitively understand to be life marked a major milestone in the evolution of complexity in our world. Here are some definitions of life:

But what is life? Like time, life is obvious to discern yet elusive to define. Although most biologists generally skirt the issue, we suggest that our very essence can be defined as follows: Life is an open, coherent, spacetime structure maintained far from thermodynamic equilibrium by a flow of energy through it – a carbon-based system operating in a water-based medium, with higher forms metabolizing oxygen (Eric Chaisson 2001).

Life does not exist in a vacuum but dwells in the very real difference between 5800 Kelvin incoming solar radiation and 2.7 Kelvin temperature of outer space. It is the gradient upon which life's complexity feeds (Margulis and Sagan 2002).

The existence of life can be traced back to the emergence of the RNA/DNA molecules (if not earlier).

Life depends on the availability of free energy. Without an input of free energy or 'negative entropy' (see below), all processes would tend to take a system towards a state of equilibrium and, entropic death. Intake of food keeps an organism alive by providing negative entropy. The complex molecules constituting food are full of free energy or negative entropy, which is derived ultimately from the Sun.

26.1 Schrödinger and life

In 1943-1944 Schrödinger wrote a little book *What is Life*. This is how Roger Penrose described this book (in 1991): ... *which, as I now realize, must surely rank among the most influential of scientific writings in this century. It represents a powerful attempt to comprehend some of the genuine mysteries of life, made by a physicist whose own deep insights had done so much to change the way in which we understand what the world is made of. ... Indeed, many scientists who have made fundamental contributions in biology, such as J. B. S. Haldane and Francis Crick, have admitted to being strongly influenced by (although not always in complete agreement with) the broad-ranging ideas put forward here by this highly original and profoundly thoughtful physicist.*

It is important to realize that when Schrödinger wrote the book, the atomic structure of DNA was not known (it was determined later by Watson and Crick). I quote Freeman Dyson (1985):

Schrödinger's book was seminal because he knew how to ask the right questions. The basic questions which Schrödinger asked were the following: What is the physical structure of the molecules which are duplicated when chromosomes divide? How is the process of duplication to be understood? How do these molecules retain their individuality from generation to generation? How do they succeed in controlling the metabolism of cells? How do they create the organization that is visible in the structure and function of higher organisms? He did not answer these questions, but by asking them he set biology moving along the path which led to the epoch-making discoveries of the subsequent forty years: to the discovery of the double helix and the triplet code, to the precise analysis and wholesale synthesis of genes, and to the quantitative measurement of the evolutionary divergence of species.

How did Schrödinger define life? He avoided giving a direct *definition*, but highlighted an important property of life by invoking the idea of *negative entropy*. He characterized living matter as that which stays alive ('evades the decay to equilibrium') by feeding on negative entropy or *negentropy*. How can entropy be negative? How does living matter feed on negentropy, and what is the source of this negentropy?

The answer to the latter question is that the Sun has been bombarding our ecosphere with low-entropy or 'high-grade' energy (Fig. 26.1). Why 'low-entropy'? Recall that differential entropy is defined as $dS = dQ/T$, and the average value of temperature T for the Sun is huge compared to that of the Earth. On entering our ecosphere, the energy of the photons coming from the Sun gets degraded to a large extent through the processes of 'thermalization,' namely dissipation into a state of much lower average temperature (and therefore a correspondingly high value for the entropy).

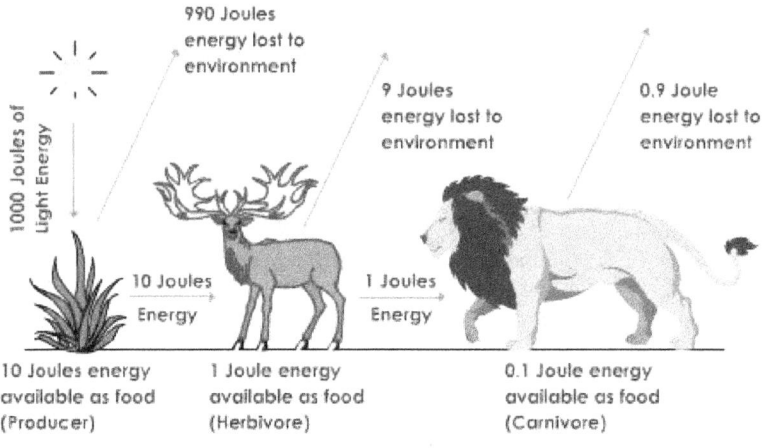

Fig. 26.1 Progressive loss of solar energy in the food chain.
http://www.tutorvista.com/content/biology/biology-iv/ecosystem/ten-percent-law.php

A small fraction of this energy, however, gets trapped as free energy. It is stored in our ecosphere in the form of simple or complex molecules. Some of the energy-rich simple molecules in which the free energy from the Sun gets stored are: H_2S, FeS, H_2, phosphate esters, HCN, pyrophosphates, and thioesters.

When food is consumed by a living organism, its processing by the organism builds up a high *information content* (negative entropy) for the organism. Boltzmann-Gibbs entropy S is the same as missing information I. This means that $-S$ (negentropy) is the same as available information.

Until 1944 most scientists were of the view that genetic information was carried by the proteins of the chromosome. Schrödinger's book, apart from invoking negative entropy for the sustenance of life, introduced new concepts for the genetic code. It inspired Watson and Crick to investigate the gene, which led to their discovery of the double-helix structure of DNA. In their 1953 paper they wrote: '*It has not escaped our notice that the specific pairing [of the two strands of DNA] we have postulated suggests a possible copying mechanism for genetic material.*' This sudden blaze of understanding laid bare the inside story of heredity, and of life itself.

What is even more important, Schrödinger argued that life is not a mysterious or inexplicable phenomenon, as some people believe, but a scientifically comprehensible process like any other, ultimately explainable by the laws of physics and chemistry.

26.2 Koshland's 'seven pillars of life'

Koshland (2002) once attended a conference focused on the vexing question of defining life. He recounts that, after considerable discussion, somebody formulated the essential characteristic of life as 'the ability to reproduce.' There seemed to be general consensus on this, till somebody said: 'Then one rabbit is dead. Two rabbits - a male and female - are alive but either one alone is dead.' Similarly, a mule must be a dead entity by this definition. Nevertheless, many of us still have, at least intuitively, an idea of what is living and what is nonliving. Koshland has come up, reluctantly, with the following short definition of life: *A living organism is an organized unit, which can carry out metabolic reactions, defend itself against injury, respond to stimuli, and has the capacity to be at least a partner in reproduction.* But he is happier giving a large set of criteria for deciding what constitutes life. He calls them the 'seven pillars of life', like the pillars of a Greek temple. Koshland's (2002) seven 'pillars' are the essential thermodynamic and kinetic principles which enable a living system to operate and propagate. The mechanisms listed by him are with reference to life as we know it on our Earth. The same seven principles may involve other mechanisms for other forms of life, or for life elsewhere. The acronym PICERAS introduced by Koshland has seven letters, corresponding to the seven principles or pillars of life: program; improvisation; compartmentalization; energy; regeneration; adaptability; seclusion.

The first pillar of life is a *program* that describes the ingredients themselves, as well as the kinetics of the interactions among the ingredients. For life on Earth, this program resides in the DNA molecules that encode the genes.

Improvisation is the second pillar of life, and is to be distinguished from another pillar, namely adaptability, discussed below. Both involve a response to change. The difference is in time scales. Adaptability is about direct response to quick changes, and does not entail a change of the genetic program of the organism. Improvisation is about gradual change (evolution) in response to long-term changes in the environment.

A living organism depends on the reaction kinetics of its ingredients. The kinetics requires certain concentrations of the chemicals involved, and that requires confinement in a 'container.' Therefore the third pillar of life is *compartmentalization*, namely the presence of a membrane or skin that confines the organism to a certain volume. In fact, large organisms have several compartments (organs), because the concentration requirements of different organs are different.

Energy is the fourth pillar of life on Earth. Without a steady input of energy, a living system would soon approach a state of equilibrium and death.

Regeneration (including reproduction) is the fifth pillar of life. There are always some thermodynamic losses and wear and tear as a living organism is sustained by the myriad chemical reactions going on inside it. Food and other intakes are one means of ensuring that the organism does not degenerate or degrade inexorably with time. Several organs of the human body (e.g., the heart) have a mechanism for tissue regeneration. However, over time, aging effects become too strong and the organism dies. To ensure that the species can still survive, Nature has evolved replication (cell division) and reproduction (which amounts to regeneration by starting all over again through the progeny) as an essential pillar of life.

A living organism may face a variety of sudden hazards. *Adaptability* is therefore the sixth pillar of life. For example, a living system must move away from an environment that is too hot for its well-being and survival.

The seventh pillar of life listed by Koshland is what he calls *seclusion*. In a living cell there exist a large number of chemical reaction pathways, all simultaneously active. Natural evolutionary processes have ensured that the enzymes catalysing the various reactions have specific shapes and reactivities, so that the different reactions do not interfere with one another. The specificity of an enzyme provides a high degree of seclusion to the relevant chemical reaction occurring in the living cell. Various feedback and feedforward mechanisms also ensure that the specificity is not completely unchangeable, but changes only under certain special signals.

The existence of many of Koshland's pillars of life can be ultimately traced back to the DNA molecule, portions of which constitute the genes. DNA is a large molecule with very high information content. Life is information. Therefore there is a direct link between life and free energy. Without an input of free energy or negative entropy, all processes would tend to take a system towards a state of entropic death. Intake of food keeps an organism alive by providing negative entropy. As Szent-Györgyi (1957) said, 'We need energy to fight against entropy.' The complex molecules constituting food are full of free energy or negative entropy, which is derived ultimately from the Sun.

27. Models for the Origins of Life

Life is an eminently active enterprise aimed at acquiring both a fund of energy and a stock of knowledge, the possession of one being instrumental to the acquisition of the other. The immense effectiveness of these two feedback cycles, coupled in multiplying interaction, is the pre-condition, indeed the explanation, for the fact that life had the power to assert itself against the superior strength of the pitiless inorganic world.

Chaisson, *Cosmic Evolution*

How did life emerge out of nonlife? It did so through a progressive and cumulative evolution of chemical complexity. In fact it is now believed that there was something *inevitable* about the emergence of life on Earth (Kauffman 1993, 2000; Margulis and Sagan 2002; England 2013), considering that the Earth has a conducive ecosphere and is an open thermodynamic system which is under a sustained bombardment of low-entropy energy from the Sun. Nevertheless, it is instructive to review some of the important theories put forward over the years for the detailed processes which made the emergence of life possible.

27.1 The early work

Life as we know it depends on the availability of certain long molecules (polymers): proteins, RNA, DNA, polysaccharides, lipids, etc. And the chemistry observed is all organic, i.e., carbon-based. We have to explain three stages in the emergence of life on Earth:

Stage I. Formation of organic molecules ('monomers') from inorganic matter.

Stage II. Covalent bonding (polymerization) of the monomers into long biomolecules.

Stage III. Emergence of the complete biological cell, ~3.5 billion years ago, capable of survival and propagation.

The first major theory for explaining Stage I was put forward by Alexander Oparin in 1924. He argued that certain organic molecules that were necessary building blocks for the emergence of life required an oxygen-free, 'reducing' ambience for their self-synthesis. He imagined a 'primeval soup' of organic molecules that got created in the oxygen-free atmosphere prevailing at that time, aided by the action of sunlight. And these molecules combined in increasingly complex ways, till they formed the so-called 'coacervate' droplets. Such droplets could 'grow' by fusion with other droplets. And they could 'reproduce' through fission into smaller droplets. This primitive metabolism underwent Darwinian-like chemical evolution in which factors favouring cell integrity survived and evolved. No DNA-type replication was involved in this earliest of 'life forms'.

Similarly, J. B. S. Haldane argued that the prebiotic oceans of the Earth were very different from what they are now, and that a 'hot dilute soup' existed in them in which organic molecules could get formed.

A famous experiment for verifying this *Oparin-Haldane hypothesis* was carried out by Miller and Urey in 1952. They created electric sparks in a mixture of water, hydrogen, methane, and ammonia. After one week a full 10-15% of the carbon in the mixture was found to have converted to organic compounds, including amino acids (the building blocks of proteins).

Recent analysis of their saved vials has revealed that as many as 23 amino acids were formed, whereas they had detected only five at that time.

Sidney Fox carried out several experiments that confirmed the formation of long organic molecules from inorganic matter. He allowed amino acids to dry out as if from a warm puddle, mimicking prebiotic conditions. The amino acids were found to form long, even cross-linked, thread-like molecules, now called 'proteinoids'. He also collected volcanic material from a cinder cone in Hawaii, and found that the temperature was over 100°C just four inches beneath the surface. He speculated that this was probably the environment in which life emerged. Biomolecules formed in such conditions could have got washed away to the seas. In an experiment, he placed lumps of lava over amino acids created from a mixture of methane, ammonia and water, sterilized the whole thing, and baked it for a few hours in a glass oven. What he got on the surface was a brown sticky substance. On drenching the lava in a sterilized water tank, a brown liquid leached out. The amino acids had formed proteinoids which, in turn, had combined to form small, cell-like spheres. These microspheres are now known as 'protobionts'. These were not biological cells; they did not contain any functional nucleic acids. But they did form lumps and chains, rather like what cyanobacteria do.

Manfred Eigen and Leslie Orgel demonstrated, independently, that a solution of nucleotide monomers can, under suitable conditions in the laboratory, give rise to a nucleic-acid polymer molecule (RNA) which replicates and mutates and competes with its progeny for survival.

For achieving this, Eigen used a polymerase enzyme, a protein catalyst extracted from a 'bacteriophage' (the synthesis and replication of the RNA depends on the structural guidance provided by this enzyme).

Orgel did something complementary to the experiment of Eigen. He made RNA grow out of nucleotide monomers by adding a *template* for the monomers to copy, but did not add a polymerase enzyme.

Thus Eigen made RNA using an enzyme but no template, and Orgel made RNA using a template but no enzyme. Modern-day living cells use both templates and enzymes for making RNA. *This work pointed to a possible parasitic development of RNA-based life in an environment created by a pre-existing protein-based life.*

There is a strong possibility that life appeared on Earth during the so-called '*thermophilic energy regime*' (I shall explain this terminology in a later chapter). This form of life comprised of microorganisms that thrived in hot conditions. Emergence of any form of life means a build-up of order, complexity, and information. For this to have become possible, there had to be conditions far from equilibrium, i.e., an energy gradient. One view is that, most probably, sunlight did not play a major role in this, and that the energy sources were of geothermal origin. Volcanic heat sources under the sea (hydrothermal vents) provided the upper end of the energy gradient, the lower end being the cold atmosphere above the seas. Under the dominant influence of the driving force provided by this energy gradient, chemical evolution and diversification of molecular structure occurred. Coming to more recent research, Kauffman (1993) proposed that such chemical evolution led to the emergence of autocatalytic reactions. His model obviates the need for the prior presence of information-rich DNA molecules for the synthesis of protein molecules, and also provides a non-random mechanism for the origin of life. Closed-loop autocatalytic reactions led to the production of life-like molecules of increasing complexity. Things progressed to a point where the forebears of DNA started appearing, which had the potential for replication. The biological prokaryotic cell emerged in due course.

There is also a viewpoint that, because of the presumed unfeasibility of the time scales involved for a terrestrial origin of life, life might have originated elsewhere in the cosmos, and brought to our Earth by meteors etc.

There is still some debate on what really happened that led to the emergence of life on Earth. Choi (2016) has listed seven theories for the origin of life. I shall consider two important models here.

27.2 The RNA-world model for the origin of life

Modern living organisms depend on both DNA and proteins. But what came first, DNA or proteins? DNA has the codes for the synthesis of proteins, but it itself needs proteins (enzymes) for its creation. A way out of this chicken-or-egg dilemma has been found by the RNA-world model for the origin of life. According to it, since RNA (a close cousin of DNA in terms of structure etc.) has been observed to act both as an information-storage (replicating) molecule and as an enzyme, life probably started as 'nude replicating RNA molecules'.

This model for the origin of life became popular after of the discovery, made in the mid-1980s by Thomas Cech and coworkers, that certain RNA sequences called *ribozymes* can act as enzymes, i.e., catalyse reactions. This dual functionality of RNA might have allowed for the existence of an '*RNA species*' that could replicate itself and thus seed the beginning of biomolecular evolution.

RNA is indeed known to be involved in a number of fundamental cell-biology processes. All biological cells contain ribosomes, which comprise of as much as ~60% ribosomal RNA (rRNA); the rest ~40% is protein. In molecular biology and genetics, the term 'translation' is used for a process in which the ribosomes inside a cell create proteins. Fig. 27.1 depicts this process schematically.

The ribosome machinery is almost identical throughout the living world; perhaps it existed almost from the beginning of life on Earth. This is an important clue because all life is believed to have had a common ancestor, a la Darwin (Fig. 27.2).

According to one version of the RNA-world hypothesis, a different type of nucleic acid, namely pre-RNA, was the first to emerge as a self-reproducing molecule, and was replaced by RNA only later. But it is also true that activated pyrimidine ribonucleotides (two of the four nucleotides constituting RNA are pyrimidines) have been synthesized in the laboratory under reasonably prebiotic conditions.

However, several scientists have expressed reservations about the RNA-world model. I quote the objections of Stuart Kauffman (1995, 2000):

- It is difficult to get RNA strands to reproduce in a test tube. 'No one has succeeded in achieving experimental conditions in which a single-stranded DNA or RNA could line up free nucleotides, one by one, as complements to a single strand, catalyse the ligation of the free nucleotides into a second strand, melt the two strands apart, then enter another replication cycle. It just has not worked'.
- Even if life did tend to originate and evolve by the RNA route, naked RNA molecules must have suffered an 'error catastrophe' during the replication processes, thus corrupting the genetic message from generation to generation. In present-day cells, such errors (mutations) are kept to a minimum by certain 'proofreading' and 'editing' enzymes. (Such considerations form part of a subject called 'systems biology' (see, e.g., Alon 2007).)

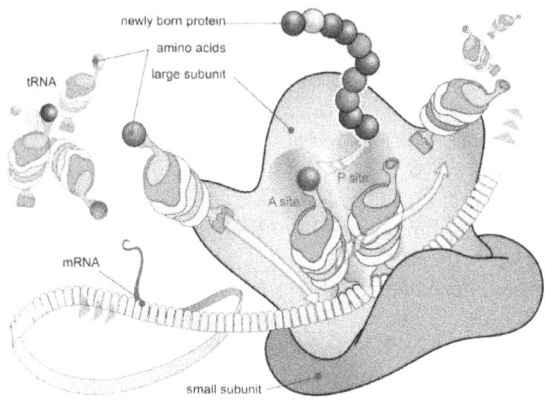

Fig.27.1 Ribosome RNA translation.
https://en.wikipedia.org/wiki/Translation_(biology)

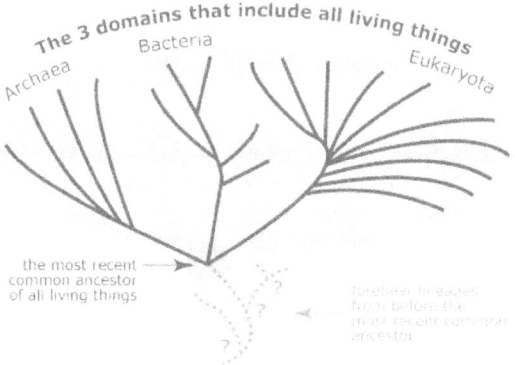

Fig. 27.2 Common ancestry of all living beings.
http://evolution.berkeley.edu/evosite/evo101/IIE2bStudyorigins.shtml

- *RNA-based life, even if it did emerge, was not complex enough to sustain itself.* In other words, it was too far from *the edge of chaos* where complexity thrives best. Why viruses do not have independent life? Why is it that the simplest free-living cells are the so-called pleuromona, and nothing less complex than them? Pleuromona are the simplest known bacteria, and they are complete with cell membrane, genes, RNA, protein-synthesizing machinery, proteins. *All free-living cells have at least the minimal molecular diversity of pleuromona.* Why nothing simpler exists that is alive on its own? The nude RNA or the nude ribozyme polymerase idea for the origin of life offers no decent explanation for the observed minimum necessary complexity of any life form.

I shall discuss an alternative model of the origin of life in the next section. It is based on the 'metabolism-first' idea (as opposed to the replication-first idea inherent in the RNA-world model). There are many variations of the metabolism-first idea; i.e., metabolism without genes. For example there is the lipids-first model (Segre *et al.* 2001): Phospholipids are known to form lipid bilayers and vesicles in water, the same structure as in cell membranes. The main idea of *the lipids-first model* invoking vesicles is that initially information was stored in the molecular composition of lipid bodies, and the evolution of more efficient information-storage molecules like RNA and DNA occurred in a later epoch.

27.3 Dyson's proteins-first model for the origins of life

> *Although it is difficult to say why the universe is so organized, the measured universal expansion since the Big Bang of space continues to provide a "sink" (a place) into which stars as sources can radiate: A progenitive cosmic gradient, the source of the other gradients, is thus formed by cosmic expansion. For the foreseeable future the geometry of the universe's expansion continues to create possibilities for functionally creative gradient destruction, for example, into space and in the electromagnetic gradients of stars. Once we grasp this organization, however, life appears not as miraculous but rather another cycling system, with a long history, whose existence is explained by its greater efficiency at reducing gradients than the nonliving complex systems it supplemented.*
>
> Margulis and Sagan (2002)

In 1985 Freeman John Dyson wrote a little book *Origins of Life*, in which he argued that metabolic reproduction and replication are logically separable propositions, and that *natural selection does not <u>require</u> replication, at least for simple creatures*. In higher-level life as seen today, reproduction of cells and replication of molecules occur together. But there is no reason to presume that this was always the case. According to Dyson, it is more likely that *life originated twice*, with two separate kinds of organisms, one capable of metabolism without exact replication, and the other capable of replication without metabolism. At some stage the two features came together. When replication and metabolism occurred in the same creature, natural selection as an agent for novelty became more vigorous.

We have seen above that Eigen and Orgel had demonstrated, by two different experiments (one involving templates and the other involving enzymes), that a solution of nucleotide monomers can, under suitable conditions in the laboratory, give rise to a nucleic-acid polymer molecule (RNA) which replicates and mutates and competes with its progeny for survival. Living cells use both templates and enzymes for making RNA. This work pointed to a possible parasitic development of RNA-based life in an environment created by a pre-existing protein-based life.

During millions of years of chemical (and later also biological) evolution, the initial primitive but living cells diversified and refined their metabolic reaction pathways. In particular, they evolved the synthesis of ATP (adenosine triphosphate) through some autocatalytic reaction mechanisms. ATP is the main energy-carrying molecule in all present-day cells. ATP-carrying primitive cells had an evolutionary advantage over other, less efficient, cells. In time, other molecules like AMP (adenosine monophosphate) emerged; or perhaps AMP came first, and then ATP.

Although ATP and AMP have similar chemical structures (Fig. 27.3), they play totally different roles in present-day cells. ATP is the universal biological currency for energy. AMP, on the other hand, is one of the *nucleotides* in the structure of the RNA molecule.

If ATP loses two of its three phosphate groups, it becomes AMP. Dyson argued that, although the primitive cells had no genetic apparatus to begin with, they were loaded with ATP molecules which could easily convert to AMP molecules. Accidentally, in one such cell which happened to be carrying AMP and other nucleotides (the 'chemical cousins' of AMP), *the Eigen experiment for synthesizing RNA happened spontaneously*. With some help from pre-existing enzymes, an RNA molecule got produced. Once created, it went on replicating itself because of the proclivity of base A to hydrogen-bond with base U, and of G to hydrogen-bond with C.

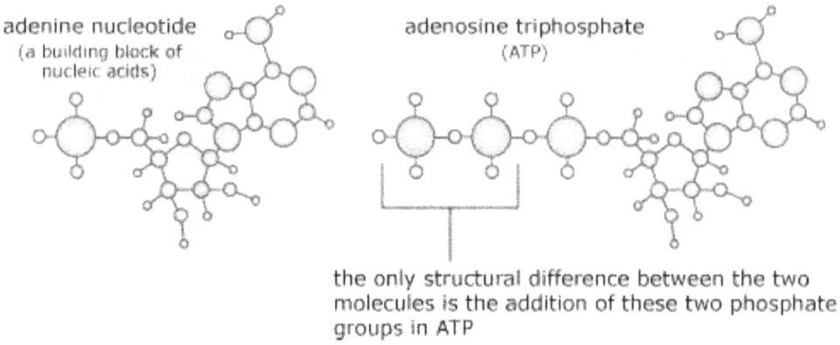

the only structural difference between the two molecules is the addition of these two phosphate groups in ATP

Fig. 27.3 Similarities between the structures of AMP and ATP.
http://evolution.berkeley.edu/evolibrary/home.php

Thus, RNA first appeared as a parasitic disease in the cell. Although most such cells died of disease, some evolved to survive the infection, à la Lynn Margulis (Margulis and Sagan 2002). In such cells, the parasite gradually became a symbiont (this will become more clear in Chapter 29). Further evolution resulted in a situation in which the protein-based life learnt to make use of the ability for exact replication provided by the chemical structure of RNA.

Is it really true that proteins emerged *before* RNA? The early evidence came from laboratory experiments done during the 1950s. The well-known experiments by Miller and others had demonstrated that amino acids form easily in a reducing atmosphere from the still simpler molecules, in the presence of ultraviolet radiation. What about nucleotides?

They are more difficult to synthesize. A nucleotide has three parts: an organic base, a sugar, and a phosphate ion. The phosphate ion occurs naturally as a constituent of rocks and sea water. The sugar (ribose) part can be synthesized with substantial efficiency from formaldehyde. And the synthesis of an organic base was demonstrated by Oró in 1960. He prepared a *concentrated* solution of ammonium cyanide in water, and just let it stand. Adenine was self-created, with a 0.5% yield. Guanine also got synthesized in a similar way. *But the catch here is that it is difficult to imagine how such high degrees of concentration of ammonium cyanide could occur in Nature*, although some possible scenarios have been suggested.

Dyson has given an updated version of his earlier ideas, in a later book *Life: What a Concept!* (2008). In his updated model there are six stages in the evolution of chemical complexity, leading to the emergence of life.

Stage 1. The early cells were just little bags of some kind of cell membrane (as I have mentioned above in the lipid-first model). This is the *garbage bag model* for Stage 1. And inside the bag there was a more or less random collection of organic molecules, with the characteristic that small molecules could diffuse in through the vesicle membrane, but big molecules, once synthesized, could not diffuse out. Thus the 'garbage bag' situation was conducive to the conversion and retention of small molecules into large molecules. The higher concentration of organic materials in the bag led to a higher efficiency of the chemical processes involved.

This evolution did not involve any replication processes. 'When a cell became so big that it got cut in half, or shaken in half, by some rainstorm or environmental disturbance, it would then produce two cells which would be its daughters, which would inherit, more or less, but only statistically, the chemical machinery inside.'

Stage 2. Parasitic RNA appeared in some of the cells in Stage 2. ATP had appeared in one of the garbage bags by a random process in Stage 1, and the cell hosting it had a metabolic advantage over other cells. Therefore many cells with large amounts of ATP got created. Then, again by chance, ATP changed to AMP in one of the cells. In due course, AMP and its chemical cousins polymerized into a primitive form of RNA. Thus there was parasitic RNA inside these cells, forming a separate form of life, which was pure replication without metabolism. To quote Dyson: 'Then the RNA invented viruses. RNA found a way to package itself in a little piece of cell membrane, and travel around freely and independently. Stage two of life has the garbage bags still unorganized and chemically random, but with RNA zooming around in little packages we call viruses carrying genetic information from one cell to another. That is my version of *the RNA world*'.

Stage 3. This stage started when the protein and the RNA systems started to collaborate. It happened after the emergence of the ribosome. Although this arrangement had the rudiments of the modern cell, the genetic information was shared mostly via viruses travelling from cell to cell. This was some kind of *open-source heredity*. The chemical inventions made by one cell could be shared with others. Evolution went on in parallel in many different cells. The best chemical devices could be shared between different cells and combined, so the chemical evolution was very rapid, as it occurred in parallel by many pathways.

Stage 4. Speciation and sex appeared in Stage 4, and that marked the beginning of the Darwinian era, when species appeared. 'Some cells decided it was advantageous to keep their intellectual property private, to have sex only with themselves or with the members of their own species, thereby defining species. That was then the state of life for the next two billion years, the Archeozoic and Proterozoic eras. It was a rather stagnant phase of life, continued for two billion years without evolving fast.'

Stage 5. Multicellular organisms appeared in Stage 5, which also involved death.

Stage 6. This is the stage when we humans appeared.

More recent work at the University of Cambridge, supporting the possible emergence of life without any need for the involvement of RNA, has been reported by Ralser and coworkers (see Keller, Turchyn and Ralser 2014). They suggest that the complex processes needed for life may have surprisingly simple beginnings. Many of the reactions that underpin life on Earth could have occurred spontaneously in Earth's early oceans, catalysed by *metal ions* rather than the *enzymes* that drive them in cells today. I quote from their paper:

'The reaction sequences of central metabolism, glycolysis and the pentose phosphate pathway provide essential precursors for nucleic acids, amino acids and lipids. However, their evolutionary origins are not yet understood. Here, we provide evidence that their structure could have been fundamentally shaped by the general chemical environments in earth's earliest oceans. We reconstructed potential scenarios for oceans of the prebiotic Archean based on the composition of early sediments. We report that the resultant reaction milieu catalyses the interconversion of metabolites that in modern organisms constitute glycolysis and the pentose phosphate pathway. The 29 observed reactions include the formation and/or interconversion of glucose, pyruvate, the nucleic acid precursor ribose-5-phosphate and the amino acid precursor erythrose-4-phosphate, antedating reactions sequences similar to that used by the metabolic pathways. Moreover, the Archean ocean mimetic increased the stability of the phosphorylated intermediates and accelerated the rate of intermediate reactions and pyruvate production. The catalytic capacity of the reconstructed ocean milieu was attributable to its metal content. The reactions were particularly sensitive to ferrous iron Fe(II), which is understood to have had high concentrations in the Archean oceans. These observations reveal that reaction sequences that constitute central carbon metabolism could have been constrained by the iron-rich oceanic environment of the early Archean. The origin of metabolism could thus date back to the prebiotic world.'

27.4 Why was evolution extremely fast for the earliest life?

> *Evolution is the fundamental physical process that gives rise to biological phenomena. Yet it is widely treated as a subset of population genetics, and thus its scope is artificially limited. As a result, the key issues of how rapidly evolution occurs and its coupling to ecology have not been satisfactorily addressed and formulated. The lack of widespread appreciation for, and understanding of, the evolutionary process has arguably retarded the development of biology as a science, with disastrous consequences for its applications to medicine, ecology, and the global environment. . . . Evolution is a problem in nonequilibrium statistical mechanics, where the key dynamical modes are collective, as evidenced by the plethora of mobile genetic elements whose role in shaping evolution has been revealed by modern genomic surveys.*
>
> Goldenfeld and Woese (2011)

Nigel Goldenfeld's (http://guava.physics.uiuc.edu/~nigel/) recent work supports many of Dyson's ideas described in the previous section (see Cepelewicz 2017), as also of Margulis (to be discussed in Chapter 29). He points out that if you look at *the last universal common ancestor (LUCA)* (the organism from which all life evolved), it becomes clear that that was not the beginning of life. There was simpler life before that, with no species and no genes. The LUCA was there about 3.8 billion years ago. And the Earth is 4.6 billion years old. So, life evolved from zero to nearly the complexity of the modern cell in considerably less that a billion years. That was very rapid indeed, considering that relatively little biological evolution has taken place since then in terms of the cellular architecture. Evolution has been slow during the last 3.5 billion years, and very fast before that. Why?

Goldenfeld's answer is that early life evolved in a different way than life in the present era. Early life communicated through gene transfer, or rather whatever was the equivalent of genes then. And it was *horizontal* gene transfer (HGT), rather than vertical gene transfer of the present era (parents give genes to their children, who give them to *their* children, and so on, down the line). Horizontal gene transfer is between unrelated organisms. It happens even today in bacteria etc. (when they develop resistance to antibiotics), with genes that are not essential to the structure of the cell. Early life was of a collective nature (a network), more of a community linked by gene transfer, than simply a sum total of collection of individuals (Goldenfeld and Woese 2011; Goldenfeld 2014). A good analogy would be that of a beehive, which operates as a superorganism because of the networking among the individual bees (see Section 18.1). According to Goldenfeld, such a postulation is the only way forward for understanding the rapid evolution of early life: horizontal gene transfer in the network, rather than vertical gene transfer, for biological evolution.

28. Genetic Regulatory Networks and Cell Differentiation

Each cell of our body carries the same genome. Then what tells some cells to become kidney cells, and others to become liver cells, and still others to become neurons (Fig. 28.1)? The term *cell differentiation* is used for this phenomenon. How does cell differentiation occur?

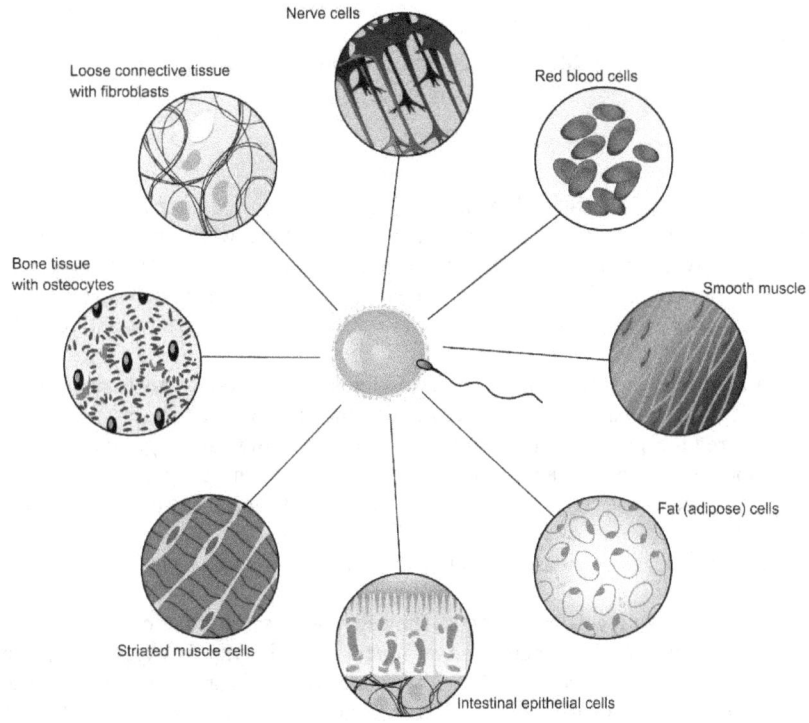

Fig. 28.1 Cell differentiation. Image credit:
http://cc.scu.edu.cn/G2S/Template/View.aspx?courseType=1&courseId=17&topMenuId=113307&menuType=1&action=view&type=&name=&linkpageID=113785

The answer has to do with 'genetic networks'. But first a word about networks in general. Use of graph theory and network theory in complexity science has paid rich dividends. The reasons are not far to see. Once we have mapped a complex system onto a network of nodes and edges (Fig. 28.2), the full power of the mathematics of graph theory and network theory can be brought to bear on the investigation of the problem. In a genetic network, the genes are the nodes of the network, and an interaction between two genes comprises an edge in the network.

Networks can be of various types. A *random network* is one in which the presence or absence of any edge is a matter of random occurrence; the edges are distributed randomly. If there are N nodes, and if there is a probability p that any pair of nodes is connected, then it can be worked out that there are $pN(N-1)/2$ edges. Random networks are the simplest imaginable. We shall encounter other types as we go along.

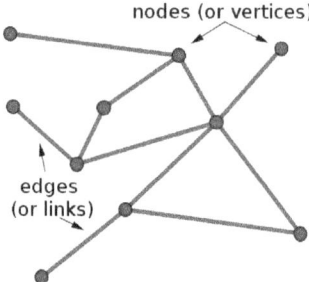

Fig. 28.2 Nodes and edges comprise a network.
http://mathinsight.org/network_introduction

28.1 Circuits in genetic networks

French scientists François Jacob and Jacques Monod were awarded (along with Andre Lwoff) the Nobel Prize for physiology or medicine for 1965 for their work on 'circuits' in genetic networks. There are thousands of genes arrayed along a DNA molecule. And the genes may be either 'on' (i.e., active) or 'off' (not active). When a gene is on, it is said to be *expressing* itself; it is doing the transcription work, directing the synthesis of the protein it has the code for.

E. coli are a kind of bacteria. When exposed to lactose, they produce an enzyme (a protein) that can digest lactose. Take away the lactose, and the enzyme production stops. Jacob and Monod discovered the so-called '*lac operon*' and the underlying *gene regulation mechanisms* responsible for the sensing of lactose and the production of the enzyme. They found that, adjacent to the gene encoding a protein called beta galactosidase, a small 'operator' DNA sequence (called 'O') bound a 'repressor protein' called 'R'. When R was bound to O, the adjacent gene for beta galactosidase could not be copied into its messenger RNA.

In more general terms, Jacob and Monod showed that a small fraction of the genes are '*regulatory genes*' which can function as *switches*. Such activity is triggered by, say, the availability of a particular hormone in the surroundings of a cell (or the presence of lactose in the case of *E. coli*). This chemical may switch-on a particular gene. The newly activated gene sends out chemical signals to fellow genes, that can switch them on or off, depending on the states they are already in. The altered state of each of these genes then releases, or stops releasing, other chemical signals, which are received by the genetic switches in the network, altering their states in turn, in a cascading manner. *This continues till the network of genetic switches settles down to a stable, self-consistent pattern.*

The term 'gene regulation' covers all factors that control gene expression. A gene, or a set of genes, is said to be expressed if the protein expected to be synthesized by its expression is found in the cell. Gene-transcription and gene-translation are the stages through which this regulation can occur. Other factors that can regulate gene expression are temperature, pH value, and the presence of certain molecules (e.g., hormones).

The work of Jacob and Monod had several implications. For example, it established DNA as not just a repository of the blueprint for the cell, telling it how to synthesize the various proteins, but also as *an engineer in charge of construction*. The DNA was established to be a molecular-scale computer that computed how the cell was to build and repair itself, and how it was to interact with the surrounding world.

This work also solved the mystery of cell differentiation. It was concluded that each type of cell corresponds to a different *pattern* of the genetic network, influenced by the presence of specific hormones etc. Although there is only a single genome involved, the genome can have many stable patterns of activation or expression, each corresponding to a different cell type (liver, kidney, brain, etc.). Thus the genome was viewed as a complex network of interacting components, which control homeostasis and differentiation through very specific control circuits among the genes. [Homeostasis is the ability of higher animals to maintain an internal consistency.]

28.2 Kauffman's work on genetic regulatory networks

A genetic regulatory network (GRN) in a cell is a set of DNA segments that interact with one another and govern the rates at which the genes in the network are transcribed into mRNA. GRNs in single-celled organisms like *E. coli* react to external stimuli to do what is good for the organism. In multicellular organisms the GRNs play other roles also, like cell differentiation (there are some 285 different types of cells in the human body).

Kauffman went many steps further than Jacob and Monod, and demonstrated that even *randomly* constructed networks of high molecular specificity can engender homeostasis and differentiation. This was a remarkable result because it meant that *highly ordered dynamical behaviour can arise even for randomly constructed genetic networks, getting just a few inputs per gene.*

He had introduced in 1969 the notion of *random Boolean networks* (RBNs). [RBNs have been reviewed recently by Drossel (2008).] We have seen above how genes in a genetic circuit may be on or off. He made the simplifying assumption that each node (gene) has two discrete (binary) states: 1 for on, 0 for off. Suppose there are N such nodes. So an RBN is a random network of N binary-state nodes (representing genes in the present case). Each gene or node was modelled as receiving K inputs ($K \leq N$) from randomly chosen 'controlling' genes or nodes, and also receiving one random 'update' function for its K inputs.

The update function prescribed the state of the gene in the next time step, given its state in the current time step, and was chosen according to some probability-distribution function. By varying N and K for these RBNs, the behaviour of a variety of such finite sequential switching automata could be investigated. At any time step, each gene or node had a value 1 or 0, and the network was a collection of these 1s and 0s, representing the 'state' of the network. This pattern of 1s and 0s served as the input, determining the pattern for the next time step of the gene, and so on.

The RBN has 2^N possible states; i.e., it has a *finite* number of states. This finiteness, coupled with the fact that the modelled dynamics is deterministic, implies that, as the RBN proceeds through a sequence of states, it must eventually return to a pattern it had at some earlier time step, and from then on it must repeat the same pattern-sequence periodically. That is, it must be trapped in a re-entrant cycle of states, or an *attractor* in phase space. Each such state cycle or attractor represents a distinct temporal mode of behaviour of the net, and was equated by Kauffman with a distinct cell type (kidney, liver, etc.). Cell types differ only in the *pattern* of gene activity; otherwise they all carry the same genome.

Kauffman focussed his attention on '*critical*' RBNs. These lie at the edge of chaos, i.e., at the boundary between *frozen* networks and *chaotic* networks. Frozen networks have very short attractors or cycle lengths. And chaotic networks have large-sized attractors that may include a substantial portion of the phase space. To quote Kauffman:

Let's talk about networks as a model of the genetic regulatory system. My claim is that sparsely connected networks in the ordered regime, but not too far from the edge (of chaos) do a pretty good job of fitting lots of features about real embryonic development, and real cell types, and real cell differentiation. And insofar as that's true, then it is a good guess that a billion years of evolution has in fact tuned real cell types to be near the edge of chaos. So that's very powerful evidence that there must be something good about the edge of chaos. So let's say the phase transition is the place to be for complex computation. Then the second assertion is something like 'Mutation and selection will get you there.'

Thus Jacob and Monod's cell types, distinguished from one another by the distinct and stable network *patterns* of gene activity, were interpreted by Kauffman as represented by different attractors in phase space. For $K = 1$ and for $K = N$ the length of the attractor cycles is very large. But for $K = 2$, i.e., when there are two inputs per gene, the lengths of the cycles are very small, roughly scaling as $\sim\sqrt{N}$ for critical networks. For example, for $N = 1000$, i.e., for 2^{1000} possible states of the network, the modelled genome was found to cycle typically among *just 30 time steps*, a remarkable result indeed.

Kauffman's work demonstrated that highly ordered dynamical behaviour is typical even for *randomly* constructed genetic networks getting just a few inputs per component. This implied that homeostasis in living complex systems is a direct consequence of the high molecular specificity among the macromolecules involved. Similarly, cell differentiation reflects the capacity of complex adaptive systems to behave in several distinct, highly localized ways.

Kauffman's work established that complex genetic networks could come into being by spontaneous self-organization, without the need for slow evolution by trial and error. After all, the whole thing had to be there together, and not partially, to function at all.

Kauffman's work, though extremely important and path-breaking, was handicapped by the limited computational power available at that time, as also the limited nature of biological data. We now know that the number of genes (N) is not proportional to the mass of DNA, contrary to what was assumed by biologists at that time; it is much smaller for higher organisms. And that, for larger N, the increase in the number of genes with N is much faster than $\sqrt{N}$. In fact, the attractor number, as also the attractor length of $K = 2$ networks, both increase with the size of the network faster than any power law.

As discussed earlier in Chapter 27, Kauffman also tackled the question of how extremely large molecules like RNA and DNA came into existence in the first place. In any case, even DNA requires the availability of certain protein molecules for its genetic role. Therefore, there must have been a mechanism which resulted in the spontaneous creation of protein molecules without the intervention of DNA or RNA. In other words, there must have been a *non-random origin of life*. There must have been another way, independent of the need to involve DNA molecules, for self-reproducing molecular systems to have got started. Kauffman carried Melvin Calvin's (1969) idea of autocatalytic reactions much further to explain how this could happen. In Kauffman's model, *life originated before the advent of RNA or DNA*. And his network model could incorporate features like reproduction, as also competition and cooperation for survival and evolution (including 'coevolution').

29. More Ideas on the Origins of Species: From Darwin to Margulis

The most inspiring way of teaching evolution is to say that it's all about the genes. It's the genes that, for their own good, are manipulating the bodies they ride about in. The individual organism is a survival machine for its genes.

Richard Dawkins, *The Selfish Gene*

We have come a long way since Darwin published his theory of evolution of species. Particularly notable has been the more recent work of Margulis, which marks a significant departure from conventional Darwinism and also 'neo-Darwinism' (Margulis and Sagan 2002).

29.1 Darwinism and neo-Darwinism

The main postulate in Darwin's theory of evolution was that a species evolves because natural selection acts on small heritable variations in the members of the species (Fig. 29.1).

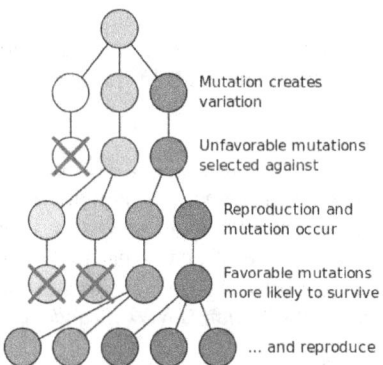

Fig. 29.1 Basic Darwinistic evolution.
http://en.wikipedia.org/wiki/Evolution

But it was argued by his opponents that, since a species is also characterized by interbreeding, such small variations should get averaged away. Darwin had no answer to counter this because the actual mechanism of inheritance was not known at that time.

The answer in fact had been provided in 1865 (i.e., during the lifetime of Darwin, but apparently unknown to him) by the work of Gregor Mendel (Fig. 29.2), the founder of the subject of genetics. We now know that the '*genotype*' or the genome of an organism is its genetic blueprint, and is present in the nucleus of every cell of the organism. The '*phenotype*', on the other hand, is the end-product (the organism) which emerges through execution of the instructions carried by the genotype. It is the phenotype that is subjected to the battle for survival and natural selection, but it is the genotype which carries the accumulated evolutionary benefits to succeeding generations. The phenotypes compete, and the fittest among them have a higher chance of exchanging genes among themselves.

What we call Mendel's laws of genetics were actually rediscovered independently by quite a few workers. One of them was the Dutch botanist Hugo de Vries, who not only rediscovered Mendel's laws for the inheritance of 'dominant' (or expressed) and 'recessive' (or suppressed) characteristics, but also discovered *genetic mutations*. These were sudden (unexplained) changes of form which were inherited by the offspring.

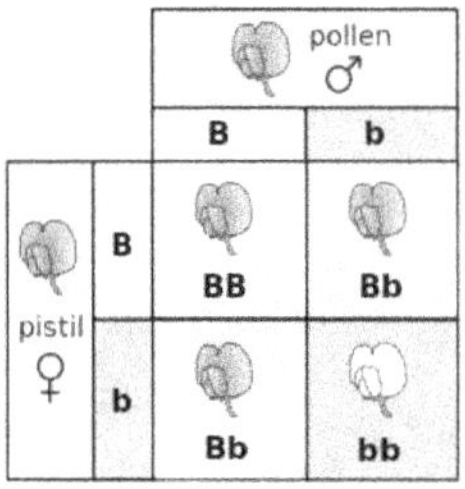

Fig. 29.2 A pea plant experiment carried out by Mendel.
https://en.wikipedia.org/wiki/Mendelian_inheritance

The present, *post-Darwinian*, picture is that the inherited characteristics of the progeny are carried by genes. In sexually reproducing organisms, each parent provides one complete set of genes to the offspring. Genes are portions of molecules of DNA, and their specificity is governed by the sequences in which their four bases (adenine (A), thymine (T), guanine (G), and cytosine (C)) are arranged. The double-helix structure of DNA, together with the restriction on the pairing of bases comprising the DNA molecule to only A-T and G-C, provides a mechanism for the *exact replication* of DNA molecules. And the DNA sequences on genes determine the sequence of amino acids in the specific proteins created by the live organism.

Genes programme embryos to develop into adults with certain characteristics, and these characteristics are not entirely identical among the individuals in the population. Genes of individuals with characteristics that enable them to survive and reproduce successfully tend to survive in the *gene pool*, at the expense of genes that tend to fail. This feature of natural selection at the gene level has consequences which become manifest at the organism or phenotype level. *Cumulative natural selection is NOT a random process.*

If like begets like (through inheritance of characteristics), by what mechanism do slight differences arise in the gene pool of successive generations so that the species evolves towards evolutionary novelty and improvement? One mechanism is that of *mutations*. Mutations, brought about by radiation or by chemicals in the environment, or by any other agents causing *replication errors*, change the sequence of the four bases in the DNA molecules comprising the genes. Most mutations are deleterious and get weeded out by the natural-selection processes, but those which happen to be beneficial to the population have a selection advantage and get further propagated in the population.

If all living beings have the same or only a few ancestors, how have the various species arisen? The Darwinistic answer lies in *isolation* and *branching*, aided by evolution. *Migrations* of populations also play a role in the evolutionary development of species. If there are barriers to interbreeding, geographical or otherwise, single populations can branch and evolve into new, distinct, species over long enough periods of time. Each such branching event is a '*speciation*': A population accidentally separates into two, and they evolve independently. When separate evolution has reached a stage that no interbreeding is possible even when there is no longer any geographical or other barrier, a new species is said to have originated.

The term *neo-Darwinism* essentially connotes a modification of the original ideas of Darwin in the light of later knowledge about the mechanism of transmittal of genetic information from one generation to the next. Margulis and Sagan (2002), whose work I shall describe in the next section, disagreed with this neo-Darwinistic view of the origin of species They summed up neo-Darwinism as follows:

'All organisms derive from common ancestors by natural selection. Random mutations (heritable changes) appear in the genes, the DNA of organisms, and the best "mutants" (individuals bearing the mutations) in competition with the others, are naturally selected to survive and persist. The unsuited offspring die – they tend to be called "unfit" – with fitness, a technical term, referring to the relative numbers of offspring left by an individual to the next generation. The most fit, by definition, produce the largest number of offspring. The mutant variations then leave more offspring, and populations evolve; that is, they change through time. When the number of changes in the offspring accumulates to recognizable proportions, in geographically isolated populations, new species gradually emerge. When sufficient numbers of changes in offspring populations accumulate, higher (more inclusive) taxa gradually appear. Over geological periods of time new species and higher taxa (genera, families, orders, classes, phyla, and so on) are easily distinguished from their ancestors.'

As emphasized by Stuart Kauffman, evolution of biological complexity is determined by *two* factors: natural selection, *and* self-organization. Self-organization creates order in any complex system. Darwinian natural selection acts on this existing order and hones it further.

29.2 Biological symbiosis and evolution

Ironically the popular evolutionist's view that organisms evolve by the accumulation of random mutation best describes the evolutionary process in bacteria. All of the larger, more familiar organisms originated by symbiont integration that led to permanent associations.

Margulis and Sagan (2002)

Not so frequently emphasized, the finding that all life is related implies the existence of a last universal common ancestor (LUCA), now known to be positioned between the Bacterial and the Archaeal/Eukaryotic branches and representing in one extreme view a single organism, or in another view, a community of associated organisms.

Nigel Goldenfeld (2014)

Biological symbiosis means a prolonged living arrangement or physical association among members of two or more different species. Levels of partner integration in symbiosis may vary in intimacy; and integration may be behavioural, metabolic, or 'genic' (i.e., of gene products).

An example of symbiosis in action are the microbes that live in a special stomach of the cow, providing the enzymes for the digestion of cellulose. The cow, in turn, provides shelter and nutrition to the microbes. We say that the microbes are the *symbionts* of the cow.

Biological cells are either prokaryotes or eukaryotes. Examples of prokaryotes are: *E. coli*; blue-green algae or cyanobacteria; and archaebacteria.

Eukaryotes can be divided into four main *kingdoms* (Fig. 29.3):

- Protoctists (algae; amoebas; ciliates; slime moulds).
- Fungi (moulds; yeasts; mushrooms).
- Plants (mosses; ferns; flowering plants).
- Animals (molluscs; arthropods; fish; mammals).

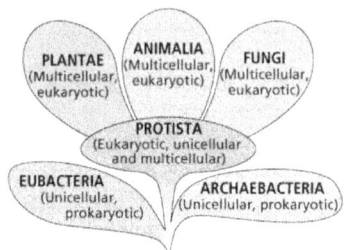

Fig. 29.3 Kingdoms among the eukaryotes.
https://www2.estrellamountain.edu/faculty/farabee/biobk/BioBookDiversity_3.html

The genealogical relationship of all living organisms has three main branches (Fig. 29.4): *bacteria*, *archaea*, and *eucarya*. However, some authors distinguish just two branches, namely bacteria and eucarya, and subdivide bacteria into eubacteria and archaebacteria. The eubacteria and the archaea or archaebacteria are prokaryotes. The eucarya are eukaryotes.

Phylogenetic Tree of Life

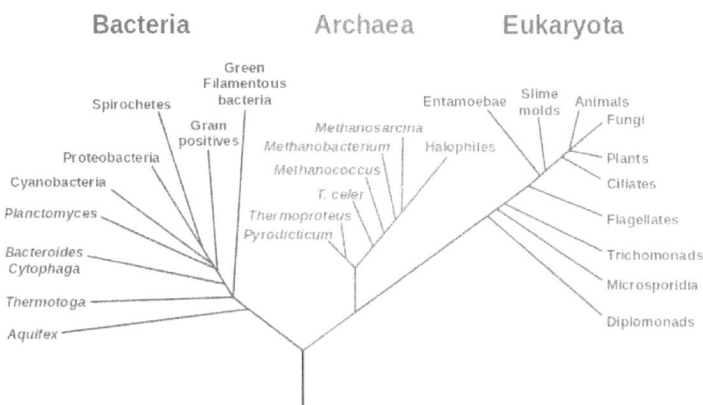

Fig. 29.4 The phylogenetic tree of life.
http://en.wikipedia.org/wiki/Phylogenetic_tree

Lynn Margulis is a major proponent of the idea that *parasitism and symbiosis were major driving forces in the evolution of cellular complexity*. She has been hammering home the point that the main components of eukaryotic cells have descended from *independent* living creatures which 'attacked' the cells from outside. In due course, the attackers and the host evolved a relationship of mutual dependence and benefit. In stages, the erstwhile invading organisms became first chronic parasites, then symbiotic partners, and finally an indispensable part of the host (Fig. 29.5).

The mitochondria are symbionts in both plant and animal cells, as are chloroplasts in plant cells. The evidence for this is that the molecular structures of mitochondria and chloroplasts are very close to certain bacteria. The emergence of a new life form from such symbiosis is called *symbiogenesis*.

Margulis marshalled evidence to argue that most of the big steps in cellular evolution were caused by parasites. And that *nucleic acids were the oldest and the most successful cellular parasites.* Thus the classical viewpoint that speciation occurs, i.e., new species arise, only as a result of the cumulative effect of mutations etc., has been strongly contested by Margulis. According to Margulis and Sagan (2002), 'No evidence in the vast literature of heredity change shows unambiguous evidence that random mutation itself, even with geographical isolation of populations, leads to speciation. Then how do new species come into being? How do cauliflowers descend from tiny, Mediterranean cabbage-like plants, or pigs from wild boars?' Their answer was that *species arise largely by the acquisition of entire genomes through symbiogenesis.*

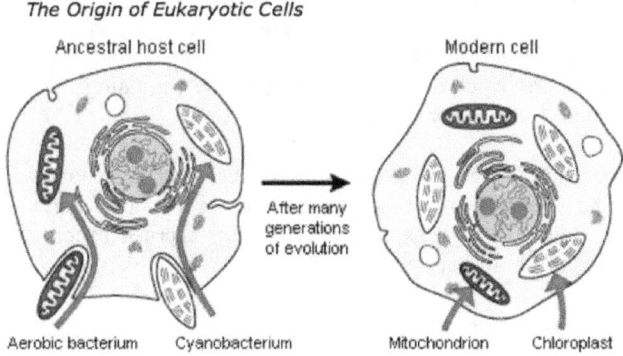

Fig. 29.5 Biological sysmbiosis.
http://www.ecocitybuilders.org/lynn-margulis-and-the-lifeenvironment-self-organizing-key-to-evolution-and-economy-continued/

Margulis's stance raised debate. Ernst Mayr wrote an appreciative foreword to the book by Margulis and Sagan (2002). But the foreword also said this: 'Speciation – the multiplication of species – and symbiogenesis are two independent, superimposed processes. There is no indication that any of the 10,000 species of birds or the 4,500 species of mammals originated by symbiogenesis.'

But contrast this with the statement of Rachel Nowak (2005): '*Symbiosis has popped up so frequently during evolution that it is safe to say that it's the rule, not the exception.*'

29.3 What is a species

In biology, studies of evolution are about tracking the changes of life through time. In particular, they are about tracking the origin and evolution of species. But what exactly is a species?

The morphological definition. Creatures belonging to a species look alike (dogs look like dogs).

The biological definition. Creatures belong to a species if they can mate and produce fertile offspring. This was introduced by zoologists, and also by botanists.

The phylogenetic definition. Groups of organisms considered to be all descended from the same ancestors are said to belong to the same species.

Margulis and Sagan (2002) rejected the biological and the phylogenetic definition, and suggested something that can accommodate only the morphological definition:

The symbiogenetic definition. 'If organism A belongs to the same species as organism B, then both are composed of the same set of integrated genomes, both qualitatively and quantitatively. All organisms that can be assigned to a unique species are products of symbiogenesis'.

Life originated with bacteria. Bacteria do not speciate. The idea of a species does not apply to them. Bacteria can pass genes back and forth (*'horizontal gene transfer (HGT)'*). There is no fixed genome to define the species of any bacteria. Bacteria are prokaryotes.

The first eukaryote emerged by the symbiogenesis of two prokaryotes. *The concept of a species can apply only to eukaryotes. It follows that the origin of species occurred long after the origin of life in the form of bacteria.* Therefore, the species of all the larger organisms (protoctists, fungi, animals, plants) originated symbiogenetically in the beginning. Nucleated organisms emerged on Earth some 1.2 billion years ago.

Symbiosis can occur only if the arrangement is beneficial to all the partners involved. In the next chapter, when I present a historical narrative of the various 'energy regimes' in the evolution of our ecosphere, I shall explain what benefits accrued to the partners when the eukaryotic cell evolved from its partner organisms.

I think there is no reason to believe that symbiogenesis is the *only* way in which new species can arise. It is a characteristic of complex systems that, often, small changes can have unexpectedly large consequences, including the emergence of new species. Effects of mutations can gradually build up to a stage wherein a sudden bifurcation occurs in phase space, and a new species arises. Speciation may well be an emergent phenomenon often. This contention of mine is in disagreement with the statement of Margulis and Sagan (2002) that 'intraspecific variation never seems to lead, by itself, to new species'.

29.4 Horizontal gene transfer in the earliest life forms

Support for the ideas of Margulis has come from the later work on horizontal gene transfer (HGT) being a predominant mechanism responsible for the very rapid pace of biological evolution in the pre-LUCA era (see Section 27.4). The question addressed by Carl Woese and Nigel Goldenfeld (2011) was: 'How could evolution have achieved so much, starting from an abiotic earth and attaining an essentially modern translational machinery in a time frame of what could be at most one billion years?'. I quote from Goldenfeld (2014):

'My early discussions with Carl centred on this issue, because Carl had the intuition that some understanding of complex dynamical systems would be pertinent to the issue. That is to say, he felt that the picture of evolution, which had emerged from the modern synthesis during the first half of the 20th century, was somehow missing an important aspect of the evolutionary process. One way to phrase this seeming inadequacy is to ask how population genetics can possibly be considered as a full explanation of the evolutionary process, when, by construction, the biological world before there were genes as such was manifestly beyond the regime of validity of the theory. Carl had already given a lot of thought to life before genes, and in the same year as the discovery of the Archaea had, in another magnificent paper also with Fox, initiated conceptual discussion on such a phase of life, which he called the "progenote." The progenote was a phase of life in which the distinction between genotype and phenotype had not yet emerged. Carl considered the emergence of the genotype–phenotype relationship to be the primary force shaping the evolution of the cell, an argument he based on the simple incisive observation that the translational apparatus is large and complex, perhaps more than any other cellular machinery. Carl attributed the size and complexity to the requirements for accuracy of the translational process, citing as inspiration a summary of theoretical work on cellular automata conducted by John von Neumann during the phase of his life devoted to the construction of the world's first modern computer.'

The HGT idea is in line with Margulis's postulation of parasitism and symbiosis as the major driving forces in the evolution of cellular complexity. The genes of the attacking parasites became, in due course, integral parts of the host cell, which means HGT.

29.5 Epigenetics

When Darwin published his theory of biological evolution, the field of genetics had not yet taken shape. Naturally, Lamarck, whose work on evolution *preceded* that of Darwin, was also unaware of the crucial role genetic mechanisms play in the evolution of species. We now have the familiar concepts of genotype and phenotype. The term genotype refers to the genetic blueprint encoded in the DNA chains. Phenotype, on the other hand, signifies the characteristics manifested by an organism; it is the structure created by the organism from the instructions in its genotype. In phase-space language, genotypes correspond to the *search space*, and phenotypes to the *solution space*.

As discussed above, biological evolution is generally believed to be Darwinian, or rather neo-Darwinian. Lamarck's evolution model, or *Lamarckism*, on the other hand, was based on two premises: the principle of use and disuse; and the principle of inheritance of *acquired* characteristics (without involving the genotype). The Lamarckian viewpoint is not acceptable in modern biology because it runs counter to the *central dogma of molecular biology* (see Chapter 24). The field of research called epigenetics has brought us close to Lamarckism in a superficial sort of way, but *without violating the central dogma of molecular biology*. It is now clear that changes other than those in the *sequence* of nucleotides in DNA, acquired during the lifetime of parents or grand-parents, can sometimes be inherited in the next few generations. Gene expression or interpretation *can* be influenced by molecules hitchhiking on gametes (germ cells) and therefore on genes. This *temporarily* heritable hitchhiking of the genes is called *epigenetic inheritance*.

Some experts use the term 'epigenetics' for the entire field of study of those chemical reactions, and those factors, which influence the activation and deactivation of parts of the genome at specific locations and at certain strategic stages during the growth and development of an organism.

Genes are portions of the long sequence of nucleic acids in the DNA chain, and contain coded information for synthesizing the various proteins. And we have known since the days of Mendel that not all genes are active all the time. Some genes are *dominant*, while others are *recessive*. One of the Mendelian laws of genetics says that it is the combination of dominant (or switched on) and recessive (or switched off) genes, inherited from the parents, that dictates the characteristics (phenotype) of an offspring. We saw in Chapter 28 how the presence of certain chemicals can influence gene expression, and once a gene has been switched on, it acts as a switch which can alter the 'on' or 'off' states of other genes. Hormones are one example of what can influence gene activity.

Can chemicals other than hormones, for example those in the diet of an organism, also influence gene expression? The answer is 'Yes'. And not just food, but even the mental state of a creature can be responsible for the secretion of chemicals which can influence gene expression. But from the point of view of genetics and transmittal of acquired characteristics, the influence on the transmittal mechanism of genes should be of an irreversible, perpetual nature; only then can it affect the future generations in a permanent manner: Then only can we have an inheritance of acquired characteristics by the progeny. *This is not found to be the case.*

Epigenetic transmission of acquired characteristics does not last forever, or even for a very large number of generations.

Epigenetic effects influence the phenotype over a few generations, without changing the sequence of nucleotides along the DNA chain.

One particular (temporarily) heritable marking of DNA that has been investigated substantially is that of *methylation*, i.e., attachment of the -CH$_3$ group to one or more nucleotides along the DNA chain. It has been found, for example, that methylation is quite frequent in cancer cells.

Methylation affects gene expression and, as a feedforward mechanism, can have serious transgenerational effects. Epigenetic changes can be passed through the germ line for a few generations. And epigenetic changes can occur throughout the life time of an individual: Methylation/demethylation can turn some genes off/on.

Richard Dawkins' phrase (also the title of a book by him) '*The Selfish Gene*' sums up the essence of neo-Darwinistic evolution; at least the Dawkins version of it: Natural selection acts only on genes, via their expression in an organism's body and behaviour. The genes are not naked. The phenotype is the vehicle that promotes their interest of propagating to the next generation. The organism is the survival machine for the genes. Dawkins spoke of the 'selfish gene' because the gene promotes its own survival, without necessarily promoting the survival of the organism, group, or even species.

There are two aspects of a gene: The specific sequence of the four nucleic-acid bases (A, T, G, C) along the backbone of DNA; and the paraphernalia of interactions that occur on the *surface* of the long DNA molecule. The backbone part does not change, except through the occasional replication error, or mutation, or during 'chromosomal crossover'. Epigenetics is all about what happens on the surface, with no influence on the backbone *sequence* of the nucleotides.

Genes propagate over many generations with a high degree of integrity because the replication process involves the strongly bonded backbone structure of DNA. By contrast, the epigenetic alterations are of a temporary (few-generations) nature. There is no evidence that the changes are permanent. In view of this, Dawkins has responded to the claims of epigenetics by replacing his phrase 'selfish gene' by '*selfish replicator*':

'*The "transgenerational" effects now being described are mildly interesting, but they cast no doubt whatsoever on the theory of the selfish gene'. . . . 'Whether [epigenetic marks] will eventually be deemed to qualify as "selfish replicators" will depend upon whether they are genuinely high-fidelity replicators with the capacity to go on forever. This is important because otherwise there will be no interesting differences between those that are successful in natural selection and those that are not.*'

Dawkins points out that if the epigenetic effects fade out within the first few generations, they cannot be said to be positively selected. True. But the feeling is growing that, for many practical purposes, epigenetic effects are far more important in terms of their consequences than was realized when they were first reported.   Klosin *et al.* (2017) have reported that epigenetic effects persisted for *14 generations*, no less, in the experiments they carried out on *C. elegans* nematodes (roundworms). Persistence for such a large number of generations can possibly have evolutionary significance, at least in some cases.

30. Coevolution of Species

Living organisms interact, not only with the inanimate surroundings, but also with organisms of the same or different species. Therefore, evolution of a species cannot occur in isolation from the evolution of other species with which it interacts. They *coevolve*. Coevolution of species is an important aspect of biological evolution.

Darwin's theory of evolution says that a species evolves to become better and better for the task of survival in a given set of conditions. Genes of individuals in the population that are not good enough for the task of survival tend to get eliminated in successive generations. All this happens through an interplay of the blind forces of Nature, but the end effect appears to be as if the genes have the thinking power of being 'selfish' at the job of survival and propagation. Richard Dawkins' (1989) phrase 'the selfish gene' sums up the situation well.

This may seem to indicate that a species will always evolve strategies (consciously or 'purely chemically') that are entirely selfish when it comes to dealing with other species. In fact, even within a species, it is conceivable that the genotype, and thence the phenotype, of an individual member may exhibit selfishness, without regard for other members of the species. But what happens in reality can actually be far from this simple-minded speculation. There can be *evolutionarily stable strategies*, involving both competition and cooperation, and there can be *evolutionary arms races*. And, as we discussed in Section 29.2, there can also be symbioses of species.

Before I discuss some of these processes, it is interesting to take note of an analysis by Douglas Caldwell (see Margulis and Sagan 2002), according to which the following terms were *never* used by Charles Darwin in his *Origin of Species*: association; affiliation; cooperate, cooperation; collaborate, collaboration; community; intervention; symbiosis. What I am now going to describe in this and the next few sections are some of the *post*-Darwinian developments.

30.1 Punctuated equilibrium in the coevolution of species

Although species tend to evolve towards a state of stable adaptation to the existing environment, it is also true that many of them are extinct, as the fossil records show. This is so because there is never a state of permanent, static equilibrium in Nature. There is a never-ending input of energy, climatic changes, terrestrial upheavals, cumulative effects of mutations, as also a coexistence with other competing or cooperating species. Rather than evolving towards a state of permanent equilibrium and adaptation, the entirety of species (in fact, the Earth as a whole) evolves towards a state of *self-organized criticality* (SOC) (see Chapter 20), and this state of complexity is poised at *the edge of order and chaos*. The never-ending inputs just mentioned result in minor or major 'avalanches' (catastrophic events of various magnitudes), which sometimes lead to the extinction of species, or the emergence of new ones.

SOC can explain the finding by Eldredge and Gould (1972) in fossil records that there are long periods of stasis, followed by quick bursts of evolutionary change; i.e., there is *punctuated equilibrium* (Fig. 30.1). Imagine an ecosystem in which the various species have reached more or less a state of equilibrium with one another. Suppose there is a random genetic crossover in one of the species that is beneficial to it and thus survives and gets propagated to other members of the species. The propagation is not linear; it is more likely to be avalanche-like or 'explosive' with time, rather like the avalanches in the sandpile experiment I described in Chapter 20. In due course, things stop changing too much, but then some other member of the population may mutate. There is thus a steady drizzle of mutations, resulting in periods of avalanches and relative equilibrium.

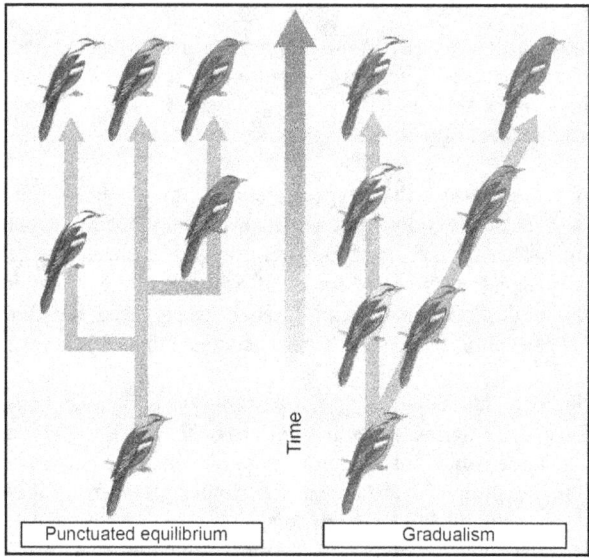

Fig. 30.1 Gradualism vs. punctuated equilibrium. Image credit: https://online.science.psu.edu/biol011_sandbox_7239/node/7290

30.2 Evolutionarily stable strategies

In intra-species and inter-species interactions there are bound to be situations of conflicts of interest. The mathematical basis for understanding such conflicts is provided formally by *game theory*. In it, one analyses how decisions are taken by individuals and groups in situations involving a diversity of individual goals and differing levels of strategic control over the environment. The basic formal aspects of game theory are given in Appendix A10.

In conventional game theory it is usually assumed that each player is a *rational* being, who expects that other players are also rational beings. Each player adopts a strategy to minimize his losses, assuming that other players will work out strategies to maximize the losses of other players as they try to minimize their own losses (the so-called *minimax strategy*). In the evolutionary context, if each member of a population did this, the end result may well be an extinction of that species (humans, please note!). The fact that a species has survived can imply that an evolutionarily *stable* strategy (ESS) must have evolved. In an ESS, the survival of an individual, though not completely subservient to the survival of the species, is determined by a strategy that ensures that contests between individuals, though leading to an improvement of the gene pool of the population, do not result in an excessive annihilation of the contest-losers. The ESS idea, which entailed a modification of conventional game theory, was put forward by Maynard Smith (1974, 1976, 1982), and it can be operative even in the *coevolution* of two or more species. In fact, the ESS concept is also relevant to the evolution of entire ecosystems.

The gist of the ESS idea is this: Whenever the best strategy of an individual depends on what others are doing, the strategy actually adopted will be an ESS, meaning that two competing species will evolve to specialize on the use of *different* resources, even though each species may tend to be a *generalist* when in isolation. 'If everyone else is eating spinach, it will pay to concentrate on cabbage; (similarly), since most forest trees put out their leaves late in spring, it pays forest herbs to put out leaves early' (Maynard Smith 1976).

As an illustration of the validity of the ESS idea *within* a species, consider the question of male-female ratio. Why is it generally close to unity (50:50)? The answer was first suggested by R. A. Fisher (1930). He assumed that genes of an individual can affect the relative number of his male and female offspring. That sex ratio will tend to dominate which maximizes the number of descendants of the individual. Suppose there are more than 50% males in the population. Then, because of selection pressure, an individual with genes which favour the birth of females will stand a better chance of contributing to the gene pool of the population (presuming that monogamy is the preponderant practice). If, on the other hand, there are more females in the population, then individuals which tend to produce more sons will have more grandchildren. The 1:1 ratio is the only *stable* ratio, and the evolutionary strategy which ensures this is called an ESS (even though no player is taking any *conscious* decisions regarding strategy).

Let us next discuss '*the logic of animal conflict*' within a species. It is observed that such conflicts are usually of the 'limited war' type. Serious injury to one or both contestants is avoided. This must be the result of an ESS. Individuals which had a tendency to violate an ESS must have got eliminated gradually by the processes of Darwinian natural selection. I give here some details of the game-theoretic 'two-player' 'non-zero-sum' version of the ESS approach applied to animal contests or conflicts by Maynard Smith and Price (1973) (details of the jargon used here are explained in Appendix 10).

Suppose there is a contest between two males of a species because there is a conflict of interest. The various possible outcomes of the contest will not be viewed in the same order of preference by the two contenders because the relevant genotype determines the pure or mixed game-theoretic strategy adopted by an animal. Suppose the first animal adopts strategy I, and the other animal adopts strategy J. We write the 'payoff' to the player with strategy I as $E_J(I)$: It is the expected gain, namely the change it will introduce in the fitness of the first player as a result of the contest. This fitness change is the increase or decrease in the contribution of this animal to the gene pool of future generations. An ESS is defined formally as follows: Suppose there is a population of individuals adopting strategies I or J with probabilities p and q respectively, with $p + q = 1$. Then the payoff or fitness of an individual adopting strategy I (we call it individual I), or strategy J, is, respectively:

Fitness of $I = p.E_I(I) + q.E_J(I)$ (30.1)

Fitness of $J = p.E_I(J) + q.E_J(J)$ (30.2)

Suppose a strategy, say I, is an ESS. Then it must have the property that a population of individuals playing I must be 'protected' against invasion by any mutant strategy, say J. If this is not the case, then I is an unstable strategy. Thus, *when I is common*, it must be fitter than any mutant strategy. Formally, I is an ESS if, for $J \neq I$,

either $E_I(I) > E_I(J)$,   (30.3)

or $E_I(I) = E_I(J)$ and $E_J(I) > E_J(J)$ (30.4)

Under these conditions, a population playing strategy I is evolutionarily stable, and no mutant can intrude successfully into the population. This is so because, for small q (i.e., when most of the members of the population adopt strategy I) the fitness of strategy I is greater than the fitness of J.

We can also state this in the language of 'mixed-strategy game theory'. An ESS consists of a square payoff matrix $\mathbf{A}$, along with a possibly mixed strategy, such that

either $\mathbf{xAx}^t \geq \mathbf{yAx}^t$ for all $\mathbf{y} \neq \mathbf{x}$,   (30.5)

or, if $\mathbf{xAx}^t = \mathbf{yAx}^t$, then $\mathbf{xAy}^t > \mathbf{yAy}^t$ (30.6)

The condition '$\mathbf{xAx}^t \geq \mathbf{yAx}^t$ for all $\mathbf{y} \neq \mathbf{x}$' says that no intruder ($\mathbf{y}$) will do better against the established population $\mathbf{x}$ than $\mathbf{x}$ does against itself. And the condition 'if $\mathbf{xAx}^t = \mathbf{yAx}^t$, then $\mathbf{xAy}^t > \mathbf{yAy}^t$' says that if $\mathbf{y}$ does as well against $\mathbf{x}$ as $\mathbf{x}$ itself, then there would be created a new environment (mainly $\mathbf{x}$, but with a small proportion of $\mathbf{y}$) in which $\mathbf{x}$ will do better than $\mathbf{y}$, thus making it an ESS.

The ESS idea is related to the notion of 'noncooperative equilibrium' in game theory, with the proviso that the choice of strategy gets interpreted genetically, rather than as a conscious strategic choice (Shubik 1993).

In the next section I illustrate some of these ideas with a case study that models the various scenarios in an analysis of the logic of animal conflict.

30.3 Of hawks and doves in the logic of animal conflicts

Let us consider a species that possesses offensive weapons capable of inflicting serious injury. Two types of tactics of conflict between individuals of the species are conceivable: *conventional* (C) which are unlikely to cause serious injury; and *dangerous* (D) which can cause serious injury if employed for long periods. C tactics may, for example, be threat displays like puffed up size, without any actual fighting.

In a game-theoretic approach to the problem of understanding such behaviour, the conflict is modelled as a sequence of moves and countermoves by two players. At each time step, the possible moves are C, D, and R, where R denotes *retreat*. Suppose player 1 employs tactic D in a move, for which there is a certain fixed probability that it would seriously injure player 2. When a player is seriously injured, he retreats, and the game ends, with the other player becoming the winner. The winner gains mates, dominance rights, etc., resulting in higher chances for the propagation of his genes into future generations.

If a player plays D in response to a move C by the opponent, we speak of a *probe* or a *provocation*. If a probe is made immediately after an opening move C of the contest, it is said to *escalate* the contest from level C to level D. If a player plays D in response to a probe, we speak of *retaliation*.

There are payoffs to each contestant at the end of a game, which are a measure of the likely reproductive success of the contestant. The payoffs depend on three considerations: (i) the advantage of winning; (ii) the disadvantage of getting seriously injured; and (iii) the disadvantage to the species as a whole of wasting time and energy in the contest.

Maynard Smith and Price (1973) modelled five possible strategies of combat:

1. Mouse. True to its name, the mouse (or the dove, if you want to call it that) never plays D. If the other player plays D, it retreats immediately, thus ending the contest. Otherwise, it goes on playing C for a preassigned number of moves. This is a 'limited war' strategy.

2. Hawk. Always plays D. Fights till the end. The end comes either when he is seriously injured, or when the opponent retreats. This is a 'total war' strategy.

3. Bully. Plays *D* if making the first move, or if the opponent plays *C*. Plays *C* in response to *D*. Retreats if the opponent plays *D* a second time.

4. Retaliator. Plays *C* if making the first move, or when the opponent plays *C*; but plays *R* if the contest has lasted a pre-assigned number of moves. Retaliates with a *D* (with a high probability) if the opponent has played *D*. A 'limited war' strategy.

5. Prober-Retaliator. A 'limited war' strategy. Plays *C* with high probability or *D* with low probability if making the first move, or if the opponent has played *C*. But plays *R* if the contest has lasted a certain fixed number of moves. Reverts to *C* after making a probing move if the opponent retaliates, but 'takes advantage' by continuing to play *D* if the opponent plays *C*. Plays *D* with high probability on receiving a probe.

Thus the Hawk strategy is a total-war strategy, and mouse, retaliator, and probe-retaliator are limited-war strategies. The main purpose of the simulation study was to see whether *individual* Darwinian selection (as opposed to the still-debated notion of *group selection*) would favour total-war strategy or one of the limited-war strategies. *They could demonstrate that individual selection does explain the observed behaviour, without any recourse to group selection.* A group-selection model explaining absence of escalated contests would go something like this: Escalated contests would result in serious injury to many members of a population, and would militate against the survival of the species.

The probabilities used in the model were as follows:

0.10 for serious injury from a single *D* play;

0.05 that a Prober-Retaliator will probe on the opening move, or after the opponent has played *C*;

1.0 for the case that Retaliator or Prober-Retaliator will retaliate against a probe (if not injured) by the opponent.

The payoffs were assigned as follows: +60 for winning; -100 for receiving a serious injury; -2 for each *D* received that does not cause a serious injury, and only causes a 'scratch'. In addition a payoff varying between 0 and +20 was awarded (to each contestant who was not seriously injured) for saving time and energy; the payoff was 0 for a contest of maximum duration, and +20 for a contest of minimal duration.

'Symmetric contests' were assumed. That is, the fighting prowess was modelled to be identical for the two players in a contest, as also the payoffs; they could differ only in the strategy adopted. An asymmetric contest, by contrast, would be one in which the fighting ability of the players can differ, and/or the value of the resources gained or lost (i.e., the payoff) is different.

The five modelled strategies define 15 types of two-player games. Maynard and Price (1973) simulated 2000 games of each type. The table below shows the average payoffs to each player in each type of play or contest. The number in a given row and column is the pay-off obtained by the row strategy when the opponent plays the column strategy. For example, the average payoff for the Mouse row and the Hawk column is 19.5; this is the payoff to the Mouse when playing against the Hawk. And the average payoff to the Hawk when playing against the Mouse is at the intersection between the Hawk row and the Mouse column, namely 80.0.

Opponent					
		Mouse	Hawk	Bully	Retaliator	Prober-Retaliator
Player receiving the payoff	Mouse	29.0	19.5	19.5	29.0	17.2
	Hawk	80.0	-19.5	74.6	-18.1	-18.9
	Bully	80.0	4.9	41.5	11.9	11.2
	Retaliator	29.0	-22.3	57.1	29.0	23.1
	Prober-Retaliator	56.7	-20.1	59.4	26.9	21.9

Which of the five strategies is an ESS for this population? Suppose it is Hawk. Such a population should consist almost entirely of Hawks. Therefore we have to look at the Hawk column in the table. We see that Mouse-Hawk and Bully-Hawk are evolutionarily more likely possibilities than Hawk-Hawk. Therefore natural selection will increasingly favour the occurrence of genes (alleles) with Mouse and Bully tendencies, at the cost of Hawk genes. *Thus Hawk or 'total war' is not an ESS.* If the species has mainly Hawks, it will wear itself out fighting.

The table shows that *Mouse is also not an ESS.* Such a population will consist almost entirely of Mouse, and the Mouse column tells us that Mouse-Mouse will fare poorly compared to other combinations in that column, meaning that natural selection will take the population away from a 'mostly Mouse' situation. A Mouse-only population will be easily dislodged from its habitat by an intruding species. Similarly, Bully is not an ESS.

It can be seen that Retaliator is an ESS (see the Retaliator column), even though Mouse-Retaliator contests do as well as Retaliator-Retaliator contests. In any case, as Maynard Smith and Price (1973) remarked, *a real population would contain young, senile, diseased and injured individuals who will adopt the Mouse strategy for nongenetic reasons.*

Prober-Retaliator is almost an ESS. Thus, individual selection can explain why potentially dangerous offensive weapons are almost never used in contests *within* a species. An ESS does, however, require retaliation: The contestants must respond to an escalated attack by escalating in return.

Such simulation studies have a bearing on how we humans can survive as a species.

30.4 Evolutionary arms races and the life-dinner principle

'If adaptation were solely to the inanimate environment, it is easy to believe that evolution will simply track Darwin's 'elements of air and water' in their random walk through time. Selection would be stabilizing until a change in the climate or an accidental geographical displacement introduced a brief interlude of directionality. Each such directional interlude would seem to be as likely to reverse as to continue the previous one. But in fact consistent directionality is introduced because the environment of any one evolving lineage includes other evolving lineages. Above all, it is because adaptations in one lineage call forth counter-adaptations in others, setting in motion the unstable evolutionary progressions we call arms races' (Dawkins and Krebs 1979).

The ESS idea has been extended substantially by the inclusion of asymmetric conflicts, involvement of multiple species, and the possibility of learning. Such conflicts often lead to 'evolutionary arms races': Adaptations in one lineage can alter the selection pressure and call forth counter-adaptations in other interacting lineages. If this occurs reciprocally, a runaway escalation of complexity of behaviour may occur, rather like a spiralling arms race between two rival nations.

Arms races take place on evolutionary time scales. It is *lineages* that evolve, not individuals.

Several factors can make an arms race asymmetric. Before I list them, I want to mention here parenthetically that we humans can also draw some lessons from this analysis for managing our own affairs (one more example of research carried out on one complex system helping in understanding another complex system). Briefly, here are the factors listed by Dawkins and Krebs (1979):

1. *The specialist vs. the generalist factor.* A predator species that is good at hunting several species of prey is unlikely to be strongly effective against any one of them, and therefore any of these prey species stands a greater chance of outrunning the predator in the evolutionary adaptation race. If, on the other hand, the predator has specialized in hunting a particular prey, any adaptively gained advantage by the prey is more likely to be matched by an evolutionary increase in the hunting capability of the predator. Of course, it is also true that a predator able to hunt several types of prey may, on the whole, make a good living: it runs many races.

2. *Unequal selection pressure.* 'The rabbit runs faster than the fox, because *the rabbit is running for his life while the fox is only running for his dinner*' (Dawkins 1979). This *life-dinner principle* can explain the asymmetry in the arms race in favour of the prey. The penalty of failure is much higher for the prey than for the predator, resulting in unequal selection pressures.

3. *General inequalities in potential rates of evolution.* Mammals evolve faster than bivalve molluscs because of the fiercer interspecific competition in the case of mammals. It may also be true that bivalves and, say, frogs have evolved slowly because they are getting on fine as they are.

4. *Learning.* Learning is similar to natural selection in that both tend to improve the performance of individuals in a population. The difference, of course, is that learning is adaptation *during* the life time of an individual, whereas in natural selection the improvement is seen only as an average over many generations of many individuals. But in a predator-prey context, learning by a predator can reduce the chances of survival of the prey, and may reduce substantially the number of generations over which the prey would have had an upper hand because of some genetically transmitted set of improved strategies. [Epigenetic factors may complicate the matter still further in some cases.]

Interspecific asymmetric arms races

An asymmetric arms race is like an attack-defence arms race, an example being the contest between parasites and their hosts; in fact, any predator-prey contest. Offensive adaptations on one side are countered by defensive adaptations on the other.

Why are hosts of cuckoos so good at detecting cuckoo eggs, but so bad at detecting and rejecting cuckoo nestlings? Dawkins and Krebs (1979) explained it by analogy with the case of the heroin addict who knows that the drug is killing him, and yet cannot stop taking it because the drug is able to manipulate his nervous system, making him crave for the drug. There is evidence that suggests that the orange gape and the loud begging calls of the cuckoo chick have a 'supernormal' (irresistible) effect on the brain of the cuckoo. It is conceivable that, in the evolutionary arms race, cuckoos have put their adaptive emphasis on two different things in the life cycle of the chick: On mimetic deception at the egg stage (the cuckoo eggs look similar to the eggs of the host); and on manipulation of the nervous system of the host at the late nestling stage. As the host species became better and better at distinguishing cuckoo eggs from its own, cuckoos responded by evolving eggs with increased similarity to the eggs of the host species. And as the hosts evolved to become less and less susceptible to the cuckoo nestlings' begging calls for food, the nestlings' calls evolved to become more and more plaintive and irresistible. The life-dinner principle is applicable here. The cuckoo *had* to do better in the evolutionary arms race for shear survival of the species. The host species, on the other hand, had no threat to its survival by of the presence of cuckoo eggs and nestlings side by side with its own.

Predator-prey dynamics was modelled as early as the 1920s by Lotka (1925) and Volterra (1926). Population densities N_1 and N_2 of two competing prey and predator species are modelled on an evolutionary time scale by the famous *Lotka-Volterra equations*:

$$\partial N_1/\partial t = N_1(r_1 - b_1 N_2), \tag{30.7}$$

$$\partial N_2/\partial t = N_2(-r_2 - b_2 N_1). \tag{30.8}$$

In these equations, r_1 is the rate of increase of the prey population when there are no predators present; r_2 is the death rate of predators in the absence of prey; b_1 denotes the rate at which the prey are eaten up by the predators; and b_2 is the ability of the predators to catch the prey. Elaboration on this is beyond the scope of this book. For more information, see http://mathworld.wolfram.com/Lotka-VolterraEquations.html, and the references therein. The Wikipedia is also a good source for further study: http://mathworld.wolfram.com/Lotka-VolterraEquations.html.

Intraspecific asymmetric arms races

Such arms races are difficult to analyse because the contestants (parents *vs.* offspring; males *vs.* females; queen ants *vs.* worker ants; etc.) belong to the same gene pool. Any advantage gained by either of the contestants goes to the same gene pool. It is an arms race between two branches of the same 'conditional strategy', rather than an arms race between two independent lineages with different gene pools. There are queen ants and there are worker ants in an ant colony (see Section 18.2); a similar situation prevails in a beehive (Section 18.1). All female honeybees, including the queen bees, develop from larvae which are identical genetically. Those fed on a certain 'royal jelly' become fertile queens, while the rest remain sterile workers. Something similar happens in ant colonies. The contest between queen and worker ants over relative parental investment is an example of arms race of this type.

Suppose there is an ant colony in which there is only one queen, who is the mother of all members of the ant colony. What determines the male-female sex ratio in this population? One possible approach for answering this question is to use a game-theoretic model in which the contestants are genetically related (Maynard Smith 1978). The term *haplodiploidy* is used in biology for the way in which the sex of the progeny is determined by external factors like selective feeding or presence or absence of fertilization of eggs. For example, in some social insects females develop from fertilized eggs, and males from unfertilized eggs. This ensures that sisters are more closely related to each other than to their own offspring, meaning that the best chance they can give their own genes of surviving is to look after each other, instead of laying eggs of their own. This ESS has evolved in Nature several times, and forms the basis of the stability enjoyed, for example, by beehives and termite mounds (Douglas 2005b).

Interspecific symmetric arms races

In a symmetric arms race, the two sides get better and better at doing the same thing. This type of arms race is unlikely to be important. The competitors in this category would rather diverge than escalate the competition. There may be, for example, evasion of competition by 'niche separation' (Lawlor and Maynard Smith 1976).

Intraspecific symmetric arms races

Symmetric arms races are more likely to occur within the same species. They are essentially of a competitive nature, and are the stuff Darwinian evolution is made of. Adaptation to male-male competition for gaining females in a species comes under this category of evolutionary arms races (Maynard Smith 1977). Even adaptation to being eaten up by predators is an arms race of this type, because individuals less competent at this task will gradually get eliminated by natural selection.

How do arms races end?

There are several scenarios:

1. One lineage may drive the other to extinction.

2. One lineage may reach an optimum, and thus prevent the other from doing so.

3. Both sides may reach a mutual local optimum (as in the flower-bee coevolution).

4. There may be no stable end, and the arms race may cycle continuously (as in the case of humans?).

31. The Various Energy Regimes in the Evolution of Our Ecosphere

By definition, dynamical systems are those that evolve with time. If a dynamical system does not receive an input of energy, it decays towards a state of equilibrium, and then stops evolving. *Energy is the real engine that drives all evolution.*

The planet we live on is a natural spaceship, which has been receiving an input of negative-entropy solar energy throughout its existence. Mainly it is this influx of energy that keeps our planet in a state away from equilibrium which, in turn, is the reason for the ever-rising levels of its complexity. All life-forms on Earth owe their existence and sustenance to this input of energy from the Sun. And the entire ecosphere can be regarded as one single, highly complex, system: *System Earth.*

In a notable book, Niele (2005) presented *the Earth's historical energy-staircase of increasing complexity*. As becomes clear on a perusal of Niele's great book, an energy-based evolutionary approach to the history of the Earth can provide major insights into why our ecosphere is what it is.

Our energy-emitting Sun came into being ~4.6 billion years ago. The energy it emits in all directions comes from the thermonuclear fusion reactions taking place in it. Some of the energy emitted by the Sun is intercepted by our Earth, and most (though not all) of it is emitted back into outer space in due course in a degraded form.

Energy has the capacity and tendency to cause change. For the change to occur, the energy has to transform to another 'quality'. According to the second law of thermodynamics, such change and dissipation of energy entails an overall state of higher disorder or entropy. This dissipation can occur only if there are *dissipative paths* available. As we shall see below, such paths are indeed there for the energy available on Earth.

Most of the solar energy received by the atmosphere surrounding the Earth escapes from it ultimately. If this were not so, the average temperature of the Earth would go on rising. What we have instead is a fairly constant average temperature.

The tiny fraction of low-entropy (or high-quality) solar energy retained by the Earth drives processes such as photosynthesis. Some other sources of energy on the Earth are: geothermal energy; cosmic rays; and the energy released by natural and artificial radioactivity. The energy flow in and around the Earth is influenced by the energy flows in the universe. A delicate balance exists among gravitation, nuclear reactions, and radiation; this balance moderates the flow of energy.

As analysed by Niele (2005), there have been *five energy revolutions* since the origin of life on Earth:

1. The *photo-energy revolution* (emergence of photosynthesis). This occurred ~3.8 billion years ago.

2. The *oxo-energy revolution* (aerobic respiration). This occurred ~2.1 billion years ago.

3. The *pyro-energy revolution* (domestication of fire by humankind). This occurred ~0.5

million years ago.

4. The *agro-energy revolution*. This occurred ~12,000 years ago.

5. The *carbo-energy revolution*. This occurred ~400 years ago.

Each energy revolution heralded a new dominant '*energy regime*': Exposure to a new (local) energy source on Earth can herald a new dominant-energy regime, provided a new path for energy dissipation is available.

Shown in Fig. 31.1 is Niele's 'historical energy staircase' for System Earth, with near-constant influx of low-entropy energy (mainly solar energy). Six ecologically dominant energy regimes have been identified by him in the history of the Earth. These are:

(i) *Thermophilic regime*. The corresponding energy period is called the *thermion period*.

(ii) *Phototrophic regime* (*photian period*).

(iii) *Aerobic regime* (*oxian period*).

(iv) *Pyrocultural regime* (*pyrian period*).

(v) *Agrocultural regime* (*agrian period*).

(vi) *Carbocultural regime* (*carbian period*).

After human beings appeared on the evolutionary scene, each energy revolution was influenced by a '*cultural trigger*' or signal:

For the pyro-energy revolution the trigger was the origin of the human species.

The agro-energy revolution was triggered by the '*Symbolisational Signal*'.

The current carbo-energy revolution was triggered by the '*Quantificational Signal*'.

The next energy revolution of the future may be triggered by the '*Macroscopical Signal*'.

These terms will be explained below, in the respective sections.

31.1 The thermophilic energy regime

As mentioned in Chapter 27, life appeared on Earth during the thermophilic regime. It probably originated through the emergence of heat-loving or thermophilic organisms. They were the *ecologically dominant* organisms in that period; hence the term 'thermophilic energy regime' for the energy regime engendered by them. They have been called *hyperthermophiles* because they were 'hard-nosed heat lovers'.

This energy regime also saw the emergence and establishment of a metabolism mechanism for the supply of energy, with ATP as the principal cellular energy currency (Smil 2003). The hyperthermophiles used nucleotides for synthesizing DNA, and amino acids for synthesizing proteins. During this regime there was practically no oxygen in the atmosphere of the Earth, although there was plenty of carbon dioxide.

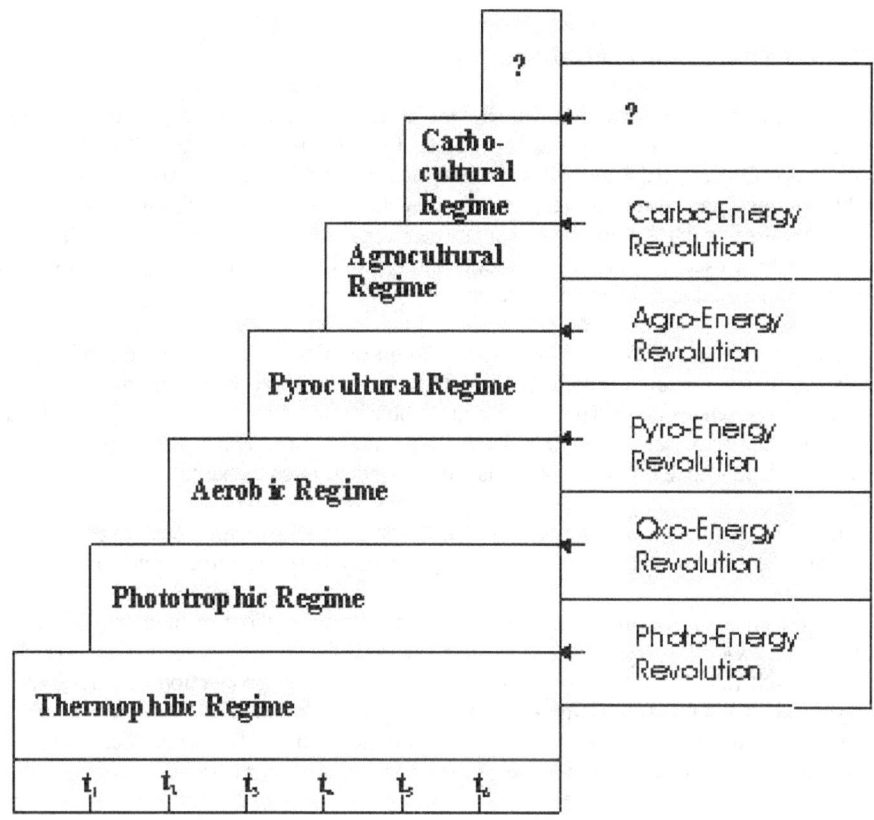

Fig. 31.1 Niele's (2005) 'historical energy staircase' for System Earth. Plotted on the x-axis in the left part of the drawing is time (not to scale). The y-axis depicts schematically the generally increasing degree of (terrestrial) complexity. Niele has identified the various energy revolutions (at times marked t_1, t_2, etc.), each such revolution heralding the onset of a specific energy regime. For example, the Photoenergy Revolution, which occurred at time t_1, marked the emergence of the Phototrophic Energy Regime. $t_1 = \sim 3.8$ billion years ago; $t_2 = \sim 2.1$ billion years ago; $t_3 = \sim 0.5$ million years ago; $t_4 = \sim 12000$ years ago; and $t_5 = \sim 400$ years ago. The time t_6 is when the next energy revolution will occur, and is a question mark at present. We can only speculate about it. One possible scenario is that t_6 will mark the emergence of a Nucleocultural Energy Regime, heralded by a Nuclear-Energy Revolution, but, as discussed in this and the next chapter, there are other possibilities also.

For higher and higher complexity to emerge, there must be an energy gradient from a local energy source to an energy sink, *and there must be an energy-dissipating pathway*. For the thermophilic energy regime:

Energy source: Primordial heat from the accretion events during the formation of the planet Earth.

Energy sink: The cold atmosphere above the seas.

Energy-dissipating pathways: Chemical evolution, autocatalytic processes/metabolism.

Chief drivers: The hyperthermophiles.

31.2 The phototrophic energy regime

After the thermophilic energy regime, the next to emerge was the phototrophic energy regime. It was dominated by solar energy as the source for the energy gradient. This came about because some of the hyperthermophiles reached the surface of the sea, where they encountered sunlight. Chemical and biological adaptation and evolution followed, enabling them to develop a new metabolism which did not depend anymore on the energy provided by the hydrothermal vents, but instead used solar energy through photosynthesis. These solar-energy-dissipating organisms established the phototrophic regime.

Two major survival tools emerged: Fixing of carbon dioxide; and the stripping of hydrogen from water (which liberated oxygen). The newly evolved microorganisms doing this were *cyanobacteria* or *blue-greens*. They produced carbohydrates from carbon dioxide and water, and gradually built up the molecular-oxygen (O_2) content of the Earth's atmosphere as a by-product. The dependence on DNA, proteins, and ATP continued as before.

The colour of the blue-greens comes from the chlorophylls in them. These pigments act as 'molecular solar panels', harvesting solar energy and converting it into chemical energy.

As taught in elementary chemistry classes, 'loss of electrons is oxidation, and gain of electrons is reduction' (LEOGER). The blue-greens strip electrons from water molecules, thus releasing hydrogen which (along with carbon dioxide) gets used in the production of carbohydrates. Photosynthesis amounts to sunlight-driven conversion of carbon dioxide and water into carbohydrates and oxygen. Since airborne carbon dioxide is the only source of carbon that the blue-greens use, we can say that they create *organic matter from inorganic matter*.

Phototrophy literally means use of light as an energy source. In the phototrophic regime, solar light was the dominant energy source for the energy-dissipating pathway for the sustenance and further evolution of life. The blue-green bacteria were the chief drivers of the biochemical cycle during this regime. The oxygen-producing ('oxygenic') photosynthesis mechanism evolved by them enabled an increase in the organic productivity by two to three orders of magnitude, compared to what was done by the hyperthermophiles in the thermophilic regime.

The Earth's atmosphere in the early thermoic era was mostly carbon dioxide, and practically no molecular oxygen. Geochemical processes buried much of the carbon dioxide as silicate-carbonates, and, in addition, biochemical processes converted this gas to bioorganic matter. Similarly, the molecular oxygen liberated by the photosynthesis processes was not available initially as atmospheric gas. Instead, much of it (~97%) was captured by rocks, volcanic gases, and upwelling oceanic iron particles. This was a slow but irreversible process. Only after it was completed (~2.2 billion years ago) did the oxygen gas start permeating the atmosphere surrounding the earth.

Within a few hundred thousand years the atmospheric oxygen levels rose from less than 1% to ~15% of present-day levels. The air became more breathable.

Thus, for the phototrophic energy regime:

Energy source: Solar light.

Energy sink: Chemical energy.

Energy-dissipating pathways: Photochemical reactions; photosynthetic life forms; other solar-energy-dissipating superstructures in the ecosphere.

<u>Chief drivers</u>: The cyanobacteria (blue-greens).

31.3 The aerobic energy regime

The release of molecular oxygen as a waste product of the photosynthetic process by the blue-greens fell into a positive-feedback loop: Abundant availability of solar light made the population of the blue-greens to grow, producing more and more free oxygen. But oxygen itself was *poison* to these organisms. That was a crisis situation indeed.

Evolutionary adaptation led to the development of a new kind of cell, namely the eukaryotic cell. Such a cell had *organelles*, which have the feature that they are enclosed in membranes. As we shall see below, emergence of the eukaryotic cell resolved the crisis mentioned above.

In due course, more complex, multi-cellular life forms emerged, and dominated this energy regime, *the aerobic regime*, in which respiration provided the main fuel-burning mechanism. Before the emergence of the eukaryotic cell, all life on Earth had existed as bacteria and archaea only (for over a billion years).

The atmospheric oxygen was conducive to the aerobes, but poison for the anaerobic blue-greens. But the blue-greens did not simply fade away in such a situation, as they were instrumental not only in the production of molecular oxygen but also food for the respiring aerobes. Therefore the build-up of oxygen in the atmosphere was a threat to *both* types of organisms: a direct threat to the anaerobes, and an indirect threat to the aerobes. How did they still manage to survive and flourish? By taking the *symbiont-integration* evolutionary route I described in Chapter 29 (courtesy Margulis and Sagan 2002), resulting in the emergence of the eukaryotic cell.

The evolution of a symbiotic 'pact' between oxygenic photosynthesis and aerobic respiration was at the heart of the oxo-energy revolution, resulting in the emergence of the aerobic energy regime (Niele 2005).

This strategic alliance between 'light eaters' and 'oxygen breathers' not only saved the light-harvesting technology of the blue-greens, it also increased by an order of magnitude the photosynthetic metabolism. The evolved eukaryotic cell design embodied sunlight-harvesting photosynthesis, and protection against oxygen toxicity. Its *highly efficient* metabolic combustion via aerobic respiration triggered the appearance of multicellular life forms, which, in turn, led to the emergence of still more complex life forms and ecosystems. Humans appeared on the scene in due course, and this was a development with unprecedented consequences.

The eukaryotic organisms have continued to coexist with the prokaryotic organisms (namely the bacteria and the archaea) in several schemes. In fact, the prokaryotes 'maintain the foundation of all functioning ecosystems on this planet'. An example is the nitrogen that bacteria make available for biological processes.

For the aerobic regime:

<u>Energy source</u>: Photosynthetic carbohydrates together with free oxygen.

<u>Energy sink</u>: Carbon dioxide plus water.

Energy-dissipating pathway: Aerobic respiration.

Chief driver: The eukaryotic cell.

Aerobic respiration produced 18 times more ATP (the cell fuel) from carbohydrates than the anaerobic processes reigning till then. The aerobic regime saw a tremendous growth in biomass, most of which ended up ultimately as fossilised minerals.

We humans emerged near the end of the aerobic energy regime. We are '*thinking reeds*'. Our emergence on the scene had a societal and cultural aspect, which impacted very strongly the otherwise purely raw, blind-forces-of-Nature evolution of the energy regimes. What is more, our existence has led to a very rapid increase in the rate of increase of complexity. It is a qualitatively different scenario now, and I continue the narrative in the first chapter of Part III of the book.

III. Humans and the Evolution of Complexity

At present there is no definite knowledge that life comparable to humans in intelligence exists anywhere else in our universe. Therefore our emergence on Earth, as a result of millions of years of evolution of cosmic complexity, can be taken as an event of cosmic importance. The reason is that the rate of increase of complexity has been increasing immensely, in fact exponentially, after we emerged. What is more, we humans have the ability and knowledge to *control* the course of evolution, both biological and nonbiological. We have the ability to create computers and other sophisticated machines which can surpass us in all our capabilities, including intelligence. These 'mind children' of ours can further enhance their own capabilities in ways which we cannot even imagine, and can colonize the universe in due course.

This part of the book covers, among other things, the evolution of artificial intelligence and robotics. Till our machines take over from us, we humans will continue to contribute to the evolution of complexity in many ways. Even afterwards, our fate will depend in a critical way on how we manage our science, technology, and world affairs at present. Needless to say, understanding complex adaptive systems well enough is the need of the hour. We ourselves are nothing but one big complex adaptive system.

The advance from the simple to the complex, through a process of successive differentiations, is seen alike in the earliest changes of the Universe to which we can reason our way back, and in the earliest changes which we can inductively establish; it is seen in the geologic and climatic evolution of the Earth; it is seen in the unfolding of every single organism on its surface, and in the multiplication of kinds of organisms; it is seen in the evolution of Humanity, whether contemplated in the civilized individual, or in the aggregate of races; it is seen in the evolution of Society in respect alike of its political, its religious, and its economical organization; and it is seen in the evolution of all those endless concrete and abstract products of human activity which constitute the environment of our daily life. From the remotest past which Science can fathom, up to the novelties of yesterday, that in which Progress essentially consists, is the transformation of the homogeneous into the heterogeneous.

Herbert Spencer
Progress: Its Law and Cause (1857), 35.

What secret it is that allows nature seemingly so effortlessly to produce so much that appears to us so complex (?)

S. Wolfram (2002)

The human mind is an indefatigable pattern seeker. In order to survive in a hostile world, we have evolved sensitivity to patterns, which we use to predict what is going to happen to us. Human beings like inventing patterns . . . Science is about digging out the secret patterns that make the universe tick. We call these patterns the laws of nature. The most effective way for human beings to think about patterns is to use mathematics. So we've convinced ourselves that the laws of nature are mathematical.

Ian Stewart, *What Shape is a Snowflake*

32. Evolution of Niele's Energy Staircase After the Emergence of Humans

Let us continue from where we left off in Chapter 31 on Niele's energy staircase of increasing complexity. Our emergence on Mother Earth marked the end of an energy regime, namely the aerobic regime. This happened because we are so different in terms of intelligence from all the existing life forms. We not only have a mind of our own, this mind has a formidable capacity for changing things, for better or for worse. We humans started what Niele (2005) calls the pyrocultural energy regime.

32.1 The pyrocultural energy regime

Humans have been instrumental in the creation of an '*anthroposphere*', comprising of the following four '*anthroposystems*' (Niele 2005):

Human knowing (leading to discovery, or new observation).

Human capacity (leading to invention, or new creation).

Human action (leading to innovation, or new practice).

Human living (leading to diffusion, or new way of living).

The anthroposphere emerged ~2.5 million years ago. Early humans noticed the hardness of stones and the sharpness of some shapes of stones. This was 'discovery' or *earth wisdom* (the first of the four anthroposystems listed above).

The next stage was 'invention', namely the creation of tools (axes, cleavers, picks) by striking stone against stone (*stone technology*).

Innovation followed invention. The invented tools or artefacts were used for procuring and processing food (*foraging and scavenging*).

All this changed the way of living, an example being the emergence of the practice of *cave dwelling*.

Man the toolmaker and cave dweller could survive and thrive through his Earth wisdom or comprehension of his surroundings. The dominance of the human species triggered the *pyro-energy revolution*, resulting in the pyrocultural energy regime. The aerobic regime had changed the face of the Earth. The new-look planet got ~20% oxygen in the atmosphere, and supported plants and animals. Niele (2005) has pointed out another important fallout of the aerobic regime, namely the appearance of *wild fire* on the scene. A new energy gradient had emerged, with wood plus oxygen serving as the energy source. The energy sink for this gradient was carbon dioxide plus water.

In due course, humans acquired mastery over fire. This was a development with far-reaching consequences. *Anthropogenic fire can be said to have marked the beginning of the human civilization*. It engendered the beginning of the pyrocultural energy regime. The new energy-dissipating structure (based on wood-burning) was *societal* in nature. Fire mastery meant several things: heating; lighting; roasting of food; scaring away animals, and most significantly, the emergent social intercourse around the fireplace.

The societal aspect of the pyrocultural energy regime had ever-spiralling fallouts. The ever-increasing energy dissipation (through burning of wood) took the System Earth farther and farther away from equilibrium, leading to the emergence of new kinds of complexity. Since the fire economy was a societal dissipative structure, the emergent phenomena were cultural by nature (Niele 2005). As people tended to assemble around the fireplace, emergent phenomena like coordination, communication, spoken languages, symbolic thinking, etc. were the result.

The driving force here was solar energy trapped in wood, and the shaping force was *human ingenuity*.

Thus, for the pyrocultural energy regime:

Energy source: Wood plus oxygen for creating controlled fire.

Energy sink: Carbon dioxide plus water.

Energy-dissipating pathway: Societal structure around the fireplace.

Chief drivers: Humans.

According to Niele (2005) the key phrases for the four anthroposystems characterizing the pyrocultural regime are: *Symbolic thinking* (discovery); *fire technology* (invention); *hunting and cooking* (innovation); and *nomadic bands* (new way of living).

Humans had observed wild fires and the burning of wood, and also experienced the heat of the fire. They soon learnt how to create and sustain this fire in a controlled manner. The fireplace became a daily practice, making cave dwelling more attractive. This had a major influence on the life style of people. They could not only hunt with their tools, they could also cook the food.

There is a book *Modernist Cuisine*: *The Art and Science of Cooking*, by Nathan Myhrvold *et al*. (2011). Among other things it explains in an appetising manner how cooking made humans smarter. Cooked food is akin to pre-digested food in certain ways. Therefore it takes the load off the intestines, thus making extra energy available for the brain. This was one of the factors leading to an increased brain size of humans, compared to the other apes.

Since humans cook their food, they spend just 5% of the day eating. Uncooked food is hard and stringy, requiring hours of chewing and still not giving the same level of nourishment. The extra time available to early humans enabled them, among other things, to look for new kinds of food, gather fruits, or lie in wait of animals for hunting.

The fireside not only resulted in the emergence of nomadic culture, it also provided the right milieu for the development of symbolic thinking: The fireside became the hub of social evolution, and its most important fallout was the uniquely human trait of symbolic thinking. *This led to the development of language*, as also an increasingly sophisticated way of looking at Nature. The coevolution of brainpower and technology, or 'memes' and artefacts, accelerated. [As I shall discuss later, 'memes' are the social equivalent of genes.] Humans even created 'non-useful' artefacts like jewellery and musical instruments.

Emergence of symbolic 'doings' like these has been viewed by Niele (2005) as a *Symbolisational Signal*, which triggered the agro-energy revolution, and the consequent agrocultural energy regime.

32.2 The agrocultural energy regime

The booming fire economy of the pyrocultural energy regime led to the emergence of agriculture. Humans observed that, when a piece of land was charred by wildfire, it got cleared of the forest and, what is more, new plants sprouted in the resurrected land, under certain conditions. They burnt forests to clear more and more land for sowing seeds of edible crops. Agriculture was a far more efficient mode for acquiring food, compared to hunting and gathering. Thus evolved *the agrocultural energy regime*.

The ever present influx of solar energy and its utilization through agriculture led to the emergence of new features like crafts, villages, a growing population, and new energy chains. 'Cooking' got elevated and diversified to other technologies like baking of bricks and making of glass, as also the processing of iron ore (see below). Alphabet and money also emerged, among several other such things. The inexorable march towards ever-increasing complexity continues to this day. For example, we now grow crystals of complex materials in the laboratory, for applications in technology. Development of carefully patterned nanocomposites is another such activity.

For the agrocultural energy regime:

Energy source: Crops (requiring seeds, water, carbon dioxide, and solar light).

Energy sink: Carbon dioxide plus water.

Energy-dissipating pathways: Various social and cultural activities of humans.

Chief drivers: Humans.

As argued by Niele (2005), this description of the agrocultural regime must be supplanted with the socio-technological description. With the emergence of agriculture, the nomadic way of life gave way to a more sedentary settled-down lifestyle, leading to farms and villages. Another life-style-changing invention was pottery. It must have been observed that materials used for making the hearth got hardened by the heat treatment. This discovery led to pottery-making. Several innovations like pots, dishes, and ovens followed. Application of the oven improved the cooking process. Use of ceramic pots for storage of various kinds of edible items increased their shelf life.

The observation of the effect of heat on material properties was the forerunner of the *evolution of the empirical sciences*. Invention of other metallurgical techniques followed. Innovations resulting from these inventions include cooking utensils, ornaments, and weapons. The increased economic diversity in goods and trade engendered barter trade.

Niele has listed the four anthroposystems of the agrocultural regime as: *practical know-how*; *agricultural technology*; *farming and bartering*; and *farms and villages*.

During this regime, humans developed the quality 'to measure reality'. According to Niele, 'a strong signal surged around circa 1250 to 1350 near the end of the Agricultural Regime: this is termed the *Quantificational Signal*'. This signal touched all the anthroposystems, prompting 'modern science, technology, business practice, and bureaucracy' (Crosby 1997).

32.3 The carbocultural energy regime

Another energy revolution, namely the carbo-energy revolution, occurred ~400 years ago when humans discovered a fuel other than wood, namely fossil fuel (coal, petroleum, natural gas). This marked the onset of the carbocultural energy regime. The fossil fuel had been created in the aerobic regime by the deposition of large volumes of dead biomass deep inside the Earth's crust. This fossilization amounted to the conversion of carbohydrates of biological origin to mineral hydrocarbons.

Discovery of this new form of fuel resulted not just in its use in place of wood for burning, but led eventually to the development of *the combustion engine*. This development had truly far-reaching consequences. The engine converted heat to mechanical movement, resulting in locomotion, electricity production, etc.

The availability of energy in a convenient form (electricity) led to a whole new set of societal energy-dissipating structures and emergent phenomena, apart from a phenomenal growth in population and economies. Niele lists some of these developments as: quantum mechanics, antibiotics, pop music, the world-wide web, man on the moon, cities, the United Nations, unions, buildings, vehicles, medicines, computer networks, mobile phones, etc.

The explosive growth in the exploitation of fossil fuels has resulted in a steady build-up of the amount of carbon emissions into the atmosphere, which is now a cause for serious concern.

For the carbo-cultural energy regime:

<u>Energy source</u>: Fossil fuel plus oxygen.

<u>Energy sink</u>: Carbon dioxide plus water.

<u>Energy-dissipating pathway</u>: Burning of fossil fuel in combustion engines.

<u>Chief drivers</u>: Human activities.

The agro-cultural regime and the carbo-cultural regime also saw the emergence of wind power, solar power, hydroelectric power, and nuclear power. But none of these has yet risen to the level of *ecological dominance* for naming an energy period after any of them.

On the socio-technological side, the carbocultural regime saw the grand alliance of science and crafts, giving rise to technology as we know it today. It is based largely on fossil fuels. Humankind progressed from 'farms and villages' to 'cities and nations'.

Niele has identified the four anthroposystems of the carbocultural as: *reductionistic science*; *conversion technology*; *manufacturing and trading*; and *cities and nations*.

Reductionistic science, while hugely successful when applied to simple or simplifiable systems, has its limitations when we try to understand complex systems. One big reason why we now have the audacity and the capability to attack research problems involving complex systems is the evolution of increasingly powerful computers. A whole new field of study called 'Computational Intelligence' has emerged, which I shall take up in the next chapter.

32.4 The green-valley approach to System Earth

As explained above, there have been three energy cultures after we humans appeared on the scene: the pyroculture, the agroculture, and the present carboculture. But now a fourth one is in the offing. Why?

Humans in the carbocultural energy regime are turning against themselves by exceeding the carrying capacity of the habitat. This is a good example of how history repeats itself sometimes, because similar things happened in the pyrocultural regime and the agrocultural regime as well. In the pyrocultural regime, when the overshooting of the carrying capacity of the habitat occurred, the Symbolisational Signal provided a new perception of reality, which enabled humans to increase the carrying capacity of the habitat by inventing agriculture. But in due course the agrocultural regime also reached a stage wherein the carrying capacity of the habitat was exceeded. Once again, another signal, namely the Quantificational Signal, provided the way out in the form of exploitation of fossil fuels, heralding the emergence of the carbocultural energy regime.

We are now in the carbocultural regime, and there is a clear signal about another overshooting of the carrying capacity of the habitat. What we are now seeing is the *Macroscopical Signal* (Niele 2005).

The meaning of the term 'macroscope' is just the opposite of that of 'microscope'. Whereas a microscope magnifies and shows detail at small length scales (a case of zooming in), a macroscope is a 'symbolic instrument' which combines data from various sources and presents the big picture in a way we can comprehend (a case of zooming out). de Rosnay (1979) introduced this conceptual tool for investigating highly complex systems. At present a variety of macroscopical signals are impinging on our consciousness, and are making us acutely aware of problems like the global warming.

Ecological footprint (more commonly called carbon footprint) is another important term in this context. It is 'the area of productive land and water that people need to support their consumption and to dispose of waste'. The Macroscopical Signal is telling us that our ecological footprint is overshooting the carrying capacity of the habitat, and this can be very dangerous.

But our response to this signal is not at all unanimous. Two broad viewpoints have emerged (Worster1994): The *'imperial view'* and the *'Arcadian view*. The former is an aggressive approach, aiming to control Nature. The latter advocates humility in the face of forces of Nature, and aims at a life of harmony and peaceful coexistence with other creatures, advocating a reduction in the size of our current ecological footprint, so that long-term sustainability can be attained.

The imperial approach was first advocated by the highly influential 16th century philosopher Francis Bacon. According to Worster (1994), 'Bacon promised to the world a manmade paradise, to be rendered astonishingly fertile by science and human management. In that utopia, he predicted, man would recover a place of dignity and order, as well as authority over all the other creatures he once enjoyed in the Garden of Eden. Where the Arcadian naturalist exemplified a life of quiet reverence before the natural world, Bacon's hero was a man of "Active Science", busy studying how he might remake nature and improve the human estate. Instead of humility, Bacon was all for self-assertiveness: "the enlargement of the bounds of Human Empire, to the effecting of all things possible". . . "The world is made for man", he announced, "not man for the world"'. I shall discuss this approach in the next section.

The Arcadian Man, on the other hand, believes that it is futile to try to conquer Nature, and that the most sensible thing to do is to live in harmony with it, and to ensure that all the other creatures with whom we share Mother Earth get their due portion of the bounty. If this requires a reversal of the clock for shrinking our current ecological footprint, then so be it. The Arcadian Man has no use for nuclear energy, nanotechnology, or genetic engineering. Even economic growth must be arrested, even reversed, if it has a deleterious effect on the ecosphere.

The Arcadian Man aims at using solar power, and emulating Mother Nature in cycling matter in (nearly) closed loops, thus taking the carbocultural regime towards the '*Green Valley*'. Four hundred years ago, at the start of the Carbian Period, the man-made emissions of carbon dioxide, resulting from the use of fossil fuels, were so small that they got readily processed and absorbed by green plants by photosynthesis. But today these emissions have reached more than 25 Gtons per year, and natural processes can fix only a part of it into solid forms. Therefore it is no longer tenable to go on following the practice of mostly 'linear' once-through conversion of natural resources into human waste. The Macroscopical Signal is loud and clear. We must resort to recycling matter in nearly-closed-loops metabolisms, so that the increasing burden on the ecosphere can be reversed. Innovative means must also be found for sequestering the carbon dioxide gas released into the atmosphere. Some possibilities are: reforestation; chemical fixation; and injection into geological formations.

But is the Green Valley approach really the best thing to do? In the next few sections I shall discuss some alternative ideas, and then describe the symbiotic approach discussed by Niele (2005). He foresees the emergence of a '*heliocultural energy regime*' as the panacea for our current and near-future ecological problems.

32.5 The imperial approach to System Earth

The scientific and technological achievements of humans have been remarkable, and we have now entered a technologically explosive phase of our evolution. Naturally, the mood is often upbeat and there is no shortage of people who are confident that we can overcome the ecological footprint problem mentioned above. Nanotechnology may provide some critical breakthroughs for this. Progress in the burgeoning field of artificial smart structures (Wadhawan 2007) may well lead to the emergence of *cyborgs* (creatures which are part human part machine) and *Robo sapiens* (robots so advanced that they would leave their creators, i.e., humans, far behind in practically all respects). Intelligent robots, cyborgs, and genetically engineered humans are expected to be not only less delicate when it comes to survival in harsher conditions, *they will actually be far more efficient consumers of energy, thus having a lower ecological footprint*.

Francis Bacon played a major role in establishing what we now call the scientific method (I outlined it in Section 2.1). He was also the original proponent of the imperial approach, namely the aggressive use of science and technology for conquering Nature and establishing the supremacy of humans in all things that matter.

Michio Kaku is, in a way, a modern-day advocate of the imperial approach. In his book *Visions: How Science Will Revolutionize the 21st Century*, he visualizes a 'phase transition' for humanity from passive observer to active choreographer of Nature. Genetic modification 'will ultimately give us the nearly God-like ability to manipulate life almost at will'. Concepts like 'gene doping' and 'synthetic life' are already being explored by biochemists. According to Gibbs (2004): '*Biologists are crafting libraries of interchangeable DNA parts and assembling them inside microbes to create programmable living machines ... Evolution is a wellspring of creativity ... But there is still plenty of room for improvement.*'

How would the energy needs be met for this scenario? Fossil fuels cannot last forever. Conventional oil and gas will be the first to go. Coal can last a little longer. Unconventional oils (oil shales, heavy oils, tar sands) can stretch the economically feasible fossil-fuels era by a century. Unconventional gas sources (methane in coal-beds and in other deposits such as the tight reservoirs and the high-pressure aquifers, as also the methane in hydrates) can also be exploited for some time, provided the necessary technology becomes available in an economically viable manner. The gas hydrates comprising of huge amounts of combustible carbon offer a potentially large source of energy, although expert opinion is divided about their economical and ecologically clean exploitation (Smil 2003).

A new dominant energy source other than fossil fuels must emerge, and according to the Imperial Man it must be *nuclear energy*.

Nuclear fission is already being exploited for power production on a commercial scale. The so-called second-generation nuclear reactors produce ~16% of the total electricity we consume (Smil 2003). Third-generation reactors, with better safety and productivity features, went into operation a few years ago. Fourth-generation reactors, based on totally new approaches, are in the pipeline. But can nuclear-fission reactors dominate the energy scene for a long time to come, resulting in the emergence of a possible nucleocultural energy regime, superseding the present carbocultural regime?

Sustained research and development work can perhaps make available a large supply of fissile nuclear fuel, which may last for centuries, if not millennia. This hope is based on the utilization of thorium, after uranium stocks have been exhausted. Breeder reactors add further to this sense of optimism. But all this cannot be taken for granted, because it is difficult to predict the course of scientific and technological development.

Nuclear *fusion*, rather than fission, offers another kind of hope for the possible emergence of a nucleocultural regime, provided certain technological hurdles can be overcome. Nuclear fusion involves the fusing together of two isotopes of hydrogen (deuterium and tritium; or deuterium and deuterium), overcoming the strong Coulombic repulsion between them by making their velocities very high. This is done by heating them to ~50 million degrees Celsius. Once the Coulombic barrier has been effectively pierced, the very strong and attractive nuclear interaction comes into play, and the end result is the formation of the very stable helium nucleus. The mass of the helium nucleus is a little less than the sum total of the masses of the two nuclei that are fused. So the excess matter/energy is released as kinetic energy or heat. This thermonuclear process is what has been going on in the Sun, and is also what makes a hydrogen bomb possible.

Tritium for this reaction must be obtained from a nuclear reaction using lithium, and the latter is available in plenty in the Earth's crust, and also in the seas. Deuterium is also available almost limitlessly in seawater.

Both fission and fusion operate with hardly any emission of greenhouse gases.

It may be possible to operate a commercial fusion reactor by the end of this century, if not earlier. But this is only an estimate. One can never be sure about such things. Whether or not a nucleocultural energy regime will emerge is a difficult question to answer. The difficulty stems from the inherently unpredictable nature of complex systems. The ecosphere is certainly a most complex system. And so are human affairs. The complexity is not only of a scientific or technological nature, but also involves socio-economic issues and political decisions.

In the public mind a major misgiving regarding nuclear reactors is about safety and long-term health hazards from radioactive waste. This needs a close, dispassionate look, and I shall devote the next section to it, and to some other aspects of the nuclear-energy option.

32.6 A nucleocultural energy regime?

No power is more expensive than no power.
Homi Bhabha

At present we are facing a major carbon-footprint problem which will only get worse with time. Nuclear energy is attractive at least on this count; there is little emission of carbon dioxide in the entire nuclear cycle from ore mining to radioactive-waste management. And if nuclear fusion also becomes available as a source of power, there is the possibility of a near-perennial solution of all our power-production woes.

How much per capita production of electric power should we humans aim for? The United Nations has worked out a *Human Development Index* (HDI) as a measure of the quality of life. Power consumption by a population is a good indicator of its technological progress and comfort level. An interesting fact is that the HDI does not go on increasing indefinitely as a population consumes more and more electricity. There is a saturation effect, although initially the HDI does increase with increasing consumption of power. Beyond ~9000 kWh per person per year there is no significant increase in the HDI.

Norway, Canada, USA, Australia and Argentina have an HDI of ~0.95 (the highest), and their power consumption ranges from ~25000 to ~2500 kWh per person per year (Fig.32.1). Some of the lowest HDIs are: 0.31 (Niger), 0.42 (Zambia), 0.55 (Pakistan), and 0.62 (India). The power consumption of these countries varies from ~0 (Niger) to ~700 kWh per person per year. For China the HDI is ~0.78, and the per capita power consumption is ~1500 kWh per annum.

The world average for the HDI is 0.741, and that for power consumption is 2490 kWh per person per year. Beyond ~6500 kWh per person per year there is no substantial increase in the HDI.

Much of the information I am giving here is from an article by an ex-Chairman (Dr. Anil Kakodkar) of the Atomic Energy Commission of India, published in the April 2012 issue of *Physics News* (India). He suggests that India should aim for 5000 kWh per person per year. If India manages to stabilize its population at 1.6 billion, it would then need ~8 trillion kWh of electric power, a full 40% of the global total generation of electricity.

Every year the world emits ~30 billion tons of carbon dioxide, of which the Indian contribution at present is ~1.7 billion tons. The present power production in India is ~0.8 trillion kWh, so India would produce 17 billion tons of carbon dioxide per annum if the power production goes up to 8 trillion kWh. The consequent effect on climate change would be unacceptably high unless we harness nuclear energy (and also solar energy) in a big way.

India has a three-stage nuclear-fission programme, which has been formulated keeping in mind the large amounts of thorium available indigenously. Thorium is a 'fertile' material, and not a 'fissile' material. This means that it has to undergo 'breeding' in a nuclear reactor to become a fissile material usable as a fuel for producing nuclear power.

Stage 1 of the Indian nuclear programme, which is currently underway, uses uranium in thermal nuclear reactors. But such reactors are not very efficient for breeding fissile materials. By contrast, a 'fast breeder reactor' can produce much more fissile material than it consumes. Such reactors comprise Stage 2 of the Indian programme. This is how large amounts of thorium will be converted to fissile material, namely U^{233}. The first such 500 MWe commercial fast-breeder reactor is now commissioned.

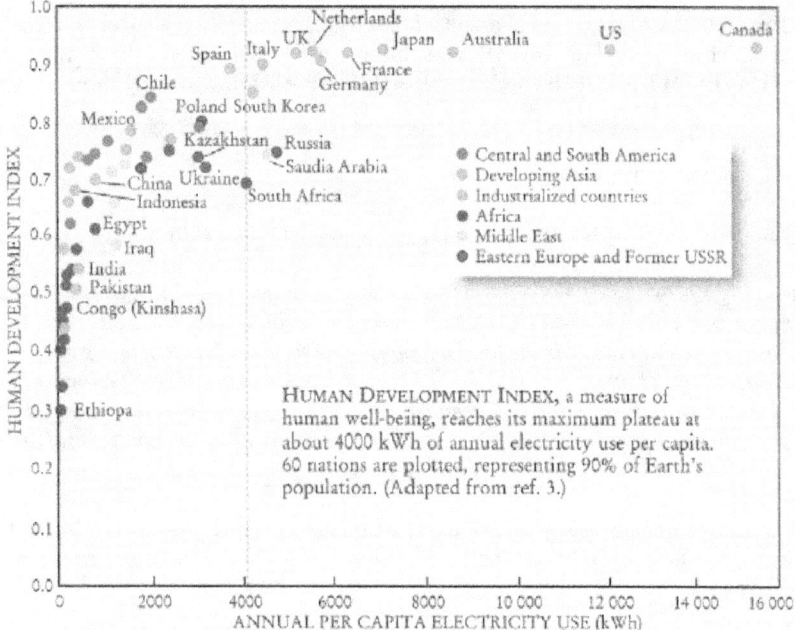

Figure 1.2. Human development index vs. per capita electricity use for selected countries. Taken from S. Benka, *Physics Today* (April 2002), pg 39, and adapted from A. Pasternak, Lawrence Livermore National Laboratory rep. no. UCRL-ID-140773.

Fig. 32.1 HDI *vs.* per capita use of electricity by some countries.
http://www.geni.org/globalenergy/library/energytrends/currentusage/index.shtml

Indian reactors in Stage 3 will use U^{233} as fuel, and will be able to meet the Indian power needs for the next 200 years.

The opposition to nuclear programmes comes from the perception of short-term and long-term safety problems, with Fukushima fresh in the public memory. But here are some facts:

1. The death rate per TWh from the various energy-production options is the least for nuclear energy.

2. No correlation has been established between the incidence of cancer and *low level* radiation. The nuclear industry follows what is called the '*linear no threshold principle*'. It is a conservatively formulated principle which deliberately ignores the fact that for *low* radiation levels the consequences do not vary linearly with level of radiation. In case of a nuclear accident, one calculates the area affected by the release of radioactivity and the number of people living in that area. If the number of people is large, the calculated consequences work out to be huge, even for very low levels of radiation. This is what the anti-nuclear lobby exploits, ignoring the fact that the linear correlation is NOT applicable for low levels of radiation. According to the American Health Physics Association, there is no risk to human life up to 100,000 microSv life-time dose of natural radiation.

3. In the Fukushima episode, no worker was exposed to a dose of radiation with a significant probability for a serious health consequence. It was the tsunami that killed more than 13000

people, with another 14000 missing. [However, it is also true that the radioactive fallout has gradually affected a large population in many parts of the world.]

4. To address the risk of diversion of nuclear materials, 4th-generation nuclear reactors are being designed which will provide cost-effective, clean, and reliable energy with minimal risk of proliferation. The Advanced Heavy Water Reactor (AHWR) designed by India is an example of that.

Above all, *nuclear power is largely green.*

I end by quoting from Dr. Kakodkar's (2012) article, mentioned above:

Nuclear energy provides 16% of the world's electricity today and it has been supplied for decades in a cost effective manner. Despite Fukushima, or Chernobyl or Three Mile Island, the real risk of nuclear energy is the lowest among various other forms of energy in commercial use. The advantage of nuclear would be seen to be much greater if one factors in additional risks associated with the predicted consequences of climate change as a result of use of fossil energy in business as usual manner. That could well be a bigger killer than several atom bombs together.

But the collective human psyche is not exactly famous for rational thinking. Should we look for options other than entirely nuclear? I think we should. In the next section I describe one such scenario.

32.7 A possible 'heliocultural' energy regime

Niele (2005) has argued in favour of drawing inspiration from what was done by the blue-greens, two billion years ago (see Section 31.2). They were the rulers of the phototrophic regime, just as we are the rulers of the current carbocultural regime. They went for a partnership, or *symbiosis*: The self-induced crisis of oxygen emission, which was poison for the blue-greens, was overcome by the evolution of a new type of cell, the eukaryotic cell, *which had organelles limited by membranes.* In the new (aerobic) regime, respiration provided the main fuel-burning mechanism: The atmospheric oxygen was conducive to the aerobes, but poison for the blue-greens. The evolution of a symbiotic 'pact' between oxygenic photosynthesis and aerobic respiration was at the heart of the oxo-energy revolution, resulting in the emergence of the aerobic regime. The eukaryotic cell design embodied sunlight-harvesting photosynthesis, as well as protection against oxygen toxicity.

A similar symbiosis can happen again, this time for saving man from the consequences of the loud and clear macroscopical signal of unsustainability. We may be heading for the emergence of '*Symbian Man*' (Niele 2005), who will effect a symbiosis of the various energy options. It will be a symbiosis of many things, *born out of the fundamental perception that System Earth is one big complex superorganism.* As envisioned by Niele (2005), it would be a symbiosis between:

- Imperial man and Arcadian man.
- Scientific reductionism and scientific holism, or simplicity and complexity.
- Knowledge of natural disciplines and knowledge of cultural disciplines.
- Anthropocentrism and ecocentrism.
- 'Nature mastery' and 'back to Nature'.
- Techno-scientific virtues and socio-ethical virtues.

Although energy from the Sun will form the backbone of this regime, there are other renewable-energy options also. To quote Smil (2003): 'Beyond the fossil fuels the world can tap several enormous renewable flows: direct solar radiation and wind energy in the accessible layer of the troposphere are both several orders of magnitude larger than the current global total primary energy supply and they can be supplemented by hydroenergy and geothermal flows'. The watchword will be: *sustainable energy* (Fig. 32.2).

The Symbian Man will consciously bring about the '*Heliocultural Energy Revolution*' (Niele 2005). The aim will be to develop '*closed-loop*' technologies enabling solar-driven *recycling* of matter. Even wind energy is of solar origin. There will be partnerships or symbiotic relationships of all kinds: from local to regional to global, and emphasis will be on *integrated and cascaded flows of renewable energy and recyclable matter*.

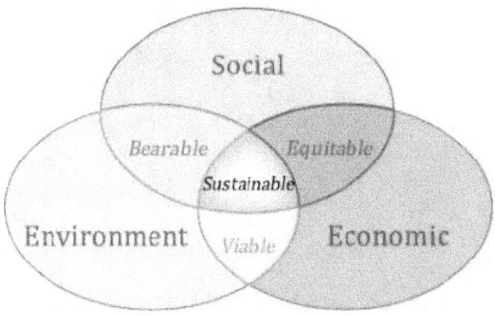

Fig. 32.2 Factors for achieving sustainable energy (Adams 2006).
http://cmsdata.iucn.org/downloads/iucn_future_of_sustanability.pdf

The energy carriers will be green electricity, solar hydrogen, and green biofuels. 'The beauty of green biofuels is that Nature looks after carbon recycling through photosynthesis, with energy storage for free. And albeit efficiencies are relatively low, residues of food and wood production nevertheless grow. In the Sun Valley, socio-metabolisms could differ regionally with ecological circumstances and heliocultures. They show optimised configurations of large-, small- and also medium-scale solutions . . ' (Niele 2005).

The Symbian approach favours distributed small-scale and decentralised medium-scale socio-metabolic sites and corresponding infrastructures.

To suppress the discharge of the greenhouse gas carbon dioxide into the environment, the Symbian Man will seek to exploit geological and chemical sequestration.

An interesting aside regarding carbon dioxide is the large amount of this and other greenhouse gases released by cattle: They emit from both ends! Their population should be reduced. In fact, there is a strong case for reduction in the use of food products of animal origin. I have written at length about this in my blog post 'Climate Justice and What We Eat' (Wadhawan 2015). Their production is very energy-intensive, with a very large carbon footprint. Humans should move towards a larger use of foods of plant origin. Not to mention the fact that a voluntary and phased reduction of the total human population will also be an important step in the right direction.

Centralized/remote production or processing of electricity, drinking water, sewage, food, and fuels results in huge transportation and loss problems. Self-sufficient *local* communes with closed-loop economy are the answer. This will also help in the direct local use of heat produced in industrial processes, instead of first converting heat to electricity, transporting the electricity over long distances, only to convert it back to heat. All this would call for Symbian partnerships between governments, NGOs, universities, and R&D companies.

So, what we have at present are three possible approaches: the Green Valley, the Nuclear Valley, and the Sun Valley. There may be an evolutionary battle between the Arcadian Man, the Imperial Man, and the Symbian Man. Better still, a symbiosis may emerge wherein the best features of the Green Valley approach and the Nuclear Valley approach are adopted and subsumed in the Heliocultural Energy Regime.

33. Computational Intelligence

33.1 Introduction

Principles of Darwinian evolution have been exploited to great advantage by carrying them over to evolution inside a computer. This is evolution carried out and controlled by humans. A further advance has been that this 'artificial' evolution has been applied even to real-life environments for the machines in which the evolution takes place; *intelligent or smart robots* is the generic term used for such machines (Wadhawan 2007).

As discussed in Chapter 29 and some earlier chapters, Lamarckian evolution does not enjoy a very respectable place in modern biology. The reason is the central dogma of microbiology, according to which information can flow only from DNA to RNA to proteins, and not in the opposite direction. But this restriction is unnecessary when evolution is occurring inside a computer. In fact, so far as such 'artificial' evolution is concerned, Lamarckian evolution can be advantageous in certain situations over Darwinian evolution.

Before discussing artificial evolution, intelligent robots, etc., let us first get familiar with what has come to be known as the field of *computational intelligence* (CI) (Konar 2005; Kruse *et al.* 2013).

The underlying approach in a conventional computation is to work through a *precise* algorithm that works on *accurate* data. But there are innumerable complex systems that cannot be adequately tackled through such an approach. One should be able to work with partially accurate or insufficient or time-varying data, requiring the use of suitably variable or tenable software. One would like to work with computational systems that are fault-tolerant, and *computationally intelligent*, making adjustments in the software intelligently and handling imperfect or 'fuzzy' data the way we humans do (Nolfi and Floreano 2000; Zomaya 2006). The subject of computational intelligence caters to this requirement.

CI consolidated as a subject in the early 90s. Zadeh (1965, 1975) had introduced the notion of *linguistic variables* for making reasoning and computing more human-like. By computing with words, rather than with numbers, one could deal with *approximate* reasoning. With increasing emphasis on the use of *biomimetics* in computational science, Zadeh's fuzzy-logic approach was clubbed with 'artificial neural networks' (ANNs), 'genetic algorithms' (GAs), 'evolutionary' or 'genetic programming' (EP or GP), and 'artificial life' (AL) to define the field of CI.

FL, ANNs, GAs, GP, and AL constitute the five hard-core components of CI (Fig. 33.1), although there are also a number of other peripheral disciplines (Konar 2005; Krishnamurthy and Krishnamurthy 2006 (in Zomaya 2006); Kruse *et al.* 2013).

Let us get a feel for each of the main topics which come under the umbrella term CI.

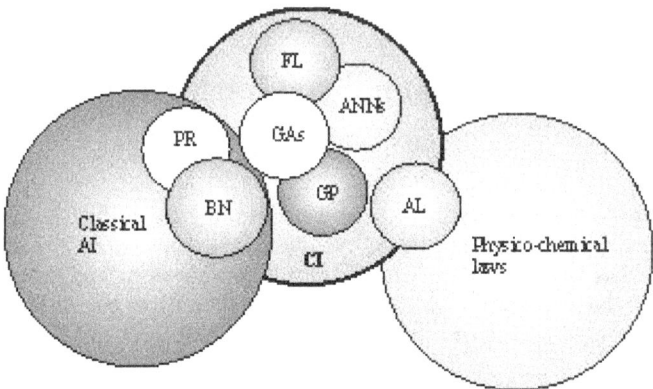

Fig. 33.1 The field of computational intelligence covers many subjects, the five main ones being FL (fuzzy logic); ANNs (artificial neural networks); GAs (genetic algorithms); GP (genetic programming); and AL (artificial life). Only some parts of old classical AI (artificial intelligence), namely PR (probabilistic reasoning) and BN (belief networks) overlap with CI. And the subject of artificial life is not all about computations; it does draw inputs from the real-world physico-chemical laws also.

33.2 Fuzzy logic

Conventional computation is based on precise logic, whereas humans are also able to process information that is not always very precise or complete. Humans are able to employ 'fuzzy logic' in their thinking and analysis. Fuzzy logic has developed as a mathematical discipline for devising computational strategies that can deal with imprecise knowledge. The imprecise nature of the information available may result from our limited capability to resolve detail, or because the data are partial, noisy, vague, or incomplete.

FL involves inference and intuition, just like the logic used by humans in certain situations. One removes the restriction that propositions can only be either true or false; instead, they are allowed to be true or false to different *degrees*. For example: if A is (HEAVY 0.8) and B is (HEAVY 0.6), then A is 'MORE HEAVY' than B. This is in sharp contrast to classical binary logic, in which A and B may be either members of the same class (both HEAVY, or both NOT-HEAVY), or of different classes.

For dealing with such human-like logic, the important notion of linguistic variables was introduced by Zadeh (1965, 1975). This made it possible to compute with words, rather than numbers, enabling approximate reasoning. *Fuzzy-rule-based systems* (FRBSs) were introduced for dealing with uncertain and vaguely defined problems: One deals with IF-THEN rules, the antecedents and consequents of which consist of fuzzy-logic statements. Representation of knowledge is enhanced by the use of linguistic variables and their *linguistic values*, which are defined by *context-dependent* fuzzy sets. These sets are specified by *gradual* membership functions. FL rules are normally of the form:

IF (*variable*) IS (*property*) THEN (*action*)

Here is how a temperature controller using a fan would be programmed:

IF temperature IS very cold THEN stop fan
IF temperature IS cold THEN turn down fan
IF temperature IS normal THEN maintain level
IF temperature IS hot THEN speed up fan

There is no 'ELSE' option because the temperature might be 'cold', 'normal' and 'hot' at the same time to different degrees.

The concept of '*expert systems*' is based on analogy with human experts, and usually has a large FL component. An expert system is a computer programme that holds and processes the information and expertise gained by humans in one or more domains. A fuzzy expert system uses a collection of fuzzy membership functions and rules, instead of Boolean logic. A typical rule for reasoning about available fuzzy data looks like this:

IF x is low and y is high, THEN z = medium

Zadeh also introduced the *f.g-generalization*. An f-generalization fuzzyfies any theory, technique, method or problem by replacing the corresponding crisp set by a fuzzy set. A g-generalization does the opposite; it *granulates* a set by partitioning its variables, functions and relations into *granules* or information clusters. f.g-generalization is a combination of these two. One ungroups an information system into components by some rules, and regroups them into clusters or granules by another set of rules. This can result in new types of information subsystems.

33.3 Neural networks, real and artificial

Artificial neural networks (ANNs) attempt to mimic the human brain for carrying out computations. Let us first familiarize ourselves with some basic ideas about how our brain functions; i.e., let us first discuss *real* neural networks.

The top outer portion of the human brain, just under the scalp, is the *neocortex*. Almost everything we associate with intelligence occurs in the neocortex, with important roles also played by the thalamus and the hippocampus. The human brain has $\sim 10^{11}$ nerve cells or *neurons* (Fig. 33.2), in the neocortex.

Most of the neurons have a pyramidal shaped central body (the *soma*), which houses the *nucleus*. Like any other biological cell, the nucleus is where the genetic material resides. In addition, a neuron has an *axon*, and a number of branching structures called *dendrites*. The axon is a signal emitter, and the dendrites are signal receivers. A connection called a *synapse* is established when a strand of an axon of one neuron 'touches' a dendrite of another neuron. The axon of a typical neuron (and there are 10^{11} of them) makes several thousand synapses. *The brain is a massively parallel computation system.*

When a sensory or other pulse ('spike') involving a particular synapse arrives at the axon, it causes the synaptic vesicles in the first (the 'presynaptic') neuron to release chemicals called *neurotransmitters* into the gap or synaptic cleft between the axon of the first neuron and the dendrite of the second (the 'postsynaptic') neuron. These chemicals bind to the receptors on the dendrite, triggering a brief local depolarization of the membrane of the postsynaptic cell. This is described as a *firing* of the synapse by the presynaptic neuron.

If a synapse is made to fire repeatedly at high frequency, it becomes more sensitive: subsequent signals make it undergo greater voltage swings or spikes. Building up of memories amounts to formation and strengthening of synapses.

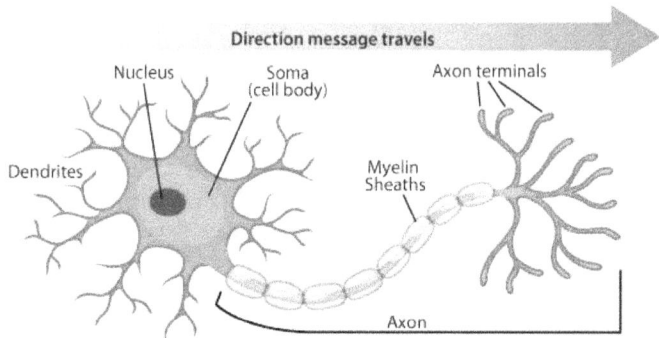

Fig. 33.2 A typical neuron. Image credit:
https://askabiologist.asu.edu/neuron-anatomy

The firing of neurons follows two general rules:

1. *Neurons which fire together wire together*. Connections between neurons firing together in response to the same signal get strengthened.

2. *Winner-takes-all inhibition*. When several neighbouring neurons respond to the same input signal, the strongest or the 'winner' neuron will inhibit the neighbours from responding to the same signal in future. This makes these neighbouring neurons free to respond to other types of input signals.

McCulloch and Pitts (1943) postulated that the brain can be modelled as *a network of logical operations* like AND, OR, NOT etc. Their paper presented the first ever model of a neural network. The McCulloch-Pitts model was the first to try to understand the workings of the brain as a case of information processing. This model was also the first to demonstrate that a network of simple *logic gates* can perform even complex computations (see Nolfi and Floreano 2000).

In 1949 Hebb published his celebrated book, *The Organization of Behaviour*. The first assumption he made was that the synaptic connections among the neurons in the brain are constantly changing, and these synaptic changes form the basis of all learning and memory. As a result of these self-regulated changes, even a randomly connected neural network would soon organize itself into a *pattern*, and the pattern is ever-changing in response to feedbacks and sensory inputs. A kind of positive feedback is operative: Frequently used synapses grow stronger, and the seldom used synapses become weaker. The frequently used synapses eventually become so strong as to be locked in as memories. The memory patterns are not localized; they are widely distributed over the brain. Hebb was among the earliest to use the term '*connectionist*' for such distributed memories.

Hebb made a second assumption: The selective strengthening of the synapses among the neuron cells by the nerve impulses causes the brain to organize itself into *cell assemblies* (Hebbian assemblies). These are subsets of several thousand neurons in which the circulating nerve impulses reinforce themselves, and continue to circulate. They were assumed to be the *basic building blocks of information*. One such block may correspond to, say, a flash of light, another to a particular sound, a third to a portion of a concept, etc. These blocks are not physically isolated. A neuron may belong to several such blocks simultaneously. Because of this overlap among blocks, activation of one cell assembly or block may trigger a response in many others blocks as well. Consequently, the blocks rapidly organize themselves into more complex concepts, something not possible for a block in isolation.

Hebb's work provided a model for the essence of thought processes. Rochester *et al.* (1956) verified Hebb's ideas by modelling on a computer what were among the first ANNs ever. Their work was also one of the earliest in which a computer was used not just for number crunching, but for simulation studies. Cell assemblies and other forms of emergent behaviour were observed on the computer screen, beginning from a random or uniform configuration.

The term 'processing element (PE)' (or '*perceptron*') is used for the basic input-output unit of an ANN which can transform a synaptic input signal into an output signal.

An ANN can be made to *evolve* and *learn*. The *training* of an ANN amounts to strengthening or weakening the connections (synapses) among the PEs, depending on the extent to which the output is the desired one. There are various approaches for doing this; e.g. *supervised learning*: A series of typical inputs are given, along with the desired outputs. The ANN calculates the root-mean-square error between the desired output and the actual output. The connection weights (w_{ij}) among the PEs are then adjusted to minimize this error, and the output is recalculated using the same input data. The whole process is repeated cyclically until the connection weights have settled to an optimum set of values.

Fig. 33.3 illustrates the idea that a neuron fires if the weighted sum of the input exceeds a threshold value.

An important feature of neural networks has been pointed out by Kurzweil (1998). It is the selective *destruction* of information. This is something so important that it can be regarded as an essence of all computation and intelligence. An enormous amount of information flow occurs in a neural network. Each neuron receives thousands of continuous signals. But finally it either fires or does not fire, reducing all the input to a single bit of information, namely the binary 0 or 1.

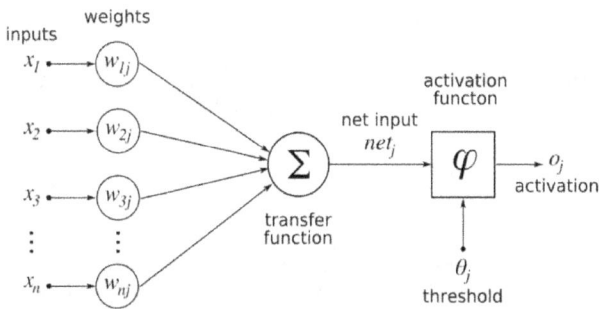

Fig. 33.3 The basic McCulloch-Pitts unit.
http://en.wikibooks.org/wiki/Artificial_Neural_Networks/Print_Version

33.4 Genetic algorithms

Holland (1975) introduced the formalism of genetic algorithms (GAs) by analogy with how biological evolution occurs in Nature. Deep down under, a computer program is nothing but a string of 1s and 0s, something like 110101010110000101001001..... This is similar to how chromosomes are laid out along the length of a DNA molecule. We can think of each binary digit as a 'gene', and a string of such genes as a digital 'chromosome'. For example, chromosome A may be 1101, B may be 0111, etc. In a GA, an individual in the population is represented schematically as the sequence of its chromosomes, say ABCDEF. Another individual may be abcdef, a third may be aBCdeF, etc.

The essence of Darwinian evolution is that, in a population, the fittest have a larger likelihood of survival and propagation. In computational terms, it amounts to maximizing some mathematical function representing 'fitness'. But it is important to remember that whereas Darwinian evolution is an open-ended and *blind* process, *GAs have a goal*. GAs are meant to solve particular pre-conceived problems.

For solving a maximization problem (e.g., for finding the maximum possible value of a complicated function (say fitness) of an artificial genome), the steps involved are typically as follows:

1. We first let the computer produce a population of, say, 1000 individuals, each represented by a randomly generated digital chromosome.

2. The next step is to test the relative fitness of each individual (represented entirely by the corresponding chromosome) regarding its effectiveness in maximizing the function under consideration; e.g., the fitness function. A score is given for the fitness, say on a scale of 1 to 10. In biological terms, the fitness is a probabilistic measure of the reproductive success of the individual. The higher the fitness, the greater is the chance that the individual will be selected (by our computer code) for the next cycle of reproduction.

3. Mutations are introduced occasionally in a digital chromosome by arbitrarily flipping a 1 to 0, or a 0 to 1.

4. The next step in the GA is to take (in a probabilistic manner) those individual digital chromosomes that have high levels of fitness, and produce from them a new generation of individuals by a process of digital sexual reproduction (which includes *crossover*). The GA chooses pairs of individuals, say ABCDEF and abcdef, and produces two new individuals (children), say ABCdef and abcDEF.

5. The new generation of digital individuals produced is again subjected to the entire cycle of gene expression, fitness testing, selection, mutation, and crossover.

6. These cycles are repeated a large enough number of times, till optimization or maximization has been achieved.

The *sexual crossover* in reproductive biology, as also in the artificial GA, serves two purposes. It provides a chance for the appearance of *new* individuals in the population that may be fitter than any earlier individual. Secondly, it provides a mechanism for the existence of clusters of genes (called *building blocks*) which are particularly well-suited for occurring *together* because they result in higher-than-average fitness for any individual possessing them.

Holland used the symbol # to denote 'does not matter'; i.e., the corresponding digital gene may be either 0 or 1. Suppose it is found that certain patterns of genes turn out to have high fitness; e.g. 111##10####1#10, or 0##1000##1111##, etc. Such patterns function as building blocks with a high potential for fitness.

What is more, since the population can shuffle its genetic material in every generation through sexual reproduction, new building blocks, as well as new combinations of existing building blocks, can arise. Thus the GA quickly creates individuals with an ever-increasing number of 'good' building blocks (the 'bad' ones get gradually eliminated by natural selection). If there is a survival advantage to the population, the corresponding individuals that have the good building blocks spread rapidly, and the GA converges to the solution rapidly (a case of positive feedback).

By the mid-1960s, Holland had proved his fundamental *schema theorem*: *In the presence of reproduction, crossover and mutation, almost any compact cluster of genes that provides above-average fitness will grow in the population exponentially.* ['Schema' was the term used by Holland for any specific pattern of genes (also see Nolfi and Floreano 2000).]

Efficient *drug design* based on evolutionary principles is an example of GAs in action. Often, for a drug to be effective, its molecular structure should be such that it can fit snugly into a relevant cleft in a relevant protein molecule. It can be very expensive to actually synthesize all those trial drugs and test their compatibility with the protein-molecule cleft. In the GA approach the computer code generates billions of random 'molecules', which it tests against the cleft in the protein. One such imaginary molecule may contain a site which matches one of, say, six sites on the cleft. This molecule is 'selected', and a billion variations of it are created, and tested. And so on.

Here is another example of application of GAs, this time from materials science. Giro, Cyrillo and Galvao (2002) designed *conducting polymers* by the GA approach. Copolymerization is one way of developing new polymers. Because of the huge number of possible configurations, and because of the ever-present possibility of structural disorder, the task becomes daunting when one wants to investigate theoretically the various ways of combining two or more monomers to design a copolymer with the desired properties. Giro *et al.* (2002) have described the successful implementation of the task of designing binary and ternary disordered copolymers of conducting polyaniline by employing GAs.

GAs have also been used for demonstrating or understanding the *evolution of strategies* in a variety of situations (see Axelrod 2006). An example is the emergence of sex as the mechanism of reproduction in practically all the large animals and plants. This has emerged in Nature in spite of the fact that sexual reproduction can be very costly. For example, in mammals the cost is that half of the members of the population, namely the males, do not give birth, thus reducing the pace of adaptive evolution for overcoming challenges. By using GAs, it was shown by Hamilton *et al.* (1990) that sexual reproduction evolved as an adaptation strategy for resisting parasites. Large mammals and plants have to overcome parasites. For this, they should be able to distinguish their own cells from the parasites that evolve to fool them. The parasites, being much smaller, reproduce much faster, and therefore evolve much faster, than the hosts. The hosts evolved sexual reproduction as a way to solve this problem: The juggling of the genes of the mother and the father made the new hosts different from both the father and the mother, and that too in every generation, thereby making it hard for the parasites to re-adapt fast enough for the offspring hosts. This idea of Hamilton needed proof, and the use of GA provided the proof.

Axelrod (2006) has listed a number of other topics that are amenable to such simulation studies: mutation; crossover; inversion; coding principles; dominant and recessive genes; gradual *vs.* punctuated evolution; population viscosity; and speciation and ecological niches.

33.5 Genetic programming: Evolution of computer programs

> *We cannot expect to find a good child-machine at the first attempt. One must experiment with teaching one such machine and see how well it learns. One can then try another and see if it is better or worse. There is an obvious connection between this process and evolution, by the identifications.*
>
> Alan Turing (1950)

Evolutionary searches through a phase space can be made, not only for finding solutions using a suitably written computer program (genetic algorithm), but also for evolving the computer programs themselves. Genetic programming (GP) is about making computers evolve their own algorithms (Fig. 33.4).

Fig. 33.4 The shape of things to come: genetic programming (courtesy the home page of Genetic Programming Inc., the brainchild of John Coza).
http://www.genetic-programming.com/

Genetic programming now routinely delivers high-return human-competitive machine intelligence. But why should this be attempted at all?

The computations occurring in the human brain are of a massively parallel nature. And ambitions about mimicking the human brain must therefore involve writing of programs with parallel computing on a *very* large scale. But this is not possible. I quote Tom Ray (http://life.ou.edu/): *The complexity of programming a massively parallel machine is probably beyond us. I don't think we will ever be able to write software that fully uses the capacity of parallelism.*

And it is not just the question of mimicking the human brain. For any highly complex system (and the modern technological world abounds in them), it is very difficult, if not well-nigh impossible, to write efficient, versatile, and *robust* computer programs. The problem becomes particularly acute when a large amount of parallel programming is essential. It appears that the only hope lies in letting computer codes *evolve*, just as the *solution* of a problem by the use of a fixed program evolves in a genetic algorithm (GA).

We have to *breed* software, which can be turned on itself, *ad infinitum*, so that it evolves to a desired end. Such breeding becomes a necessity when we are trying to solve some extremely difficult problems in technology development. An example is the development of a multimillion-line code for automatic flying of an aircraft. This is an example of wanting our machines to solve problems we do not know how to *solve*, but merely know how to *state*. It is impossible to make such huge programs completely bug-free. And unless adequate safety features are built in, just one bug may be enough to cause the plane to crash. We do not want that to happen.

Learning from ant colonies ('ant algorithms') is one way to proceed (Chapter 18). But a more general way is to attempt GP. Here are some general characteristics of GP for massively parallel computer codes:

- There is *an absence of central command*. After all, no human is doing the coding. Only the problem is stated, and some 'boundary conditions' or initial conditions are specified.

- The massiveness and *distributedness* of such codes ensures that even if there is a bug or a local crash, the system carries on, after some minor readjustment.

- There is *incremental* expansion, followed by testing at each such 'modular' stage.

- Parasites, viruses, adversaries are deliberately self-introduced at the testing stage, so that there is a built-in experience and therefore the ability for tackling such eventualities. This is the equivalent of *vaccination* in living beings.

Naturally, such codes develop the all-important *robustness* feature. This is crucial when one is trying to develop, say, a code for flying a fully computer-controlled aircraft. We cannot afford a programming failure. [Think of the beehive (Chapter 18). There is enough *redundancy* that even if a portion of the hive or population gets decimated, the complex adaptive superorganism makes adjustments, and carries on regardless.]

Typically, one begins with a *population* of competing computer programs. Variations (including mutations) are introduced, as also the equivalent of crossover in biological reproduction. The variations are made heritable. Fitness criteria are defined, and the individual programs are made to compete for survival. The fitter programs are assigned a larger chance of surviving and reproducing in the next cycle of the evolutionary operation.

Each of the competing algorithms is normally a *sequence* of commands. What is being attempted, in effect, is that during the process of evolution the sequence of commands is shuffled and the commands are tempered with in other ways. This process is continued till the right combination of everything is achieved.

You might be wondering how can we tinker with a computer program and still hope that it would be functional at all? Even the slightest of errors in writing the code may make it impossible to run. Yes; conventional algorithms are *fragile*. Even a minor crossover or *mutation* of any command in an algorithm can really be a *mutilation*, resulting in a nonsensical or illogical programming statement, making the computer program to crash. In principle one could get over this problem by assigning zero fitness to all such illegal algorithms, but then most of the members of the population would be declared unfit, and the evolution process would come to a halt because it would not have the all-important benefit of continual *variety*.

One way to make computer programs less fragile is to give them a *tree structure*, rather than a sequential structure. The technical definition of a tree structure is given in Appendix A9 on network theory. But even the commonsense meaning of the word 'tree' is enough for our purpose here. The tree structure can make the algorithm robust enough to withstand the depredations of crossovers and mutations etc. with a certain degree of resilience. More details can be found in Sipper (2002), and Konar (2005).

Fig. 33.5 is a program for the function f(X) = (2.2 − (X/11)) + (7 * cos(Y)). And Fig. 33.6 is an example of how crossover between two parent GPs is carried out.

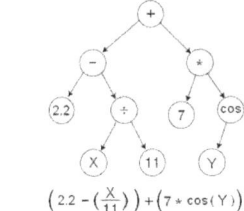

$$\left(2.2 - \left(\frac{X}{11}\right)\right) + \left(7 * \cos(Y)\right)$$

Fig. 33.5 Example of a tree structure depicting a function.
http://en.wikipedia.org/wiki/Genetic_programming

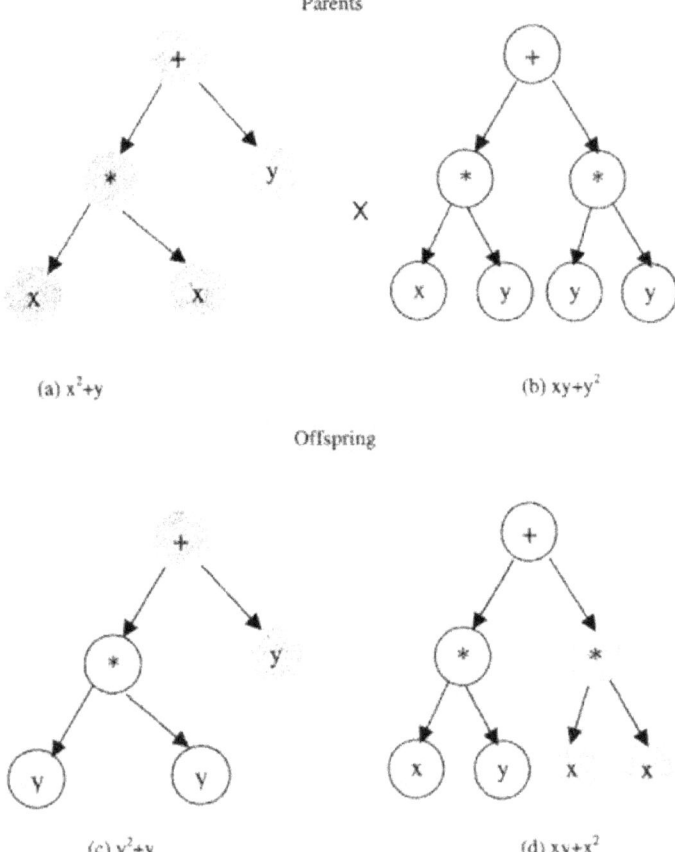

Fig. 12.15: Crossover between 2 genetic programs (a) and (b) yields new programs (c) and (d).

Fig. 33.6 Crossover between two parent GPs giving rise to two new programs (c) and (d).
Image credit: Sipper (2002).

Such evolutionary or 'bio-inspired' computing has also been used for evolving sophisticated analogue electrical circuits (Koza *et al*. 1997).

'Gene-expression programming' is an important variation on GP (Ferreira 2006).

33.6 Artificial life

> *With the advent of artificial life, we may be the first creatures to create our own successors. . . If we fail in our task as creators, they may indeed be cold and malevolent. However, if we succeed, they may be glorious, enlightened creatures that far surpass us in their intelligence and wisdom. It is quite possible that, when conscious beings of the future look back on this era, we will be most noteworthy not in and of ourselves but rather for what we gave rise to. Artificial life is potentially the most beautiful creation of humanity.*
>
> Doyne Farmer and Alletta Belin (1992)

Christopher Langton is the main originator of the subject of artificial life. The term artificial life (AL, or Alife) was coined by him around 1970: AL is '. . *an inclusive paradigm that attempts to realize lifelike behaviour by imitating the processes that occur in the development or mechanics of life.*'

In the more familiar field of artificial intelligence (AI) one uses computers to model neuropsychology. Likewise, in the field of AL one uses computers to model the basic biological mechanisms of evolution and life (Heudin 1999). In abstracting the basic life processes, the AL approach emphasizes the fact that life is not a property of matter per se, but the *organization* of that matter. The laws of life must be laws of dynamical form, independent of the details of a particular carbon-based chemistry that just *happened* to arise here on Earth. It attempts to explore other possible biologies in new media, namely computers and robots.

The idea is to view *life-as-we-know-it* in the context of *life-as-it-could-be*. In conventional biology one tries to understand life phenomena by a process of *analysis*: We take a living community or organism, and try to make sense of it by subdividing it into its building blocks. By contrast, AL takes the *synthesis* or bottom-up route. We start with an assembly of very simple interacting units, and see how they evolve under a given set of conditions, and how they change when the environmental conditions are changed.

One of the most striking characteristics of a living organism is the distinction between its *genotype* and *phenotype*. The genotype can be thought of as a collection of little computer programs, running in parallel, one program per gene. When activated, each of these programs enters into the logical fray by competing and/or cooperating with the other active programs. And, collectively, these interacting programs carry out an overall computation that is the phenotype. The system *evolves* towards the best solution of a posed problem.

By analogy, the term GTYPE is introduced in the field of AL to refer to any collection of low-level rules. Similarly, PTYPE means the structure and/or behaviour that results (*emerges*) when these rules are activated in a specific environment.

What makes life and brain and mind possible is a certain kind of balance between the forces of order and the forces of disorder. In other words, there should be an *edge-of-chaos existence*. Only such systems are both stable enough to store information, and yet evanescent enough to transmit it.

Life is not just *like* a computation; *life literally is computation* (Chapter 14). And once we have consciously made a link between life and computation, an immense amount of computational theory can be brought in. For example, the question 'Why is life full of surprises?' is answered in terms of *the undecidability theorem of computer science*, according to which, unless a computer program is utterly trivial, the fastest way to find out what it would do (does it have bugs or not?) is to actually run it and see. This explains why, although a biochemical machine or an AL machine is completely under the control of a program (the GTYPE), it still has surprising, spontaneous behaviour in the PTYPE. It never reaches equilibrium, and there is perpetual novelty.

The computational aspect of the AL approach invokes the theory of *complex dynamical systems* (Wadhawan 2010). Such systems can be described at various levels of complexity, the global properties at one level emerging from the interactions among a large number of simple elements at the next lower level of complexity. The exact nature of the emergence is, of course, unpredictable because of the extreme nonlinearities involved.

Here are some websites devoted to artificial life:

- http://biology.kenyon.edu/slonc/bio3/AI/A_LIFE/a_life.html
- http://www.biota.org/nervegarden
- http://www.digitalspace.com/avatars
- http://www.biota.org/

34. Adaptation and Learning in Complex Adaptive Systems

I discussed complex adaptive systems (CASs) in Section 7.2. John Holland, whose genetic-algorithm (GA) formalism was described in Section 33.4, realized that a GA by itself was not an adaptive agent. *An actual adaptive agent is playing games with its environment, which amounts to prediction (or thinking ahead) and feedback* (Waldrop 1992). He came up with a model for this.

34.1 Holland's model for adaptation and learning

The ability of a CAS for thinking ahead, or prediction, requires the emergence and constant revision of a *model* of the environment. And this is not a prerogative of the brain alone. It occurs in all CASs all the time (Holland 1995, 1998): They all evolve models that enable them to anticipate the near future.

That a brain is not needed for doing this is illustrated by the case of many bacteria that have special enzymes that enable them to swim along directions along which there is an increasing concentration of glucose. Effectively, these enzymes seem to model a world in which chemicals diffuse outwards from their source. There is also the implicit prediction that if you swim towards regions of higher concentration, then more of something nutritious may be obtained. This ability has evolved through processes of Darwinian natural selection. Individuals which had even a slight tendency towards this behaviour had an evolutionary advantage over those which were lacking it, and over time the ability became stronger and stronger through processes of natural selection and inheritance. A case of *frozen memory* so to speak; frozen in the genes.

How do such models of the environment arise, even when there is no 'conscious' thinking involved? Holland's (1998) answer was: *Through feedback from the environment.* Holland drew inspiration from Hebb's (1949) neural-network model I described in Section 33.3. The neural network learns not only through sensory inputs, but also through internal feedbacks. Such feedbacks are essential for the emergence of the resonating cell assemblies in the neural network.

A second ingredient Holland put into his simulated adaptive agent was the IF-THEN rules used so extensively in expert systems. This enhanced the computational efficiency of the artificial adaptive agent. Holland argued that an IF-THEN rule is, in fact, equivalent to one of Hebb's cell assemblies. And there is a large amount of overlap among different cell assemblies. Typically a cell assembly involves ~1000 to 10000 neurons, and each neuron has ~1000 to 10000 synaptic connections to other neurons. Thus, activating one cell assembly is like posting a message on something like an '*internal bulletin board*', and this message is 'seen' by most of the other cell assemblies overlapping with the initially activated cell assembly. Those of these latter assemblies that are properly and sufficiently overlapping with the initial assembly would take actions of their own, and post *their* messages on the bulletin board. And this process will occur again and again.

What is more, each of the IF-THEN rules constantly scans the bulletin board to check if any of the messages matches the IF part of the rule. If it does, then the THEN part becomes operative, and this can generate a further chain of reactions from other rules and cell assemblies, each posting a new message on the internal bulletin board.

In addition to the role of cell assemblies and IF-THEN rules, some of the messages on the bulletin board come directly from sensory input data from the environment. Similarly, some of the messages can activate actuators, or emit chemicals, making the system affect the environment. Thus Holland's digital model of the adaptive system was able to get feedback from the environment, as well as from the agents constituting the network; it also influenced the environment by some of its outputs.

Having done all this, the third innovation introduced by Holland was to ensure that even the *language* used for the rules and for the messages on the metaphoric internal bulletin board was not in terms of any human concepts or terminology. For this he introduced certain rules called '*classifiers*'. He wrote: '*GAs offer robust procedures that can exploit massively parallel architectures and, applied to classifier systems, they provide a new route toward an understanding of intelligence and adaptation*'.

The rules and messages in Holland's model for adaptation were just bit-strings, without any imposed interpretation of what a bit string may mean in human terms. For example, a message may be 1000011100, rather like a digital chromosome in his GAs. And an IF-THEN rule may be something like this: If there is a message 1##0011####10 on the board, then post the message 1000111001 on the board (here # denotes that the bit may be either 0 or 1). Thus this abstract representation of IF conditions *classified* different messages according to specific patterns of bits; hence the name 'classifiers' for the rules.

In this classifier system, the meaning of an abstract message is not something defined by the programmer. Instead, it emerges from the way the message causes one classifier rule (or a sensor input) to trigger another message on the board. *Apparently, this is how concepts and mental models emerge in the brain in the form of self-supporting clusters of classifiers which self-organize into stable and self-consistent patterns.*

There are thousands or tens of thousands of mutually interacting IF-THEN rules, cell assemblies, and classifiers. This may lead to conflicts or inconsistencies regarding action to be taken, or regarding the state a neuron can be in. Instead of introducing a conflict-resolution control from the outside (as in a typical top-down approach), Holland decided that even this should emerge from within. The control must be *learnt*, emerging from the bottom upwards. After all, this is how things happen in real-life systems. He achieved this by introducing some more innovations into his model, as I shall describe in the next section.

34.2 The bucket brigade in Holland's algorithm

In Holland's model for adaptation and learning in CASs, the bulletin-board messages must *compete* for domination and survival. Moreover, since in the game for survival in real life, certain players form *syntheses* or *alliances* for mutual benefit, Holland added this feature in his model by first assigning to each classifier a certain *plausibility or probability factor*. The classifiers with high plausibility stood a higher chance of being selected for bulletin-board display in a given cycle (this gave them more 'visibility' or prominence); the rest were likely to be ignored or eliminated. The Hebbian reinforcement and feedback mechanism was used for determining the plausibility factor for a classifier. The plausibility was determined by performance. If the selection of a classifier resulted in better performance, as indicated by a positive feedback from the environment, its plausibility factor was enhanced. If the performance was poor, the plausibility factor was diminished, just like the weakening of the synaptic strength in the Hebbian model of the brain.

Next, Holland took note of the fact that since a classifier interacts with many others, and is influenced by them, the credit for doing well (if it does well) is not entirely its own; a chain of classifier actions leads to the successful action of the final classifier. To factor-in this, Holland deviated from the general Hebbian approach and, instead, drew inspiration from how a market economy functions. The market is driven by the powerful *profit motive*. Let us see what this marketplace metaphor is.

In a market, there is buying and selling. If a product (say a raw material, or a service) is good, it fetches a good price from the buyer. The act of buying depletes the funds of the buyer, but the buying is done with the expectation that, by doing value-addition to the product purchased, the buyer can sell it in the market at a good price, and thus not only recover the cost of the original product, but also make a profit. There is a chain of buyers and sellers at various levels, leading to the manufacture of the end product, ready for buying and use by the consumer, and the profit motive operates at each level of transaction. Thus, for a successful end product, although the payment of price is made by the consumer only to the final seller, all the players in the game get rewarded. And if an end product fails, nobody buys it and the loss (negative profit) percolates down the line to all the persons or firms involved.

Holland used similar ideas for his computer simulation of adaptation for *evolving* the plausibility factors for the classifiers. The messages competing for being posted on the bulletin board are like the goods and services for sale, and the classifiers generating those messages are like the firms that produce the goods and services. The classifiers not only sell, they also buy, because they scan all the messages on the board, and get activated if the corresponding IF condition is satisfied by a displayed message. When a classifier buys, it is like a firm making a bid, and the highest bidder wins the bid. And how does the classifier 'pay' for the purchase? By transferring some of its plausibility strength to the supplying classifier. And since the classifiers are a highly interconnected system, the profits and losses get dispersed, just like in the market economy. Since the environment (likened to the consumer of the end product) provides a Hebbian reinforcement for good performance by feedback, the 'understanding' evolves at all levels that if you make a good intermediate product, you get rewarded with a profit, and if your product in not good you stand to lose. Finally the plausibility factor of each classifier evolves to a value matching its true worth to the environment, *and this happens without any top-down instructions*.

Holland called this part of his modelled CAS the '*bucket brigade*' algorithm: It passes the reward from each classifier to the previous classifier down the chain. [The term 'bucket brigade' was used by analogy with how fire-fighting is done efficiently by making people stand in a row from the source of water to the site of the fire, so that each person takes the bucketful of water from the neighbour and passes it on to the neighbour on the other side; thus crucial time is saved because nobody has to run carrying water from one end to the other.]

The idea is similar to how synapses get strengthened in the Hebbian model of the brain. Something similar also happens when reinforcement of synaptic weights occurs when an artificial neural network (ANN) is trained for achieving the desired performance via gradual learning.

Exploitation *vs.* exploration

The bucket-brigade algorithm was still lacking in one additional aspect of evolution of learning. It lacked the *novelty* feature. It strengthened the classifiers that the system already possessed, but lacked the ability to explore new possibilities. It had the *exploitation* feature, but not the *exploration* feature. This was easy to handle. All that Holland had to do was to bring in his genetic algorithms (GAs). Classifiers evolved via the GA approach. The crossover feature of GAs ensured that new possibilities were explored and, once in a while, fitter classifiers arose which could replace weaker ones.

To summarize the description so far in this chapter, the rule-based system, along with the bucket-brigade approach and the GA, not only learned from experience, but was also innovative and creative. The bottom-up formalism eliminated the need for a *deductive* approach wherein rules have to be imposed for deciding what is consistent and desirable, and what is not. Instead, processes of inference, learning, and discovery emerged by *induction* through the bottom-up route. This was probably the first successful model of *emergence* in a CAS.

A number of successful computer simulations using Holland's model have been reported. The one that impressed Holland the most in those days was the work of his student Goldberg (1989). Goldberg's Ph. D. thesis work demonstrated how Holland's adaptive-systems formalism could be used successfully for controlling a simulated gas pipeline, an extremely complex problem.

Holland's formalism holds the promise of a theory of cognition. Various attempts have been made for formulating theories of cognition. An overview of social cognition theory (Fig. 34.1) has been given by Pajares (2002). For an overview of Piaget's theory of cognitive development, visit the URL https://en.wikipedia.org/wiki/Piaget%27s_theory_of_cognitive_development.

34.3 Langton's work on adaptive computation

We discussed Neumann's self-reproducing cellular automata (CA) in Section 16.3. Langton (1989) extended the CA approach to his work on artificial life (AL) (see Section 33.6) by introducing *evolution* into the Neumann universe.

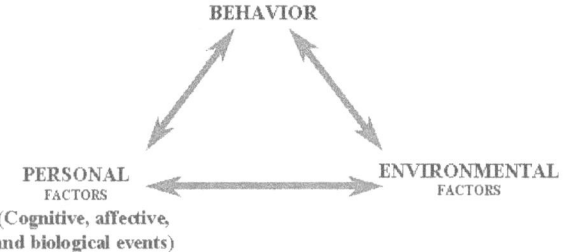

Fig. 34.1 Modelling social foundations of thought and action.
Image credit: Pajares (2002).
http://www.uky.edu/~eushe2/Pajares/eff.html

In the self-reproducing CA created by Langton, a set of rules (the so-called GTYPE in the field of AL) specified how each cell interacted with its neighbours, and the overall pattern that resulted was called the PTYPE. The local rules could *evolve* with time, rather than remaining fixed. His pioneering work was a fine example of *evolutionary computation* or *adaptive computation* (Fig. 34.2). For more on adaptive computation, visit
https://mitpress.mit.edu/search/mitpress_search/adaptive%20computation?page=1.

Langton also correlated his work on AL with Wolfram's four classes of CA (Section 16.4). We saw in Chapter 15 on chaos how the dynamics described by Lorenz by his equations for modelling weather phenomena changes drastically with even small changes in the values assigned to the three adjustable or 'control' parameters. In particular, the dynamics is chaotic only for a certain range of values of the control parameters, and nonchaotic or not fully chaotic for other values. Small values of control parameters give nonchaotic behaviour, similar to the dynamics described by Wolfram's Class 1 and Class 2 CA. And sufficiently large values of the control parameters result in totally chaotic dynamics, which corresponds to Class 3 CA.

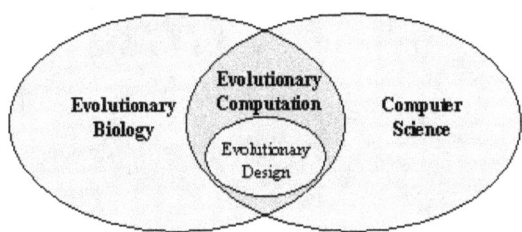

Fig. 34.2 Evolutionary computation and evolutionary design
(see Bentley: http://www0.cs.ucl.ac.uk/staff/ucacpjb/wc3paper.html).

Langton investigated the introduction of a similar (only one) parameter into the rules controlling CA behaviour to examine the above analogy more clearly, and particularly to investigate the connection between Class 4 CA on one hand, and partially chaotic systems on the other. After a number of trials, he came upon a parameter (λ) for the CA rules to correspond to the control parameter in an equation governing chaotic behaviour.

This λ was defined as *the probability that any cell in the CA will be 'alive' after the next time step*: In Chapter 16 we chose the colours black and white for distinguishing the two states of a CA cell, which we can now relate to 'alive' and 'dead'; just like what was done for the Game of Life. For example, if $\lambda = 0$ in the rule governing the evolution of a particular set of CA, all cells would be white or dead after one time step. The same would be true if $\lambda = 1$. In fact, the CA dynamics is symmetrical about the value $\lambda = 0.5$, except for the fact that the colours black and white get interchanged.

In his computer experiments Langton found that, as expected, $\lambda = 0$ corresponds to Wolfram's Class 1 rules. The same was true for very small nonzero vales of λ.

As this control parameter was increased gradually, Class 2 features started appearing at some stage, with characteristic oscillating behaviour. With increasing values of the control parameter, the oscillating pattern took longer and longer to settle down.

Taking $\lambda = 0.5$ resulted in totally chaotic behaviour, typical of the Wolfram Class 3. Langton found that clustered around the critical value $\lambda \approx 0.273$ were Class 4 CA. Thus, as the control parameter increases from zero onwards, we see a transition from 'order' to 'complexity' to 'chaos' (Fig. 34.3).

The variation of a control parameter like temperature can cause phase transitions to occur in a system such as water (from steam to liquid water to ice, on cooling). Langton realized that his control parameter λ plays a similar role in determining the dynamics of CA. It is like temperature. At low temperatures a material (such as H_2O) is solid; it is in a crystalline state, an *ordered* state. At high temperatures we have a fluid state (liquid or vapour), which signifies *chaos* or disorder. Langton drew the analogy with such phase transitions for describing the Class 4 behaviour in CA which sets in for values of λ around 0.273.

The solid-fluid transition in water is actually a *first-order* or discontinuous phase transition. But a crystalline material can also undergo one or more solid-to-solid phase transitions. Such transitions may be of first order or second order. As Langton pointed out, a second-order phase transition is actually better for drawing the analogy with Class 4 CA.

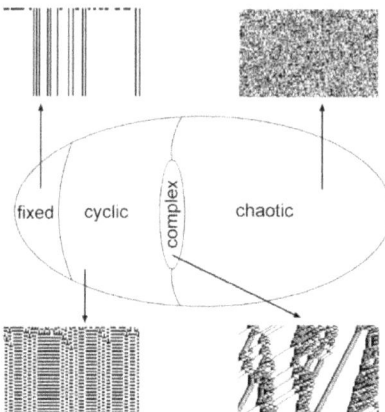

Fig. 34.3 With an increase in the value of the control parameter from zero onwards, there is a transition from 'order' to 'complexity' to 'chaos'. Image credit: http://www.noyzelab.com/research/ulamizer2.html

A second-order phase transition is characterized by the occurrence of '*premonitory phenomena*' (Wadhawan 2000). That is, even before the phase-transition temperature T_c is reached, the material exhibits 'critical phenomena' and regions of the new phase start appearing and disappearing in the old phase. This is unlike a first-order phase transition, which generally occurs sharply (e.g. at the melting point of ice), and there are generally no critical fluctuations or premonitory phenomena (even though there is a range of temperatures in which the parent phase and the daughter phase coexist).

Langton's control parameter λ for CA gives Class 4 behaviour not only for the value 0.273, but for a certain *range* of values around that number. And even for a specific value of the control parameter in this range, coherent structures may exist indefinitely, or over an arbitrarily large number of cells of the cellular automaton. *Langton gave the phase boundary in the Neumann universe the name edge of chaos.*

34.4 The edge-of-chaos existence of complex adaptive systems

The 'edge-of-chaos' nomenclature was introduced for the boundary in phase space that separates ordered behaviour from chaotic behaviour, *with complex behaviour sandwiched between these two extremes*. In 2-dimensional phase space we need a *line* to separate the two regimes. In 3-dimensional space we need a *plane* or a membrane (a 2-dimensional object) for demarcating this separation. In n-dimensional phase space we need an $(n$-1$)$-dimensional hyper-membrane for separating ordered behaviour from chaotic behaviour. The phrase 'edge of chaos' should be understood in this spirit.

We should also remember that this 'edge' is not a sharp one. It is more like a thin or thick membrane in phase space, with chaotic behaviour on one side, and ordered behaviour on the other. There is a gradation from chaos to complexity to order across the membrane (Fig. 34.4). Complex behaviour is at its most creative within the membrane.

This is true of so many complex systems. As we saw in Section 16.2, the Game of Life, a simple algorithm, is independent of the computer used for running it and exists in the Neumann universe, just like other Class 4 CA. They are capable of information processing and data storage etc. Being a mixture of coherence and chaos, they have enough stability to store information, and enough fluidity to transmit signals over arbitrary distances in the Neumann universe. There are many analogies, not only with computation, but with life, economies, and social systems. After all, they are all just a series of computations. Life, for example, is nothing if it cannot process information. And it strikes a right balance between too static a behaviour and excessively chaotic or noisy behaviour.

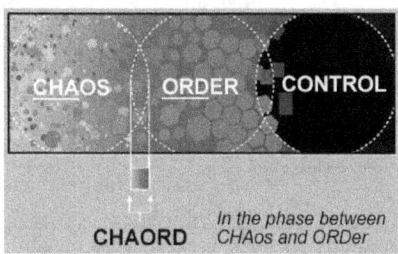

Fig. 34.4 The transition from chaos to order is separated in phase space by a membrane of a certain thickness, and complex behaviour thrives best in this membrane region. Image credit: http://www.innovativecommunities.org/how-we-work-together

In fact, the occurrence of complex phase-transition-like behaviour in the edge-of-chaos domain is something very common in all branches of human knowledge. Kauffman (1969), for example, recognized it in genetic regulatory networks (see Chapter 28). In his work on such networks in the 1960s, he discovered that if the connections were too sparse, the network would just settle down to a 'dead' configuration and stay there. If the connections were too dense, there was a churning around in total chaos. Only for an optimum density of connections did the stable state cycles arise. Similarly, in the mid-1980s, Farmer, Packard and Kauffman (1986) found in their autocatalytic-set model (Chapter 25) that when parameters such as the supply of 'food' molecules, and the catalytic strength of the chemical reactions etc. were chosen arbitrarily, nothing much happened. Only for an optimum range of these parameters did a 'phase transition' to autocatalytic behaviour set in quickly.

Many more examples can be given (Farmer *et al.* 1986b): Coevolutionary systems, economies, social systems, etc. A right balance of defence and combat in the coevolution of two antagonistic species ensures the survival and propagation of both. Similarly, the health of economies and social systems can be ensured only by a right mix of feedbacks and regulation on the one hand, and plenty of flexibility and scope for creativity, innovation, and response to new conditions, on the other. *The dynamics of complexity around the edge of chaos is ideally suited for evolution that does not destroy self-organization.*

Based on the work of Kauffman, Langton, and many others, one can compile a number of analogies or correspondences from a variety of disciplines. The table below does just that. There is a gradation from (a) excessive order, to (b) complex creativity and progress, to (c) total anarchy and failure. In each case there is some kind of a 'phase transition' to the complexity regime as a function of some control parameter (like temperature in the case of second-order phase transitions in crystals).

How do complex systems move towards the edge-of-chaos regime, and then manage to stay there? For complex *adaptive* systems, Holland (1975, 1995, 1998) provided an answer in terms of Darwinian evolution. Complex systems that are capable of sophisticated behaviour have an evolutionary advantage over systems that are too static and conservative or too turbulent. Therefore both the latter categories would tend to evolve towards the middle-course of complexity, if they are to survive at all. And once they are in that regime, any deviations or fluctuations would tend to be reversed by evolutionary factors.

Cellular automata classes	1 & 2 → 4 → 3
Dynamical systems	Order → complexity regime → chaos
Matter	Solid → phase-transition regime → fluid
Computation	Halting → undecidability regime → nonhalting
Life	Too static → life / intelligence → too noisy
Genetic networks	No activity → stable-state cycles → chaotic activity
Autocatalytic sets of reactions	No autocatalysis → autocatalysis → no autocatalysis
Coevolution	No coevolution → coevolution → no coevolution
Economies, social systems	No creativity → health and happiness → anarchy

The next question is: What do complex systems do when they are in the edge of chaos. Holland's (1998) assertion about the occurrence of 'perpetual novelty' in a CAS essentially amounts to saying that the system *moves around* in the edge-of-chaos membrane in phase space. But that is not all. The moving around actually takes the system to states of higher and higher sophistication of structure and complexity. Learning and evolution not only take a CAS towards the edge-of-chaos membrane, they also make it move in this membrane towards states of higher complexity (Farmer *et al.* 1986 (also see https://en.wikipedia.org/wiki/Self-organization)). The ultimate reason for this, of course, is that our universe is ever expanding, and there is therefore a perpetual input of free energy or negative entropy into it.

Farmer gave the example of *the autocatalytic-set model* (which he had proposed along with Packard and Kauffman) to illustrate the point. When certain chemicals can collectively catalyse the formation of one another, their concentrations increase by a large factor spontaneously, far above the equilibrium values. This implies that the set of chemicals as a whole emerges as a new 'individual' in a far-from-equilibrium configuration. Such sets of chemicals can maintain and propagate themselves, *in spite of the fact that there is no genetic code involved*. In a set of experiments, Farmer and colleagues tested the autocatalytic model further by allowing occasionally for novel chemical reactions. Mostly such reactions caused the autocatalytic set to crash or fall apart, but the ones that crashed made way for a further evolutionary leap. New reaction pathways were triggered, and some variations got amplified and stabilized. Of course, the stability lasted only till the next crash. Thus a succession of autocatalytic metabolisms emerged. Apparently, *each level of emergence through evolution and adaptation sets the stage for the next level of emergence and organization.*

35. Smart Structures

Smart or adaptronic structures are defined as structures with an ability to respond adaptively in a pre-designed useful and efficient manner to changing environmental conditions, including any changes in their own condition. The response is adaptive in the sense that two or more stimuli or inputs may be received and yet there is a single response function as per design. The structure is designed to ensure that it gives optimal performance under a variety of environmental conditions.

Smart bridges, smart surfaces, smart wings of aircraft, smart cars, and robots are some examples of smart structures. In all probability there is no clear distinction between life and nonlife. As I have explained in an earlier book on smart structures (Wadhawan 2007), this fact will become more and more apparent as we humans make progress in the field of smart structures.

35.1 The three main components of a smart structure

Any smart structure, biological or artificial, typically has a host structure or 'body', which has an interface with a source of energy. The body houses or supports one or more sensors (e.g. 'nerves') and one or more actuators (e.g. 'muscles'). The sensors and the actuators interface with a control centre or 'brain' (Fig. 35.1). Since both sensors and actuators interface with the control centre, they interface with each other also, albeit indirectly. In a good smart structure there is extensive and continuous feedback and communication of information among the various subunits.

The basic action plan of a smart structure is essentially as follows: The input of data from the sensors is analysed by the control centre. If the course of action is clear, the control centre signals the actuators to take the action. If the course of action is not clear, the control centre directs the sensors to collect additional data. This goes on cyclically till the course of action is clear, and then the action is taken. The action taken depends on the overall purpose or objective.

To consider the example of a human (obviously a smart structure), if a person puts his / her hand on a hot surface, the tactile sensors send a signal to the brain, which then immediately directs the muscles to take the hand away from the hot surface. The purpose of the smart structure here is to survive.

Since the smartest structures around us are those designed by Nature through aeons of trial and error and evolution, it makes sense for us to emulate Nature when we want to design artificial smart structures. In fact, an alternative definition of a smart structure can be given as follows:

Smart structures are those which possess characteristics close to, and, if possible, exceeding, those found in biological structures.

Sensors

Sensing involves measurement followed by information-processing. The sensor may be a system by itself, or it may be a subsystem in a larger system (e.g., a complete smart system). In artificial smart structures, optical fibres constitute the most versatile sensors. A variety of ferroic materials also serve as sensors. Some notable examples are piezoelectric materials like quartz and PZT, and relaxor ferroelectrics like PMN-PT.

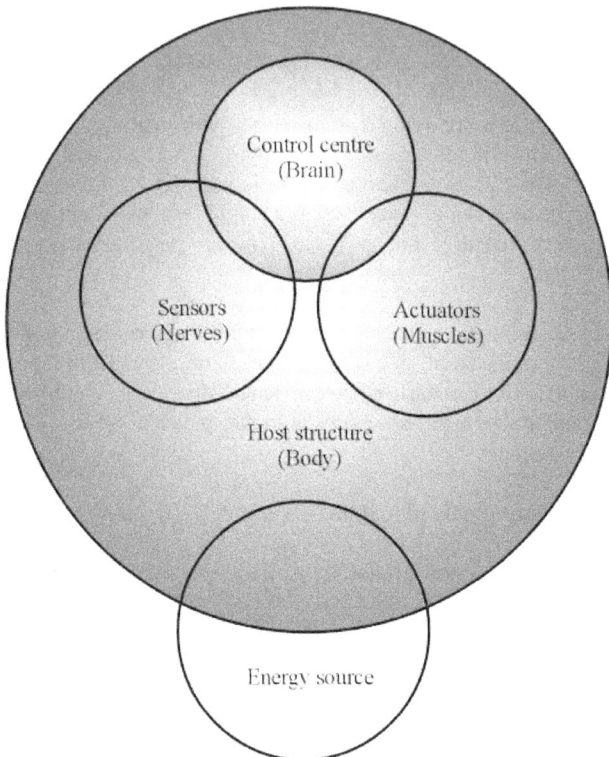

Fig. 35.1 The main components of a typical smart structure, and the interactions among them
(Wadhawan 2007).

The concept of '*integrated sensors*' has been gaining ground for quite some time. Typically, there is a microsensor integrated with signal-processing circuits on a single package. This package not only transduces the sensor inputs into electrical signals, but may also have other signal-processing and decision-making capabilities. There are several advantages of integrated sensors: better signal-to-noise ratio; improved characteristics; and signal conditioning and formatting.

Actuators

An actuator creates controllable mechanical motion from other forms of energy. Materials used for sensors and actuators in smart structures fall into three main categories: ferroic materials; nanostructured materials; and soft materials.

Microactuators are the current rage. At present it is usually necessary that they be compatible with the materials and processing technologies used in silicon microelectronics. The two main types of microactuators are '*mechanisms*' and '*deformable microstructures*'. The former provide displacement through rigid-body motion. The latter do this by mechanical deformation or straining.

Control systems

The use of computers is necessary for developing sophisticated control systems for smart structures (e.g., robots) which can learn and take decisions. There are various approaches to computational intelligence, as also to evolutionary robotics. The evolution of *distributed intelligence* should be emphasized in this context, and will be taken up later in the book.

Microelectromechanical systems (MEMS)

MEMS are the current choice for integrated smart structures, as they involve a high degree of miniaturization and integration of sensors, actuators, and control systems. Miniaturization and integration have many advantages: lower cost; higher reliability; higher speed; and capability for a higher degree of complexity and sophistication. Some authors equate smart structures with MEMS (Varadan *et al.* 2006).

Applications of MEMS include those in biomedical engineering, wireless communications, and data storage. Some of the more integrated and complex applications are in microfluidics, aerospace, and biomedical devices.

Silicon tops the list of materials used for MEMS because of its favourable mechanical and electrical properties, and also because of the already entrenched IC-chip technology. More recently, there have been advances in the technology of using multifunctional polymers for fabricating 3-dimensional MEMS (Varadan, Jiang and Varadan 2001). Organic-materials-based MEMS are also conceivable now, after the invention of the organic thin-film transistor. In the overall smart systems involving MEMS, there is also use for ceramics, metals, and alloys, as also a number of ferroic or multiferroic materials.

35.2 Reconfigurable computers and machines that can evolve

The design of smart structures is all about drawing inspiration from biological systems. Reconfigurable computers and evolvable machines work on the same idea:

'*The malleability of configurable processors means that they can adapt at the hardware level. Does the term adaptation ring a bell? Over the past few years, researchers working on bio-inspired systems have begun using configurable processors; after all, they reasoned, what could be better suited to implement adaptive systems than soft hardware? And what do you get when you marry configurable processors with evolutionary computation? Evolvable hardware*' (Sipper 2002).

In a typical device having a microprocessor, e.g. a washing machine or a digital camera, the control part comprises hardware and software. The hardware is usually fixed and unchangeable. The software has been written by somebody and is therefore fixed (although new software can be loaded), and the user can only key-in his/her preferences about some operational parameters. Computational science has been progressively moving towards a scenario in which the hardware is no longer absolutely hard; it is changeable, even evolvable.

Evolutionary ideas have permeated a whole host of scientific and engineering disciplines, and computational science is no exception. I discussed the evolutionary aspects of software in Section 33.5 on genetic programming. On-the-job changeability and evolution of *hardware* configurations offers exciting possibilities.

Conventional general-purpose stored-program computers have a fixed hardware, and they are programmable through software. Their microprocessors can be led through just about any conceivable logical or mathematical operations by writing a suitable set of instructions. Since their general-purpose, low-cost, hardware configuration is not fine-tuned for any specific task, they tend to be relatively slow.

On the other hand, for major specialized jobs involving a very large amount of number-crunching, it is more efficient to design *application-specific ICs* (ASICs), but then the overall long-term cost goes up, particularly if there is a need for upgradation or alteration from time to time.

An intermediate and very versatile approach is to have *reconfigurable* computers (RCs). A computer has a memory and a processor. The software is loaded into the memory; and the processor normally has a fixed, unalterable, configuration. By 'configuration' we mean the way the various components of the processor are interconnected, and the logical and mathematical operations they perform. If we can make these interconnections and operations (*gate arrays*) alterable, *without having to physically change them by hand*, we get an alterable or reconfigurable hardware. This is equivalent to having customized hardware which can be reconfigured at will (within limits), without incurring additional costs.

In RCs one makes use of '*field-programmable gate arrays*' (FPGAs), the logic structure of which can be altered and customized by the user. The generic architecture of such RCs has four major components connected through a *programmable* interconnect: multiple FPGAs; memories; input/output channels; and processors.

FPGAs are highly tuned hardware circuits that can be altered at almost any point during use. They comprise of arrays of reconfigurable logic blocks that perform the functions of logical gates. The logic functions performed within the blocks, as well as the connections among the blocks, can be changed by sending control signals. A single FPGA can perform a variety of tasks in rapid succession, reconfiguring itself as and when instructed to do so.

Villasenor and Mangione-Smith (1997) made a single-chip video transmission system that reconfigures itself four times per video frame: It first stores an incoming video signal in the memory; then it applies two different image-processing transformations; and then becomes a modem to send the signal onward. FPGAs are ideally suited for algorithms requiring rapid adaptation to inputs.

Such softening of the hardware raises visions of *autonomous* adaptability, and therefore *evolution*. It should be possible to make the *soft hardware* evolve to the most desirable (*fittest*) configuration. That would be a remarkable Darwinian evolution of computing machines, a marriage of adaptation and design.

Self-healing machines

Biological systems are robust because they are fault-tolerant and because they can heal themselves. Machines have been developed that can also heal themselves to a certain extent. The field of *embryonic electronics* (Mange *et al.* 2004) has drawn its sustenance from ontogeny (i.e., the development of a multicellular being from a single fertilized cell, namely the zygote) and embryogenesis observed in Nature. One works with a chessboard-like assembly of a large number of reconfigurable computer chips described above. This is the equivalent of a multicellular organism in Nature. The chips or *cells* are blank-slate cells to start with. We specify a task for the assembly; say, to show the time of the day. We also wish to ensure that the artificial organism is robust enough to heal itself; i.e., it should repair itself if needed. Such a BioWatch has indeed been built (see Sipper 2002):

Ontogeny in Nature involves cell division and cell differentiation, with the all-important feature that *each cell carries the entire genome*. The BioWatch borrows these ideas. The genome, of course, comprises of the entire sequence of instructions for building the watch. One starts by implanting the genome (the zygote) in just one cell. Cell division is simulated by making the zygote transfer its genome to the neighbouring cells successively. When a cell receives the genome from one of its neighbours, information about its relative location is also recorded. In other words, each of the cells knows its relative location in the assembly.

This information determines how that cell will specialize by extracting instructions from the relevant portion of the genome. Thus each cell or chip, though specialized ('differentiated') for doing only a part of the job, carries information for doing everything, just as a biological cell does. The BioWatch is now ready to function. Its distinctive feature is that it can *repair* itself. How?

Suppose one of the cells malfunctions, or stops functioning. There are kept some undifferentiated (and therefore unused) cells in the same assembly. Repair or healing action amounts to simply ignoring the dead cell after detecting its relative position, and transferring its job to one of the fresh cells which already has the complete genome, and has to only undergo differentiation for becoming operational.

One can do even better than that by making the system *hierarchical*. Each cell can be given a substructure: Each cell comprises of an identical set of 'molecules'. When one of the molecules malfunctions, its job is transferred to a fresh molecule. Only when too many molecules are non-functional does the entire cell become dead, and the services of a fresh cell are requisitioned.

Hardware you can store in a bottle

Adamatzky and coworkers (2005) have been developing chemical-based processors that are run by ions rather than electrons. At the heart of this approach is the well-known *Belousov-Zhabotinsky* or *BZ reaction* (see Section 25.5). It is a repeating cycle of three sets of chemical reactions. After the ingredients have been brought together, they only need some perturbation (e.g. a catalyst, or a local fluctuation of concentration) to trigger the first of the three sets of reactions. The products of this reaction initiate the second set of reactions, which then set off the third reaction, which then restart the first reaction. And so on, cyclically. The BZ reaction is *self-propagating*. Waves of ions form spontaneously and diffuse through the solution, inducing neighbouring regions to start the reactions.

Adamatzky has been developing *liquid logic gates* (for performing operations like NOT and OR) based on the BZ reaction. It is expected that an immensely powerful parallel processor (*a liquid robot brain*) can be constructed, which will be a blob of jelly, rather than an assembly of metal and wire. Such a system would be highly reconfigurable and self-healing.

A possible host material for this *blobot* is a jelly-like polymer called PAMPS, an electroactive gel. It expands or contracts when an electric field is applied. BZ waves can travel through it without getting slowed down substantially, and the waves can be further manipulated interactively by internal or external electric fields. One day we may end up having an intelligent, shape-changing, crawling blob based on such considerations.

Who would have thought that the FPGA-chip idea would one day find a direct analogue in the way the human brain has evolved to be? But that is exactly what has happened. As explained in a recent and highly successful theory of the human brain (Kurzweil 2012), a child starts out with a huge number of 'connections-in-waiting' to which the 'pattern recognition modules' can hook up. We shall discuss such things later in this book.

Some details of the future of intelligence will be given in the next two chapters. As Kurzweil

(2006)   predicted,   we   are   approaching   a   'technological   singularity'
(http://www.singularity.com/), beyond which robots and other smart structures will overtake
us in all abilities, and technological progress will be so rapid as to outstrip our ability to
comprehend it. And we shall transform ourselves and augment our minds and bodies with the
help of genetic alterations, MEMS, NEMS (nanoelectromechanical systems), and *true* machine
intelligence. That would mark a complete blurring of the distinction between the 'living' and
the 'nonliving'.

36. Robots and Their Dependence on Computer Power

Robots are a subset of smart structures. The evolution of robots (our 'mind children') is being brought about by us at a rapid pace. This is artificial evolution. Historically speaking, the evolution of terrestrial complexity had gone on increasing till we humans emerged, but our activities are resulting in a much faster increase of complexity on a variety of fronts, an important one being the evolution of artificial intelligence. Intelligent robots have artificial intelligence.

There are two main types of robots: *industrial* robots, and *autonomous* robots. Industrial robots do useful work in a structured or pre-determined environment. They do repetitive jobs like fabricating cars, stitching shirts, or making computer chips, all according to a set of instructions programmed into them.

Autonomous or smart robots, by contrast, are expected to work in an *unstructured* environment. They *move around* in an environment that has not been specifically engineered for them, and do useful and 'intelligent' work. They have to interact with a dynamically changing and complex world, with the help of sensors and actuators and a brain centre.

There have been several distinct or parallel approaches to the development of machine intelligence (Nolfi and Floreano 2000). The classical artificial-intelligence (AI) approach attempted to imitate some aspects of rational thought. Cybernetics, on the other hand, tended to adopt the human-nervous-system approach more directly. And evolutionary or adaptive robotics embodies a convergence of the two approaches. Thus the main routes to the development of autonomous robots are:

- behaviour-based robotics;
- robot learning;
- artificial-life simulations (in conjunction with physical devices comprising the robot);
- evolutionary robotics.

We discussed artificial life in Section 33.6. Let us begin with behaviour-based robotics here.

36.1 Behaviour-based robotics

In the traditional AI approach to robotics, the computational work for robot control is decomposed into a chain of information-processing modules, proceeding from overall sensing to overall final action. By contrast, in behaviour-based robotics (Brooks 1986; Arkin 1998), the designer equips the robot with a set of simple basic behaviours. A parallel is drawn from how coherent intelligence ('swarm intelligence') emerges in a beehive or an ant colony from a set of very simple behaviours. In such a vivisystem, each agent is a simple device interacting with the world with sensors, actuators, and a very simple brain.

In Brooks' '*subsumption architecture*', the decomposition of the robot-control process is done in terms of behaviour-generating modules, each of which connects sensing to action *directly*. Like an individual bee in a beehive, each behaviour-generating module directly generates some part of the behaviour of the robot. The tight (proximity) coupling of sensing to action produces an intelligent network of simple computational elements that are broad rather than deep in perception and action.

There are two further concepts in this approach: '*situatedness*', and '*embodiment*'. Situatedness means the incorporation of the fact that the robot is situated in the real world, which directly influences its sensing, actuation, and learning processes. Embodiment means that the robot is not some abstraction inside a computer, but has a body which must respond dynamically to the signals impinging on it, using immediate feedback. This makes evolution of intelligence in a robot more realistic than the artificial evolution carried out entirely inside a computer.

In Brooks' (1986, 1990) approach, the desired behaviour is broken down into a set of simpler behaviours ('layers'), and the solution (namely the control system) is built up incrementally. Simple basic behaviours are mastered first, and behaviours of higher levels of sophistication are added gradually, layer by layer. Although basic behaviours are implemented in individual subparts or layers, a coordination mechanism is incorporated in the control system, which determines the relative strength of each behaviour in any particular situation.

Coordination may involve both competition and cooperation. In a competitive scenario, only one behaviour determines the motor output of the robot. Cooperation means that a weighted sum of many behaviours determines the robot response.

In spite of the progress made in behaviour-based robotics, the fact remains that autonomous mobile robots are difficult to design. The reason is that their behaviour is an *emergent* property (Nolfi and Floreano 2000). By their very nature, emergent phenomena in a complex system (in this case the robot interacting with its surroundings) are practically impossible to predict, even if we have all the information about the sensor inputs to the robot and the consequences of all the motor outputs. The major drawback of behaviour-based robotics is that the trial-and-error process for improving performance is judged and controlled by an outsider, namely the designer. It is not a fully self-organizing and evolutionary approach for the growth of robotic intelligence. And it is not easy for the designer to do a good job of breaking down the global behaviour of a robot into a set of simple basic behaviours. One reason for this difficulty is that an optimal solution of the problem depends on *who* is describing the behaviour: the designer or the robot? Accordingly, the description can be either *distal* or *proximal* (Nolfi and Floreano 2000): The proximal description of the behaviour of the robot is a description from the vantage point of the sensorimotor system that describes how the robot reacts to different sensory situations. The distal description is from the point of view of the designer or the observer. In it, the results of a sequence of sensorimotor loops may be described in terms of high-level words like 'approach' or 'discriminate'. Such a description of behaviour is the result of not only the sensory-motor mapping, but also of the description of the environment. It thus incorporates the dynamical interaction between the robot and the environment, and that leads to some difficult problems. The environment affects the robot, and the robot affects the environment, which in turn affects the robot in a modified way, and so on. This interactive loop makes it difficult for the designer to break up the global behaviour of the robot into a set of elementary or basic behaviours that are simple from the vantage point of the proximal description. Because of the emergent nature of the behaviour, it is difficult to predict what behaviour will result from a given control system. Conversely, it is also difficult to predict what pattern of control configurations will result in a desired behaviour.

This problem is overcome in *evolutionary robotics* by treating the robot and the environment as a single system, in which the designer has no role to play. After all, this is how all complexity has evolved in Nature.

36.2 Evolutionary robotics

Evolutionary robotics is a technique for the automatic creation of autonomous robots (Nolfi and Floreano 2000). It works in analogy with the Darwinian principle of selective reproduction of the fittest.

I described genetic algorithms (GAs) in Section 33.4. Typically, a population of artificial chromosomes or digital organisms (strings of command) is created in the computer controlling the robot. Each such chromosome encodes the control system (and even the morphology) of a robot. The various possible robots (actual, or just simulations) then interact with the environment, and their performance is measured against a set of specified criteria (the 'fitness functions'). The fittest robots are given a higher chance of being considered for the creation of next-generation robots (specified by a new set of chromosomes drawn from the most successful chromosomes of the previous generation). Processes such as mutations and crossover etc. are also introduced, by analogy with biological evolution. The whole process of decoding the instructions on the chromosomes and implementing them is repeated, and the fitness functions for each member of the new robot population are computed again. This is repeated for many generations, till a robot configuration having a pre-determined set of performance parameters is obtained.

The goal here is to evolve robots with creative problem-solving capabilities. This means that they must *learn* from real-life experiences so that, as time passes, they get better and better at problem-solving. The analogy with how a small child learns and improves as it grows is an apt one. The child comes into the world equipped with certain sensors and actuators (eyes, ears, hands, etc.), as well as a brain. Through a continuous process of experimentation (unsupervised learning), as well as learning from parents and teachers etc. (supervised learning), and also learning from the hard knocks of life, including rewards for certain kinds of action (reinforced learning), the child's brain performs *evolutionary computation*. This field of research involves a good deal of simulation and generalization work. Apart from GAs, other techniques of computational intelligence are also employed.

In evolutionary robotics, the key question to answer is: Under what conditions is artificial evolution likely to select individual robots which develop *new* competencies for adapting to their environment? Nolfi and Floreano (2000) discuss the answer in terms of three main issues: (i) generating incremental evolution through interspecific and intraspecific competition; (ii) leaving the system free to decide how to extract supervision from the environment; and (iii) bringing the genotype-to-phenotype mapping into the evolutionary process itself.

Incremental evolution

In biological evolution, natural selection is governed by the ability to reproduce. But there are many examples of improvements (sophisticated competencies) over and above what is essential for sheer survival and propagation of a species. Emergence of the ability to fly is an example of this. A more complex example is that of the emergence of speech and language in humans. Such competencies, which bestow a survival and propagation advantage to a species, do arise from time to time. As discussed in Chapter 33, Holland used GAs in his model for ensuring an element of *perpetual novelty* in the adapting population. He also incorporated, among other things, the notion of classifier systems. Classifier systems were also introduced by Dorigo and Colombetti (1997), who used the term *robot shaping*, borrowing it from experimental psychology wherein animals are trained for eliciting pre-specified responses.

A simple approach to evolving a particular ability or competency in a robot is to attempt selection through a suitable fitness function. This works well for simple tasks. For a complex task, the requisite fitness function turns out to be such that all individuals fail the test, and evolution grinds to a halt. This is referred to as *the bootstrap problem* (Nolfi and Floreano 2000). Incremental evolution is a way out of this impasse. Define a new task and the corresponding fitness function or selection criterion that is only a slight variation of what the robot is already able to do. And do this repeatedly. The sophisticated competency can thus be acquired in a large number of incremental stages (Dorigo and Colombetti 1997).

But such an approach has the potential drawback that the designer may end up introducing excessive constraints. A completely self-organized evolution of new competencies, as happens in biological evolution, has to invoke phenomena like the *coevolution of species* and the attendant *arms races* within a species or between species (Chapter 30). Coevolution can lead to progressively more complex behaviour among competing robots. For example, predator and prey both improve in their struggle for survival and propagation. In such a scenario, even though the selection criterion may not change, the task of adaptation becomes progressively more complex.

Extracting supervision from the environment through lifetime learning

A human supervisor can help a robot evolve new competencies. But it is more desirable if the robot can autonomously extract high-level supervision from the environment itself. A large amount of information is received by the robot through its sensors, and it adapts to the environment through the feedback mechanisms. As the environment changes, new strategies are evolved for dealing with the change. The evolving robot also gets feedbacks on the consequences of its motor actions. Thus, the robot can effectively extract supervision from the environment through a lifetime of adaptation and learning about what actions produce what consequences.

One makes a distinction between *ontogenetic adaptation* and *phylogenetic adaptation*. The former refers to the adaptation resulting from a lifetime of learning, in which the individual is tested almost continuously for fitness for a task. The latter involves only one testing for fitness, namely when it comes to procreating the next generation of candidate robots. Not much progress has been made yet regarding the exploitation of phylogenetic adaptation in evolutionary robotics.

Development and evolution of evolvability

In a GA the chromosomes code the information, which must be decoded for evaluating the fitness of an individual. Thus there is a mapping of information from the genotype to the phenotype, and it is the phenotype which is used for evaluating the fitness. Genotypes of individuals with fitter phenotypes are given a larger probability for selection for procreating the next generation. The problem is that a one-to-one genotype-to-phenotype mapping is too simplistic a procedure. In reality, i.e., in real biological evolution, a phenotype may be influenced by more than one genes. The way genetic variation maps onto phenotypic variation in evolutionary biology is a highly nontrivial problem, and is commonly referred to as the *representation problem*. The term *evolvability* is used for the ability of random variations to result in improvements sometimes. As emphasized by Wagner and Altenberg (1996), evolvability is essential for adaptability. And incorporating evolvability into evolutionary robotics amounts to tackling the representation problem. Efforts have indeed been made to incorporate more complex types of mapping than a simple one-to-one mapping of one gene to one robotic characteristic (Nolfi and Floreano 2000).

36.3 Evolution of computer power per unit cost

Let us next take a look at developments regarding the control centres of robots, namely computers. There are three parameters to consider for the computational underpinnings of robotic action: the processing power (or speed) of the computer; its memory size; and the price for a given combination of processing power and memory size.

The processing power or computing power can be quantified in terms of *million instructions per second* or MIPS. And the size of the memory is specified in megabytes. By and large, the MIPS and the megabytes for a computer cannot be chosen independently: Barring some special applications, they should have a certain degree of compatibility (per unit cost), for reasons of *optimal performance in a variety of applications.*

An analysis by Hans Moravec (in 1999) revealed that, for general-purpose or *'universal computers'*, the ratio of memory (the megabytes) to speed (the MIPS) has remained remarkably constant during almost the entire history of computers. A 'time constant' can be defined here as roughly the time it takes for a computer to scan its own memory once. One megabyte per MIPS gives one second as the value of this time constant. This value has remained amazingly constant (till recently) as progressively better universal computing machines have been developed over the decades, as depicted in Moravec's (1999) 'evolution slide' (Fig. 36.1).

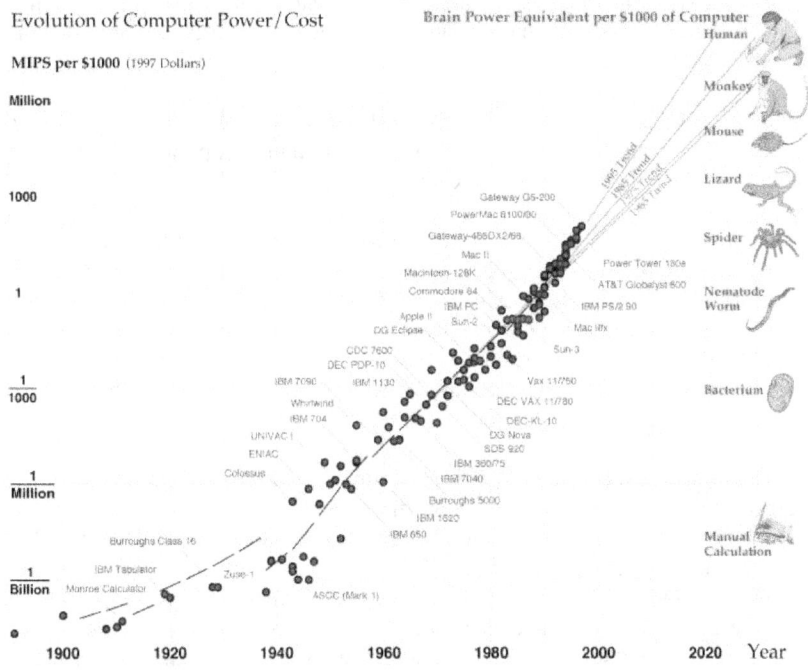

Fig. 36.1 Moravec's famous *evolution slide*, graphically depicting the progress of the MIPS-to-megabytes ratio for general-purpose or 'universal' computers over their entire history. Image credit: http://singularitybookreviews.com/2012/01/

Machines having too much memory for their speed are too slow (for their price), even though they can handle large programs. Similarly, lower-memory higher-speed computers are not able to handle large programs, in spite of being fast. Therefore, special jobs require *special computers* (rather than universal computers), entailing higher costs, as also a departure from the above universal time constant. For example, IBM's Deep Blue computer, developed for competing with the chess legend Garry Kasparov (in 1996-97) had more speed than memory (~3 million MIPS and ~1000 megabytes, instead of the universally optimal combination of, say, 1000 MIPS and 1000 megabytes). Similarly, the MIPS-to-megabytes ratio for running certain aircraft is also skewed in favour of MIPS. Examples of the other kind, namely slow machines (less MIPS than megabytes), include time-lapse security cameras and automatic data libraries.

Moravec estimated in 1999 that the most advanced supercomputers available at that time were within a factor of 100 of having the power to mimic the human brain. But then such supercomputers came at a prohibitive cost. Costs must fall if machine intelligence is to make much headway. Although this has indeed been happening for a whole century, what about the future? How long can this go on? The answer is: For quite a while, provided technological breakthroughs or new ideas for exploiting technologies keep coming.

An example of the latter is the use of *multicore processors*. Multicore chips, even for PCs, are already in the market. Nvidia introduced a chip, GeForce 8800, capable of a million MIPS speed, and low-cost enough for use in commonplace applications like displaying high-resolution videos. It has 128 processors (on a single chip) for specific functions including high-resolution video display. In a multicore processor, two or more processors on a chip process data in tandem. For example, one core may handle a calculation, a second one may input data, while a third one sends instructions to an operating system. Such load-sharing and parallel functioning improves speed and performance, and reduces energy consumption and heat generation.

Nanotechnology holds further promise for the next-generation solutions to faster and cheaper computation. DNA computing (Adams 1998) is an alternative approach being investigated; this technique has the potential for massive parallelism.

Quantum computing (Hooker 2012) is one more exciting possibility. Extreme miniaturization is another. Lovat *et al.* (2017) have reported the creation of a *single-molecule* room-temperature transistor from 14 atoms. This is part of an effort towards atomically precise, reproducible transistors made from single molecules and operating at room temperature. They created a two-terminal transistor with a diameter of about 0.5 nm and core consisting of just 14 atoms. The device can reliably switch from insulator to conductor when charge is added or removed, one electron at a time.

Single-molecule-level data storage (with 100 times higher data densities than current technologies) is an exciting possibility in terms of cost, space, and energy saving, as also higher speeds. A step in that direction seems to have been taken by Goodwin *et al.* (2017). They have reported molecular magnetic hysteresis at 60 K (i.e., above liquid-nitrogen temperature) in dysprosocenium. Molecular-level data storage can result in much smaller hard drives that require less energy. This means that data centres could be smaller, lower-cost, and very energy-efficient. One should be able to store more than 25 terabytes of data in a device the size of a U.S. quarter.

A factor hindering rapid progress in robotics has been the high costs incurred on sensors and actuators. Progress in nanotechnology (e.g., the development of MEMS (see, e.g., Wadhawan 2007a)) is resulting in continuously falling costs of sensors and actuators. It is now far less expensive to incorporate GPS (Global Positioning System) chips, video cameras and array microphones, etc., into robots.

Bill Gates (2007) announced the development of universally applicable software packages by his company, Microsoft, that would further facilitate the use of ordinary PCs for controlling and developing robots of ever-increasing sophistication. Many robots already have PC-based controllers. It is anticipated that a large-scale move of robotics towards universally applicable PC-based architecture will cut costs and reduce the time needed for developing new configurations of autonomous robots. In the 1970s the development of Microsoft BASIC had provided a common foundation that made it possible to use software written for one set of hardware to run on another set. Something similar has been happening in robotics.

One of the challenging problems faced in robotics was that of *concurrency*, namely how to process *simultaneously* the large amount of data coming in from a variety of sensors, and send suitable commands to the actuators of the robot. The approach adopted in the beginning was to write a 'single-threaded' program that first processes all the input data and then decides on the course of action, before starting the long loop all over again. This was not a happy situation because the action taken on the basis of one set of input data may be too late for safety etc., even though subsequent input data has already indicated a drastically different course of action. To solve this problem, one must write 'multi-threaded' programs that can allow data to travel along many paths. This tough problem has been tackled by Microsoft by developing what is called CCR (concurrence and coordination run-time). Although CCR was originally meant to exploit the advantages of multicore and multiprocessor systems, it may well be just the right thing needed for robots also. The CCR is basically a sequence of library functions that can perform specific tasks. It helps in developing multithread applications quickly, for coordinating a number of simultaneous activities of a robot. Of course, competing approaches already exist, not only to CCR, but also to the so-called 'decentralized software services' (DSS).

Although low-cost *universal robots* will be run by universal computers, their proliferation will have more profound consequences than those engendered by low-cost universal computers alone. Computers only manipulate symbols; i.e., they basically do '*paperwork*' only, although the end results of such paperwork can indeed be used for, say, automation. A sophisticated universal robot goes far beyond mere paperwork. It goes into perception and action in real-life situations. There is a far greater diversity of situations in the real world than just paperwork, and in far greater numbers. There would thus be a much larger number of universal robots in action than universal computers. This, of course, will happen only when the cost per unit capability falls to low levels.

When will robotic intelligence go past human intelligence? Kurzweil has made an optimistic prediction (https://futurism.com/kurzweil-claims-that-the-singularity-will-happen-by-2045/):

'2029 is the consistent date I have predicted for when an AI will pass a valid Turing test and therefore achieve human levels of intelligence. I have set the date 2045 for the 'Singularity' which is when we will multiply our effective intelligence a billion fold by merging with the intelligence we have created'.

[The 'Singularity' is that point in the evolution of complexity when the advances in technology, particularly in artificial intelligence (AI), will lead to machines that are smarter than human beings. We shall continue the discussion on machine intelligence in the next chapter.]

I conclude this chapter by providing some updates on Moravec's 'evolution slide' (Fig. 36.1 above) prepared in 1999. This information is from:
http://www.donbot.com/Futurebot/NewTech/NT01360MoravecsGraphFirstModification.html
On this website an extension of the data has been given up to ~2010 (Fig. 36.2).

I quote from the description accompanying the updated slide:

'Here is presented a modified version of Moravec's graph to add more recently available computers and to clarify which area corresponds to MIPS per $1000 and which area corresponds to MIPS. If the graph's projections were correct and if we had the software available today, we would be able (to) purchase a computer to simulate the human brain in 2010 for $1000.

'Notice that if we could perform at 10^3 in 2000 and will make it to 10^9 by the end of 2010, then we can project that a rate of improvement of at least 1 million times (10^6) per decade is now in effect. This should accelerate as we continue to use more intelligent robots to help us design more intelligent robots.

'I have updated the left hand part of the graph by adding the black dots. These points were determined by looking up the MIPS ratings on the latest computers on Wikipedia.

'As to the right-hand side of the slide, my question was always, "Why doesn't Hans Moravec update his Evolution Slide?" Perhaps it is because the estimate for how many Millions of Instructions Per Second (MIPS) would be needed to simulate the human brain seems to have been off by a factor of between 1000 and 10,000,000. This is still a matter of debate. Marvin Minsky thinks we already have enough computer power to simulate a human brain now, others think it will not happen until 2050.'

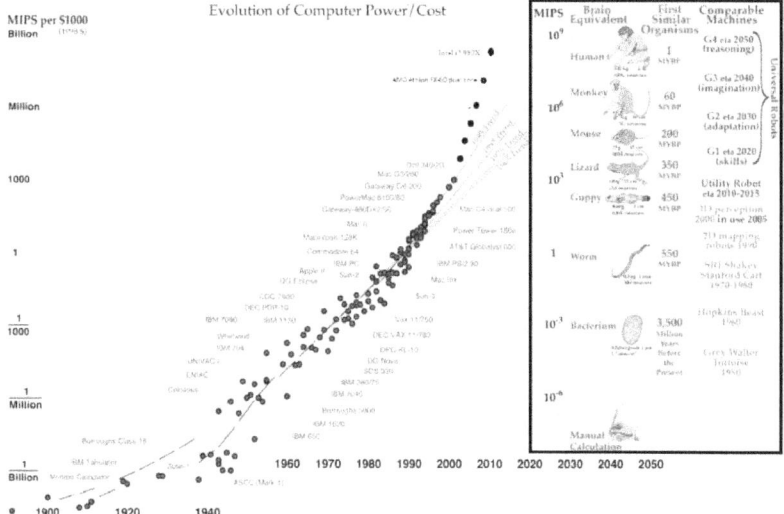

Fig. 36.2 Updation of Fig. 36.1 with data up to ~2010.
http://www.donbot.com/Futurebot/NewTech/NT01360MoravecsGraphFirstModification.html

37. Machine Intelligence

The idea that humans will always have a unique ability beyond the reach of non-conscious algorithms is just wishful thinking.

Yuval Noah Harari (2016)

Learning, smartness, and intelligence are terms we have been normally associating with living beings. But if our machines are going to be helping us in truly worthwhile ways, we must make them intelligent. And the best way to do this is to make their intelligence *evolve*. Intelligence is an *emergent* property, possible only in complex adaptive systems. Human intelligence arises from the complex interactions among a huge assembly of neurons. Intelligence is a property of the neural network as a whole; each neuron is as dumb as can be. Machine intelligence will also evolve on similar lines.

37.1 Artificial distributed intelligence

People learn to make sense of the world by talking with other people about it

J. Kennedy (2006)

Swarm intelligence (the kind we see in beehives and ant colonies) is the archetypal example of how intelligence emerges, namely by interactions among a large number of dispersed individuals. '*Distributed perceptive networks*' are the artificial equivalents of such vivisystems.

A variety of small and low-cost smart sensor systems are available these days. Therefore a large number of them ('smart dust') can be easily distributed over the area to be monitored, or spied upon (Ananthaswamy 2003). Similarly, very simple computers and operating systems can be procured in large numbers and linked by radio transceivers and sensors to form small autonomous nodes called *motes*. They run on specially designed operating systems, like the one called TinyOS (which was developed by a graduate student).

Thus each mote can link up and communicate with its neighbours. Although each unit by itself has only limited capabilities (like a bee in a beehive), a system comprising of hundreds or thousands of them can spontaneously emerge as a *perceptive* network (O'Hare and Jennings 1996). The overall approach in the construction of such networks is based on the following considerations: cut costs; conserve power; conserve space (miniaturize); have wireless communication (networking) among the nodes or agents; let collective intelligence emerge, like in a beehive; ensure robust, efficient, and reliable programming of a large and distributed network of motes; incorporate regrouping and reprogramming; and include redundancy of sensor action to increase the reliability of the motes, keeping in view the fact that they may have to operate in hostile environmental conditions.

Each mote has its own TinyOS (like the tiny brain of a bee). This software runs on microprocessors that require very little memory (as little as eight kilobytes). The so-called *multi-hop networking* approach is used for saving power: Each mote is given an extremely short-ranged radio transmitter. A *multi-tiered* network is established, with motes in a particular tier or layer communicating only with those in the next lower and the next upper layer. Thus information hops from one mote to another mote only by single one-layer steps. This hierarchy also makes room for parallelism. If, for example, a particular mote stops functioning, there is enough redundancy and parallelism in the network that other motes reconfigure the connectivity to bypass that mote.

To update or replace the software for a network of motes, the method used is similar to how viruses or worms spread in PCs via the Internet. The new software is placed only in the 'root' mote, which then 'infects' the neighbouring motes with it, and so on, up the hierarchy.

The brain, the sensors, and the actuators of an artificial intelligent structure can be situated in *different* locations, employing what is called *pervasive computing* (Hansmann 2003). Add to this the fact that the sensors in this network can be *different* from the human sensory system, and something truly remarkable can emerge. Why? Human intelligence is based on *pattern formation* based on the sensors and computing capability available to it. A different set of sensors and computing architecture can result in a different *kind* of intelligence, which can see patterns we humans have not and cannot (Hawkins 2004).

Pervasive computing will have truly far-reaching consequences for applications concerning, say, terrestrial phenomena like weather forecasting, animal and human migrations, etc. Suppose we let loose weather sensors with a certain degree of artificial intelligence all over the globe, all communicating with local brain centres, and indirectly with a centralized brain. Since these are intelligent artificial brains, they will form a worldview with the passage of time, as more and more sensory data are received, processed, and generalized. And since the sensors need not be like those of humans, the pattern formation in the artificial superbrain will be different from that in our brains. This machine-brain will recognize local patterns and global patterns about winds etc., and there will be a perspective about weather patterns at different time scales: hours, days, months, years, decades. *The crux of the matter is that the intelligent machine-brain will see patterns we humans have not and cannot.*

With the ready availability of technologies like the Microsoft 'decentralized software services' (DSSs) for easily writing the programs needed for distributed robotic applications, it is now possible for a network of wireless robots to tap into the power of desktop PCs for carrying out tasks like recognition and navigation (Gates 2007) (https://en.wikipedia.org/wiki/Microsoft_Robotics_Developer_Studio). The DSSs enable the creation of applications in which the various services operate as *separate* processes that can be orchestrated, just like we can aggregate images, text, and information from different servers on a Web page. The DSSs need not reside entirely on the robot. They can be distributed over many PCs.

One can link wireless domestic robots or *personal robots* (PRs) to PCs. This makes it possible, for example, to keep a long-distance tab of what is being done by the PRs. The robots, of course, communicate with one another also.

An example of a practical application of what has been called '*particle swarm intelligence*' is that of optical network optimization (Durand *et al.* 2013). Dynamic optimization enables increased network flexibility and capacity.

It goes without saying that artificial intelligence of any kind can go only as far as the power of the computers behind it. We shall now take a look at this aspect of machine intelligence.

37.2 Evolution of machine intelligence

> *Our intelligence, as a tool, should allow us to follow the path to intelligence, as a goal, in bigger strides than those originally taken by the awesomely patient, but blind, processes of Darwinian evolution. By setting up experimental conditions analogous to those encountered by animals in the*

*course of evolution, we hope to retrace the steps by which human intelligence
evolved. That animals started with small nervous systems gives confidence
that today's small computers can emulate the first steps toward humanlike
performance.*

Moravec (1999)

Moravec (2000, 2003) estimated that artificial smart structures in general, and robots in particular, will evolve millions of times faster than biological creatures, and will surpass the humans in intelligence long before the present century is over. This evolution in machines is being assisted by us, and this makes two big differences, both of which can accelerate its speed: (i) unlike 'natural' Darwinian evolution (which is not goal-directed), artificial evolution has goals set by us; (ii) Lamarckian evolution is not taboo in artificial evolution (in biological evolution it is banned by the central dogma of molecular biology).

Sophisticated machine intelligence, successfully modelled on the human neocortex, should be only a few decades away. As Moravec pointed out in 1999, already machines read text, recognize speech, and even translate languages. Robots drive cross-country, crawl across Mars, and trundle down office corridors. He also discussed how the music composition program EMI's classical creations had pleased audiences who rated it above most human composers. The chess program Deep Blue, in a first for machinekind, won the first game of the 1996 match against Gary Kasparov. It appears certain that we shall indeed be able to evolve machine intelligence comparable to human intelligence in sophistication. As argued by Moravec, a fact of life is that biological intelligence has evolved from that of, say, insects to humans. There is a strong parallel between the evolution of robot intelligence and biological intelligence that preceded it. The largest nervous systems doubled in size every fifteen million years or so, since the Cambrian explosion 550 million years ago. Robot controllers double in complexity (processing power) every year or two. They are now at the lower range of vertebrate complexity, but should (hopefully) catch up with us within half a century. This will happen so fast because artificial evolution is being assisted by the intelligence of humans, and not just determined by the blind processes of Darwinian evolution.

In due course, probably well within this century, intelligent robots would have evolved to such an extent that they would take their further evolution into their own hands. The scenario beyond this crossover stage has been the subject of much debate. For example, the books by Moravec (1999) and Kurzweil (2005) continue to invite strong reactions.

I mentioned motes earlier in this chapter. Perceptive networks using motes will be increasingly used, and not just for spying. Such distributed supersensory systems will not only have swarm intelligence, they will also undergo *evolution* with the passage of time. Like in the rapid evolution of the human brain, both the gene pool and the 'meme pool' (a term introduced by Dawkins (1989), and to be discussed later in this book) will be instrumental in this evolution of distributed intelligence. This ever-evolving superintelligence and knowledge-sharing will be available to each agent of the network, leading to a snowballing effect. However, estimates about the speed with which machine intelligence will evolve continue to be uncertain. For example, this is what Moravec wrote in 2003:

'Before mid-century, fourth-generation universal robots with humanlike mental power will be able to abstract and generalize. The first ever AI programs reasoned abstractly almost as well as people, albeit in very narrow domains, and many existing expert systems outperform us. But the symbols these programs manipulate are meaningless unless interpreted by humans. For instance, a medical diagnosis program needs a human practitioner to enter a patient's symptoms, and to implement a recommended therapy. Not so a third-generation robot, whose simulator provides a two-way conduit between symbolic descriptions and physical reality.

Fourth-generation machines result from melding powerful reasoning programs to third-generation machines. They may reason about everyday actions with the help of their simulators (as did one of the first AI programs, the geometry-theorem prover written in 1959 at IBM by Herbert Gelernter. This program avoided enormous wasted effort by testing analytic-geometry "diagram" examples before trying to prove general geometric statements. It managed to prove most of the theorems in Euclid's "Elements," and even improved on one). Properly educated, the resulting robots are likely to become intellectually formidable, besides being soccer stars'.

However, there are some who believe that things may not move that fast. There may be a reason for this pessimism, as best expressed by Nolfi and Floreano (2000): 'The main reason why mobile robots are difficult to design is that their behaviour is an emergent property of their motor interaction with the environment. The robot and the environment can be described as a dynamical system because the sensory state of the robot at any given time is a function of both the environment and of the robot's previous actions. The fact that behaviour is an emergent property of the interaction between the robot and the environment has the nice consequence that simple robots can produce complex behaviour. However it also has the consequence that, as in all dynamical systems, the properties of the emergent behaviour cannot easily be predicted or inferred from a knowledge of the rules governing the interactions. The reverse is also true: it is difficult to predict which rules will produce a given behaviour, since behaviour is the emergent result of the dynamical interaction between the robot and the environment'.

But there are some inveterate enthusiasts as well, who are expecting an incredibly rapid narrowing of the distinction between humans and robots, and even a merging of the two identities. Here is what Kurzweil wrote (in 2012): 'My sense is we're making computers in our own image and we'll be merging – we already have – with that technology. We're going to use those tools to make ourselves more intelligent'. The bold title of his book, *How to Create a Mind: The Secret of Human Thought Revealed*, published in November 2012, says it all; he is using the word 'mind', not 'brain'. The intelligence of machines and humans will evolve together, in a symbiotic way.

37.3 The future of intelligence and the status of humans

> *Technology, hailed as the means of bringing nature under the control of our intelligence, is enabling nature to exercise intelligence over us.*
>
> Dyson (1997)

We are orders of magnitude more intelligent than any other species around. But can we become even more intelligent as a species? Stephen Hawking made an interesting suggestion. To the extent that our typical level of intelligence has a correlation with our typical brain size, there is scope for manipulation and improvement. Our present average brain size is limited by the size of the orifice through which a newborn baby has to pass at birth. This limitation can be got over by going for conception, followed by the entire growth of the foetus, *outside* the womb.

But it is more likely that the future of our intelligence will be influenced by developments underway in robotics (Moravec 2009). Progress in machine intelligence will have absolutely mind-boggling effects on our intelligence. How, and why?

At present we make a distinction between two kinds of intelligence: biological or organic intelligence, and machine or inorganic intelligence. A composite (organic-inorganic, or man-machine) intelligence will evolve in the near future. It appears inevitable that, aided by human beings, an empire of *inorganic life* (intelligent robots) will evolve, just as biological or organic life has evolved. We are about to enter a *post-biological world*, in which machine intelligence, once it has crossed a certain threshold, will not only undergo Darwinian and Lamarckian evolution *on its own*, but will do so millions of times faster than the biological evolution we are familiar with so far. The result will be intelligent structures with a composite, i.e., organic-inorganic or man-machine, intelligence (Moravec 2000).

Moravec's (1999) book, '*Robot: Mere Machine to Transcendent Mind*', sets out a possible scenario. He expects robots to model themselves on successful biological forms, not human forms always. One such form - used by trees, the human circulatory system, and basket starfish - is a network of ever-finer branches.

A 'bush robot' is another likely development in Moravec's scheme of things: 'Twenty-five branchings would connect a meter-long stem to a trillion fingers, each a thousand atoms long and able to move about a million times per second.' Medical applications are one among the many likely uses of the bush robot listed by Moravec: 'The most complicated procedures could be completed almost instantaneously by a trillion-fingered robot, able, if necessary, to simultaneously work on almost every cell of a human body.'

When will all this happen, and what will be its possible bearing on our intelligence? Estimates vary widely (Moravec 1999b). The most optimistic ones are those of Kurzweil (1999, 2001, 2013). This is what he wrote in 1999: 'Sometime early in the next century, the intelligence of machines will exceed that of humans. Within several decades, machines will exhibit the full range of human intellect, emotions and skills, ranging from musical and other creative aptitudes to physical movement. They will claim to have feelings and, unlike today's virtual personalities, will be very convincing when they tell us so. By 2019 a $1,000 computer will at least match the processing power of the human brain. By 2029 the software for intelligence will have been largely mastered, and the average personal computer will be equivalent to 1,000 brains.'

There were reasons for the optimism exuded by Kurzweil: 'We are already putting computers – neural implants – directly into people's brains to counteract Parkinson's disease and tremors from multiple sclerosis. We have cochlear implants that restore hearing. A retinal implant is being developed in the U.S. that is intended to provide at least some visual perception for some blind individuals, basically by replacing certain visual-processing circuits of the brain. Recently scientists from Emory University implanted a chip in the brain of a paralyzed stroke victim that allows him to use his brainpower to move a cursor across a computer screen.'

In his book *The Age of Spiritual Machines*, Kurzweil (1999) enunciated his *Law of Accelerating Returns*, which simply paraphrases the occurrence of positive feedback in the evolution of complex adaptive systems in general, and biological and artificial evolution in particular. The law embodies *exponential* growth of evolutionary complexity and sophistication: ' . . . advances build on one another and progress erupts at an increasingly furious pace. . . . As order exponentially increases (which reflects the essence of evolution), the time between salient events grows shorter. Advancement speeds up. The returns – the valuable products of the process – accelerate at a nonlinear rate. The escalating growth in the price performance of computing is one important example of such accelerating returns. . . The Law of Accelerating Returns shows that by 2019 a $1,000 personal computer will have the processing power of the human brain – 20 million billion calculations per second. . . Neuroscientists came up with this figure by taking an estimation of the number of neurons in the brain, 100 billion, and multiplying it by 1,000 connections per neuron and 200 calculations per second per connection. By 2055, $1,000 worth of computing will equal the processing power of all human brains on Earth (of course, I may be off by a year or two).'

The law is similar to Moore's law, except that it is applicable to all 'human technological advancement, the billions of years of terrestrial evolution' and even 'the entire history of the universe'.

Kurzweil's book *How to Create a Mind* (2012) puts forward a 'pattern recognition theory' for how the brain functions, similar to Jeff Hawkins' theory published in his famous book *On Intelligence: How a New Understanding of the Brain will Lead to the Creation of Truly Intelligent Machines* (2004). [Kurzweil's book has been summarized in considerable detail at http://newbooksinbrief.com/2012/11/27/25-a-summary-of-how-to-create-a-mind-the-secret-of-human-thought-revealed-by-ray-kurzweil/.] According to Kurzweil, our neocortex contains 300 million very general pattern-recognition circuits which are responsible for most aspects of human thought, and a computer version of this design can be used to create artificial intelligence more capable than the human brain. As computational power grows, machine intelligence would represent an ever increasing percentage of total intelligence on the planet. Ultimately it will lead (by 2045) to the 'Singularity', a merger between biology and technology. 'There will be no distinction, post-Singularity, between human and machine intelligence . . .'. It is only a matter of time before we merge with the intelligent machines we are creating.

Stephen Hawking expressed the fear that humanity may destroy itself if there is a nuclear holocaust, and suggested the escape of at least a few individuals into outer space as a way for preserving the human race. But, for all our bravado, our bodies are delicate stuff which can survive only in a very narrow range of temperatures and other environmental conditions. But our robots will not suffer much from that handicap, and will be able to withstand high radiation fields, high temperatures, low temperatures, near-vacuum conditions, etc. Such robots (or even cyborgs) will be able to communicate with one another, with the inevitable possibility of developing distributed intelligence. And when each such robot is already way ahead of us in intelligence, a *distributed superintelligence* will emerge, capable of further evolution, of course. The ever evolving superintelligence and knowledge will benefit each agent in the network, leading to a snowballing effect.

Further, the superintelligent agents may organize themselves into a hierarchy, rather like what occurs in the human neocortex. Such an assembly would be able to see incredibly complex patterns and analogies which escape our comprehension, *leading to a dramatic increase in our knowledge and understanding of the universe*. Moravec (2000) expressed the view that this superintelligence will advance to a level where it is more mind than matter, suffusing the entire universe. We humans will be left far behind, and may even disappear altogether from the cosmic scene.

An alternative though similar picture was painted by Kurzweil (2005), envisioning a *coevolution* of humans and machines via neural implants that will enable an uploading of the human carbon-based neural circuitry into the then-prevailing hardware of the intelligent machines. Humans will simply merge with the intelligent machines. The inevitable habitation of outer space and the further evolution of distributed intelligence will occur concomitantly. Widely separated intelligences will communicate with one another, leading to the emergence of an *omnipresent superintelligence*. You want to call that 'God'? Don't. That omnipresent superintelligence would be *our* creation; a triumph of *our* science and technology; a result of what we humans can achieve by adopting the scientific method of interpreting data and information.

These ideas have been further updated in a book by Max Tegmark (2017): *Life 3.0: Being Human in the Age of Artificial Intelligence*. Tegmark defines life as simply as a process that can retain its complexity and replicate. What is replicated is information or software (comprised of bits) specifying how the atoms of the hardware part are arranged.

He delineates the evolution of three *stages* of life: Life 1.0; Life 2.0; and Life 3.0. A bacterium is an example of Life 1.0. When a bacterium makes a copy of its DNA, a new set of atoms are arranged in the same pattern as the original, thereby copying the information. So, Life 1.0 (the *biological stage*) evolves both its hardware and software.

We humans are Life 2.0. Ours is the *cultural stage* of evolution. In this stage, life's hardware is through the processes of evolution (it is an unchangeable given), but it designs much of its software. We are complex adaptive systems par excellence, with a life time of learning (software) at our disposal. And we are using this software for developing and for being a part of Life 3.0: robots, cyborgs, distributed superintelligence.

Life 3.0 (the *technological stage*) designs both its hardware and software (unlike Life 2.0, which cannot design its hardware). As Kurzweil writes: 'After 13.8 billion years of cosmic evolution, development has accelerated dramatically here on Earth: Life 1.0 arrived about 4 billion years ago, Life 2.0 (we humans) arrived about a hundred millennia ago, and many AI researchers think that Life 3.0 may arrive during the coming century, perhaps even during our lifetime, spawned by progress in AI'.

'We are the brothers and sisters of our machines' (Dyson 1997)

.

38. Evolution of Language

A mostly Lamarckian process whereby evolution of a transformational nature proceeds via the passage of acquired characters, cultural evolution, like the stellar evolution before it, involves no DNA chemistry and perhaps less selectivity than biological evolution. Culture enables animals to transmit survival kits to their offspring by nongenetic routes; the information gets passed on behaviourally, from brain to brain, from generation to generation, the upshot being that cultural evolution acts much faster than biological evolution.

Eric Chaisson (2002), *Cosmic Evolution*

If there had been no speech, then right and wrong, truth and falsehood, good and bad, attractive and unattractive would not have been made known. Speech makes known all this. Worship speech.

Chandogya Upanishad VII-2-1

As far as humans are concerned, language has got to be the ultimate evolutionary innovation. It is central to most of what makes us special, from consciousness, empathy and mental time travel to symbolism, spirituality to morality.

Kate Douglas (2005a)

Somewhere in the last 100,000 years or so, human beings hit upon language. Human language must have seemed an odd-sounding innovation to the other animals around. But by allowing the expression of arbitrarily complicated concepts, human language allowed people to process information in a highly distributed fashion. The distributed nature of human information processing in turn allowed people to cooperate in new ways, forming groups, associations, societies, companies, and so on. Some of these new forms of cooperation proved strikingly effective, as various forms of distributed information processing, such as democracy, communism, capitalism, religion, and science, took on a life of their own, propagating themselves and evolving over time. It is the richness and complexity of our shared information processing that has brought us this far.

Seth Lloyd (2006)

Story-telling or spoken language was the first major invention of humans that enabled them to represent ideas with distinct utterances. When written language was invented, we developed distinct shapes to symbolize our ideas. The evolution of language, speech, and culture are some of the causative factors in the rapid evolution of the size and capacity of the human brain. The emergence of human language has been a major milestone in the relentless evolution of complexity on our planet.

We 'know' what our thoughts and memories mean. But if we want to share them with others, they have to be translated into language. Our neocortex accomplishes this using what Kurzweil (2012) calls '*pattern recognizers*', which have been trained with patterns that we have learnt for the purpose of using language.

According to Kurzweil (2012), language is highly hierarchical and it evolved to take advantage of the hierarchical nature of the neocortex, which in turn reflects the hierarchical nature of reality. Noam Chomsky wrote about the innate ability of humans to learn the hierarchical structures in language. Hauser, Chomsky and Fitch (2002) cited the attribute of *'recursion'* as accounting for the unique language faculty of the human species. Recursion, according to Chomsky, is the ability to put together small parts into a larger chunk, and then use that chunk as a part in yet another structure, and so on, iteratively and hierarchically. That is how we are able to build the elaborate structures of sentences and paragraphs and sections and chapters from a limited set of words.

According to Richard Dawkins (1989), 'most of what is unusual about man can be summed up in one word: "culture".' Of course, one must make a distinction between 'culture' and 'society'. 'A *society* refers to an actual group of people and how they order their social relations. A *culture* . . . refers to a body of socially transmitted information' (Barkhow 1989). The term 'culture' encompasses 'all ideas, concepts and skills that are available to us in society. It includes science and mathematics, carpentry and engineering designs, literature and viticulture, systems of musical notation, advertisements and philosophical theories – in short, the collective product of human activities and thought' (Distin 2005).

It is notable that, on the biological evolutionary time scale, there has been an exceptionally rapid expansion of brain capacity in the course of evolution of *one* of the ape forms (chimpanzees) to *Homo sapiens*, i.e., ourselves. This has happened in spite of the fact that the genome of humans is incredibly close to that of chimpanzees. The evolution of language, speech, and culture are believed to be some of the causative factors for this rapid evolution of the human brain. Let us see how.

Homo sapiens was preceded by *Homo heidelbergensis*, which also had a fairly large brain, but was not very effective as a hunter. He was not able to establish *ecological dominance* over other animals, even after two million years of evolution. Our human advantage is believed to have arisen from the emergence of language. 'No topic is more intriguing and more difficult to address concretely than the evolution of language, but ... [it] is almost a kind of sixth sense, since it allows people to supplement their five primary senses with information drawn from the primary senses of others. Seen in this light, language becomes a kind of "knowledge sense" that promotes the construction of extraordinarily complex mental models, and language alone may have provided sufficient benefit to override the cost of brain expansion' (Klein and Edgar 2002). [The reference to 'the cost of brain expansion' here is to the fact that in humans the brain takes up ~20% of the metabolic resources of the body, and the brain tissue requires 22 times more energy than a comparable piece of muscle at rest.]

Deacon (1997) emphasizes the big difference between human language (talking) on one hand and the various modes of communication among other live entities: 'Although other animals communicate with one another, at least within the same species, this communication resembles language only in a very superficial way – for example, using sounds – but none that I know of has the equivalents of such things as words, much less nouns, verbs, and sentences. Not even simple ones.'

Deacon continues: 'Though we share the same earth with millions of living creatures, we also live in a world that no other species has access to. We inhabit a world full of abstractions, impossibilities, and paradoxes ... We tell stories about our real experiences and invent stories about imagined ones, and we even make use of these stories to organize our lives. In a real sense, we live our lives in this shared virtual world. ... The doorway into this virtual world was opened to us alone by the evolution of language, because language is not merely a mode of communication, it is also the outward expression of an unusual mode of thought - *symbolic representation*. Without symbolization the entire virtual world is . . . out of reach:

inconceivable . . . symbolic thought does not come innately built in, but develops by internalising the symbolic process that underlies language'.

Homo heidelbergensis had a big brain. But was he also a great symbolic thinker? Probably not. Deacon argues that probably a single symbolic innovation triggered a coevolution of language and brain-size. Greater brain power resulted in a greater capacity to symbolise, speak, think. The cascading effect led to more complex languages and more complex brains. But all this required *social* interaction and support: 'Language is a social phenomenon. ... [and] ... the relationship between language and people is symbiotic'.

Deacon traces the evolution of social complexity by assuming that the early humans were dual-parenting. Since their sense of smell was not very acute (thus ruling out a role for chemical signalling through pheromones), some other type of sexual signalling evolved between the male and the female. This is how *social communication* originated and evolved as a kind of social hormone.

Other than sex, availability of food is the major factor determining the survival of a species. Male humans had to cooperate with one another for hunting. Deacon again: 'Males must hunt cooperatively; females cannot hunt because of their ongoing reproductive burdens; and yet hunted meat must get to those females least able to gain access to it directly (those with young), if it is to be a critical subsistence food. It must come from males ... [who] ... must maintain constant pair-bonding relationships'. This need for hunting in groups resulted in the evolution of a social structure implying a symbolisational solution to the problem of survival. This is because symbolic reference, as also speaking and thinking, are basically of a social nature. There was naturally a concomitant evolution of the speech organ (voice box).

Grooming

According to Robin Dunbar (1998): 'One of the most important ways that primate allies show their affection to each other is by grooming. Grooming not only gets rid of lice and other skin parasites, but it is also soothing. Primates turn grooming into a social currency that they can use to buy the favour of other primates. But grooming takes a lot of time, and the larger the group size, the more time primates spend grooming one another. Gelada baboons, for example, live on the savannahs of Ethiopia in groups that average 110, and they have to spend twenty per cent of their day grooming one another. ... If we had to bond our groups of 150 the way primates do, by grooming alone, we would have to spend about 40 to 45 per cent of our total daytime in grooming'.

The primates in the savannahs also had to find food, and therefore such a large investment of time in grooming would have caused a non-sustainable work *vs.* life balance. *Language emerged as a better way of bonding.*

Emergence of a word-speaking species

Humans began with sound language, gradually increasing the vocabulary. But there is a severe limit to how many sound calls you can have which still sound distinct. The next step in the evolution of language was a stringing together of sounds into specific sequences, namely *words*. A word-speaking species naturally had an evolutionary advantage.

Sentences syntaxing words were the next level of evolving complexity. Brain size increased concomitantly to understand and remember words, syntax, grammar, and sentences (Zimmer 2006): 'A syntax-free language beats out syntax when there are only a few events that have to be described. But above a certain threshold of complexity, syntax became more successful.

When a lot of things are happening, and a lot of people or animals are involved, speaking in sentences wins … Something about the life of our ancestors became complex and created a demand for a complex way in which they could express themselves … A strong candidate for that complexity, as Dunbar and others have shown, was the evolving social life of hominids'.

This social evolution of complexity is the advantage we humans have over other animals. We have the capacity to introduce and expand complexity in social life, and development of language is both a cause and an effect of this capacity. As Kate Douglas (2005) said, 'In a sense, language is the last word in biological evolution. That's because this particular evolutionary innovation allows those who possess it to move beyond the realms of the purely biological. With language, our ancestors were able to create their own environment – we now call it culture – and adapt to it without the need for genetic changes'.

Whereas humans and chimpanzees have many genes in common, the *expression* of certain genes is more common in the human brain. Interestingly, the brains of newborn humans are far less developed than those of newborn chimpanzees, and the neural networks of human babies are developed over *many years of exposure to a linguistic environment*. Through a continuous process of unsupervised learning, supervised learning, and reinforced learning, the child's brain performs evolutionary computation (see Section 34.3).

As we shall see in the next chapter, with language came the possibility of emergence of 'memes'. Language coevolved with memes.

39. Memces and Their Evolution

We are different from all other animals because we alone, at some time in our far past, became capable of widespread generalized imitation. This let loose new replicators – memes – which then began to propagate, using us as their copying machinery much as genes use the copying machinery inside cells. From then on, this one species has been designed by two replicators, not one. This is why we are different from the millions of other species on the planet. This is how we got our big brains, our language and all our other peculiar 'surplus' abilities.

Susan Blackmore (2000)

'*Internet memes*' is a familiar term these days. An Internet meme is an idea, quotation, style or action which spreads from person to person via the Internet. The good memes get viral, and ordinary or bad ones do not propagate because they are ignored by the netizens.

What evolves in any type of evolution is information content. The two most basic aspects of evolution are *replication of information* (also ensuring preservation of the replicated information), and *the mode of transmittal of information*. Genes preserve biological information, and they use DNA for this. What about culture?

Similar to the gene, which is the unit of biological inheritance, Richard Dawkins (1976/2006) introduced the notion of the *meme*, which is the *unit of cultural inheritance*. A meme may be some good idea, a witty saying, a soul-stirring tune, a logical piece of reasoning, or a great philosophical concept. Dawkins visualized that two different evolutionary processes must have operated in tandem: the classical Darwinian evolution, and another one centred around intelligence, language, and culture. Memes are, roughly speaking, the cultural analogues of genes.

The genes that exist in many copies in a population are those that are good at surviving and replicating. Through a reinforcement effect, genes whose carriers in a population are good at cooperating with one another stand a better chance of surviving. Similarly, the fittest set of cooperating memes has a better chance of surviving to form the *meme pool* of the population. They replicate themselves by imitation or copying (Blackmore 1999), and also by a variety of other mechanisms (Distin 2005). Cultural evolution and progress occurs through a selective propagation of the fittest set of cooperating memes.

Memes evolve, just as genes evolve. [In fact, any entities that can replicate, and that have a variation both in their specific features and in their reproductive success, are candidates for Darwinian selection or natural selection.] And the coevolution of gene pools and meme pools (through language etc.) resulted in a rapid enlargement of the brain size of *Homo sapiens*. A large brain size, once attained, resulted in several other capabilities as well.

An important *difference* between memes and genes is that the speed of cultural evolution (development of ideas, customs, etc.) is far higher than the speed of genetic evolution. Nevertheless, there are several proposed analogies between the two. How far can we carry the gene analogy for understanding the nature of memes? This continues to be a subject of debate. Distin (2005) has listed several characteristics of memes. One of them is *the essential particulate nature of memes*. The most efficient methods of replicating complexity are *hierarchical* (or modular or particulate). If variation were to be permitted in every element of a complex structure, then copying processes would lose much of their stability. In genetics, Mendel's work established the particulateness of genes, namely the clear presence or absence of the effects of these replicators on the world. Something similar is necessary for memes in their role in the cultural evolution of complexity. This means that memes must be able to fit into established *cultural assemblies* without their own informational content being lost or blended in the process. That is, memes must have a certain degree of particulateness, so that the results that they produce are generally of a fixed nature. Their identity should be such that they are discernible *packets of information* (like the genotype). But, whereas the genotype is distinct and clearly definable, the phenotype (which is a manifestation of the genotype) in biological systems possesses a certain degree of flexibility and variability. Likewise, the manifestations of memes have a certain degree of flexibility that enables their effects to be produced in a variety of cultural contexts. Copied in these ways, information is given the stability to grow and develop in complexity. The breadth and depth of human culture is thus explained by the cumulative replication of *particulate information.*

In both genetics and memetics, the replicators carry information about the effects that they control. In the case of genes, their independence is maintained via the medium of DNA, which preserves biological information in a form that is replicable and can produce its effects in a variety of contexts. In the case of memes, this role is performed by what is called the '*representational content,*' which is the memetic or cultural equivalent of DNA: As *representations* of a portion of information, memes have a certain *content*. A representation in the human mind is some piece of our 'mental furniture' that carries information about the world. For example, a thought that 'the object on my desk is a book' is a mental representation of a bit of the world (namely that book). Therefore 'representational content' refers to the information that is included in the content of our representations.

It is representational content which accounts for the mechanisms of memetic heredity and for the influence of memes over their phenotypic effects. Distin (2005) uses the term *memetic DNA* for the representational content. It provides the mechanism for memetic evolution, just as DNA provides the mechanism for genetic evolution.

How is the representational content fixed in our brains? Replicators preserve and copy specific portions of information. For memes, we should be able to identify precisely which bits are carried in each replicator. This means pinpointing the exact content of any representation, and this is something determined partly by the various properties of the object or situation being represented. Yet representational content is determined by other factors as well; e.g. by the capabilities and history of the organism doing the representing.

Some organisms are capable of forming representations the content of which is determined by a combination of the relevant properties of that which is represented, and the organism's own individual and social learning capacities. Such organisms are able, in other words, both to preserve information and to transmit it among themselves.

In the case of complex representations, which have links not only externally to perceptions and behaviour but also internally to other representations, the resultant behavioural flexibility can enable us to track down their content more completely. It should be possible to test all the links, by altering the associations that the organism encounters, and observing the effects on its behaviour. *Only representations with this determinacy of content can count as memes, since a crucial aspect of any replicator is the preservation of given information.*

Thus memes are representations which preserve their content in a way that can be copied between generations. As representations, they are specifically those bits of our mental furniture which control our behaviour in response to the information they carry. In other words, the basis of memes in representational content is precisely what accounts for their ability to exert executive effects on the world.

Representations gain meaning from their context within a *representational system* (RS), and the uniquely human capacity that lies at the heart of culture is our ability to copy and develop RSs, as well as adding individual representations to our repertoire: the ability, in other words, to *meta-represent*. Natural languages, as also systems of mathematical and musical notation, are some examples of cultural RSs, and each is peculiarly appropriate to its particular cultural area. Human minds acquire replicators on an ongoing basis throughout their lives, and this means that they can acquire novel RSs as well as novel representations. Among these various RSs, the natural languages have primacy: they alone benefit from an innate device for their acquisition. Yet they benefit, too, from the innate ability to meta-represent – and it is this that allows us to develop nonlinguistic RSs also, the diverse rules and structures of which are realized in media other than speech. Once these sorts of RSs have been taken into account, it becomes more clear that there are many concepts that are not available to us until the RS that supports them has been developed.

According to Distin, humans are born with a degree of 'mindedness' that includes, for example, the '*representation instinct*': an ability and tendency to learn and manipulate vast numbers of representations, as well as the various systems in which they are embedded. And this innate mental potential of an infant is realized as a result of exposure to the cultural environment.

Genes preserve and replicate biological information by building *vehicles* for their own propagation and protection. The effects of the genes are found in the machines that they build for their survival, and their replication also depends ultimately on this same machinery. Memes depend for their replication on a faculty of the human mind that is ultimately of a *genetic* nature, namely the representation instinct. Organisms, as well as minds, develop via interaction between the innate potential and the environment, and in the case of the mind a crucial part of that environment comprises of the memes.

A human mind is thus partly a product of the memes, but only because it has the innate potential to interact with and develop in response to these memes. And culture is the product of human minds, although the preservation of information in representational content ensures that the culture we see today is mostly the result of memes produced by human minds of long ago. The development of human minds depends on a combination of two types of processes: their innate potential is the result of an interaction between genes and the physical environment, and that potential is fulfilled as a result of interaction with memes.

The selfish meme?

'Memes are best thought about not by analogy with genes but as new replicators, with their own ways of surviving and getting copied' (Blackmore 1999).

Dawkins (1976) described the essence of his 'selfish gene' theory as the insight 'that there are two ways of looking at natural selection, the gene's angle and that of the individual'. Similarly, the essence of his selfish *meme* hypothesis is the insight that there are two ways of looking at cultural change, the meme's angle and that of the human individual.

One of the most significant implications of Dawkins' selfish-*gene* theory is that the individual organism was not an inevitable outcome of genetic evolution: it so happens that genes have banded together to build *survival machines*, but the only crucial feature of any form of evolution is the replicator – the unit of selection. Although organisms clearly exist, and have a perspective from which the world of genes is irrelevant to their everyday lives, fundamentally their lives and evolution are determined by that world. According to Distin (2005), no analogous insight arises from the theory of the selfish *meme*, because memes do not build survival machines. Their replicative mechanisms, and the means of their variation and selection, lie in genetically determined human faculties, and not in vehicles that they themselves build.

However, Blackmore (1999) takes the view that we are *meme machines*, just as we are *gene machines*. Consequently, she argues that 'there is no conscious self inside' those machines; and that 'a complex interplay of replicators and environment' is all there is to life. Our sense of self may not be illusory, but our sense of control over the collective products of our minds may well be. Although our minds provide the mechanisms of memetic evolution, there is a very real sense in which the directions of that evolution are independent of us.

If I have to choose between Distin's and Blackmore's theses, I would go for Blackmore's. The grip of the basic tenets of some of the major organized religions of today over the collective outputs of the minds of their followers is the most glaring example of the point made by Blackmore.

Here is an interesting aside in the context of memes. The idea of free will has generated a lot of debate. Free will is probably an illusion. Kurzweil (2012) is of the interesting view that the free-will idea is a useful *meme* (irrespective of whether it has a rational basis or not). For example, it enables us to fix responsibility for an act of crime.

For more on memes, here is an interesting online article:
http://www.lycaeum.org/~sputnik/Memetics/day.life.html.

40. Evolution of the Human Brain, and the Nature of Our Neocortex

. . . our intelligence is not disembodied, but is instantiated in physical objects: our brains. Their structure is due to the long process of evolution, and their operations are governed by the laws of physics. Since they are physical entities, our brains run without being told how to run.

Douglas Hofstadter, *Gödel, Escher, Bach*

The emergence of humans on the scene has sharply accelerated the rise of the overall complexity of our Earth. This has happened, and is still happening at an ever-increasing pace, because of the evolution of cultural complexity. A major reason for this is the very high level of intelligence possessed by humans. I shall discuss the nature of intelligence in this and the next few chapters.

The human brain is a physical organ, governed by the laws of physics. The mind is what the brain and the rest of the body does.

There is a network-theory approach to the question of intelligence. It asserts that our intelligence is no different from *swarm intelligence* (as in a beehive or an ant colony), the swarm here being that of the large number of strongly interacting neurons.

There is a belief that the transition from intelligence to what many people call 'consciousness' needs the acquisition of a human language. Our language has a *hierarchical* structure, from the alphabet to words to sentences to paragraphs to chapters, etc. This is rather like Minsky's (1986) 'society of mind' (to be discussed in the next chapter), comprising of 'communities' of large numbers of interacting neurons, emerging as a hierarchical structure, so typical of any complex adaptive system.

[As an aside I may mention here that even in scientific circles one often comes across the statement that 'there is a deep connection between the mind and the body, something we are just beginning to fathom'. I want to point out that what we really have is a *mindbody* (a *single* entity), and not a mind *and* a body. It is mindbody, and not even mind-body. An analogy with the introduction of the term *spacetime* by Einstein (Chapter 5) would help understand the difference in emphasis that can ensue when carefully chosen words are used for describing a reality. The recognition that spacetime is a single entity made a huge difference in our understanding of the true nature of the cosmos. Of course, the terms 'space' and 'time' are still used when the space aspect or the time aspect of spacetime is more relevant in a given context. Similar is the case for the use of the terms mindbody, mind, body, and mind-body, depending on what is more relevant or appropriate in a given situation. But one should not lose sight of the fact that the real entity is *always* mindbody. Treating mindbody as a single entity can make a *qualitative* difference in the way we theorize about this most important of questions about ourselves. Fig. 40.1 is an Internet meme I created on Facebook for emphasizing the fact that 'mindbody' does *not* have the same connotation as mind-body.]

MINDBODY
Einstein explained how space and time are not two separate entities,
and introduced the term *spacetime* as an improvement over 'spacetime'. Similarly the term *'mindbody'* must be introduced as an
improvement over 'mind-body' for acknowledging and emphasizing
the inseparable nature of the mind-body entity.

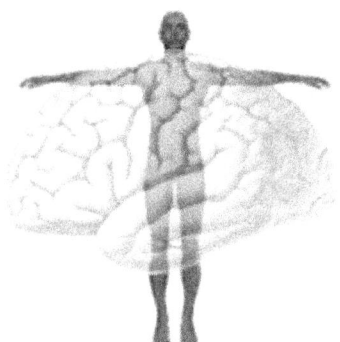

Image credit: http://www.simplypsychology.org/mindbodydebate.html

Fig. 40.1 Mindbody, not mind-body.

40.1 Evolution of the brain

The fundamental law 'Nature abhors gradients' has resulted in an accumulation of enormous amounts of structure and order in our ecosphere, as also in the rest of the universe. Any living entity exploits this structure and order of its surroundings to ensure its survival and reproduction. Consider a single-celled organism (a bacterium) in a pond. On its surface are molecules that can 'detect' (are influenced by) the presence of nutrients. There is usually a gradient of the nutrient concentration, so that it is higher on one side of the organism than on the other. The single-celled organism has chemical sensors which can detect this gradient (Robson 2011). Biological evolution has programmed it to propel itself in the direction of increasing concentration of nutrient. One of the attributes of intelligence is the problem-solving capacity of the system; other important attributes are prediction and memory capabilities. As Hawkins (2004) pointed out, both prediction and memory are involved here. The prediction is that, by moving in the direction of increasing concentration of nutrient, more nutrient will be found. This is not something the organism has 'learnt' and 'remembered' in its lifetime. The 'memory', evolved over many generations of evolution, is embedded in its DNA.

Plants also exploit the existing order and structure (constancy or sameness over reasonably long time scales) by employing memory and prediction. The memory frozen in the genes of a tree tells it that it will find greater sunshine by sending its branches and leaves towards the sky. And that it will find water and minerals by sending its roots down into the soil. These actions are automatic, and there is no 'thinking' involved, just as there is no thinking involved in the actions of a bacterium mentioned above.

At a certain stage in the evolutionary history of plants, more complex behaviour emerged in the form of *communication systems* among the various parts of a plant, based mainly on chemical signals. Suppose an insect damaged some part of a tree, and this led to the slow transmittal of a chemical through the vascular system to other parts of the tree. This triggered a defence mechanism; e.g., the making of a toxin for the insect.

Neurons evolved in due course, as a faster way of communicating information to different parts of an organism. The electrochemical spikes in a neuron travel much faster than the diffusion speed of chemicals. In due course, the synaptic connections between neurons became modifiable. A neuron may or may not send a signal, depending on what happened in the past. This rudimentary nervous system had elements of both memory and learning.

The evolutionary advantage of this to the creature was huge, and *qualitatively* different. Instead of depending on just 'genetic memory' and instinct coded in DNA, the creature could now learn from experience during its own lifetime, and modify its behaviour for achieving better survival and propagation possibilities. In particular, if the environmental structure and order changed rather suddenly, the animal could still make a generally adequate response, instead of having to depend only on the somewhat outdated (and therefore inadequate) genetic memory and instinct. Such plastic nervous systems entailed a huge evolutionary advantage, and there was a burst of new species from fish to snails to mammals, including humans.

Why is it that intelligence evolved mainly in the animal kingdom, but not so much in the plant kingdom? According to Hans Moravec the difference has arisen because animals are mobile and plants are generally not. The mobility of animals presents to them an ever-changing environment. Therefore intelligence is an important prerequisite for survival and propagation: An animal can survive only if it has a large repertoire of solutions to the continuous stream of problems it faces in a changing environment.

40.2 The human neocortex

The human brain, like the brain of any other mammal, has something distinctly additional compared to the brain of reptiles from which it evolved, namely the *neocortex*. Our brain has two main parts: the 'old brain' or the reptilian brain or the R-brain or the 'primitive' brain; and the neocortex. These two parts, along with the spinal cord, comprise *the central nervous system*. The top outer portion of the brain, just under the scalp, is the neocortex (or *cortex* for short). It covers most of the R-brain, and has a crumpled appearance, with many ridges and valleys. The R-brain is rather similar in reptiles and mammals, and has a number of parts, including the thalamus and the hippocampus (Fig. 40.2). But practically everything we associate with conscious memory and intelligence occurs in the neocortex, although the thalamus and the hippocampus also play important roles.

In the evolutionary history of life on Earth, sophisticated sensory and actuation organs had evolved in reptiles, and their behaviour was controlled by the old brain, with no cortex. The evolution of the cortex in one of the offshoots of the reptiles, along with the availability of a stream of sensory inputs into it which it could remember and analyse much better than the reptiles could, gave the mammals an evolutionary advantage: When they found themselves in situations they remembered to have faced earlier, their much-improved memory and analysis power told them what to expect next, and how to respond effectively.

Humans are special compared to other mammals because of their very large *prefrontal cortex* (or frontal lobe). The prefrontal cortex (particularly the upper two-thirds of it, including the dorsolateral prefrontal cortex) can be regarded as the rational centre of the brain; or the *rational brain*. The rest of it is the *emotional brain*.

I have described many details about the ~10^{11} neurons in the human neocortex, including the massive networking among them, in Section 33.3. The human cortex, if stretched flat, is the size of a large napkin, and ~2 mm thick. It has six layers, each roughly the thickness of a playing card. There is a branching hierarchy among the layers. Layer 6 is at the bottom of the hierarchy, and Layer 1 is at the top. The inputs from the various sensory organs are received in

Layer 6, and then interpreted and correlated. Then more and more abstract and generalized versions of the information are sent up the hierarchical layers. There is a very high degree of feedback and feedforward among the layers, as also cross-correlations.

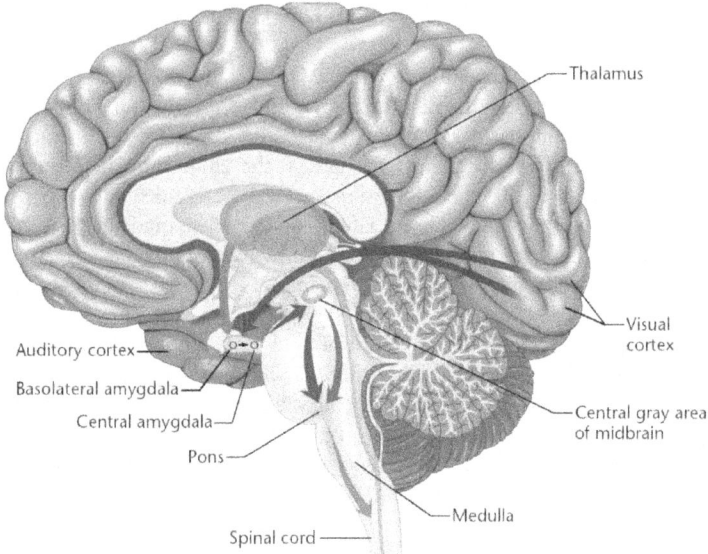

Fig. 40.2 The human brain. Image credit:
http://northofneutral.wordpress.com/2013/05/

Portions of the cortex can be identified as different *functional areas* or *regions*. For example, a portion of the frontal lobe is the *motor cortex*. It controls movement and other actuator functions of the body. The *visual cortex* is another example.

The cortical tissue can be functionally divided into vertical units or *columns*. Neurons within a column respond in a similar manner to external signals with a particular attribute.

The functionality of the cortex is arranged in a branching hierarchy. The primary sensory regions constitute the lowest rung of the hierarchy (Layer 6). The sensory region for, say, vision (called V1) is different from that for hearing etc. V1 feeds information to higher layers called V2, V4 and IT, and to some other regions (Fig. 40.3). The higher they are in the hierarchy, the more abstract they become. V2, V4 etc. are concerned with more specialized or abstract aspects of vision. The higher echelons of the functional region responsible for vision have the visual memories of all sorts of objects. Similarly for other sensory perceptions.

In the higher echelons are areas called *association areas*. They receive inputs from several functional regions. For example, signals from both vision and audition reach one such association area.

Although the primary sensory mechanism for, for example, vision is not the same as for hearing, what reaches the brain at higher levels of the hierarchy is qualitatively the same. The axons carry neural signals or spikes which are partly chemical and partly electrical, but their nature is independent of whether the primary input signal was visual or auditory or tactile. Finally *they are just patterns*.

Creation of *short-term memory* in the brain amounts to a stimulation of the relevant synapses, which is enough to temporarily strengthen or sensitize them to subsequent signals.

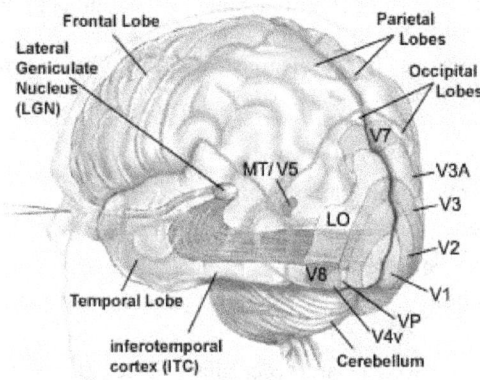

Fig. 40.3 The sensory regions for vision, V1, V2, etc. Image credit:
http://thebrain.mcgill.ca/flash/a/a_02/a_02_cr/a_02_cr_vis/a_02_cr_vis.html#3

This strengthening of the synapses becomes permanent in the case of *long-term memory*. This involves the activation of genes in the nuclei of postsynaptic neurons, initiating the production of proteins in them. Thus *learning* requires the synthesis of proteins in the brain within minutes of the training. Otherwise the memory fades away.

Information meant to become the higher-level or generalized memory, called *declarative memory*, passes through the hippocampus, before reaching the cortex. The hippocampus is like the principal server on a computer network. It plays a crucial role in consolidating long-term memories and emotions by integrating information coming from sensory inputs with information already stored in the brain.

40.3 The history of intelligence

As the most important phenomenon in the universe, intelligence is capable of transcending natural limitations, and of transforming the world in its own image. In human hands, our intelligence has enabled us to overcome the restrictions of our biological heritage and to change ourselves in the process. We are the only species that does this.

Kurzweil (2012)

Exactly three years ago, on January 13, 2011, humans were dethroned by a computer on the quiz show JEOPARDY! A year later, a computer was licensed to drive cars in Nevada, after being judged safer than a human.

What's next? Will computers eventually beat us at all tasks, developing superhuman intelligence?

I have little doubt that this can happen: our brains are a bunch of particles obeying the laws of physics, and there's no physical law precluding particles from being arranged in ways that can perform even more advanced computations.

Max Tegmark (13 January 2014)
http://www.kurzweilai.net/humanity-in-jeopardy

Ray Kurzweil is an inventor and futurologist (among other things) who has not only been making predictions about the future of artificial intelligence and other things (see https://en.wikipedia.org/wiki/Predictions_made_by_Ray_Kurzweil), but has also been instrumental in making many of his predictions come true. This 'ultimate thinking machine' has been the principal inventor of:

- the first CCD flat-bed scanner;
- the first omni-font optical character recognizer;
- the first print-to-speech reading machine for the blind;
- the first text-to-speech synthesizer;
- the first music synthesizer capable of re-creating the grand piano and other orchestral instruments; and
- the first commercially marketed large-vocabulary speech recognizer.

[For an update on Kurzweil's 'most exciting predictions', see https://futurism.com/ray-kurzweils-most-exciting-predictions-about-the-future-of-humanity/. For more on his predictions, see http://www.sciencealert.com/rey-kurzweil-s-most-exciting-predictions-about-the-future-of-humanity.]

In his book *How to Create a Mind* (2012) he describes our current understanding of how the brain functions and how the mind emerges from the functioning of the brain. I believe that we can now say that we are finally on the right path when it comes to getting the hang of what intelligence is all about. The latest model of the functioning of the brain, presented by Kurzweil, is so successful that I have little doubt that we can reverse-engineer the human brain and create a truly worthy artificial analogue of it.

I shall describe Kurzweil's model after doing some groundwork in this and the next few chapters, but here is the gist of what the model says (Kurzweil 2012):

*'Modern neuroscience has proved that the neocortex operates according to a relatively straightforward **pattern-recognition scheme**. This scheme is hierarchical in nature, such that lower-level patterns representing discrete bits of input combine to trigger higher-level patterns that represent more general categories. The hierarchical structure is innate, but the specific categories and meta-categories are filled in by way of learning. Also, the direction of information travel is not only from the bottom up, but also from the top down, such that the activation of higher-order patterns can trigger lower-order ones, and there is feedback between the various levels'.*

This model or theory has been called the *Pattern Recognition Theory of the Mind* (PRTM).

Consider the following statements:

1. A long tiresome speech delivered by a frothy pie topping.
2. A garment worn by a child, perhaps abroad an operatic ship.
3. Wanted for a twelve-year crime spree of eating King Hrothgar's warriors; officer Beowulf has been assigned the case.
4. It can mean to develop gradually in the mind or to carry during pregnancy.
5. National Teacher Day and Kentucky Derby Day.
6. Wordsworth said they soar but never roam.
7. Four-letter word for the iron fitting on the hoof of a horse or a card-dealing box in a casino.

8. In act three of an 1846 Verdi opera, this Scourge of God is stabbed to death by his   lover, Odabella.

Each of these is the answer to a specific question. For example, (8) is the answer to the question 'What is Attila the Hun?'. These and other such questions were asked by the anchor in the popular American TV show *Jeopardy*! (https://en.wikipedia.org/wiki/Jeopardy%21). In this game the contestant who answers a question correctly, before any other contestant can, wins that round. The questions are spoken out in English by the anchor and heard by each contestant.

Now the pleasant shocker. In 2011 the IBM computer named *Watson* competed against the two best human players in the world (Ken Jennings and Brad Rutter), and won! (see https://en.wikipedia.org/wiki/Watson_%28computer%29). The questions for the above eight answers are:

1. meringue harangue;
2. pinafore;
3. Grendel;
4. gestate;
5. May;
6. skylark;
7. shoe;
8. what is Attila the Hun?

For the 8th query Watson replied 'What is Attila?'. The host responded by saying, 'Be more specific'. Watson said (correctly): 'What is Attila the Hun?'.

As Kurzweil (2012) writes, 'It should be noted that not only did Watson read and "understand" the subtle language in the *Jeopardy*! query (which includes such phenomena as puns and metaphors), but it obtained the knowledge it needed to come up with a response from understanding hundreds of millions of pages of natural-language documents including Wikipedia and other encyclopaedias on its own. It needed to master virtually every area of human intellectual endeavour, including history, science, literature, the arts, culture, and more'. The technique used by such machines is the so-called '*hierarchical hidden Markov model*' (http://www.cs.princeton.edu/courses/archive/spr06/cos598C/papers/FineSingerTishby1998.pdf). Kurzweil was among the people who developed this technology in the 1980s and 1990s.

How has such performance become possible? It is because now we have the PRTM, which seems to be describing correctly the basic algorithm of our neocortex. What makes me very optimistic about the future of research in artificial intelligence (AI) is the fact that the work on the PRTM and the work on the further development of projects like Watson are going on hand in hand. *Insights obtained in one get tested by direct application on the other.* Things look very very impressive indeed.

How come we humans are so intelligent as to be able to understand the secrets of our own intelligence, and are even able to put that knowledge into machines that may soon overtake (and also enhance) our own intelligence? A look at the history of intelligence is in order here.

The first episode in the story of intelligence in our universe occurred when our universe got created, with its given set of fundamental constants of Nature. The rather tautological anthropic principle is the reason why the fundamental constants are what they are, namely just right for the emergence and existence of atoms and molecules and life forms, including ourselves (with a neocortex capable of hosting intelligence). The nature of our universe is such that it is capable of encoding information. Moreover, since it has been expanding ever since the Big Bang, there is a perennial creation of gradients which, in turn, lead to the evolution of complexity because of the second law of thermodynamics for open systems. So the lesson from the first part of the story of intelligence is that our universe is based on information, and the information evolves with time (see Chapter 12).

I have already recounted a whole succession of episodes in the story of intelligence:

- How atoms emerged was described in Chapter 22. The step from atoms to molecules was discussed in Chapter 23. Of particular importance in this context has been the availability of the carbon atom, which has a valence of 4, and forms an immense variety of information-rich molecules.

- As discussed in Chapter 25, chemical adaptation and chemical evolution occurred over the millennia, resulting in the emergence of the DNA molecule in due course (Chapter 27). This molecule encoded information which could be passed on to succeeding generations in a precise manner. What is more, the coded information acted like an algorithm for the growth and development of the progeny (requiring translation from genotype to phenotype). As Kurzweil (2012) narrates, in due course the organisms evolved communication and decision networks called nervous systems, which facilitated survival in the increasingly complex environment. The neurons aggregated into brains capable of intelligent behaviour.

- The next development was uniquely human. Our brains are capable of *hierarchical thinking*. I quote Kurzweil (2012): 'We are capable of hierarchical thinking, of understanding a structure composed of diverse elements arranged in a pattern, representing that arrangement with a symbol, and then using that symbol as an element in a yet more elaborate configuration. . . . we are able to call these patterns *ideas*. Through an unending recursive process we are capable of building ideas that are ever more complex. We call this vast array of recursively linked ideas *knowledge*'.

There is another thing unique to humans: An opposable thumb (see S. Kemmer: http://www.ruf.rice.edu/~kemmer/Evol/opposablethumb.html) which enables us to build sophisticated tools. We have used our intelligence and our ability to build tools to develop *technology*. This has enabled our knowledge base to grow without limits. One of our creations, namely computers, now hold the promise of creating artificial intelligence which will excel our own. I gave glimpses of this prospect in Chapter 37.

It appears that ours is the only species in our universe which possesses intelligence of an order high enough to make us aware of the grandeur of the universe. We are not just 'star stuff' (a la Carl Sagan); we are unique. Therefore we should lead lives worthy of our stature in the cosmos.

41. Minsky's and Hawkins' Models for how Our Brain Functions

Three models of how our brain functions are discussed in this book, and this chapter takes up the first two, namely Minsky's model and Hawkins' model. Kurzweil's more recent model will be taken up in a later chapter.

41.1 Marvin Minsky's 'Society of Mind'

'Our minds did not evolve to serve as instruments for observing themselves, but for solving such practical problems as nutrition, defence, and reproduction' (Marvin Minsky 2006).

Efforts at developing machine intelligence have resulted in deep insights into how the human brain functions. Marvin Minsky was a pioneer in this field. In 1986 he published the great book *The Society of Mind*, in which he formulated his ideas about human cognition. His next book, *The Emotion Machine*, published in 2006, reflected the progress made at that time in understanding the workings of the human mind via the machine-intelligence approach.

Minsky's 'society' of mind comprises of *agents* or *resources*, which are the simplest individuals that populate the brain. Each agent or resource can be visualized as a typical component of a computer program, like a simple subroutine or data structure. The agents can get connected and composed into larger systems called *agencies* or *societies of agents*. The agencies self-organize into still larger conglomerates that can perform still more complex functions, and so on into higher and higher levels of self-organization and complexity, ultimately leading to the *emergence* of abilities we attribute to the mind. There is thus a hierarchical structure and organization, like in any CAS.

The idea of hierarchical levels of organization was well documented in an earlier publication of Minsky (1981): '*One could say but little about "mental states" if one imagined the mind to be a single, unitary thing. But if we envision a mind (or brain) as composed of many partially autonomous "agents"—a "society" of smaller minds—then we can interpret "mental state" and "partial mental state" in terms of subsets of the states of the parts of the mind. To develop this idea, we will imagine first that this mental society works much like any human administrative organization. On the largest scale are gross "divisions" that specialize in such areas as sensory processing, language, long-range planning, and so forth. Within each division are multitudes of subspecialists—call them "agents"— that embody smaller elements of an individual's knowledge, skills, and methods. No single one of these little agents knows very much by itself, but each recognizes certain configurations of a few associates and responds by altering its state*'.

As is the case with any CAS, we cannot predict with certainty the properties of the mind-system in terms of the laws of physics applied to the constituent agents, nor can we start from the observed complexity of the brain and work our way downwards all the way to understand why the increasing complexity took a particular route in phase space [please note that 'deterministic' and 'unpredictable' are not mutually exclusive propositions in physics]. To quote Minsky (1990): '*the functions performed by the brain are the products of the work of thousands of different, specialized sub-systems, the intricate product of hundreds of millions of years of biological evolution. We cannot hope to understand such an organization by emulating the techniques of those particle physicists who search for the simplest possible unifying conceptions. Constructing a mind is simply a different kind of problem — of how to synthesize organizational systems that can support a large enough diversity of different schemes, yet enable them to work together to exploit one another's abilities'.*

Here is Minsky's (1986) take on consciousness: '*In this book, the word (consciousness) is used mainly for the myth that human minds are "self-aware" in the sense of perceiving what happens inside themselves. I maintain that human consciousness can never represent what is occurring at the present moment, but only a little of the recent past - partly because each agency has a limited capacity to represent what happened recently and partly because it takes time for agencies to communicate with one another. Consciousness is peculiarly hard to describe because each attempt to examine temporary memories distorts the very records it is trying to inspect'.*

Minsky described 'free will' as a myth, the myth that human volition is based upon some third alternative to either causality or chance.

The 'Single-Self' concept

Some people still subscribe to the concept that there is creature (or a set of creatures) inside us that does all the feeling or thinking for us, and makes all the important decisions for us. It is our 'identity' or 'self'. Even our legal system distinguishes between deliberate wilful murder, and murder that was not pre-planned. This Single-Self concept may be useful as a meme, but apparently has no scientific basis at a fundamental level.

Why do humans entertain such fiction? It may be partly because it makes life look pleasant, 'by hiding from us how much we're controlled by all sorts of conflicting, unconscious goals'. According to Minsky: '*That image makes us efficient, whereas better ideas might slow us down. It would take too long for our hardworking minds to understand everything all the time. However, although the Single-Self concept has practical uses, it does not help us to understand ourselves — because it does not provide us with smaller parts we could use to build theories of what we are. When you think of yourself as a single thing, this gives you no clues about issues like these: What determines the subjects I think about? How do I choose what next to do? How can I solve this difficult problem? Instead, the Single-Self concept offers only useless answers like these: My Self selects what to think about. My Self decides what I should do next. I should try to make my Self get to work'.*

He goes on to say that: '*Whenever you think about your "Self" you are switching among a huge network of models, each of which tries to represent some particular aspects of your mind — to answer some questions about yourself'.*

41.2 Can we make decisions without involving emotions?

> *What sorts of 'rules' could possibly capture all of what we think of as intelligent behaviour however? Certainly there must be rules on all sorts of different levels. There must be many 'just plain' rules. There must be 'metarules' to modify the 'just plain' rules; then 'metametarules' to modify the metarules, and so on. The flexibility of intelligence comes from the enormous number of different rules, and levels of rules. The reason that so many rules on so many different levels must exist is that in life, a creature is faced with millions of situations of completely different types. In some situations, there are stereotyped responses which require 'just plain' rules. Some situations are mixtures of stereotyped situations - thus they require rules for deciding which of the 'just plain' rules to apply. Some situations cannot be classified - thus there must exist rules for inventing new rules ...*

and on and on. Without doubt, Strange Loops involving rules that change themselves, directly or indirectly, are at the core of intelligence. Sometimes the complexity of our minds seems so overwhelming that one feels that there can be no solution to the problem of understanding intelligence - that it is wrong to think that rules of any sort govern a creature's behaviour, even if one takes 'rule' in the multilevel sense described above.

Douglas Hofstadter, *Gödel, Escher, Bach*

We cannot take decisions without involving emotions. This conclusion of modern psychology goes against the grain of what was believed to be the case about the nature of rational behaviour for most of the 20th century. The conventional picture was that at the bottom of the hierarchical complexity of the human brain is the *brain stem*, which controls bodily functions like heartbeat, breathing, and body temperature. At the next higher level is the *diencephalon*, which regulates hunger pangs and sleep cycles etc. Then comes the *limbic region*, which generates and controls emotions (violence, lust, impulsive behaviour, etc.). These three levels of brain complexity are common to all mammals, including humans. Lastly there is the prefrontal cortex, predominantly responsible for our reasoning power and intelligence etc. Although it enables us to suppress emotions to a small or large extent, it is wrong to think that this 'rationality' portion of our brain can completely overpower or overrule what the three hierarchically lower parts of the brain tend to do. In other words, *it is impossible for us to make decisions that are completely dispassionate or 'reasoned'.*

It is also true that a substantial portion of the prefrontal cortex is involved in our emotional behaviour. How do we 'manage' our emotions? We do so *by thinking about them*, and the thinking is done mainly by the prefrontal cortex. The term '*metacognition*' is used for the capacity of our prefrontal cortex to contemplate about our own mind (Fig. 41.1). The frontal cortex knows when we are, say, angry. In fact, practically every emotional state comes with self-awareness attached to it. This enables us to figure out or 'think' why we are feeling the way we are feeling. Thus we are able to exercise a certain degree of control over our emotions by what is commonly called 'rational thinking'. This is also how we make decisions. The emotional brain is constantly sending out signals about its likes and dislikes. The prefrontal cortex monitors these emotional outputs and tries to decide which signals to take seriously and which ones to overrule. Although the rational brain cannot silence emotions, it can help figure out which ones should be followed.

A highly readable account of the role of intuition and emotions in our decision-making process has been given in the book *How We Decide* by Jonah Lehrer (2009).

Unlike other regions ('columns') of the cortex, which specialize in processing specific types of stimuli, the neurons of the prefrontal cortex can process *whatever kind of data they need to process*. This enables our brain to look at a given problem from a variety of vantage points, and even come out with creative solutions.

How does the prefrontal cortex accomplish this? The answer has to do with its special kind of memory called the *working memory* (Baddeley and Hitch 1994; Smith and Jonides 1999; Diamond 2013). It is a short-term memory, but it has a *persistence* feature. It is a meeting ground, and also a melting pot, of information from various sources. Neurons in this part of the brain fire in response to a stimulus, *and then keep on firing for several seconds after the stimulus has disappeared*. This allows the brain to make creative associations. This is the so-called *restructuring phase of problem-solving*: Information is mixed together in new ways and overlapping of ideas occurs, leading to new insights. The resultant novel neural wiring enables you to identify the answers you were looking for. This is an important feature of human intelligence.

Fig. 41.1 Metacognition: thinking about thinking, or, being aware that one is aware.
http://schoolofthinking.org/2012/02/metacognition-whats-the-fuss/

The emotional brain is important too

Excessively rational thinking can backfire, because it often amounts to suppressing what the primitive brain is trying to tell us (so show healthy respect for your 'gut feeling'!). This problem arises because the rational brain is not an infinitely powerful supercomputer; i.e., rational analysis cannot always provide the best solution to a complicated problem. The cumulative wisdom buried in the (much larger) primitive brain must also be used.

The psychologist George Miller (1956) demonstrated in his essay 'The Magical Number Seven, Plus or Minus Two' that the conscious brain can only handle about seven pieces of data at any one moment. *The computational circuitry of the rational part of our brain is only a tiny fraction of the total capacity of the brain, 'just a few microchips within the vast mainframe of the mind.'* As a result, too many choices, or too much data, can overwhelm the prefrontal cortex, leading to bad decisions. The trick lies in *learning* when to trust your intuitions more than your reasoning power. 'Because working memory and rationality share a common cortical source – the prefrontal cortex – a mind trying to remember lots of information is less able to exert control over its impulses. The substrate of reason is so limited that a few extra digits can become an extreme handicap' (Lehrer 2009).

The fact of life is that the rational part of our brain (which is really a very recent novelty on the evolutionary time scale) has a rather slow and small, even erratic, CPU. Too much information can interfere with understanding. *When the prefrontal cortex is overwhelmed, correlation is confused with causation, and people tend to make theories out of coincidences.*

And yet, excessive dependence on the emotional brain can be risky too. The ideal situation is that exemplified by, say, a champion chess player. Through an unhurried analysis of the games he won or lost, he builds up experience (*turning mistakes into educational events*) which gets 'internalized' into his emotional brain. In due course, it becomes 'second nature' for him to make the right moves, not having to consciously analyse the consequences of too large a number of prospective moves. *The emotional brain is a huge supercomputer, with massive parallel-processing capabilities.*

41.3 Hawkins' model for intelligence and consciousness

Jeff Hawkins, in his 2004 book *On Intelligence*, proposed the so-called *memory and prediction theory* of how human intelligence arises. The basic idea of Hawkins' theory of intelligence, in his own words, is as follows: *The brain uses vast amounts of memory to create a model of the world. Everything we know and have learnt is stored in this model. The brain uses this memory-based model to make continuous predictions of future events. It is the ability to make predictions about the future that is the crux of intelligence.*

Kurzweil's (2012) theory (or model) of the how our brain functions has many ideas in common with those in Hawkins' book. Moreover, there is also some discussion over credit sharing. I quote from Kurzweil (2012): 'The pattern recognition theory of mind that I present here is based on recognition of patterns by pattern recognition modules in the neocortex. These patterns (and the modules) are organized in hierarchies. I discuss below the intellectual roots of this idea, including my own work with hierarchical pattern recognition in the 1980s and 1990s and Jeff Hawkins (born in 1957) and Dileep George's (born in 1977) model of the neocortex in the early 2000s'.

Hawkins (2004) pointed out in his book that the neocortical memory differs from that of a conventional computer in four ways:

1. The cortex stores *sequences* of patterns. For example, our memory of the alphabet is a sequence of patterns. It is not something stored or recalled in an instant, or all together. That is why we have difficulty saying it backwards. Similarly our memory of songs is an example of *temporal* sequences in memory.

2. The cortex recalls patterns *auto-associatively*. The patterns are associated with themselves. One can recall complete patterns when given only partial or distorted inputs. During each waking moment, each functional region is essentially waiting for familiar patterns or pattern-fragments to come in. Inputs to the brain link to themselves auto-associatively, filling in the present, and auto-associatively linking to what normally flows next. We call this chain of memories, *thought*.

3. The cortex stores patterns in an *invariant form*. Our brain does not remember *exactly* what it sees, hears, or feels; the brain remembers the important relationships in the world, independent of details.

4. The cortex stores patterns in a *hierarchy*.

Storing sequences, auto-associative recall, and invariant representation are the necessary ingredients for predicting the future, based on memories of the past. How this happens is the subject matter of Hawkins' book. According to him, making such predictions is the essence of intelligence.

Hawkins takes the view that perhaps consciousness is simply what it feels like to have a neocortex. He suggests that the self-awareness aspect of consciousness is synonymous with the formation of *declarative memories*. These are memories we can recall and talk about.

Hawkins, while formulating his theory of intelligence, took very seriously the so-called *Mountcastle's hypothesis*. Since the same types of layers, cell types and connections exist in the entire cortex, Mountcastle (1978) had put forward the following hypothesis:

There is a common function, a common algorithm, that is performed by all the cortical regions.

What makes the various functional areas different is the way they are *connected*. He went further to suggest that the reason why the different functional regions *look* different when imaged is because of these different connections only.

Hawkins suggests that, although hearing, touch, vision etc. are processed by the same algorithm in the neocortex, they are handled differently in the R-brain: *'Hearing relies on a set of audition-specific subcortical structures that process auditory patterns before they reach the cortex. Somatosensory patterns also travel through a set of subcortical areas that are unique to somatic senses. Perhaps qualia, like emotions, are not mediated purely by the neocortex. If they are somehow bound up with subcortical parts of the brain that have unique wiring, perhaps tied to emotion centres, this might explain why we perceive them differently, even if it doesn't explain why there is any sort of qualia sensation in the first place'.*

The structure of the inputs (i.e., the spatio-temporal information pattern) is qualitatively different for, say, the auditory nerve and the optic nerve. The optic nerve has a million fibres, and the auditory nerve has only thirty thousand. The optic nerve caries information that is more spatial than temporal, and the auditory nerve carries information that is more temporal than spatial. This may have a bearing on why is red red and green green. No matter how consciousness is defined, memory and prediction play crucial roles in creating it.

Here is how Hawkins answers why our thoughts appear to be independent of our bodies:

'To the cortex our bodies are just part of the external world. Remember, the brain is in a quiet and dark box. It knows about the world only via the patterns on the sensory nerve fibres. From the brain's perspective as a pattern device, it doesn't know about your body any differently than it knows about the rest of the world. There isn't a special distinction between where the body ends and the rest of the world begins. But the cortex has no ability to model the brain itself because there are no senses in the brain. Thus we can see why our thoughts appear independent of our bodies, . . '.

Kurzweil (2002) points out that his model differs from Hawkins' model in some important aspects: '. . . as the name implies, Hawkins is emphasizing the temporal (time-based) nature of the constituent lists. In other words, the direction of the lists is always forward in time. His explanation for how the features in a two-dimensional pattern such as the printed letter "A" have a direction in time is predicated on eye movement. He explains that we visualize images using saccades, which are very rapid movements of the eye of which we are unaware. The information reaching the neocortex is therefore not a two-dimensional set of features but rather a time-ordered list. While it is true that our eyes do make very rapid movements, the sequence in which they view the features of a pattern such as the letter "A" does not always occur in a consistent temporal order. (For example, eye saccades will not always register the top vertex in "A" before its bottom concavity.) Moreover, we can recognize a visual pattern that is presented for only a few tens of milliseconds, which is too short a period of time for eye saccades to scan it. It is true that pattern recognizers in the neocortex store a pattern as a list and that the list is indeed ordered, but it may also represent a spatial or higher-level conceptual ordering . . .'

Dileep George was a coworker of Kurzweil. His doctoral dissertation presents a more up-to-date description of the hierarchical temporal memory method [Dileep George, 'How the Brain Might Work: A Hierarchical and Temporal Model for Learning and Recognition' (PhD dissertation, Stanford University, June 2008)].

42. Inside the Human Brain

For developing a credible model for how our brain functions, the key requirement is the availability of detailed knowledge about its anatomy, connections, and processes. Naturally, the full force of the progress made by science and technology in developing the probing and peering techniques has been brought to bear on this task.

42.1 Probing the human Brain

If we want to reverse-engineer the human brain, the most important thing to do first is to probe its structure and function with experimental tools that have the highest possible spatial and temporal resolution. I give here a very brief historical account of the progress made in achieving this objective. This information is based largely on the work of Kurzweil (2005, 2012).

1. At the beginning of the 20th century, crude tools were developed for examining the physical processes inside the brain. In 1928, E. D. Adrian measured the electrical output of nerve cells, thus demonstrating that *there are electrical processes occurring inside the brain.* To quote Adrian: 'I had arranged electrodes on the optic nerve of a toad in connection with some experiments on the retina. The room was nearly dark and I was puzzled to hear repeated noises in the loudspeaker attached to the amplifier, noises indicating that a great deal of impulse activity was going on. It was not until I compared the noises with my own movements around the room that I realized I was in the field of vision of the toad's eye and that it was signalling what I was doing'. As Kurzweil (2005) remarks, 'Adrian's key insight from this experiment remains a cornerstone of neuroscience today: the frequency of the impulses from the sensory nerve is proportional to the intensity of the sensory phenomena being measured. For example, the higher the intensity of light, the higher the frequency (pulses per second) of the neural impulses from the retina to the brain'.

2. Horace Barlow, a student of Adrian, provided another crucial insight, namely 'trigger features' in neurons. He discovered that the retinas of frogs and rabbits have single neurons that trigger on 'seeing' specific shapes, directions, or velocities. This meant that *perception involves a series of stages, with each layer of neurons recognizing more sophisticated features of the image.* Even today, electroencephalography (EEG) is a common investigative and diagnostic tool that records the electrical activity occurring along the scalp. It measures the voltage fluctuations resulting from the ionic currents flowing within the neurons (see below). The spectral content of an EEG can provide information for, for example, the epileptic activity in the brain of a patient. This technique can provide millisecond-level temporal resolution.

3. In 1939, A. L. Hodgkin and A. F. Huxley began developing an idea of how neurons perform, namely by accumulating their inputs and then producing a *spike in membrane conductance*: There is a sudden increase in the ability of the neuron's membrane to conduct a signal and the corresponding voltage along the axon of the neuron. This was described by Hodgkin and Huxley as the axon's '*action potential*' (voltage). They actually measured the action potential on an animal neuron in 1952. Squid neurons were chosen by them for this, because of their large size and accessibility.

4. Building on the work of Hodgkin and Huxley, W. S. McCulloch and W. Pitts worked out in 1943 a simple model of neurons and neural nets. I described their model in Section 33.3, under the title 'Neural networks, real and artificial'. This very basic model was further refined by

Hodgkin and Huxley in 1952. Whether in the brain or in a computer simulation, the model introduces the idea of a neural 'weight' which represents the strength of the neural connection (synapse), and also a *nonlinearity* (firing threshold) in the neural cell body (the soma).

5. As described in Section 33.3, another breakthrough idea was put forward in 1949 by Donald Hebb. His theory of neural learning (the *Hebbian response theory*) said that if a synapse is stimulated repeatedly, it becomes stronger. Over time this conditioning produces a learning response. Such *connectionist* ideas flourished during the 1950s and 1960s, and led to much research on artificial neural nets.

6. There is another form of Hebbian learning, namely a loop in which the excitation of a neuron feeds back on itself, causing reverberation (a continued reexcitation of the neurons in the loop). Hebb suggested that this type of reverberation could result in short-term memory: 'Let us assume that the persistence or repetition of a reverberatory activity (or 'trace') tends to induce lasting cellular changes that add to its stability. . . When an axon of cell A is near enough to excite a cell B and repeatedly or persistently takes part in firing it, some growth process or metabolic change takes place in one or both cells such that A's efficiency, as one of the cells firing B, is increased'. This form of Hebbian learning is well captured by the popular phrase '*cells that fire together wire together*'. *Brain assemblies* can create new connections and strengthen them, based on their own activity. The actual development of such connections by neurons has been seen in brain scans.

In Hebb's theory the central assumption is that the basic unit of learning in the neocortex is the neuron: a *single* neuron. But the current theory, put forward by Kurzweil (2012), of how the brain functions is based, not on the neuron itself, but rather on an *assembly* of neurons (I shall describe the theory in next chapter). This basic unit of learning comprises of ~100 neurons. According to Kurzweil (2012) 'the wiring and synaptic strengths *within* each unit are relatively stable and determined genetically. . . . Learning takes place in the creation of connections *between* these units, not within them, and probably in the synaptic strengths of those interunit connections'. As we shall see in the next chapter, experimental evidence has indeed been obtained for the existence of 100-neuron thick modules as the basic units of learning.

7. The connectionist movement suffered a temporary setback in 1969 when Marvin Minsky and Seymour Papert published the book *Perceptrons*. This book included a theorem which demonstrated that the most common neural net used at that time (namely *Rosenblatt's Perceptron*) was unable to answer whether or not a line-drawing was fully connected.

8. The neural-net or connectionist movement staged a resurgence in the 1980s when the *back propagation* method was invented. In it, the strength of each simulated synapse is governed by a learning algorithm that adjusts the synaptic weight or the strength of the output of each artificial neuron after each training trial, thus enabling the net to learn to match the right answer more correctly. This type of *self-organization* has helped solve a whole range of pattern-recognition problems. But back propagation is not a feasible model for the training occurring in real mammalian biological neural nets.

9. Spectacular progress continues to be made in developing experimental techniques for peering into the brain. According to Kurzweil (2005) the resolution of noninvasive brain-scanning devices has been doubling every 12 months or so (per unit volume). There is also a comparable improvement in the speed of brain scanning image reconstruction.

A commonly used brain-scanning technique is fMRI (functional magnetic resonance imaging). This technique is based on the fact that cerebral blood flow and neuronal activation are coupled. When an area of the brain is in use, blood flow to that region increases. fMRI provides a spatial resolution of ~1 mm, and a time resolution of ~1 second (or 0.1 second for a thin brain slice). It measures blood-oxygen levels, and is an indirect technique for recording neuronal activity. Another such indirect technique is PET (positron emission tomography). It measures the regional cerebral blood flow (rCBF).

Both fMRI and PET reflect local synaptic activity, rather than the spiking of neurons. They are particularly reliable for recording the relative *changes* in the state of the brain, for example when a particular task is being carried out by the subject.

10. Another brain-scanning technique is MEG (magnetoencephalography). It measures the magnetic fields outside the skull, coming mainly from the pyramidal neurons of the neocortex. It can achieve millisecond-level temporal resolution, but has a very poor spatial resolution (~1 cm).

11. Optical imaging is an invasive technique capable of providing high spatial and temporal resolution. It involves removing a part of the skull, staining the brain tissue with a dye that fluoresces during neural activity, and imaging the emitted light.

12. When it is feasible to destroy a brain for the purpose of scanning it, immensely high spatial resolutions become feasible. It has been possible to scan the nervous system of the brain and body of a mouse with a resolution better than 200 nm.

42.2 Peering into the human brain

As Ray Kurzweil (2005) has kept on emphasizing for many years, the law of accelerating returns (LOAR) is always operative in the evolution of all information-based technologies, resulting in their *exponential* growth (Moore's law is just one example of that). Naturally, progress in the development of better and better brain-probing technologies is no exception to this. In the previous section I gave you a summary of Kurzweil's account of the early history of progress in the development of experimental techniques for probing the human brain. Our present capabilities are already mind-boggling, and much more will be coming in the near future. And needless to say, experiment and theory will go hand in hand. Two examples will illustrate my point.

In Chapter 41 we learnt about the path-breaking Mountcastle hypothesis, which says that *there is a common function, a common algorithm, that is performed by all the cortical regions.* Kurzweil (2012) has rightly emphasized the fundamental importance of this insight: 'A critically important observation about the neocortex is the extraordinary uniformity of its fundamental structure. This was first noticed by American neuroscientist Vernon Mountcastle (born in 1918). In 1957 Mountcastle discovered the *columnar organization* of the neocortex. In 1978 he made an observation that is as significant to neuroscience as the Michelson-Morley ether-disproving experiments of 1887 were to physics. That year he described the remarkably unvarying organization of the neocortex, hypothesizing that it was composed of a single mechanism that was repeated over and over again, and proposing the *cortical column* as that basic unit'.

Another basic insight is that the basic module of learning is a module of dozens of neurons (~100). Support for this postulate has come from the work of Henry Markram. His ambitious *Blue Brain Project* (http://bluebrain.epfl.ch/) aims to both model and simulate the human brain, including the entire neocortex, as also the old-brain (R-brain) regions such as the hippocampus, amygdala, and cerebellum: 'Reconstructing the brain piece by piece and building a virtual brain in a supercomputer — these are some of the goals of the Blue Brain Project. The virtual brain will be an *exceptional tool* giving neuroscientists a new understanding of the brain and a better understanding of neurological diseases'.

This project is using a scanning-technology tool called the *automated patch-clamp robot* (Kodandaramaiah *et al*. 2012), with which researchers are 'measuring the specific ion channels, neurotransmitters, and enzymes that are responsible for the electrochemical activity within each neuron'. It is 'an automated system with one-micrometre precision that can perform scanning of neural tissue at very close range without damaging the delicate membranes of the neurons'. The scanning technology has been already used for simulating a single neuron (in 2005), a neocortical column consisting of 10,000 neurons (in 2011), and a neural mesocircuit consisting of 100 neocortical columns (in 2011) (Fig. 42.1).

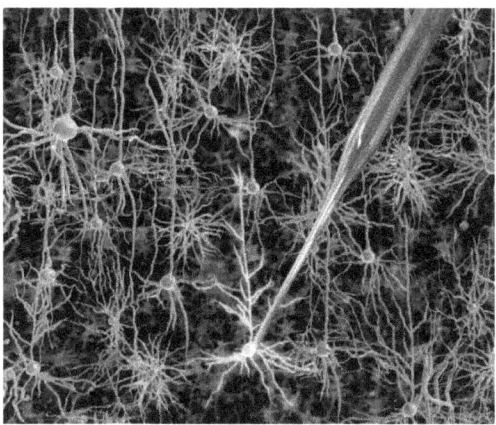

Fig. 42.1 Peering into the inner workings of the brain cells with robots.
http://web.mit.edu/newsoffice/2012/robots-recording-neurons-0507.html

The scientists developed this method to automate the process of finding and recording information from neurons in the living brain. It has been shown that a robotic arm guided by a cell-detecting computer algorithm can identify, and record from, neurons in the living-mouse brain with better accuracy and speed than a human experimenter. The automated process eliminates the need for months of training, and provides the long-sought information about the activity of living cells. Using this technique, scientists could classify the thousands of different types of cells in the brain, map how they connect to each other, and figure out how diseased cells differ from normal cells. To quote the authors (Kodandaramaiah *et al*. 2012): 'Whole-cell patch-clamp electrophysiology of neurons is a gold-standard technique for high-fidelity analysis of the biophysical mechanisms of neural computation and pathology, but it requires great skill to perform. We have developed a robot that automatically performs patch clamping *in vivo*, algorithmically detecting cells by analysing the temporal sequence of electrode impedance changes. We demonstrate good yield, throughput and quality of automated intracellular recording in mouse cortex and hippocampus'.

As quoted by Kurzweil (2012), Markram wrote in a 2011 paper that while he was 'search[ing] for evidence of Hebbian assemblies (collections of neurons that are arranged together) at the most elementary level of the cortex', what he found instead were 'elusive assemblies [whose] connectivity and synaptic weights are highly predictable and constrained'. He concluded that 'these findings imply that experience cannot mould the synaptic connections of these assemblies', and speculated that 'they serve as innate, Lego-like building blocks of knowledge for perception and that the acquisition of memories involves the combination of these building blocks into complex constructs'.

Here is more from Markram: 'Functional neuronal assemblies have been reported for decades, but direct evidence of clusters of synaptically connected neurons . . . has been missing. . . . Since these assemblies will all be similar in topology and synaptic weights, not moulded by any specific experience, we consider these to be innate assemblies . . . Experience plays only a minor role in determining synaptic connections and weights within these assemblies . . . Our study found evidence [of] innate Lego-like assemblies of a few dozen neurons . . Connections between assemblies may combine them into super-assemblies within a neocortical layer, then in higher-order assemblies in a cortical column, even higher-order assemblies in a brain region, and finally in the highest possible order in the whole brain . . . Acquiring memories is very similar to building with Lego. Each assembly is equivalent to a Lego block holding some piece of elementary innate knowledge about how to process, perceive and respond to the world. . . When different blocks come together, they therefore form a unique combination of these innate percepts that represents an individual's specific knowledge and experience'. Whoa! Now that's what I call progress.

Further evidence for the regular structure of connections across the neocortex was published in the March 2012 issue of the journal *Science* by Van J. Wedeen *et al*. They wrote: 'Basically, the overall structure of the brain ends up resembling Manhattan, where you have a 2-D plan of streets and a third axis, an elevator going in the third dimension'.

As Wedeen (http://blogs.discovermagazine.com/notrocketscience/2012/03/29/the-brain-is-full-of-manhattan-like-grids/#.WPWpuoVOJy1) said in a *Science* magazine podcast, 'This was an investigation of the three-dimensional structure of the pathways of the brain. When scientists have thought about the pathways of the brain for the last hundred years or so, the typical image or model that comes to mind is that these pathways might resemble a bowl of spaghetti – separate pathways that have little particular spatial pattern in relation to one another. Using magnetic resonance imaging, we were able to investigate this question experimentally. And what we found was that rather than being haphazardly arranged or independent pathways, we find that all of the pathways of the brain taken together fit together in a single exceedingly simple structure. They basically look like a cube. They basically run in three perpendicular directions, and in each one of these three directions the pathways are highly parallel to each other and arranged in arrays. So, instead of independent spaghettis, we see that the connectivity of the brain is, in a sense, a single coherent structure'.

A very precise form of scanning technology was used for revealing the grid-like structure of the connections, involving a variety of noninvasive scanning technologies, including new forms of MRI, magnetoencephalography, and diffusion tractography (a method to trace the pathways of fibre bundles in the brain).

This is incredible stuff! A great triumph of modern science, and of the scientific method! I re-quote: '*. . . we find that all of the pathways of the brain taken together fit together in a single exceedingly simple structure. They basically look like a cube. They basically run in three perpendicular directions, and in each one of these three directions the pathways are highly parallel to each other and arranged in arrays*' (Fig. 42.2).

As Kurzweil (2012) explains, 'Whereas the Markram study shows a module of neurons that repeats itself across the neocortex, the Wedeen study demonstrates a remarkably orderly pattern of connections between modules. *The brain starts out with a very large number of "connections-in-waiting" to which the pattern recognition modules can hook up.* Thus if a given module wishes to connect to another, it does not need to grow an axon from one and a dendrite from the other to span the entire physical distance between them. It can simply harness one of these connections-in-waiting and just hook up to the ends of the fibre. As Wedeen and his colleagues write, "The pathways of the brain follow a base-plan established by . . . early embryogenesis. Thus, the pathways of mature brain present an image of these three primordial gradients, physically deformed by development". In other words, as we learn and have experiences, the pattern recognition modules of the neocortex are connecting to these preestablished connections that were created when we were embryos' (emphasis added).

Fig. 42.2 Pathways of the brain. Image credit:
http://www.ineffableisland.com/2012_03_01_archive.html

This is rather like the field-programmable gate arrays (FPGAs) I described in Section 35.2. We humans developed the technology of FPGAs, not knowing that our own brains have evolved to have a similar configuration and working principle!

Fig. 42.3 gives a pictorial summary of the present status of tools for imaging the human brain.

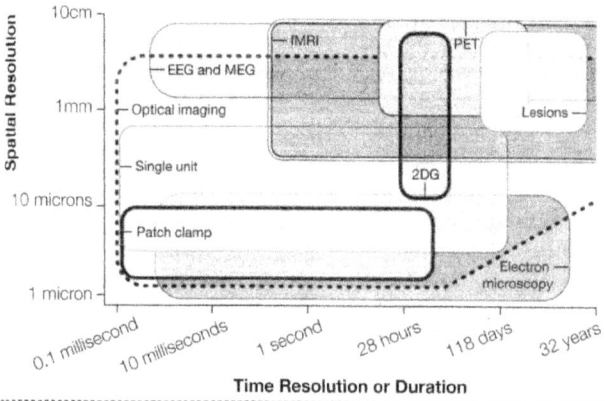

Fig. 42.3 Tools for imaging the human brain (from Kurzweil 2012).

Wedeen, whose work I mentioned above, is also involved in the truly ambitious Human Connectome Project, which aims at mapping the wiring diagram of the entire living human brain (http://www.humanconnectomeproject.org/).

Scientists have even captured videos of brains in real time (Nöbauer *et al.* 2017). They used light for peering into the brain of a mouse, and captured live neural activity of hundreds of individual neurons in a 3-D section of tissue.

43. Kurzweil's Pattern-Recognition Theory of Mind

I had stated the gist of Kurzweil's (2012) pattern recognition theory of mind (PRTM) in Section 40.3. The theory is based on a certain model of the neocortex, which I shall outline here. I believe that the model is bound to be largely correct, for two reasons. One is the immensely graphic and detailed experimental data we now have about the structure of the brain, as I outlined in the previous chapter. The other reason is the spectacular success already achieved in creating an artificial brain based on the PRTM. The success of, for example, IBM's Watson is proof of that (see Section 40.3).

The human neocortex is an essentially 2-dimensional, ~2.5-mm thick, structure, comprising of six layers. The layers are numbered from I (the outermost layer) to VI (see Section 41.3).

The PRTM says, in essence, that all of the many wonders of the neocortex can be reduced to a single type of thought process, involving hierarchical thinking. This is lent credence to by the structure of the neocortex itself. Its fundamental structure and function has an extraordinarily high degree of uniformity, *a la* Vernon Mountcastle (Section 41.3). And this structure is hierarchical in nature. Mountcastle had also postulated the existence of *cortical columns* along the thickness of the neocortex. The six layers and the cortical columns in them together imply the existence of a *grid structure*, which has been confirmed by experiment.

Kurzweil hypothesizes that the basic uniform unit of action in the entire neocortex is the so-called *pattern recognizer* (PR); it is the fundamental component of the neocortex. Deviating a bit from Mountcastle's model, Kurzweil's PRTM stipulates that the PRs are not separated by specific physical boundaries; rather they are placed closely one to the next in an interwoven fashion. *A cortical column is simply an aggregate of a large number of PRs.*

The PRs wire themselves to one another throughout the course of a lifetime. Therefore the elaborate connectivity between modules that exists in the neocortex is not specified much by the genetic code; rather it gets created to embody the patterns we actually learn over our lifetime.

Kurzweil estimates that there are ~500,000 cortical columns in the human neocortex, each being ~0.5 mm wide and ~2.5 mm long. Each contains ~60,000 neurons. Since each PR within a cortical column contains ~100 neurons, it follows that there are ~500,000 x 60,000 / 100, or ~300 million PRs in our neocortex.

How many patterns can the human neocortex store? With as many as 300 million PRs available, our brain can indulge in a huge amount of redundancy, resulting in our fantastic pattern-recognition capability, which is far in excess of what any computer system has been able to attain so far. [However, let us also remind ourselves that computer processes are millions of times faster than the electrochemical processes occurring in our brains.]

Here is an example of the redundancy with which our brain stores patterns. The face of a loved one is not stored just once, but thousands of times. Some are just repetitions, but most are different perspectives of the face, differing in lighting, facial expressions, etc. And none of these repeated patterns are stored as 2-dimensional arrays of pixels. They are stored as 1-dimensional *lists* of features, but hierarchically: The constituent elements of a pattern are themselves patterns, and so on.

Even our procedures and actions comprise of patterns, and are stored in the neocortex in a likewise manner.

Kurzweil's estimate of the total capacity of the human neocortex is on the order of low hundreds of millions of patterns, which is similar to the number of PRs, namely ~300 million.

The structure of a pattern

The PRTM says that patterns are recognized by pattern-recognition modules in the neocortex, and that the patterns and the modules are organized in hierarchies. When a pattern is recognized, there are three parts to this process. To make the description concrete, let us take the example of an APPLE, and also the word 'APPLE' we use for referring to this physical entity. The details given here are from Kurzweil (2012).

Part one is the input, consisting of the lower-level patterns that compose the main pattern. The descriptions of each of these lower-level patterns do not need to be repeated for each higher-level pattern that references them. The letter 'A' appears in the pattern for the word APPLE and also in a large number of other words. Each of these patterns need not repeat a description for the pattern of A, but can use a common description stored somewhere. All that is required is a neural connection to that location. There is an axon from the 'A' pattern recognizer that connects to multiple dendrites, one for each word that uses 'A'.

Part two of each pattern is the *name* of the pattern. This 'name' is simply the axon that emerges from each pattern processor. When the axon fires, its corresponding pattern has been recognized. It is as if the pattern recognizer is shouting: 'Hey guys, I just saw the written word "apple"'.

Part three of each pattern is the set of higher-level patterns that it, in turn, is a part of. For the letter 'A' it is all words that include 'A'.

For the example of apple the object and apple the word, just like the hierarchy for the storage and recognition of the word 'apple', another part of the cortex has a hierarchy of pattern-recognizers processing the actual *images* of objects. If you are looking at an apple, the corresponding pattern recognizer will fire its axon, saying in effect: 'Hey guys, I just saw an actual apple'. Similarly, if somebody utters the word 'apple', the corresponding auditory pattern-recognizer will be triggered.

Information flows *down* the conceptual hierarchy as well as *up*. To quote Kurzweil (2012): 'If, for example, we are reading from left to right and have already seen and recognized the letters "A", "P", "P", "L", the "APPLE" recognizer will predict that it is likely to see an "E" in the next position. It will send a signal *down* to the "E" recognizer saying, in effect, "Please be aware that there is a high likelihood that you will see your 'E' pattern very soon, so be on the lookout for it"'. The 'E' recognizer then adjusts (lowers) its threshold or action potential for the firing of the neuron which would potentially declare that 'E' has been seen. *Even if an incomplete or smudged image of 'E' appears, it would be recognized correctly because it was expected'*.

This prediction feature is one of the primary reasons why we have a neocortex at all. Our brain is making predictions all the time, and at all levels of abstraction.

The nature of data flowing into a pattern recognizer

What does the data for a pattern look like? Suppose the pattern is a face, an essentially 2-

dimensional set of data. But, as can be seen from the structure of the neocortex, the pattern inputs are only 1-dimensional lists. All the experience in the creation and functioning of artificial pattern-recognition systems also confirms that one can represent 2- or higher-dimensional data streams as 1-dimensional lists. Our memories are patterns organized as lists (that is why we have trouble reciting the alphabet backwards). And, what is more, each item in the list is another pattern, and so on, hierarchically. We have learnt these lists, and we recognize them when an appropriate stimulus is present. Memories exist in the neocortex in order to be recognized.

Autoassociation and invariance

We can recognize a pattern even if it is incomplete. This ability to associate a pattern with a part of itself is called *autoassociation.*

Often we are able to recognize patterns that are distorted, or when aspects of them are transformed. This ability is called *invariance*, and the brain deals with it in four ways.

The first way is through global transformations that are effected before the cortex receives the sensory data.

The second takes advantage of the redundancy in the storage of memory. The memory has many perspectives or variations stored away.

The third is the ability to combine two or more memory lists. That is how we understand metaphors and similes.

The fourth method derives from the 'size parameters' that allow a single module to encode multiple instances of a pattern.

Learning

As Kurzweil writes: 'Our neocortex is virgin territory when our brain is created. It has the capability of learning and therefore of creating connections between its pattern recognizers, but it gains those connections from experience. . . Learning and recognition take place simultaneously. We start learning immediately, and as soon as we've learned a pattern, we immediately start recognizing it. . . . patterns that are not recognized are stored as new patterns and are appropriately connected to the lower-level patterns that form them'.

The language of thought

At the heart of the pattern-recognition theory of mind (PRTM) is the neocortical pattern-recognition module, the inputs to and the outputs from which are shown Fig. 43.1.

The brain starts out with a very large number of 'connections-in-waiting' to which the pattern-recognition modules can hook up. As we learn and have experiences, the pattern recognizing modules of the neocortex are connecting to preestablished connections that were created when we were embryos. Kurzweil (2012) has summarized his PRTM as follows:

'(a) Dendrites enter the module that represents the pattern. Even though patterns may seem to have two- or three-dimensional qualities, they are represented by a one-dimensional sequence of signals. The pattern must be present in this (sequential) order for the pattern recognizer to be able to recognize it. Each of the dendrites is connected ultimately to one or more axons of pattern recognizers at a lower conceptual level that have recognized a lower-level pattern that constitutes part of this pattern. For each of these input patterns, there may be many lower-level pattern recognizers that can generate the signal that the lower-level pattern has been recognized. The necessary threshold to recognize the pattern may be achieved even if not all of the inputs have signalled. The module computes the probability that the pattern it is responsible for is present. This computation considers the "importance" and "size" parameters (see (f) below).

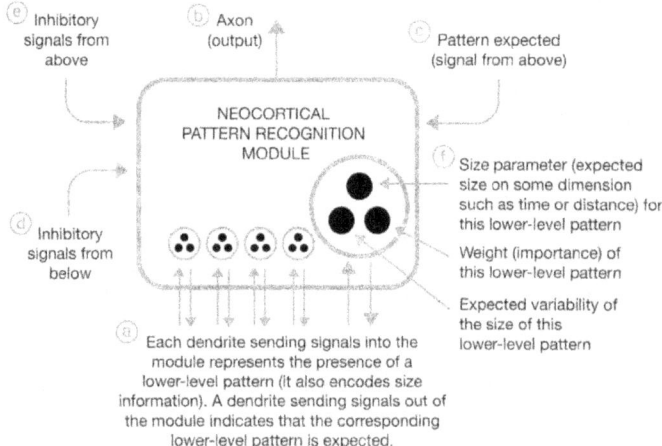

Fig. 43.1 The neocortical pattern-recognition module (from Kurzweil 2012).

'Note that some of the dendrites transmit signals into the module and some out of the module. If all of the input dendrites to this pattern recognizer are signalling that their lower-level patterns have been recognized except for one or two, then this pattern recognizer will send a signal down to the pattern recognizer(s) recognizing the lower-level patterns that have not yet been recognized, indicating that there is a high likelihood that that pattern will soon be recognized and that lower-level recognizer(s) should be on the lookout for it.

'(b) When this pattern recognizer recognizes its pattern (based on all or most of the input dendrite signals being activated), the axon (output) of this pattern recognizer will activate. In turn, this axon can connect to an entire network of dendrites connecting to many higher-level pattern recognizers that this pattern is input to. This signal will transmit magnitude information so that the pattern recognizers at the next higher conceptual level can consider it.

'(c) If a higher-level pattern recognizer is receiving a positive signal from all or most of its constituent patterns except for the one represented by this pattern recognizer, then that higher-level recognizer might send a signal down to this recognizer indicating that its pattern is expected. Such a signal would cause this pattern recognizer to lower its threshold, meaning that it would be more likely to send a signal on its axon (indicating that its pattern is considered to have been recognized) even if some of its inputs are missing or unclear.

'(d) Inhibitory signals from below would make it less likely that this pattern recognizer will recognize its pattern. This can result from recognition of lower-level patterns that are inconsistent with the pattern associated with this pattern recognizer. . . .

'(e) Inhibitory signals from above would also make it less likely that this pattern recognizer will recognize its pattern. This can result from a higher-level context that is inconsistent with the pattern associated with this recognizer.

'(f) For each input, there are stored parameters for importance, expected size, and expected variability of size. The module computes an overall probability that the pattern is present based on all of these parameters and the current signals indicating which of the inputs are present and their magnitudes. A mathematically optimal way to accomplish this is with a technique called *hidden Markov models* (https://en.wikipedia.org/wiki/Hidden_Markov_model). When such models are organized in a hierarchy (as they are in the neocortex or in attempts to simulate a neocortex), we call them *hierarchical hidden Markov models*.'

Triggered patterns trigger other patterns. Incomplete patterns send signals down the conceptual hierarchy. Complete patterns send signals up the hierarchy. These patterns are the language of thought. Like language they are hierarchical, but they are not always language per se, although language-based thoughts are also possible.

There can be two modes of thinking: nondirected and directed. In the former, thoughts trigger one another in a nonlogical way. Dreams are examples of nondirected thoughts. Directed thinking is what we use when we are trying to solve a problem, or when we formulate an organized response.

Thus, according to the PRTM, our intelligence is the result of '*self-organizing, hierarchical recognizers of invariant self-associative patterns with redundancy and up-and-down predictions*' (Kurzweil 2012).

It is rightly claimed in Kurzweil's book that the model ' . . is an incredible synthesis of neuroscience and technology and provides a road map for the future of human progress'. The operating principle of the neocortex (explained by the PRTM) 'is arguably the most important idea in the world, as it is capable of representing all knowledge and skills as well as creating new knowledge'.

44. The Knowledge Era and Complexity Science

The great unexplored frontier is complexity...I am convinced that nations and people that master the new science of complexity will become the economic, cultural, and political superpowers of the next century.

Heinz Pagels (1939-1988)
The Dreams of Reason: The Computer and the Rise of the Science of Complexity (1988)

The complexity-science concepts I have brought out in this book have been in the making for a long time, although the subject got a major boost only during the 1990s, and the progress has been exponentially rapid since then. The work done by scientists for understanding natural phenomena is, by and large, in full public glare, and their results often engender applications in many other fields of human activity. In this chapter I give you a glimpse of the spin-offs of complexity-science research in other domains of the present knowledge era.

44.1 The wide-ranging applications of complexity science

The conference series *Complex Systems* is organized regularly in the UK by the Wessex Institute and others. The announcement brochure for 'Complex Systems 2018', to be held in May 2018 (http://www.wessex.ac.uk/conferences/2018/complex-systems-2018), makes interesting reading, and I list the conference topics to help you get a feel for how so many fields other than hard-core complexity-science have benefitted from the ideas originating from complexity science:

Complexity Science
- Emergence
- Emergent Intelligence
- Adaptation in response to disruptive events
- Resilience to extreme events
- Autocatalytic properties
- Self-organisation
- Evolution

Digital Ecosystems
- Urban digital ecosystems
- Healthcare digital ecosystems

Complex Adaptive Business
- Self-organisation versus control
- Real-time logistics
- Real-time supply chains
- Adaptive business processes
- Enterprise ontology
- Mining big data (extracting knowledge from data)
- Resilience to electronic attacks and fraud

Complex Adaptive Healthcare
- Adaptation in human biology
- Self-organisation of healthcare resources

Complex Networks
- The Internet of documents
- The Internet of people
- The Internet of things
- Social media

Complexity in Transportation Systems
- Self-organisation in air traffic management
- Adaptive aircraft lifecycle
- Adaptive space logistics
- Intelligent drones
- Real-time scheduling of multi-modal transport
- Swarming

It is a long list indeed, but still not exhaustive. I shall consider just a few such topics here.

44.2 Econophysics

A financial market is a complex system in which a large number of traders interact with one another, and also react to information, and determine the 'best' price for a given item for buying or selling. The time evolution of the price and the number of transactions of a traded item is generally unpredictable. The time series indicating the price variation of an item is found to be *essentially indistinguishable* from a stochastic or random process. Like other complex systems, financial markets are open systems with many interacting subunits, and the subunits interact nonlinearly.

In the study of stock markets, the so-called *efficient-market hypothesis* (EMH) serves as a useful benchmark. In the 'strong' version of this hypothesis, an efficient market is defined as one in which all the available information is processed *instantly* when it reaches the market, and in which this fact is *immediately* reflected in new values of prices of the traded assets (Fig. 44.1).

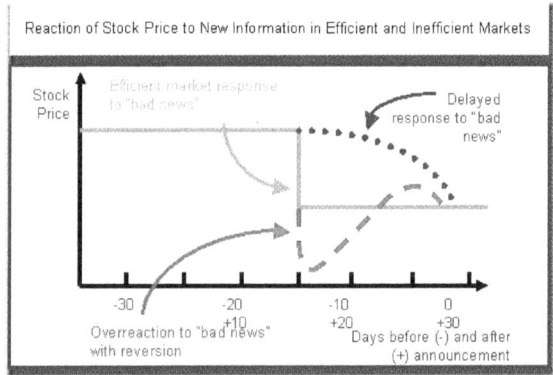

Fig. 44.1 Reaction of a stock price to new information. Image credit:

The EMH makes the idealized stipulation that any market is highly efficient in determining the most *rational* price of a traded item or asset. It was originally formulated in the 1960s. There are two assumptions involved here: (i) the market is efficient; and (ii) the behaviour of traders is strictly rational. As we shall see below, these are only idealizations.

Why does the time series of the returns appear to be random? It is because it carries so much information that there are no readily discernible regularities in it. It is, by and large, a 'non-redundant' time series, meaning that the information it carries is almost irreducible, or algorithmically incompressible, for most practical purposes. The EMH requires that the concerned time series for market prices has a 'dense' amount of non-redundant information. Since there are limits on the speed and capacity of our computers, a time series carrying this information is almost indistinguishable from a totally random time series. Of course, an analysis of the deviation from a totally random time series is a good way of testing the degree of validity of the EMH in a given situation.

In an economic activity, there can be both a law of diminishing returns and a law of increasing returns, depending on the nature of the activity.

The law of diminishing returns

Suppose there is good demand for a commodity because of its attractive existing price. Naturally, the price will increase. This will then reduce the demand. And a reduced demand will entail a lowering of the price, and so on, till the demand and the price have reached a state of *equilibrium*. Thus *negative feedbacks* tend to stabilize an economy, as per *conventional* economic theory. This law of diminishing returns (Fig. 44.2) implies *a single equilibrium point* for an economy, and such situations are amenable to analytical control.

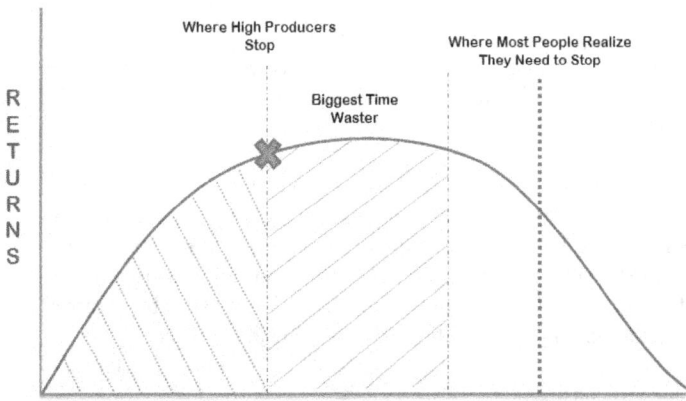

Fig. 44.2 The law of diminishing returns in an economy. Image credit: http://elizabethgatlin.com/analysis/sustainability-and-the-bottom-line/

By and large, *resource-based* economic activities (e.g., agriculture and mining) tend to follow the law of diminishing returns. By contrast, *knowledge-based* parts of an economy are generally governed by the law of *increasing* returns or positive feedback.

The law of increasing returns

As demonstrated by the pioneering studies of Brian Arthur during the 1990s, positive feedbacks often occur in an economy, with the resultant *multiple equilibrium points* (Fig. 44.3). Small shifts in the economy can get amplified, rather than smothered out. The economy evolves like any open, nonlinear complex system. There can be multiple bifurcations in phase space, and it is difficult, if not impossible, to predict which bifurcation branch will be chosen by the market forces. What is more, once the random events select a particular branch or path in phase space, the choice tends to get *locked-in*, regardless of the advantages of the alternatives.

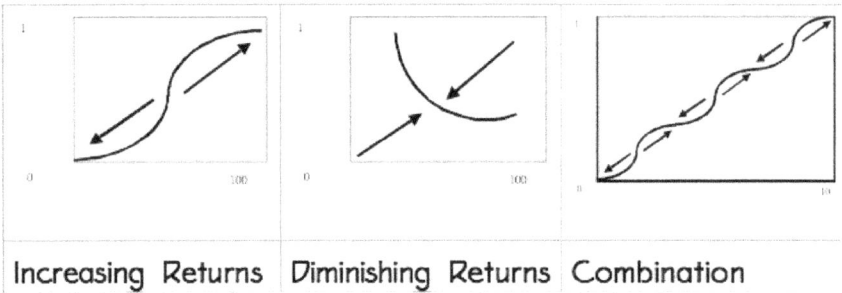

Increasing Returns Diminishing Returns Combination

Fig. 44.3 The possibility of multiple equilibrium points in an economy.
http://en.wikibooks.org/wiki/Transportation_Economics/Positive_externalities

An example of this is the history of the VCR industry. The market started out with two competing formats, VHS and Beta, selling at about the same price. (It appears, in hindsight, that Beta was technically superior.) In the beginning there were increasing returns for each format as their market shares increased. For example, a large number of VHS recorders in the hands of consumers motivated the vendors to stock more pre-recorded tapes in the VHS format. This encouraged more people to buy VHS recorders. The same law of increasing returns operated for the Beta format also. In the beginning there were *fluctuations* in the fortunes of the two competing brands, attributable to factors such as external circumstances, 'luck', and corporate manoeuvring. Then, increasing returns on early gains by VHS (reduced production costs per unit on increased volumes of production) tilted the game in favour of VHS, driving the other technology out of the market. This is something which could not have been predicted in the beginning.

The law of increasing returns can go beyond the product with which a company started (Arthur 1990): 'Not only do the costs of producing high-technology products fall as a company makes more of them, but the benefits of using them increase. Many items such as computers or telecommunications equipment work in networks that require compatibility. When one brand gains a significant market share, people have a strong incentive to buy more of the same product so as to be able to exchange information with those using it already.'

Path dependence in phase space

In a positive-feedback economy, although the individual transactions are small and essentially random events, they can accumulate by the positive (nonlinear) feedbacks. A number of characteristics or historical antecedents of positive-feedback economies can be listed:

1. In a particular industry, there is often a clustering of firms in a specific geographical location. A different location would have been better, but there is a kind of freezing of historical accidents in what has actually happened. Why? The first firm chooses a location for some logical (or even illogical or whimsical) reason. The choice of the second firm depends not only on the (real or perceived) merits of that region, but also on the fact that it is profitable to be near the first firm. There is a cascading effect because the third firm may be influenced more by the presence of the first two firms in that region, than by the absolute merit of that region;

and so on.

2. Railroad gauges are what they are at present because, once a particular choice was made (even arbitrarily), it was economical to stick to that choice everywhere in that region. There was a self-enforcement effect operating here.

3. The initial advantage possessed by a country or a multinational corporation can snowball into total dominance at the global level, until a better and / or cheaper product overcomes the monopoly. This highlights the importance of industrial research in any knowledge-based economy. Another important factor is the *timing* of release of a product.

In the language of evolution of a phase-space trajectory, what we are seeing here are random bifurcations in phase space. Once a branch of a bifurcation gets selected for further time-evolution, there is no going back; there is only a locked-in trajectory along a particular path in phase space (see Fig. 4.6 in Chapter 4). Thus the evolution of a positive-feedback economy has a strong *path dependence*. This can cause even hitherto successful economies to become locked into inferior paths of development. There is always a danger that a sound technology, with good *long-term* potential, may get rejected just because it has a long gestation period and slow initial growth. Similarly, when two new technologies compete, the one with a better *initial* acceptance by people may oust the other from the market, even when the other technology is inherently better (as shown by later events). Early superiority or 'selectional advantage' is no guarantee of long-term fitness. Arthur (1990) cites the example of how the U.S. nuclear-power programme got 'phase-locked' into the light-water-cooled reactors option, even though the high-temperature, gas-cooled, reactor designs may be inherently superior.

The bottom line is that, unlike negative-feedback economies, positive feedback economies do not head for a unique equilibrium; their phase-space trajectory is not path-independent. Like in a chaotic system (Chapter 15), even identical-looking initial conditions can lead to divergence in trajectories, simply because even small events or errors may get hugely amplified as time passes. Then long-term accurate forecasting becomes difficult, if not impossible.

Kurzweil (2005) has made an interesting point about the long-term predictability of stocks of companies dealing with information technologies. Such technologies are highly influential in just about every industry now. 'With the full realization of the GNR revolutions in a few decades, every area of human endeavour will essentially comprise information technologies and thus will directly benefit from the law of increasing returns'. [Here GNR stands for Genetics-Nanotechnology-Robotics.] This should mean that it makes sense to make long-term investments in the stocks of companies dealing with such technologies (provided the companies are managed well!).

44.3 Application of complexity-science ideas in management science

> *There may not be a predictable future but there is still a need to engage in futuring—continually constructing a future. We need ways to decide whether we are engaged in the most appropriate activities and relating in the most appropriate way to our stakeholders so to contribute to our organization's resilience and optimization.*
>
> John Vogelsang (2002)

The most popular idea from formal complexity science that has been picked up in management-science research is that of *complex adaptive systems* (CASs) (see Kelly and Allison 1999; Anderson 1999). And in complexity science a much investigated CAS is the beehive. In Section 18.1 I discussed the emergence of swarm intelligence in a such a system. In particular, I described how decisions are made collectively and democratically about the selection of a new nesting site. It turns out that the beehive can teach us a thing or two about decision-making by groups of individuals, particularly the compromise between good decisions and swift decisions. Swift decisions may be necessary at times, even at the risk of some mistakes. Seeley *et al.* (2006) have pointed out some instructive features of how the bees do it:

The first thing to note is that the foraging bees are self-organized in a way that promotes diversity of information. There is no 'leader' to snuff out dissent. The decision-making process is spread over all the members of the group in a decentralized fashion. Diverse information about all kinds of nesting sites is brought to the hive, without bias.

Secondly, the bees are autonomous, with no inclination or pressure for blindly imitating other bees. There is fair competition among the possible nesting sites. On seeing a waggle dance, a bee goes to the suggested site *to check for itself* the merits of that site. This independence of action helps prevent propagation of errors in site selection.

Thirdly, the quorum-sensing approach allows aggregation of diversity and independence of information, but only long enough to ensure a low probability of decision error.

In management science a much cited paper has been that of Dooley (1997). Hints of learning from beehives are apparently there in his work. I quote from his analysis of a CAS model of organizational change: 'A truly complex adaptive organization would appear best suited in semi-turbulent and turbulent environments where change is imminent and frequent. The art of designing such systems successfully is that convergent and divergent forces must be balanced, not in a linear, additive way, but in an organic fashion.' He came up with the following design principles for complex adaptive organizations:

'(a) create a shared purpose;
(b) cultivate inquiry, learning, experimentation, and divergent thinking;
(c) enhance external and internal interconnections via communication and technology;
(d) instil rapid feedback loops for self-reference and self-control;
(e) cultivate diversity, specialization, differentiation, and integration;
(f) create shared values and principles of action; and
(g) make explicit a few but essential structural and behavioural boundaries.'

He gave the example of the business organization VISA which 'has been developed, from scratch, using the principles of CAS. . . . VISA was developed using five principles:

'(a) it must be equitably owned by all participants (the banks),
(b) power and function must be distributive,
(c) governance must be distributive,
(d) it must be malleable but durable, and
(e) it must embrace diversity and change.

'Its transactions exceed \$650 billion annually, and yet runs with a core staff of 3000 distributed in 13 countries; its present communication system was developed in 90 days for less than \$25,000. This success story indicates: (a) CAS theory can be used effectively to design a world-class business, and (b) designing such systems anew may be easier than changing existing organizational structure and practice.'

That was the perception in 1997; it has only been strengthened since then.

44.4 Cultural evolution and complexity transitions

A vast number of our social institutions are shaped by the principles of self-organization. In recent times, the relentless evolution of the Internet, and the increased capacity for creating new information and spreading existing or new information, has propelled us into a new era of social organization. For example, the dynamics of information exchange creates new social trends organically, *without any top-down directionality* (Facebook is an example). The lessons gained from studying the emergence of such complex institutional entities from the voluntary actions of individuals acting out of self-interest can be harnessed towards achieving efficiency in corporate management, or when drafting public policy. This is one more example of how investigation of one type of complex system can provide insights into what may be happening in other complex systems.

As an aside I may mention here that the entire human civilization can be regarded as a single complex adaptive system (Bar-Yam 1997), though a hurdle to the investigation of such a system is that it is one of a kind; there is nothing similar to compare it with.

In the evolution of complex systems one can often identify the so-called '*complexity transitions*' (Bar-Yam 1997), which usually have far-reaching consequences. Evolution of life from nonlife (Maynard Smith and Szathmary 1998) was one such transition. Another one, which is still not complete, is that from monarchy or dictatorship to democracy. [In Chapter 4 I had explained that 'bifurcations in phase space' is a more general and appropriate term in many situations than 'phase transitions'. Complexity transitions are a type of bifurcations in phase space.]

In the corporate sector, and also in many other human organizations, there is currently occurring a transition away from hierarchical control. In a hierarchical organization it is presumed, or implicit, that the degree of complexity of the controlling individual is more than that of the organization. But as the complexity of the subsystems and the number of available communication channels, as also the level of interdependence, increases, such a management approach becomes untenable. The way out is increasingly turning out to be 'horizontally' interacting subsystems, rather than top-down control systems.

Our ever-increasing cultural complexity has also resulted in a highly networked global economy, another case of complexity transition from hierarchical control to networked transactions. Shown in Fig. 44.4 is a (still ongoing) complexity transition in human organizations.

As depicted in this figure, the ever-increasing complexity of human organizations has resulted in a diffuse complexity transition:

(a) A single person (king / dictator / big boss) takes all the decisions and directs the behaviour of all persons under his domain. The actions of the controlled persons are simple at both the individual and the collective level.

(b) But as the complexity of options and behaviours increases, intermediate layers of hierarchical control have to emerge. The intermediate layers filter the information reaching the top layer and, also, elaborate on the nature of the commands down the line. This can work only if the collective behaviour can be simplified in an effective manner.

(c) A complexity transition occurs when the maximum degree of complexity of an individual becomes less than, the collective complexity. Then the filtering of the information way up, and the elaboration of the directives on the way down, become ineffective, or not good enough.

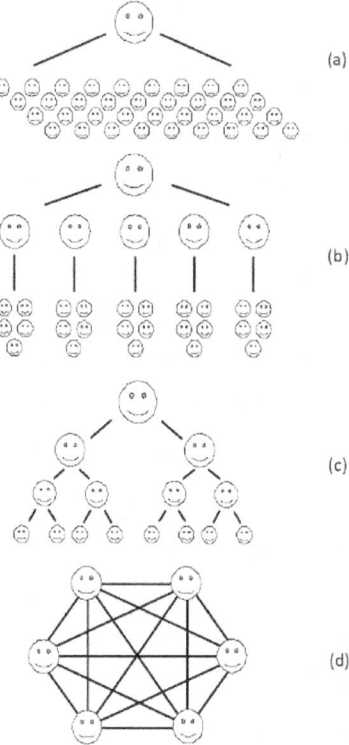

Fig. 44.4 Complexity transition in human organizations. See text for details.
From Bar-Yam (1997).

(d) Ultimately there is a *network* of individuals, in which everybody can communicate with everybody else *directly*. This results in qualitatively new emergent behaviour and characteristics. An analogy with the neural network of the human brain immediately points to the possibility of emergence of *supra-human* intelligence in the human network.

Consequences of such a complexity transition in our civilization

Prior to the transition, the complexity of the various organized structures was less than the complexity of a typical human being. After the transition the opposite is the case. There is now practically a weakening of the central control. This has consequences for the individual, as well as for the more complex environment in which the individual must function (Bar-Yam 1997).

The individual in an organization was, till recently, the most complex single organism. But now the environment is more complex than what was the most complex so far. An analogy with the rest of the animal kingdom can help us understand the response of the human individual. All other animals are less complex than the environment. They survive as species by reducing their interaction with the complex environment (e.g. by creating for themselves certain *ecological niches*), and also by reproducing excessively. The humans have also been striving for more and more *specialization*, so that they can sell their skills competently, and survive. This effort also helps them limit their exposure to the highly complex modern civilization. Specialization also helps tackle the problem of dealing with the ever-increasing mass of information and knowledge in our knowledge era.

The individual may tend to develop a sense of insecurity when exposed to the environment more complex than him / her. But the situation is mitigated by the fact that, since the entire system is one big complex system, this superorganism has the usual tendencies like the motivation to survive. This purpose is served better if the superorganism (namely the human civilization as a whole) attempts to protect and nourish its components, namely the human beings. An example is the better health services available to us, resulting in a higher life-expectancy.

44.5 Complexity leadership theory in the knowledge era

The top-down, bureaucratic, leadership models of the previous century were well-suited for *physical-production* oriented economies, but not for *knowledge-based* economies. Ideas from complexity science have made major inroads into how knowledge-oriented businesses should be run. The complexity leadership theory (CLT) formulated by Uhl-Bien, Marion and McKelvey (2007) is an example of the modern approach. Three entangled leadership roles are identified: adaptive leadership, administrative leadership, and enabling leadership (Fig. 44.5). Presuming that one is dealing with a CAS here, leadership is seen, not only as a position of authority, but also as an *emergent, interactive dynamic*, from which a collective impetus for action and change comes as an emergent phenomenon.

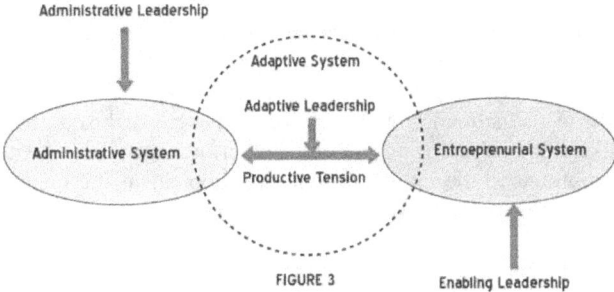

Fig. 44.5 The entangled roles of administrative leadership, enabling leadership, and adaptive leadership in CLT. From Gunawardana (2015):
http://www.dailymirror.lk/88693/complexity-leadership-theory

Rejecting the earlier approach of organizations trying to *simplify* their structures, the CAS approach advocates that organizations should *increase* their complexity to the level of the environment (the *Law of Requisite Complexity*): 'Requisite complexity enhances a system's capacity to search for solutions to challenges and to innovate because it releases the capacity of a neural network of agents in pursuit of such optimization'.

In keeping with the CAS metaphor, CLT presumes hierarchical structuring, as also differing enabling and adaptive functions across levels of hierarchy. The boundaries of the CAS are identified as open systems. And the leadership only exists in, and is a function of, the interactions. The CLT stipulates that the role of managers should not be limited to aligning worker preferences with centralized organizational goals; rather they should try to enable *emergence* of novel solutions and new contexts, even new organizational goals if necessary.

44.6 Complexity science in everyday life

Our day-to-day life is already influenced in a variety of ways by a conscious or unconscious application of complexity-science principles. A highly readable book I commend in this regard is Len Fisher's (2009) *The Perfect Swarm: The Science of Complexity in Everyday Life*. After doing a lot of statistical analysis about our behaviour in terms of the complexity metaphor, he has come up with a number of useful tips or 'rules' for deciding our course of action in real-life situations. Here is a glimpse:

1. Develop the virtues of swarm intelligence in family, community, and business environments by providing a platform that encourages people to see themselves as stakeholders rather than share-holders. You can also develop swarm intelligence in your group by using cell phones to turn yourselves into a smart mob.

2. When networking, find, use, or establish those few long-range links that bring clusters together into a small world with only a few degrees of separation.

3. If you want to give yourself the best chance of choosing the very best option in a situation that doesn't allow you to go back to the options that you have rejected, look at 37 percent of those available, then choose the next one that is better than any of them. This will give you a 1 in 3 chance of finding the best option, and a very high chance of finding one in the top few percent . . . .

4. If you want to persuade a large group of people, or even start a craze, don't rely on persuading someone with influence to pass the message on. It is far better to try for a critical mass of early adopters – people who will take the idea or product up after a single exposure.

5. Follow the example of those who have succeeded, but make sure that you have the same qualities that they do!

6. If your only clue in a situation is how others are behaving, use the quorum response, which will make your likelihood of choosing the best option increase steeply with the number of people already committed to the option. If you can supplement the quorum response with additional information, this will help your chances.

7. Avoid the perils of groupthink by escaping temporarily from the group environment, doing some independent thinking, and committing yourself to the conclusions of that thinking before returning to the group.

8. When trying to bring an issue to the notice of a group or the public as a whole, don't be a one-hit wonder; plan to bring different aspects to the fore over time.

9. Win-stay, lose-shift. In recurring situations that require cooperation, always cooperate at first, but then do in the second encounter whatever the other person did in the previous encounter (i.e., cooperate or don't cooperate).

10. When networking, choose and use hubs.

11. One final message: use such rules wisely. Remember that life is complex and that emergent patterns can't always be predicted from simple rules, even though such rules lead to them.

45. Epilogue

I think the next century will be the century of complexity
Stephen Hawking (January 2000)

The two central features of complex systems are self-organization and emergence. Self-organization occurs because of the second law of thermodynamics for open systems. This law is the mother of all organizing principles. In fact, it is central to the understanding of practically all natural processes.

The second law is an emergent law. Even the causality principle is an emergent principle.

Natural phenomena involve the evolution of complexity. Understanding the nature of this evolution has enabled us to take on even the Big Questions that have perplexed people over the millennia (Sanders 2003; Jogalekar 2013). Such questions have been largely avoided by conventional, i.e., reductionistic, science. Three Big Questions I listed in Chapter 1 were:

- How can the universe, with all its structure and organization, emerge out of nothing?
- How can life emerge out of nonlife?
- How can intelligence emerge out of nonintelligence?

Complexity science, empowered by the progress made by conventional science and by technology, is able to provide credible answers to these and many other big and small questions. In this book I have tried to give these answers in as simple a language as possible.

I am a scientist and I take pride in the fact that I value *the scientific method* (Wadhawan 2014), which is the only valid method for understanding natural phenomena. Sometimes we hear statements like the following: 'Certain questions are beyond science'. Except for some questions which have to do with ethics or morality (which are really *policy matters*, for the individual or for the state), nothing occurring in Nature can be beyond science. If not the scientific method, then what other method can we possibly have for investigating and understanding natural phenomena? None that I know of.

For understanding the origin of our universe we have to begin by reminding ourselves that all phenomena are governed by the laws of quantum mechanics. Laws of quantum mechanics are highly counter-intuitive, but there is nothing we can do about that. Laws of physics in our universe have existed long before we humans emerged on the scene. As Marvin Minsky said, our brains evolved for solving such practical problems as nutrition, defence, and reproduction. Our intelligence has made us the dominant species on Earth, but it has not been necessary for our brains to evolve to give us the intuition to such an extent, and in such a way, that the laws of quantum mechanics do not appear counter-intuitive to us. One has to understand the basics of quantum field theory to get a feel for how our universe emerged out of 'nothing' *without violating the law of conservation of energy* (with the term 'energy' including the mass part also). We are still some way from developing a widely accepted theory of quantum gravity, but the essence of the basic idea for rationalizing the birth of our universe out of 'nothing' is rather simple: We have 'something rather than nothing' because this 'quantum nothing' is something unstable, and therefore undergoing fluctuations all the time (Krauss 2012). These quantum fluctuations can occur, but only within the time limits specified by the Heisenberg uncertainty principle: Energy can emerge out of nothing provided it exists only for a time limited by the uncertainty principle. One such energy fluctuation (at the Big Bang) never got a chance to revert back to zero because it was immediately followed by *inflation*. It doubled the size of the baby universe every 10^{-34} s. This means that 100 doublings occurred within 10^{-32} s after the Big Bang. The inflation was so rapid that the quantum fluctuation just 10^{-20} the size of a proton could grow to a size of 10 cm in just 15×10^{-33} s.

The next question is: The universe has gone on expanding, and more and more mass and energy are emerging; is that not a violation of the energy-conservation principle? No. The total energy of the universe has a positive contribution and a negative contribution, and the two add up to zero. The positive part comes from all the mass and energy we have around us, and the negative part is the attractive gravitational potential energy. The negative (and increasing) gravitational energy of the ever-expanding universe is the reason why an equal amount of positive mass and energy can keep emerging out of the quantum nothing.

The Big Bang model for the origin of our universe has held sway for a long time. Recently there are strong murmurs that the model may need revisions, even drastic ones. But that does not bother me one bit. The beauty of the scientific method is that its conclusions are self-correcting. Tomorrow if there is a better model for our universe than the Big Bang model, then so be it.

Another important model or theory in cosmology is M-theory. Some of the best brains in science have been working on it, but it still requires a lot of validation. I do hope it gets confirmed. What I like best about it is that the anthropic principle emerges out of it as a natural corollary. Of course, the anthropic principle is valid even if the M-theory is not. This is because the multiverse idea can still survive, via the cosmic-inflation theory.

Apart from quantum mechanics and the principle of conservation of energy/mass, the other big idea one must get the hang of for learning the basics of complexity science is the second law of thermodynamics. The law says that for an *isolated* system the entropy always increases. People have no trouble understanding that, but the problem starts when we are dealing with thermodynamically *open* systems, rather than isolated systems, and most systems are indeed open systems. An open system is one through which mass and/or energy can flow; an isolated system is one for which this is not the case. For an open system it is meaningless to speak only in terms of entropy for stating the second law. We must bring in the concept of free energy, and the generalized second law of thermodynamics, applicable even to open systems, says that *free energy always tends to get minimized.* Nothing too profound or difficult about that; tendency towards free-energy minimization is nothing more than tendency to evolve from instability to stability. A related idea is regarding the direction of increasing time, or the *arrow of time*. Time increases in the same direction in which irreversible processes occur (Chapter 11).

Typically, the free energy of a system has two contributions, namely those from internal energy and from entropy: $F = U - TS$. So F can decrease either if the entropy term TS increases, or if the internal-energy term U decreases. Consider the example of crystal growth, say of ice from water. The ice crystal has a higher degree of order compared to liquid water, so this is not a case of increase of entropy. But this phenomenon (emergence of order out of disorder) occurs because atoms in ice crystals are more tightly bound to one another than they are in liquid water, resulting in a large fall in the U term. And the trade-off in the changes in the U term and the TS term is such that there is a net lowering of the free energy F, as demanded by the second law for open systems.

This example illustrates that order *can* emerge out of disorder if there is an appropriate flow (input/output) of energy and/or mass to or from the system. The emergence of life out of nonlife is nothing more than the culmination of a very large number of events and processes, each of them controlled by the second law for open systems.

Consider a system that is not in equilibrium. Let S_1 be its entropy. Naturally, it will tend to attain a state of equilibrium. When it has succeeded in doing so, let its entropy be S_0. Since entropy tends to increase for the system and the surroundings as a whole, we have $S_0 > S_1$. Also, for most practical purposes, what matters is the *change* in entropy, rather than its absolute value. Therefore there is no harm in shifting the entropy scale such that we associate zero entropy with the state of equilibrium; i.e., we assume that $S_0 = 0$. If we do that, we can say that, for book-keeping purposes, the entropy associated with a state in disequilibrium is *negative*.

For similar reasons, whereas entropy is a measure of absence of information, negative entropy is a measure of available information. Let us consider our ecosphere. It is an open system which continuously receives energy, mainly from the Sun. Since input of energy pushes any system away from equilibrium (a state of negative entropy), we can say that the Sun has been pumping information or negative entropy into our ecosphere. Some of this information enters the structure and function of biomolecules and other complex molecules. The structure of any biomolecule carries a very large amount of information compared to simple molecules like O_2, CO_2, N_2, etc. How and why did the biomolecules evolve out of simple molecules? They did so because, ultimately, some of the negative entropy, or information, being pumped into our ecosphere by the Sun got embedded in the structure of the biomolecules. This is how life-originating and life-sustaining molecules emerged on Earth. This is how life emerged out of nonlife (a case of 'more order' emerging locally out of 'less order' in an open system).

A game-changing idea in biology is that of Darwinian evolution. Two factors guided Darwin's formulation of the theory of evolution. One was *Malthusian ideas*: If resources are limited, the fitter individuals in a population stand a better chance of hogging more of them, and such individuals are more likely, not only to survive, but also to procreate. The other factor that influenced Darwin was *the power of gradual change*, evidenced, for example, by how even gigantic creations like the Grand Canyon can emerge simply because enough time has been available for water and winds to run their course, chiselling away one grain of rock at a time. Similarly, over adequately large time scales, biological evolution resulted in the appearance of fitter and fitter species, and even new species.

Thus, life emerged very gradually out of nonlife through chemical evolution, and evolved further because of biological evolution.

How can intelligence emerge out of nonintelligence? The beehive provides a hint for an answer. Each bee hardly has any intelligence to speak of. Its genetic information enables it to sense pheromones, and it is genetically programmed to react to the behaviour of other bees in the hive in certain simple ways. And yet the beehive as a whole is a veritable *superorganism*, able to take intelligent decisions. The key to this swarm intelligence is in the *interaction network* of the bees. This is also what happens in the human brain. Each neuron is as dumb as can be, but the complex adaptive system (CAS) comprising of billions of neurons and the trillions of interactions among them is capable of developing formidable levels of intelligence.

A grand achievement of the human mind is that, by adopting the scientific method, we have developed science and technology to a level whereby we have been able to probe the human brain to a fantastic degree of detail and accuracy. This information has enabled us to understand the mechanism of our intelligence. What is more, we are already on our way to developing artificial brains, comparable in sophistication to the human brain.

The progress in artificial intelligence (AI) research is occurring at an exponential rate, and it is certain that in the next few decades a *tipping point* will be reached when artificial intelligence will equal human intelligence. What will happen beyond that tipping point (the '*singularity*') is absolutely mind-boggling for a number of reasons:

•   The natural evolutionary processes are so slow that human intelligence can hardly change significantly over the next few decades or centuries. Moreover, our intelligence would increase only if there is some evolutionary pressure or reason for it to increase. By contrast, artificial intelligence will continue to increase at an explosively exponential rate. Therefore, beyond the tipping point the systems incorporating AI will gradually become more, much more, intelligent than us.

•   At present our pattern-recognition capability is superior to that of artificial brains. But artificial brains are bound to catch up soon.

•   Processes in artificial brains are millions of times faster than those in the human brain.

•   The human body and brain is too fragile for interstellar travel. But this is not a serious handicap for artificial brains and the robots embodying them.

•   Progress in computer science is the reason why our own intelligence will soon be enhanced by artificial intelligence, developed in a digital cortical brain.

I must make special mention here of Wolfram Alpha (http://www.wolframalpha.com/), which is an *answer engine* (rather than a *search engine* like Google Search). It *computes* answers (https://en.wikipedia.org/wiki/Wolfram_Alpha), rather than directing you to websites where the answers (or the recipes for obtaining the answers) may be available. It consists of ~15 million lines of the Mathematica code (http://www.wolfram.com/mathematica/), and computes answers from more than 10 trillion bytes of data curated by the Wolfram Research staff. The ever-increasing power of Wolfram Alpha will be available to our children and to our robots (our 'mind children') in a very routine sort of way. Look at what the scientific method, coupled with human ingenuity, has done to our lives and to our future!

In the present century itself there will be a cosmic network of immensely intelligent robots and computer-brains, communicating with one another and infusing the cosmos with a pervasive *superintelligence created by humans.*

Apart from the three Big Questions, a host of other questions about natural phenomena get answered in a satisfying way by adopting the complexity-science approach. The degree of complexity of everything taken together can be taken to be zero at the Big Bang. This big-bang event heralded the start of a system, namely the cosmos, which is perpetually far from equilibrium. Therefore there has been a dynamical evolution towards states of higher and higher information-content and complexity. The elementary particles evolved towards states of lower free energy by forming atoms, and then molecules. With further passage of time, more complex molecules emerged. As far as is known at present, there is only one planet in our universe, namely our Earth, on which chemical evolution proceeded to such an extent, and along such a route in phase space, that life forms appeared. After this happened, biological evolution became a more dominant natural process compared to chemical evolution. As Charles Darwin wrote in his *Origin of Species*:

Thus, from the war of nature, from famine and death, the most exalted object which we are capable of conceiving, namely, the production of the higher animals, directly follows. There is grandeur in this view of life, with its several powers, having been originally breathed into a few forms or into one; and that, whilst this planet has gone cycling on according to the fixed law of gravity, from so simple a beginning endless forms most beautiful and most wonderful have been, and are being, evolved.

After humans appeared on the scene, language and speech emerged, and then occurred a whole range of other cultural-evolution processes. Language, speech, writing, printing, and telecommunication etc. have resulted in the emergence of a new kind of complex adaptive system (CAS), in which we human beings are the interacting agents, interacting among ourselves and with the ecosphere of the Earth.

Just as interactions among neurons in a brain result in the emergence of intelligence, the interactions among the humans have led to the emergence of a certain superintelligence which is still evolving: It is the World Wide Web, or the Internet. Whereas the internet does embody supra-human intelligence, there is nothing supernatural about it. Its entire functioning can be understood in terms of the known laws of science.

To bring out some more aspects of complexity science, I shall now re-narrate the sequence of events regarding the evolution of complexity by taking a somewhat different route (even some repetition would be worthwhile, as it would help emphasize certain points).

As explained in the book, chemical evolution preceded biological evolution. Molecules of increasing complexity (or information content) evolved with the passage of time. In due course, metabolism and self-replication properties appeared (either together or separately), and the emergence of 'life' was simply inevitable. Life just *had* to appear in the conditions prevailing on Earth, and, after it had appeared, biological evolution did the rest. There is nothing 'miraculous' about that.

The ever-present expansion of the universe provides gradients of various types. The cosmic evolution of complexity occurs because Nature abhors gradients. The reason why the ever-present expansion of the universe is resulting in the increase of complexity is that free energy is being added to the universe as it expands.

Complex systems are open systems, and a flow of energy/information/mass through them drives them to states away from equilibrium. At appropriate conditions far from equilibrium, the system makes a critical *symmetry-breaking bifurcation* in phase space, with the attendant emergence of order. A succession of such symmetry-lowering bifurcations or 'phase transitions' can occur, leading to an increase of order, regularity, or degree of complexity.

Flow of energy (meaning both inflow and outflow of energy) through a system is essential for the emergence and increase of complexity in it. For the evolution of complexity in our ecosphere, the Sun is the main source of a steady energy influx, which serves as a steady influx of information, some of which gets stored in various chemical and other forms.

Complex systems can be divided into two categories: complex adaptive systems (CASs), and complex nonadaptive systems. CASs have the distinctive feature that they can undergo processes like learning and biological evolution (or biology-like evolution). They do not just operate in an environment created for them initially, but develop the capability to even change their environment. By contrast, galaxies and stars and other such complex objects and materials are examples of nonadaptive complex systems. They are inanimate systems which evolve with time, but within the unchanging constraints provided by the initial conditions and by the environment.

A CAS acquires information about the surroundings and about itself. It identifies regularities in that information, and condenses those regularities into a *schema* or conceptual model. It acts in the real world on the basis of that schema.

There can be many competing schemata, and the most suitable ones survive and evolve, based on the feedbacks received from the interactions with the environment.

Each agent in a CAS is constantly reacting to what other agents are doing. Therefore the environment is changing all the time.

The control of a CAS is highly dispersed. No one is really in command. Coherent behaviour or order in a CAS arises from competition and cooperation among the agents themselves.

A CAS may have many levels of self-organization. Agents at one level serve as the building blocks for agents at the next higher level.

In the light of new experience, CASs are constantly adjusting and rearranging their building blocks. This forms the basis of all learning, evolution, or adaptation in CASs.

CASs are thus characterized by *perpetual novelty*.

A CAS has an ever-changing internal model of the world, and it is always making predictions based on this model.

CASs have a certain *dynamism* not present in nonadaptive complex systems. And yet this dynamism is far from being total chaos or randomness. CASs have the ability to establish a balance between order and chaos. This balance point is referred to as *the edge of chaos*. It represents the coexistence of order and chaos.

Life signifies both stability and creativity, something that becomes possible in the vicinity of the edge of chaos. Complex systems tend to move towards the edge of chaos and order for solving complex problems.

There continues to be an ever-increasing fallout of complexity-science ideas into other fields of human activity, and I touched on some of the topics in Chapter 44. The most important notion from formal complexity science that has been lapped up by planners, administrators, managers, and strategists is that of *complex adaptive systems* (along with the *edge-of-chaos paradigm*, since CASs operate best in the edge-of-chaos regime). Here are some more examples (not discussed in Chapter 44) of how the CAS paradigm has been picked up and elaborated upon by them: strategic planning (Vogelsang 2002); exploiting the cyberspace for net-centric warfare (Phister 2010); military analysis (Green 2011); and strategy and organization research (Baumann 2015). Some of the complexity-science insights highlighted and exploited in such work are as follows:

- Exchange of information among members of the CAS must be encouraged at all levels.

- In a CAS, no one is in command, and all ideas must be judged on merit.

- The overall situation can keep changing, so the system should be ever ready to change strategy, if needed.

- The future can never be predicted with certainty, so there should be frequent real-time reviews of methods and missions. 'Misunderstandings and miscues offer variable ways of interacting   and   opportunities   to   reshape   the   assumptions and expectations that have become global patterns' (Vogelsang 2002).

• Organizations should attempt to operate and coevolve at the edge of chaos, because that is where maximum creativity, novelty and resilience are likely to occur.

• Organizations should consider cultivating multiple strategies, many of which can operate in parallel.

I am a huge fan of the great futurologist, Ray Kurzweil (http://www.kurzweilai.net/). He has made many predictions about the progress of technology, and seems to be doing a good job of it, so far. One of his predictions is that in the 2020s your life-span will increase by more than a year for every year that you live.

He has also emphasized time and again that the well-known Moore's law regarding the size and capability of IC chips is just one of the examples of how *progress occurs exponentially in all knowledge-related technologies*. He has founded a 'Singularity' University, which promulgates his philosophy that progress in biotechnology and artificial intelligence is accelerating and will rapidly, radically change what it means to be human.

The Singularity University is currently offering a week-long 'immersive course' for executives which examines how the key converging technologies will shape our future, and explores 'ethical leadership' in a world of rapid change (https://www.technologyreview.com/s/603433/for-14000-a-weeklong-firehose-of-silicon-valley-kool-aid/).

Not only executives running big businesses, but all of us, will have to prepare ourselves for the exponential speeds with which the knowledge-related technologies are advancing. In this book I have given you a feel for how Darwinian evolution occurred over the millennia. The unnerving thought is that the future of evolution is now largely in *our* hands. We, with the help of our computers and robots, are now able to achieve an exponentially increasing rate of (artificial) evolution. By comparison, Darwinian evolution is such a slow process that nothing significant will change in our brains and bodies over the next few decades. It is time to take some proactive action and formulate some carefully conceived action plans for our future. And for that to happen in a meaningful way, it is important that a large fraction of the human population understands the basics of how complex adaptive systems behave. This book has been written with that as its principal objective.

IV. Appendices

This last part of the book is mainly for those who are not uncomfortable with mathematical equations. The mathematics is not at an advanced level, but is enough to enable me to do justice to some of the very basic ideas in the physics of complex systems.

A1. Equilibrium Thermodynamics and Statistical Mechanics

Classical thermodynamics was initially formulated to understand what it takes to efficiently convert heat into mechanical work. This subject has evolved and expanded to cover the study of interconversion of all forms of energy. And some more.

A1.1 Equilibrium thermodynamics

Energy. We define energy as the capacity for doing mechanical work. The *first law of thermodynamics* states that energy is always conserved. This law also takes note of two types of energy, namely heat energy Q, and work energy W. The law says that if we supply an incremental amount of heat energy dQ to a system, then

$$dQ = dE + dW \tag{A1.1}$$

That is, a part of dQ may the used up for doing mechanical dW, and the rest is stored as internal energy dE; there is no loss or gain of energy. Thus energy can be converted from one form to another, but it cannot be created or destroyed. [Under certain conditions, interconversion of mass and energy can occur: one of them disappears and, concomitantly, an equivalent amount of the other appears.]

Entropy. Clausius made the observation (in 1865) that, for an *isolated* system, the ratio of the heat content to the absolute temperature always increases. [An isolated system is defined as one which cannot exchange either energy or matter with the surroundings.] He called this ratio *entropy*:

$$dS = dQ/T \tag{A1.2}$$

The reason why the entropy always increases can be understood as follows. Consider the isolated system as having two parts, one at temperature T_1, and the other at T_2, with $T_1 > T_2$. When heat dQ flows from the first part to the second, the change of entropy is

$$dS = dQ/T_2 - dQ/T_1 \tag{A1.3}$$

Since $T_1 > T_2$, and heat always flows from the hotter part to the colder part, we must have $dS > 0$. This is the *second law of thermodynamics* for an isolated system that is not in equilibrium. With the passage of time, $T_1 \rightarrow T_2$, and therefore $dS \rightarrow 0$. Finally, when $T_1 = T_2$ we get $dS = 0$, and *equilibrium* is said to have been reached. Now no useful work can be done by the isolated system; i.e., no heat can be converted to mechanical work. In other words, for the system to function as a *heat engine*, we must have $T_1 > T_2$.

Entropy is thus a measure of the fraction of energy unavailable for doing useful work.

If an isolated system is in thermodynamic equilibrium, it is implied that its entropy has reached the maximum value. It also means that *the entropy has a lower value when the system is away from equilibrium*. Such a system tends to increase its entropy by heading towards equilibrium.

We have defined an *isolated* system as one which can exchange neither energy nor matter with the surroundings. If a system can exchange energy, but not matter with the surroundings, it is called a *closed* system (Nicolis and Prigogine 1977). If both energy and matter can be exchanged with the surroundings, we are dealing with an *open* system. So we can have isolated systems, closed systems, or open systems.

Free energy, or thermodynamic potentials. The term '*potential energy*' is familiar to many. Imagine an object, say a stone of mass m, lying on the surface of the Earth. It is at rest, so its kinetic energy is zero. Left to itself, it will do nothing, and just stay there. Now suppose I lift it to a height h. By lifting it I am doing work against the gravitational pull of the Earth. The work I invest in the stone gets stored in it as *gravitational potential energy*, equal to mgh (force mg multiplied by distance h); here g is the acceleration due to gravity. This energy is available for doing work. For example, the stone, when released from my grip, will fall back to the Earth. And at the moment it touches the Earth, it has kinetic energy exactly equal to the potential energy it had at the height h (if we can ignore frictional losses etc.). Thus, in this example, the gravitational potential energy is the *free energy*, or the energy free or *available* for doing work. Remember, in general, not all energy is available for doing work; some part of it is trapped as heat energy (i.e., energy of chaotic motion or vibration or rotation of the molecules).

Potential energy per unit volume is just called 'potential', and a number of different *thermodynamic potentials* can be defined, relevant in different situations. For example, suppose $dQ = 0$ in the equation embodying the first law of thermodynamics (Eq. A1.1). Then $dW = -dE$. This means that the entire energy of the system (e.g., the potential energy of the stone raised to the height h) is free energy or available energy for doing mechanical work (provided there are no other dissipative or irreversible processes in operation). Processes for which $dQ = 0$, i.e., processes for which heat Q is a constant, are called *adiabatic processes*.

Thus, for adiabatic processes the internal energy E per unit volume plays the role of a thermodynamic potential.

Chemical reactions govern all life processes, as also many other processes in our universe. Since these systems tend towards ever-increasing levels of complexity, it is important to generalize the criteria for their equilibrium-seeking tendencies by incorporating parameters like pressure, volume, temperature, and composition, apart from entropy, into the definitions of appropriate thermodynamic potentials. Consider a cylinder with a piston. Suppose there are chemicals in it which undergo a reaction, resulting in the liberation of a gas. The gas does work by pushing against the atmospheric pressure p. Helmholtz introduced a quantity H, called the *enthalpy*:

$$H = E + pV \tag{A1.4}$$

Here E is the internal energy, and V is the volume of the system. The term pV represents the work done by the gas when it expands against the ambient pressure.

Gibbs generalized this further by incorporating a term (TS) which takes care of the fact that, if the reaction is exothermic or endothermic, it will exchange heat with the surroundings (assuming that we are dealing with a *closed* system, rather than an isolated system):

$$G = E + pV - TS = H - TS \tag{A1.5}$$

G is called the *Gibbs potential*, or the *Gibbs free energy* per unit volume. The system is assumed to be in contact with a '*thermal bath*', which has such a large thermal capacity that a small amount of heat exchanged by it with the system does not significantly change the temperature of the bath.

Next, suppose the system exchanges not only heat but also matter with the surroundings; i.e., it is an *open* system. Then Eq. A1.5 for the Gibbs free energy has to be further generalized as follows:

$$G = E + pV - TS + \sum_j \mu_j N_j \tag{A1.6}$$

Here N_j is the number of molecules of type j in the system. And μ_j is the *chemical potential* (chemical free energy per unit volume) for molecules of type j.

The occurrence of a chemical reaction results in a change $\Delta S_{universe}$ in the entropy of the universe:

$$\Delta S_{universe} = \Delta S_{system} + \Delta S_{bath} \tag{A1.7}$$

If the reaction is, say, endothermic, it would absorb heat ΔH_{system} from the bath, and therefore

$$\Delta S_{bath} = -\frac{\Delta H_{system}}{T} \tag{A1.8}$$

Substituting this in the previous equation, we get

$$\Delta S_{universe} = \Delta S_{system} - \frac{\Delta H_{system}}{T} \tag{A1.9}$$

Since the universe includes everything, there is nothing outside it. It is therefore an *isolated* system (rather than being just a closed system), and therefore, in accordance with the second law for an isolated system:

$$\Delta S_{universe} \geq 0 \tag{A1.10}$$

We can now obtain a generalized version of the second law as follows. We first note that Eq. A1.9 can be rewritten as

$$T\Delta S_{universe} = \Delta S_{system} - \Delta H_{system} = \Delta G_{system} \tag{A1.11}$$

Eqs. A1.9 and A1.10 can be combined to get

$$-T\Delta S_{system} + \Delta H_{system} \leq 0 \tag{A1.12}$$

Since $G = H - TS$, we can write

$$\Delta G_{system} = \Delta H_{system} - T\Delta S_{system} \tag{A1.13}$$

Here we have assumed that $\Delta T_{system} \approx 0$. Eliminating ΔH_{system} between Eqs. A1.12 and A1.13, and taking note of Eq. A1.11, we finally get

$$\Delta G_{system} = -T\Delta S_{universe} \leq 0 \qquad\qquad (A1.14)$$

Thus, for any spontaneous process occurring in a system at constant pressure and temperature, the Gibbs free energy of the system can never increase. This is the generalized statement of the second law of thermodynamics applicable to an *open* (rather than an *isolated*) system.

The *Helmholtz potential* is $H = E + pV$. If variations in the pV term are not significant, we can define the relevant thermodynamic potential as

$$F = E - TS \qquad\qquad (A1.15)$$

Eq. A1.14 then reduces to

$$\Delta F_{system} = -T\Delta S_{universe} \leq 0 \qquad\qquad (A1.16)$$

A1.2 Statistical mechanics

The above formulation of the first and the second laws of thermodynamics arose from the desire to understand the reasons that limit the amount of work one can extract from heat in a heat engine. An alternative, though equivalent, formulation was given in terms of statistical mechanics. Statistical mechanics is formulated in terms of *probabilities*, and this is what Ludwig Boltzmann did in the 19th century. The purpose was to explain the observed *macrostates* of a system in terms of its underlying *microstates*. In this formulation, entropy S is a measure of lack of information, loosely described as *disorder*. As derived by Boltzmann,

$$S = k_B \ln W \qquad\qquad (A1.17)$$

Here k_B is what is now called *the Boltzmann constant*, and W is the number of microstates of the system under consideration. In this formulation for entropy, all the microstates were assumed to be equally probable (thus W has the same meaning as the symbol N_s used in Appendix A2 below on probability theory). The W here should not be confused with the same symbol used above for mechanical work.

The assumption about the equiprobability of the microstates was removed by Gibbs during the 1870s to get:

$$S = -k_B \sum_{i=1}^{W} P_i \ln P_i \qquad\qquad (A1.18)$$

Here P_i is the probability that the system is in microstate i. For the equiprobability case, we have $P_i = 1/W$ for all values of the microstate index i.

A1.3 The ergodicity hypothesis

It is impossible to have a totally isolated system. There is always at least the effect of contact with the surroundings (the 'thermal bath'). This contact introduces random thermal fluctuations. The dynamical equations determining the time evolution of a system result in the evolution of a trajectory in phase space, and the effect of the random thermal fluctuations is that the dynamical trajectories get mixed up randomly.

Consider a system characterized by N microscopic variables $\{s_i\}$, $(i = 1, 2, .. N)$. Any *observable A* must be a function of these variables. Its *expectation value* is:

$$< A >= \lim_{t \to \infty} (1/t) \int_0^t dt' \, A[s(t')] ,$$

(A1.19)

with $s = \{s_1, s_2, ... s_N\}$. We usually assume that enough time has been given to a system in equilibrium to 'visit' all its microscopic states many times. The justification given for making this *ergodicity hypothesis* is that, given long enough observation time, the thermal fluctuations can cause a thorough mixing of the dynamical trajectories, so that all microscopic states become accessible. The relative frequency of the visits to the various states is determined by the *probability distribution function* of the states, i.e. some function $P(s_1, s_2, ... s_N)$. Under the ergodicity hypothesis, the time average embodied in Eq. A1.19 can be replaced by an *ensemble average*:

$$< A >= \int ds_1 ds_2 .. ds_N \, A[s] P(s_1, s_2, .. s_N) ,$$

(A1.20)

subject to the constraint

$$\int ds_1 ds_2 .. ds_N \, P(s_1, s_2, .. s_N) = 1$$

(A1.21)

A1.4 The partition function

If we impose the constraints that the energy E must be conserved and that the free energy F must be minimized, it can be shown that the probability P for any microscopic state labelled by, say, α, is given by *the Boltzmann distribution function*:

$$P_\alpha = (1/Z) \exp(-\beta H_\alpha)$$

(A1.22)

Here Z is the *partition function*; β is inverse temperature ($\beta = 1/T$); and H_α is the Hamiltonian for state α:

$$Z = \sum_\alpha \exp(-\beta H_\alpha)$$

(A1.23)

$$E \equiv < H >= \sum_\alpha P_\alpha H_\alpha$$

(A1.24)

The partition function is a measure of the volume occupied by an ensemble in phase space (see Plischke and Bergersen 1994).

Once the partition function is known, one can calculate the various thermodynamic properties from it.

A1.5 Tsallis thermodynamics of small systems

The Boltzmann-Gibbs formulation of thermodynamics outlined above is based on the following three assumptions:

1. The microscopic interactions are short-ranged (compared to the linear size of the system). This amounts to assuming that the component states of the system are not strongly correlated. [In a strongly correlated system it may happen that, for example, the probability for the occurrence of a microstate depends strongly on the occurrence or otherwise of another microstate.]

2. The time range of the microscopic 'memory' of the system is short in comparison to the observation time. This amounts to excluding systems which are far from equilibrium.

3. The system evolves with time in a Euclidean-like spacetime.

A system with these properties is called a *thermodynamically large system* (or just a 'large system' for short). It has thermodynamic *additivity* or *extensivity*. In particular, the entropy S is an *extensive state parameter*: The concept of entropy is invoked for understanding how and why one form of energy changes (or does not change) to another. It affects the maximum energy available for doing useful work. It is also a measure of order and disorder, as expressed by the Boltzmann-Gibbs equation. So defined, entropy is an extensive state parameter; i.e., its value is proportional to the size of the system. For two independent systems, the entropy for the combined system is simply the sum of the entropies of the two individual systems.

This formulation for entropy has had several notable failures, and has therefore been generalized by Tsallis (1988, 1995a, b, 1997, 2009; also see Cartwright 2014). Tsallis highlighted the three basic assumptions (stated above) on which Boltzmann-Gibbs thermodynamics is based. Boltzmann-Gibbs thermodynamics fails whenever any of those assumptions is unjustified, i.e., whenever the system we are dealing with is not a large system.

Tsallis (1988) generalized Boltzmann-Gibbs thermodynamics by introducing two postulates, the first generalizing the definition of entropy, and the second generalizing that of internal energy. An *entropic index q* was introduced, and the generalized entropy was defined as

$$S_q = k_B \frac{1 - \sum_{i=1}^{W} P_i^q}{q-1},$$
(A1.24)

$$\sum_i P_i = 1$$
(A1.25)

This *Tsallis entropy* is still non-negative, but it is *nonextensive*. Its limiting value for $q = 1$ is the standard (extensive) entropy, defined by Eq. A1.18.

Tsallis entropy has the *pseudo-additivity* property. If A and B are two *independent* systems, i.e. if $P_{ij}^{A+B} = P_i^A P_j^B$, then

$$\frac{S_q(A+B)}{k_B} = \frac{S_q(A)}{k_B} + \frac{S_q(B)}{k_B} + (1-q)\frac{S_q(A)}{k_B}\frac{S_q(B)}{k_B}$$
(A1.26)

Thus $(1-q)$ is a measure of the *nonextensivity* of a system. And the entropy is greater than the sum of two or more entropies for $q < 1$, and less than the sum for $q > 1$. We say that the system is *extensive* for $q = 1$, *superextensive* for $q < 1$, and *subextensive* for $q > 1$.

Tsallis generalized the definition for internal energy as follows:

$$E_q \equiv \sum_{i=1}^{W} P_i^q \varepsilon_i \qquad (A1.27)$$

Here $\{\varepsilon_i\}$ is the energy spectrum of the microstates.

The canonical-ensemble equilibrium distribution is obtained by first defining the generalized partition function as

$$Z_q \equiv \sum_{i=1}^{W} [1-(1-q)\beta\varepsilon_i]^{1/(1-q)}, \qquad (A1.28)$$

where

$$\beta = 1/(k_B T) \qquad (A1.29)$$

S_q is then optimized under the constraints $\sum_i p_i = 1$ and $U_q \equiv \sum_{i=1}^{W} p_i^q \varepsilon_i$ as follows:

$$p_i = \frac{1}{Z_q} = [1-(1-q)\beta\varepsilon_i]^{1/(1-q)} \qquad (A1.30)$$

This is the generalized version of the conventional Boltzmann weight or Boltzmann factor, namely $e^{-\beta\varepsilon_i}$. [The Boltzmann factor is the probability of occupation of a microstate of energy E, relative to the probability of occupation of a state of energy 0.]

Eq. A1.30 embodies a very important result. It tells us that:

The Boltzmann factor need not be an exponential factor always. It can be a power law *as well.*

[Remember, power laws rule the world of complexity, as discussed in Section 20.2.]

Consider stagnant water. For it the entropy is extensive because the possible number W of microstates increases exponentially with the number n of water molecules ($W \sim 2^n$), and $s = k \log W$.

Next, consider a whirlpool. The molecular motions are now no longer totally random; they are correlated in such a manner that there is an overall whirlpool effect. Consequently, W no longer increases exponentially with n, and may have a power-law dependence like, say $W \sim n^2$. Correlations can now affect the probability of occurrence of some or many microstates.

Thus in nonextensive systems, the correlations among individual constituents do not decay exponentially with distance, but may rather obey a *power-law* dependence. This has important implications for the occurrence of complex behaviour in many systems. A number less than unity, raised to a power less than unity, becomes larger. For example, $0.4^{0.3} = 0.76$. Suppose a certain process or event in a system with $q < 1$ is somewhat rare, with p equal to, say, 0.4. If $q = 0.3$, the effective probability of the occurrence of that event becomes larger (0.76). A tornado or a cyclone is an example of how low-probability events can grow in intensity for nonextensive systems. Unlike the air molecules in normal conditions, the movements of air molecules in a tornado are highly *correlated*. Trillions and trillions of molecules are turning around in a correlated manner in a tornado. A vortex is a very rare (low-probability) occurrence, but when it is there, it controls everything because it is a nonextensive system.

Doubts have been expressed about the generality, or even the validity, of the Tsallis formulation (see, e.g., http://bactra.org/notebooks/tsallis.html). But the fact remains that the dynamics of an enormous number of systems has been understood with notable success in terms of the Tsallis thermodynamics of small systems. Extensive updates are available at http://tsallis.cat.cbpf.br/biblio.htm. Here is a partial listing of the experimental verifications and applications (https://en.wikipedia.org/wiki/Tsallis_entropy):

1. The distribution characterizing the motion of cold atoms in dissipative optical lattices, predicted in 2003 and observed in 2006.
2. The fluctuations of the magnetic field in the solar wind enabled the calculation of the q-triplet (or Tsallis triplet).
3. The velocity distributions in driven dissipative dusty plasma.
4. Spin glass relaxation.
5. Trapped ion interacting with a classical buffer gas.
6. High energy collisional experiments at LHC/CERN (CMS, ATLAS and ALICE detectors) and RHIC/Brookhaven (STAR and PHENIX detectors).

And here are some important theoretical results which support the Tsallis formalism (https://en.wikipedia.org/wiki/Tsallis_entropy):

1. Anomalous diffusion.
2. Uniqueness theorem.
3. Sensitivity to initial conditions and entropy production at the edge of chaos.
4. Probability sets which make the nonadditive Tsallis entropy to be extensive in   the thermodynamic sense.
5. Strongly quantum entangled systems and thermodynamics.
6. Thermostatistics of overdamped motion of interacting particles.
7. Nonlinear generalizations of the Schrodinger, Klein-Gordon and Dirac equations.

A2. Probability Theory

Behavioural psychologists have found that, when faced with incomplete information, we often base our decision on similarity with past experience. Probability does not even come into it. How can we know probabilities if we have incomplete information?

Edgar Peters (1999), *Patterns in the Dark*

A2.1 The notion of probability

Probability theory is the formal study of chance occurrences. The probability function P encodes what we know about the likelihood that a particular event (or a set of events) will happen. It is a number between 0 and 1, with $P = 0$ representing our expectation that there is no chance that the event will happen, and $P = 1$ expressing our certainty that the event would indeed happen. The intermediate values of P are either measures, or expectations, of the likelihood or probability of the occurrence of that event.

Suppose we flip an unbiased coin 10 times ($N = 10$), and it is 'heads' 7 times, and 'tails' 3 times. Heads and tails constitute the two (mutually exclusive) *classes* into which we can divide the possible occurrences of the event in this case. The *relative frequency* of the 'heads' class has been found to be 7/10 or 0.7 in this experiment, and that of the 'tails' class is 0.3; and $0.7 + 0.3 = 1$:

Relative frequency = (number of occurrences of an outcome) / (number of opportunities of occurrence)

Since we are working with an unbiased coin, we expect that a head and a tail have the same chance or probability P_i of showing up in a particular experiment i, i.e., the relative frequency is *expected* to be 0.5 for each of them. It is simply that our *sample size* of just 10 experiments (flipping of the coin for $N = 10$ times) is statistically not large enough to be representative of the true nature of the population of events. As $N \to \infty$, we expect the relative frequency for the occurrence of both heads and tails to approach the value 0.5. This *limiting relative frequency* for an event belonging to a class (namely the class 'head' *and* the class 'tail' in our example) is the probability P. In this case the probability of a head appearing or a tail appearing is a constant number: $P_i = 0.5$, $i = 1,2,...10$.

A2.2 Multivariate probabilities

The probability function we have considered above involves just one variable, say x: $x = 1$ if the throw of the coin shows up a head, and $x = 0$ if it does not show up a head. *Multivariate probabilities* are also relevant in a variety of situations. For example we may flip *two* coins at the same time, or one after another. Now there are four possible outcomes, each with a probability 0.25: (head, head), (tail, tail), (head, tail), (tail, head). I briefly mention four types of multivariate probabilities here: joint probability; marginal probability, conditional probability, and sequence probability (Williams 1997).

Joint probability. It is the probability of the joint occurrence of two or more specified events. The joint occurrence may be either simultaneous or sequential. Suppose we throw a pair of dice (A and B) a 100 times. Let x denote what appears on die A, and y what appears on die B. The joint probability in this case can be denoted by $P(x, y)$. Suppose die A shows up a 1 (this is a *given*). And suppose die B shows up a 1 only three times in 100 trials. Then the joint probability $P(1,1)$ (or rather the relative frequency) is 3/100 or 0.03.

Since $x = 1,2,3,4,5,6$ and $y = 1,2,3,4,5,6$, we can show the results of the 100 throws in the form of a 6×6 matrix. The 36 elements of this matrix are the joint probabilities for the 36 possible outcomes of the experiment. Since any outcome is bound to be one of these 36, the sum total of the joint probabilities must be unity:

$$\sum_{i=1}^{6}\sum_{j=1}^{6} P(x_i, y_j) = 1 \tag{A2.1}$$

Marginal probability. It is the probability of getting a particular variable in a joint distribution of probabilities, irrespective of the values of the other variable(s). In the above hypothetical experiment, suppose $P(1,1) = 0.03$, $P(1,2) = 0.02$, $P(1,3) = 0.01$, $P(1,4) = 0.01$, $P(1,5) = 0.02$, and $P(1,6) = 0.01$. The sum of these six probabilities is 0.10. Thus the marginal probability $P(x_1)$ that die A will show a 1, irrespective of what die B shows, is 0.10:

$$P(x_1) = \sum_{j=1}^{6} P(x_1, y_j) = 0.10 \tag{A2.2}$$

Conditional probability. It is the probability that a particular event will happen, given that one or more other events have happened already. Knowledge about the prior occurrence of these earlier events enables us to refine or reassess the estimate of the likelihood of occurrence of the event under consideration. For example, what is the probability that Mr. X weighs more than 70 kg, given that he is six feet tall? The known height of Mr. X makes it more likely that he weighs more than 70 kg.

In our experiment above we have listed the probabilities for $P(1,1)$, $P(1,2)$, .. $P(1,6)$, adding up to $P(x_1) = 0.10$. What is the conditional probability that $y = 5$, given that $x = 1$? Since $P(1,5) = 0.02$, the conditional probability is 0.02/0.10, or 0.2. Ten out of the 100 throws of the two dice showed $x = 1$, irrespective of what y was. And 2 out of these ten showed $y = 5$; thus the conditional probability that $y = 5$, given that $x = 1$, is 2/10, or 0.2.

We denote by $P(y_j | x_i)$ the conditional probability that y_j would occur, given that x_i has occurred. It is clear from the above example that

$$P(y_j \mid x_i) = \frac{P(x_i, y_j)}{P(x_i)} \tag{A2.3}$$

Sequence probability. It is a somewhat non-standard term, used by Williams (1991). It is the joint probability of certain successive events in time; it is the probability that a given sequence of events will take place. For the simple example of two events x_i and y_j considered above, the sequence probability $P(x_i, y_j)$ is given by Eq. A2.3 as

$$P(x_i, y_j) = P(x_i).P(y_j \mid x_i) \tag{A2.4}$$

In a chaotic system, as a control parameter is changed gradually, the event trajectory may undergo repeated bifurcations in phase space. Sequence probability is the probability that a particular *sequence* of events would take place; i.e., the probability for the tracing of a particular trajectory in phase space. As the above defining equation indicates, it is the product of the 'ordinary' probability $P(x_i)$ that a particular trajectory segment characterising the event x_i in phase space has been chosen by the system and the conditional probability that the next segment of the trajectory, or the next event, would be y_j. In general, there is a whole succession of event-segments. Sequence probability is the joint probability of getting particular values of a variable over successive events in time.

A2.3 Determinism and predictability

Classical microscopic laws are characterized by not only time-reversal symmetry, but also determinism. Determinism means that if the position and the momentum of a particle are known at any time $t = t_0$, and the boundary conditions are known, then the laws of classical mechanics determine the position and momentum at all instants of time, *both future and past*. The success of, for example, particle-accelerating machines demonstrates the applicability of the deterministic equations of motion to *simple* (or simplifiable) systems (in contrast to complex systems). Simple systems have the *linearity* feature: The inevitable imprecision in our knowledge of the physical parameters of such a system does not lead to runaway consequences in our predictions about the mechanics of the system.

By contrast, chaotic systems, though deterministic, are governed by strongly *nonlinear* equations of motion, and consequently we cannot predict their behaviour far into the future. Chaos is an example of the fact that determinism does not necessarily imply predictability.

A3. Information and Uncertainty

The natural dynamics of a physical system can be thought of as a computation in which a bit not only registers a 0 or 1 but acts as an instruction: 0 means 'do this' and 1 means 'do that'. The significance of a bit depends not just on its value but on how that value affects other bits over time, as part of the continued information processing that makes up the dynamical evolution of the universe.

Seth Lloyd, *Programming the Universe*

Constituents of a complex system interact with one another, as also with the surroundings. This interaction is basically *communication of information*.

Weiner (1948) developed a statistical theory of 'amount of information', in which *the unit amount of information* was defined as that transmitted by a single decision between equally probable alternatives. And Shannon (1948) provided a quantitative or numerical measure of information.

A3.1 Information theory

The term 'information' can be assigned a numerical measure by defining it as the uncertainty in the outcome of an experiment yet to be carried out (Shannon and Weaver 1949). The uncertainty may be high either because only one of a large number (N_s) of outcomes is possible, or/and because the probability of a particular outcome is inherently low.

Eq. 6.7 in Chapter 6 provides a numerical measure of information I:

$$I = c\log(1/P) = -c\log P \tag{A3.1}$$

Here P is the probability that an experiment would give a certain stipulated output, and c is an arbitrary constant, which can be chosen depending on the context. One possible choice for defining a numerical measure of information is to take $c = 1$, and take the logarithm to the base 2. Eq. A3.1 then reads

$$I = \log_2(1/P) = -\log_2 P \tag{A3.2}$$

This choice of the logarithm is relevant for the binary system of counting, in which a bit can be either 0 or 1.

Consider the example of flipping a coin. Suppose we get 'heads'. What is the information measure I of this experiment? Since $P = 1/2 = 0.5$, Eq. A3.2 gives $I = 1$.

Similarly, if there are four possible equiprobable outcomes of an experiment (i.e., $N_s = 4$; $P = 0.25$), we get $I = 2$ from Eq. A3.2 as the information gained from any particular outcome of the experiment. Similarly, $I = 3$ when $N_s = 8$.

Eq. A3.2 must be generalized to include cases which are not equiprobability cases. Instead of a constant probability P, we may have a set of probabilities P_1, P_2, .. P_{Ns}, one for each of the N_s outcomes of an experiment. This means that *the information gainable when a particular outcome is obtained will also depend on how likely or unlikely that outcome is going to be when a test experiment is carried out.* Suppose $P_1 = 0$; then the corresponding output will never occur, and its contribution to the information would be zero. Eq. A3.2 must be modified to reflect this fact. Such considerations lead to the following natural generalization:

$$I = \sum_{i=1}^{N_s} P_i \log_2(1/P_i) = -\sum_{i=1}^{N_s} P_i \log P_i , \qquad\qquad (A3.3)$$

subject to the constraint that

$$\sum_{i=1}^{N_s} P_i = 1 \qquad\qquad (A3.4)$$

That is, we have weighted each term in Eq. A3.3 by the probability of occurrence of that outcome, and summed over all possibilities of the outcome.

A further conceptual generalization can be introduced in Eqs. A3.2 and A3.3 by interpreting P_i as not just the 'ordinary' probability, but also any of the other probability types I described in Appendix A2: joint probability, marginal probability, conditional probability, sequence probability (Williams 1997).

A3.2 Shannon's formula for a numerical measure of information

> *The significance of a bit of information depends on how that information is processed. All physical systems register information, and when they evolve dynamically in time, they transform and process that information. If an electron 'here' registers a 0 and an electron 'there' registers a 1, then when the electron goes from here to there, it flips its bit.*
> Seth Lloyd, *Programming the Universe*

Eq. A3.3 is essentially the famous *Shannon formula* for information. Let us see how.

Shannon (1948) took the bit as the unit of information, and asked the question: *How many bits are needed to express any integer Ω?* Let the symbol I denote the *information function* introduced by Shannon for defining the quantity of information needed (or *missing*) to specify an arbitrary integer Ω. It was defined by him as $I = \log_2 \Omega$ bits:

$$I = \log_2 \Omega = \frac{\log_e \Omega}{\log_e 2} = 1.442695 \ln \Omega \text{ bits,} \qquad\qquad (A3.5)$$

or

$$I = K \ln \Omega, \qquad\qquad (A3.6)$$

where $K = 1.442695$ bits. The information function so defined is not an integer in general; one has to round it to the next higher integer.

To get a feel for the logic behind this formulation, let us first take $\Omega = 8 = 2^3$. Its binary representation is 1000, i.e., four bits are needed to specify it. In general, if $\Omega = 2^n$, then $n+1$

bits are needed to specify this integer which is an exact power of 2. Shannon generalized this to numbers which are not exact powers of 2. One needs n bits to express an integer $2^n > \Omega \geq 2^{n-1}$. The logarithm was introduced by Shannon for defining I in Eq. A3.5 for reasons similar to those given in Section A1.2 (Appendix A1) for defining entropy: If Ω is equal to, say, 8, we get $I = n = 3$ [also see the similar reasoning used for writing Eq. A1.16].

Shannon also addressed the question: *What is the amount of missing information before we perform an experiment the result of which we cannot predict with certainty?*

For this he first calculated the missing information, I_N, for N independent performances of the same experiment. Suppose P_i is the probability that, in a single performance, a particular result i will be obtained. For N performances of the experiment, and for large N, the fraction of times the result i is obtained will become close to P_i. But some information is still missing, because we do not know the *sequence* in which the various results will appear as we perform the experiment N times. Shannon showed that this missing information is given by the formula

$$I_N = K \ln \Omega, \tag{A3.7}$$

where

$$\Omega = \frac{N!}{n_1! n_2! n_3! ... n_i!}, \quad n_i \equiv N P_i \tag{A3.8}$$

On using Sterling's approximation ($\ln(n_i) \approx n_i(\ln n_i - 1)$), and dividing both sides of Eq. A3.7 by N, one gets Shannon's formula for the *missing information* before the performance of a single experiment:

$$I = -K \sum_i P_i \ln P_i \tag{A3.9}$$

Take the simple example of flipping a coin ($P_i = \frac{1}{2}$). For this case,

$$I = -K[\frac{1}{2}\ln(\frac{1}{2}) + \frac{1}{2}\ln(\frac{1}{2})] = 1 \text{ bit} \tag{A3.10}$$

Indeed, one bit of information was missing before the experiment was performed.

Eq. A3.9 defines what has come to be known as *Shannon information*.

A3.3 Shannon entropy and thermodynamic entropy

Shannon's ideas are also applicable to statistical thermodynamics. Shannon information is similar to thermodynamic entropy. In the Shannon formulation, information, or rather the *lack* of it (i.e., ignorance), occupies centre stage. And in Eq. A3.9, -I is a measure of '*information entropy*'. To establish the correspondence between the two formalisms, consider an ensemble of N identical systems. Let S_N denote the thermodynamic entropy of a macrostate of the system. We do not have information about the microstates comprising this macrostate. This is the *missing* information. A comparison of Eqs. A1.17 and A3.7 gives

$$I = S \tag{A3.11}$$

Thus, Shannon information (or rather Shannon *ignorance*) is the same as Boltzmann-Gibbs entropy. In information theory, one works with *Shannon entropy*, rather than Boltzmann-Gibbs entropy of thermodynamics. The two terms describe the same quantity, except that Shannon entropy has a larger domain of applicability:

$$S = I = -k_B \sum_{i=1}^{W} P_i \ln P_i \tag{A3.12}$$

Thus, W is Ω, S_N is I_N, and k_B is K. Both I and S are measures of *ignorance*, or *missing information*, or *lack of information*. When we say that entropy is a measure of disorder, what we mean is that, since a disordered state requires more parameters for specifying it, there is a larger amount of missing or unavailable information, compared to the situation for an ordered state.

By this line of reasoning, $-S$ (negative entropy, or *negentropy*) can be taken as a measure of *available* information, for a suitably shifted scale for measuring S. In real-life processes what matters mostly is the *change* of entropy, rather than its absolute value, so shifting of the point $S = 0$ on the scale is of no consequence.

We can work out the relationship between information and energy in terms of the actual value of the Boltzmann constant k_B: Since $k_B = K$, and $K = \dfrac{1}{\ln 2}$ bits, we must have

$$k_B = 1.442695 \ \text{bits} \tag{A3.13}$$

Since $k_B = 1.3806 \times 10^{-23}$ Joules/Kelvin, we get

$$1 \text{bit} = 0.95696 \times 10^{-23} \ \text{Joule/Kelvin} \tag{A3.14}$$

Thus 1 bit of information is approximately equivalent to 10^{-23} Joule/K (see Avery 2003). *This is the smallest thermodynamic entropy change one must associate with a measurement yielding one bit of information.*

A3.4 Uncertainty

Let us dwell some more on Shannon's quantification of information. His important contribution to information theory was that he established a measure of information that was independent of the method used for producing that information. In particular, in the context of *communication theory* this means that the information content of a message is independent of the technique used for sending the message.

Suppose a question Q is asked about a system, and we are seeking the correct answer. *But this can be a meaningful exercise only if we are already aware of all possible answers* (with different or same probabilities defined by our prior knowledge of the system). Let X denote the totality of prior knowledge we have about Q, so that probabilities P_i can be assigned to the various conceivable answers. If we know that some answer is impossible, we assign $P = 0$ to it. Similarly, assigning $P = 1$ for an answer means that we already know the answer, and the question Q need not be asked. In general, P has a set of values between 0 and 1. Thus available prior knowledge about the possible answers to a question can be encoded as a set of probabilities. Information acquired through measurement has the effect of causing an adjustment in this probability assignment. *The Shannon information is a measure of how much is expected to be learnt about the answer to a question when all that is known beforehand is a set of probabilities for the various possible answers.*

To emphasize the fact that the Shannon information depends both on how well we define the question Q and how correct is our prior knowledge X about the possible answers, we should represent it as, say, $I(Q \mid X)$, so that Eq. A3.9 can be rewritten as

$$I(Q \mid X) = -K \sum_i P_i \ln P_i \qquad (A3.15)$$

The summation in this equation is over all possible answers to the question Q.

Since the information obtained by making a measurement depends on prior knowledge X, the question of choosing a zero for the scale of information is also relevant. Suppose we receive new information or knowledge (for example, a message is communicated to us) which changes X to X'. Naturally, I will change to, say, I'. Shannon defined the information content of the message as

$$\Delta I = I' - I = I(Q \mid X') - I(Q \mid X) \qquad (A3.16)$$

It is a measure of the change in the knowledge of the recipient of the message. The zero for the scale of information is defined by putting $I(Q \mid X) = 0$ in the above equation. That is, with reference to the question Q asked about a system, the zero of the information level corresponds to a situation wherein no knowledge or information is available about the answer to the question Q.

Shannon's formulation was originally meant for designing better communication channels. An interesting aspect of modern communication theory is that the merit of a particular communication-channel design lies not just in how well the actual message is sent, but also in how well the channel could have sent all the other messages it might have been asked to convey (Tribus and McIrvine 1971).

Peters (1999) recognized two main types of uncertainty: vagueness, and ambiguity. *Vagueness* implies an inability to define something precisely by establishing a definite cutoff for classification purposes. *Ambiguity* means lack of sufficient information about the possible outcomes of our actions, thus making unclear the choice between two or more objects or actions. It is ambiguity (rather than vagueness) which is relevant in the context of complex systems.

A3.5 Algorithmic information theory

The basic idea of algorithmic information theory (AIT) is that the information carried by a bit string is better defined, not in terms of the length of the bit string (as in the older information theory), but by the length of the most efficient *algorithm* needed for generating that bit string (Chaitin 2001). If the given bit string is totally random, then the algorithm or program is no better (i.e., it is not more compact) than the bit string itself. But if there is some order or regularity or pattern in the bit string, then its information content is lower, because a compact program can be written which exploits or takes note of this order. In this case the bit string is said to be *compressible*. Otherwise it is incompressible.

How to find the most efficient or compact program for 'explaining' a given bit string? Chaitin (2001) introduced the symbol $H(x)$ to denote the size in bits of the smallest program that can calculate x. $H(x)$ is thus a measure of the *program-size complexity* (often referred to as

Kolmogorov complexity).

He also introduced a number omega (Ω), called *the halting probability*, to quantify the degree of logical randomness of any system. Suppose the term 'program' means the sum total of the computer program and the data to be read by the program Consider an ensemble of all such possible programs. Ω is the probability that a program chosen at random from this set will ever halt; i.e., it will complete a computation and then halt.

Since we must consider all possible programs, any succession of bits is a possible program for testing its halting behaviour. We can flip a coin repeatedly to get a random sequence of bits. We go on adding random bits, one at a time, till the sequence of bits is a program which can be tested for halting or otherwise.

Chaitin's AIT can be summarized in terms of the following three basic ideas:

1. The programs must be *self-delimiting* so that information is additive. This means that the complexity of a pair of systems cannot be greater than the sum of complexities of the two constituent parts:

$$H(x,y) \leq H(x) + H(y) + c \tag{A3.17}$$

Since the candidate programs are generated by the successive tossing of a fair coin, we have no way of predicting the result of a particular flip of the coin; each bit is randomly generated. The self-delimiting requirement for the program means that the computer should first test the program already available to it, and the next random digit should be given to it (by tossing a coin) only when it asks for it. Thus the computer decides whether or not it has enough bits which can constitute a whole program. The program must be self-delimiting for a correct definition of the halting probability Ω.

2. To understand the properties of the program-size complexity $H(x)$, one needs to look at the *algorithmic probability* $P(x)$. It is the probability that a program generated by tossing a coin repeatedly calculates x. Since programs are assumed to be self-delimiting, this probability is a well-defined quantity. What is more, $H(x)$ is essentially the logarithm of the inverse algorithmic probability:

$$H(x) = -\log_2 P(x) + O(1) \tag{A3.18}$$

3. The *relative complexity* $H(y\,|\,x)$ of y, given x, is defined as the size of the smallest program needed for calculating y if the minimum-size program for calculating x is a given. Only $H(x)$ is given directly, not x. The following decomposition theorem is then obtained:

$$H(x,y) = H(x) + H(y\,|\,x) + O(1) \tag{A3.19}$$

The theorem states that combining a program for getting x with a program for getting y from x gives the best possible program for the pair, give or take a fixed number of bits.

A4. Thermodynamics and Information

You cannot get something for nothing, not even an observation.

Dennis Gabor

A4.1 Entropy and information

To restate the overlap of the entropy/disorder/uncertainty concept in information theory and in thermodynamics, we integrate Eq. A1.2 in Appendix A1 to get

$$S'-S = \int_{X}^{X'} \frac{dQ}{T} \tag{A4.1}$$

This formulation for entropy states that, when heat is added incrementally to a system at temperature T, resulting in an increase of the information about the system from X to X', the entropy change ($S'-S$) of the system is as defined above.

The concept of entropy is relevant in a wide variety of disciplines. Consequently, although the basic definition is the same, and has been given above, the contextual interpretations can vary in emphasis (Williams 1997; Lloyd 2006).

The first thing to note is that, although the equations defining Boltzmann-Gibbs entropy (Eq. A1.16) and Shannon entropy (Eq. A4.15) are the same, the former is limited to *statistical mechanics only*, whereas the Shannon formulation is applicable to *any field of enquiry*, and *for any of the various types of probability* described in Appendix A2.

Another important thing to realize is that entropy is a statistic that characterizes an ensemble of probabilities, *and therefore changes if the probability distribution changes* (irrespective of any physics associated with the probability distribution). It has therefore been described as the *entropy of the probability distribution P_i*.

In the context of complexity science, *lack of information* is the most relevant interpretations of entropy. Entropy is a measure of missing information. And both entropy and information are defined in terms of a probability distribution. Just as entropy is high for a disordered system, it is equivalently high for equally probable events. Disorder (high entropy) implies that the system may be in any of a large number of possible microstates. Entropy is the information *required* (but not available) for specifying the random motions of atoms and molecules. It is the invisible portion of the information contained in a physical system.

Entropy is thus a measure of *uncertainty*. When the uncertainty is low, i.e., when we are quite sure about the outcome of a probing experiment, the information gained on carrying out the experiment is low. Correspondingly the entropy is also low. By contrast, the entropy is high when the uncertainty is high.

Here are some other ways to look at the meaning of entropy (Williams 1997):

1. *Freedom of choice*, i.e., a large number of microstates or possible outcomes of an experiment, is also a way of looking at entropy.

2. *Surprise content* is another interpretation of entropy. Uncertainty, surprise content, and

entropy are all proportional to one another.

3. Another interpretation of entropy is the amount of information needed to specify the state of a system to a particular *accuracy*. If this required or missing amount of information is large, the entropy is large.

Conditional entropy and joint entropy

Eq. A4.15 is very similar to the Boltzmann expression for entropy (Eq. A1.16):

$$S = -k_B \sum_{i=1}^{N_s} P_i \log P_i \tag{A4.2}$$

Here k_B is the Boltzmann constant, and P_i is the probability that a microstate of the system is in one of N_s possible states in phase space. It follows that we can define, for example, *conditional entropy* $S_{Y|X}$ by working with the conditional probability given by Eq. A2.3, rather than with ordinary probability $P(x_i)$.

Similarly, just as joint probability $P(x_i, y_j)$ is the probability (or relative frequency) that two independent events x_i and y_j will happen together, *joint entropy* S_{XY} is a measure of the average amount of information obtained (or the average amount of uncertainty reduced) by individual measurements of two independent systems X and Y. By using Eq. A2.4 as the definition of probability entering the definition of entropy in Eq. A4.2, we get

$$S_{X,Y} = S_X + S_{Y|X} \tag{A4.3}$$

Similarly, by reversing the roles of X and Y, we get

$$S_{Y,X} = S_Y + S_{X|Y} k \tag{A4.4}$$

A4.2 Kolmogorov-Sinai entropy

Information is a measure of complexity, and chaos has the largest (but finite) degree of complexity. Williams (1997) has given an insightful description of the various interpretations of entropy since the introduction of this term by Rudolf Clausius in 1865. The one best suited for identifying chaos, and for providing a quantitative measure of it, is the so-called *Kolmogorov-Sinai (K-S) entropy*. K-S entropy requires the computation of *sequence probabilities* (i.e., the probabilities for all the various routes that the system will follow over time; see Appendix A2). Moreover, it represents a *rate*, and is also a *limiting value*. It is defined as follows:

$$S_{KS} = \lim_{\varepsilon \to 0} \lim_{t \to \infty} \left[\sum_{i=1}^{N_r} -P_S \log P_S \right] / time = \lim_{\varepsilon \to 0} \lim_{t \to \infty} (S_t - S_{t-1}) \tag{A4.5}$$

Here the subscript s in P_S denotes the fact that we are now dealing with *sequence* probabilities.

Detailed computations show that, for the example of the logistic equation (see Appendix 8 on chaos theory), for which $k < 3$ corresponds to a fixed-point attractor, the K-S entropy is always zero. It turns out to be zero for the period-two attractor also. In fact, *it is zero for all nonchaotic attractors (of any period)*. Such attractors represent systems which do not evolve with time on a long-term basis. The entropy finally remains constant with time (i.e., the rate of change is zero). No new information is gained over time. By contrast, in a chaotic regime there is continuous evolution with time, *at all times*. At any future time the system can be in a totally unpredictable state, embodying a steady supply of *new* information. Therefore the K-S entropy for a chaotic system is always found to be some positive constant. For example, for $k = 3.7$ for the logistic equation, we get the K-S entropy as ~0.47-0.48 (see Williams 1997).

The K-S entropy also illustrates the difference between chaotic data and random or noisy data. By 'random' we mean that determinism, if any, is practically negligible. The K-S entropy for a random system works out to be infinite (at least for uniformly distributed data), unlike the finite positive values it has for chaotic systems.

A4.3 Mutual information and redundancy of information

One makes a distinction between *self-information* and *mutual information*, or, equivalently, between *self-entropy* and *mutual entropy*. Self-information is what we have mainly discussed so far. It pertains to the system under discussion, without reference to other systems. Mutual information, by contrast, involves at least two systems. If two systems do not interact with each other, they have no information about each other. And if they do interact, then each has some information about the other. Consider two systems X and Y which are mutually related in some way. Prior knowledge about X can help reduce the uncertainty in the knowledge of Y. The uncertainty in the knowledge of Y has two contributions: Y by itself has an uncertainty, a measure of which is the self-entropy S_Y; secondly, there is an uncertainty of Y, given that a measurement of X has been already performed. The latter uncertainty is given by the conditional entropy $S_{Y|X}$. Overall, the uncertainty S_Y about Y is lessened by an amount $S_{Y|X}$. Mutual information $I_{Y;X}$, defined as the net lack of uncertainty about Y is therefore

$$I_{Y;X} = S_Y - S_{Y|X} \tag{A4.6}$$

Substituting from Eq. A4.4,

$$I_{Y;X} = S_Y + S_X - S_{X,Y} \tag{A4.7}$$

Thus, mutual information is the sum of the two self-entropies minus the joint entropy. It can also be interpreted as the amount of information that one measurement of one variable gives about a later measurement of the same variable.

We can express the entropies in Eq. A4.7 in terms of probabilities to arrive at the following expression for mutual information, which does not refer to entropies at all (Williams 1997):

$$I_{Y;X} = \sum_{i=1}^{N_s} \sum_{j=1}^{N_s} P(x_i, y_j) \log_2 \frac{P(x_i, y_j)}{P(x_i)P(y_j)} \tag{A4.8}$$

Redundancy

In the context of chaos theory, the notion of redundancy extends the idea of mutual information or mutual entropy to more than two systems. It is the mutual information of three or more systems or, in the context of chaos, three or more *dimensions*. Redundancy R is the sum of the self-entropies minus the joint entropy. It is a measure of the average amount of information common to several systems or variables.

Consider two variables x and y which are related, and for which the relationship is known to us exactly and completely. If we measure x, we know y also. Therefore, measurements on y are unnecessary or superfluous or *redundant*. In this example, the mutual information is the maximum. The redundancy is also the maximum because all measurements of y are wholly unnecessary or redundant.

At the other extreme, x and y may have no relationship at all (to the best of our knowledge). Then the mutual information is zero. The redundancy is also zero, because there is nothing redundant about measuring y, even when we know x.

In intermediate situations, knowledge of x provides some, but not all, information about y.

For a two-system case, mutual information is equal to redundancy. But the situation can change when we are dealing with three or more systems (Weigend and Gershenfeld 1993). However, the terms 'mutual information' and 'redundancy' are often used interchangeably, for any number of dimensions. In the context of chaos, let X in Eq. A4.7 stand for the system at time step t, so we change the symbol from X to X_t. Let Y denote the *same* system at a later time $t+m$, so we change the symbol from Y to X_{t+m}. Eq. A4.7 can then be written as

$$I(X_t; X_{t+m}) = S(X_t) + S(X_{t+m}) - S(X_t, X_{t+m}) \tag{A4.9}$$

Generalizing from two dimensions to D dimensions, and equating mutual information I with redundancy R, we get

$$R(X_t; X_{t+m}; ... X_{t+(D-1)m}) = \sum_{i=1}^{N} S(X_i) - S(X_t, X_{t+m}, ... X_{t+(D-1)m}) \tag{A4.10}$$

where

$$\sum_{i=1}^{N} S(X_i) = S(X_t) + S(X_{t+m}) - S(X_t, X_{t+m}) + ... + S(X_t, X_{t+(D-1)m}) \tag{A4.11}$$

Redundancy, as defined by Eq. A4.10, is a measure of the average amount of common information among several systems or variables.

Predictability or incremental redundancy

Incremental redundancy is the change of redundancy when the dimension is changed by unity. It is denoted by $R_{D+1} - R_D$, or $R_D - R_{D-1}$. It is a measure of the average amount of information that several successive measurements of x give about the *next x*. It can also be interpreted as a measure of *predictability*. It tells us about the average amount of information that several sequential measurements of x give about the next x (Williams 1997).

A good source of additional information on the relationship between thermodynamics and information theory is a Wikipedia article at
http://en.wikipedia.org/wiki/Entropy_in_thermodynamics_and_information_theory.

A5. Systems Far from Equilibrium

*I have always been convinced that an understanding of the dynamical origin
of dissipative structures, and more generally of complexity is one of the most
fascinating conceptual problems of contemporary science.*

Prigogine (1996)

For a long time after the initial formalization of the subject of thermodynamics, only stable
systems at or near equilibrium were considered as coming under the purview of
thermodynamics. Only during the last few decades the subjects of nonequilibrium
thermodynamics and the dynamics of unstable systems (beginning with the notion of chaos)
have come into their own.

A5.1 Emergence of complexity in systems far from equilibrium

Complexity cannot change or emerge in systems at *stable* equilibrium. If the system is an
isolated one, its entropy tends to its maximum value. If it is not an isolated system but a closed
system kept at a fixed temperature (meaning that it can exchange energy, but not matter, with
the surroundings), then its free energy F ($F = E - TS$) tends to the minimum value. In either
case, fluctuations or perturbations tend to die out as the system approaches equilibrium.

For *unstable* equilibrium (see Fig. 4.3b in Chapter 4) the internal energy is at its *maximum*
value at the equilibrium point. Consequently, any perturbation takes the system further away
from equilibrium because the system can lower its free energy by doing so.

For an isolated system the entropy S increases with time till it reaches its maximum value at
equilibrium:

$$\frac{dS}{dt} \geq 0 \tag{A5.1}$$

This statement of the second law of thermodynamics is valid for both equilibrium and
nonequilibrium situations, with the equality sign holding for equilibrium.

For an *open* system, the entropy change in a time interval dt has two contributions (Prigogine,
Nicolis and Babloyantz 1972):

$$dS = d_e S + d_i S \tag{A5.2}$$

The first contribution is the influx of entropy due to exchanges of energy and matter with the
surroundings. And the second contribution is from the production of entropy internally, due to
processes such as diffusion, conduction, chemical reactions, etc. The second contribution is
always nonnegative (second law of thermodynamics):

$$d_i S \geq 0 \tag{A5.3}$$

But the first contribution to entropy *can* have a negative sign. In particular, it can be large
enough in magnitude to make the total change of entropy negative:

$$\Delta S = \int dS < 0 \qquad\qquad\qquad\qquad\qquad\qquad (A5.4)$$

If such a situation is to be maintained indefinitely, it should be possible for the system to be in a *steady state* (a steady state, by definition, is one in which the state parameters do not change with time). Thus, for $dS = 0$ we get

$$d_e S = -d_i S < 0 \qquad\qquad\qquad\qquad\qquad\qquad (A5.5)$$

Thus the internal entropy of the system can *decrease* if there is an influx of negative entropy from outside. *This influx can be effective only under nonequilibrium conditions.* If there is equilibrium, then $d_i S = d_e S = 0$.

A decrease of internal entropy implies a creation of order or coherence. Thus, nonequilibrium *can* be a source of order.

Lower entropy means lower missing information, or higher available information content. The steady influx of energy from the Sun has been making the Earth evolve towards states of higher and higher information content or complexity. A live organism has a much higher information content and order, compared to its dead (equilibrium) state. No wonder order and design emerges, even though there is no designer involved.

A5.2 Nonequilibrium classical dynamics

In classical statistical mechanics the initial state of a system having N particles is determined by the 'generalized' positions q and the 'generalized' momenta p of all the particles in it (we have used here a shorthand notation, in which q stands for the positions of *all* the particles, and p for the momenta of all the particles). The initial state of the system can be represented as a point (q_0, p_0) in phase space. The phase space here is a hyperspace defined by $6N$ coordinate axes; $3N$ for the coordinates, and $3N$ for the momenta. The dynamics of the classical system is determined by Newton's equations, or some other equivalent dynamical equations. Governed by these equations, the point (q_0, p_0) traces a well-defined *trajectory* in phase space with the passage of time (see Fig. 3.1).

Instead of dealing with a single system with a given set of initial conditions, we can follow Einstein and Gibbs and examine a collection or *ensemble* of systems, if we wish to investigate the population dynamics of physical systems which differ only in their initial conditions. Each member of the ensemble corresponds to a different point in phase space, and the entire ensemble is therefore represented by a collection or *cloud* of points (Fig. A5.1). The shape of the cloud changes as the system evolves with time.

The density of points in this cloud in phase space is represented by the *density function* $\rho(q, p, t)$. It is the probability per unit volume of phase space for finding at time t a representation point in the neighbourhood of the point (q, p).

A *trajectory* in phase space (like the one shown in Fig. 3.1) corresponds to a special case of the cloud depicted in Fig. A5.1: For this special case the density function is zero everywhere except at the initial point (q_0, p_0), and with time this point moves along a *sharp* curve. The Dirac delta function can be used for defining the density function for this case at $t = 0$:

$$\rho = \delta(q - q_0)\delta(p - p_0) \qquad\qquad\qquad\qquad\qquad (A5.6)$$

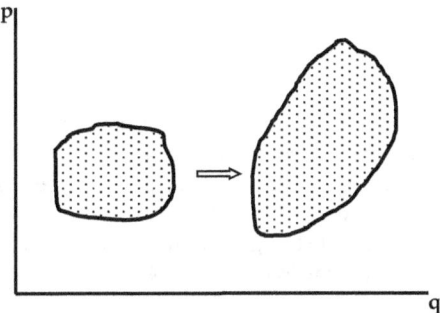

Fig. A5.1 Phase-space representation of the Gibbs ensemble, and its evolution with time. The shape of the cloud of points on the left (say at time $t = 0$) changes to the shape on the right at a later time t (Prigogine 1996).

For simple, *stable*, classical systems, the individual point of view (in terms of trajectories) and the statistical point of view (in terms of probabilities) are equivalent. [Stable dynamical systems are those for which small changes in initial conditions produce correspondingly small effects. By contrast, unstable systems are those for which small changes in initial conditions are amplified as time passes.]

For unstable systems, the equivalence to the trajectory approach is no longer there. For them, probabilities acquire a new dynamical meaning. Introduction of the density function or the probability distribution function ρ enables us to obtain a new dynamical description which can be used to predict the future evolution of the ensemble. New solutions for the density function are obtained that are *irreducible* because they do not apply to individual trajectories (Prigogine 1996). For unstable systems the initial condition is no longer a *point* in phase space, but rather some *region* described by the density function at time $t = 0$. We are thus led to a *nonlocal* description. The trajectories, though still there, are the results of stochastic processes.

Integrable *versus* nonintegrable dynamical systems

It was believed for quite some time that if the initial conditions of a system are known with infinite precision, then the evolution of its trajectory in phase space can be determined with arbitrarily high precision. At the end of the 19th century Poincaré showed that this is not really the case because of the occurrence of *resonances* among the degrees of freedom, which lead to the occurrence of powerful *instabilities* (Poincaré 1993).

Poincaré argued that a dynamical system is specified by the kinetic energies of the particles in it, plus the potential-energy term arising from the interactions among the particles. The Hamiltonian $H(p, q)$ represents this sum. Once the Hamiltonian of a system is formulated, one can derive the equations of motion, known as the *canonical equations of motion*.

A particularly simple case would be that in which the particles are free, i.e., noninteracting. The potential-energy term is then zero, and the system is fully *integrable*. The solution of the dynamical equations is a trivial matter in this case, and there is a well-defined trajectory in phase space. *Integrable systems correspond to a deterministic world described by the free Hamiltonian $H(p)$, and there are no surprises.*

Poincaré showed that real dynamical systems are largely *nonintegrable*: It is mostly not possible to choose suitable variables that would eliminate the potential-energy term. He also identified the reason for this: It is the occurrence of resonances (*Poincaré resonances*) among the degrees of freedom.

Poincaré resonances

Consider a harmonic oscillator. For small deviations from equilibrium, the oscillator moves back and forth around the equilibrium configuration with a certain *natural frequency*. Let us superimpose an external force on the oscillator. The external force has a frequency that we can vary. When the frequency of the external force bears a simple numerical ratio with the natural frequency of the oscillator, there is a strong coupling between the two, and the amplitude of oscillation increases by a large amount. We speak of a resonance in such a situation.

We can generalize from this. In fact there is no need for an external perturbation for encountering a resonance situation. Consider a system with *two* natural frequencies, ω_1 and ω_2. We have a resonance whenever two nonzero integers n_1 and n_2 exist such that

$$n_1\omega_1 + n_2\omega_2 = 0 \tag{A5.7}$$

That is, a resonance exists if the two frequencies are related by a rational fraction:

$$\omega_1 / \omega_2 = -n_2 / n_1 \tag{A5.8}$$

We speak of a Poincaré resonance in such a situation. Poincaré emphasized that such resonances in a dynamical system lead to the occurrence of small or 'dangerous' denominators, like in $1/(n_1\omega_1 + n_2\omega_2)$. The existence of any equation like Eq. A5.7 makes it very difficult, or rather impossible, to calculate a precise trajectory in phase space, *even if the initial conditions are known with infinite precision*. We speak of a *Poincaré nonintegrability* or a *Poincaré divergence* in such a situation.

Unlike integrable systems (and there are not many around), *nonintegrable systems have the potential for indeterminism and surprises, characteristic of complex systems*. According to Prigogine (1996), 'For us, Poincaré divergences are an opportunity. Indeed, . . . we can show that nonintegrability paves the way, as does chaos, for a new *statistical* formulation of the laws of dynamics'.

Poincaré resonances are ubiquitous in Nature. Here are some examples (Prigogine 1996): emission or absorption of light; the approach to equilibrium of a system of interacting particles; interactions among fields; etc. Resonances are *nonlocal* events, as they do not occur at a specific point in space or time. Therefore the Newtonian-trajectory description is inadequate for them, and one has to take recourse to a statistical or ensemble-level description.

The KAM theory

It took sixty years after the work of Poincaré for nonintegrability to be understood as a new starting point for dynamics. This was made possible by the theory of <u>K</u>olmogorov, <u>A</u>rnold, and <u>M</u>oser, or the KAM theory (see Taber 1969). The theory deals with the effect of resonances on trajectories in phase space. Here are some salient features and results of this theory:

- Since the frequencies ω depend on the coordinates and the momenta, the frequencies take on different values at different points in phase space.

- Some of the points in phase space are characterized by resonances, and others are not. For a system in the chaotic regime, this results in highly complex behaviour.

- As a result, there are two types of trajectories: 'nice' deterministic trajectories, and 'random' trajectories, associated with resonances, which shift erratically in certain regions of phase space.

- Energy plays a role similar to that of the control parameter in the logistic equation in chaos theory.

- For low values of energy, the fraction of the phase space wherein random behaviour prevails is small. Chaos sets in at a certain critical value of the energy; this is when neighbouring trajectories start diverging exponentially with time.

- When energy is large enough so that chaotic behaviour is fully developed, the cloud of points created by a trajectory results in *diffusion* in phase space.

- Diffusion implies an approach to uniformity in the *future*. Diffusion is an irreversible process which creates entropy.

The time paradox

As the KAM theory shows, starting with time-symmetric classical-dynamical equations, we end up in a breaking of time-reversal symmetry in certain dynamical situations. This is referred to as the *time paradox*. Because of it, time acquires a direction; there is an *arrow of time* distinguishing the past from the future. Poincaré resonances and the resulting nonintegrability are the basic reasons for the time paradox. The paradox persists even when we switch over to the quantum-mechanical description of Nature. Thus, in general, *the laws of dynamics must be formulated in terms of probability distributions.*

A5.3 When does the Newtonian description break down?

I state a few general results here from the work of Prigogine (1996) for answering this question:

- For *transient interactions* (e.g., when a beam of particles is scattered by an object and is able to escape from the influence of the scattering object), the diffusive terms can be neglected, and the trajectory description is quite valid.

- For *persistent interactions* on the other hand (e.g., when a beam of particles is trapped by an object), the phase-space diffusion term plays a dominant role. The appearance of diffusive terms marks the breakdown of the Newtonian, deterministic formalism.

- I discussed equilibrium thermodynamics of large systems in Appendix A1. Such systems are those which can be assumed to be in the *thermodynamic-limit* regime. In this limit, both N (the number of particles in the system) and V (the volume occupied by the system) are very large, but with the proviso that the ratio N/V is finite. The existence of a large number of particles in a system marks the breakdown of the deterministic Newtonian description, and new properties can emerge, i.e., *emergent phenomena* can arise. *Even the appearance of a new phase across a phase transition in a large system is an example of emergence.* One cannot even define the phase of a system if there are only a few particles in it. This is akin to a situation involving Poincaré resonances. We cannot take only a part of a system involving persistent interactions and consider it in isolation. This is why time-symmetry and determinism break down at the global level for a large population of interacting particles.

- Thus, irreversibility and probability considerations dominate at the macroscopic level. It is only for *transient* interactions that a Newtonian description holds.

- Thermodynamic considerations become inescapable for large, nonintegrable systems, and trajectory descriptions of dynamical problems have to be replaced by probability considerations.

A5.4 Generalization of Newtonian dynamics

I outline here a generalization of Newtonian dynamics to incorporate nonequilibrium situations, following the work of Prigogine and coworkers (Prigogine 1945, 1996). We start by asking the question: *Is a steady-state nonequilibrium system stable?* For near-equilibrium situations, also called *linear* nonequilibrium situations, the answer is 'yes'.

For dealing with deviations from a steady state, one introduces the *rate* of entropy production:

$$P = dS / dt \tag{A5.9}$$

At equilibrium this rate is zero; there is no production of entropy (Prigogine 1945). For near-equilibrium situations around the equilibrium point, P is a minimum (Fig. A5.2). Thus, fluctuations dampen and die out in this case also, just as they do for stable-equilibrium situations. But something else also happens for the near-equilibrium system. The work of Prigogine (1996) and coworkers established that *such a system may develop order (and therefore increased complexity) as a result of the irreversible processes involved.*

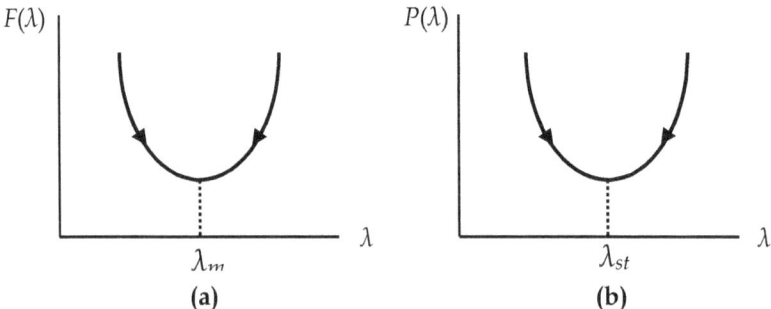

Fig. A5.2 Equilibrium (a) *versus* steady state (b) for a system kept at a fixed temperature. In (a) it is the free energy F that is minimum at equilibrium ($\lambda = \lambda_m$). In (b) it is the rate of entropy production, $P(\lambda)$, that is minimum in a steady state ($\lambda = \lambda_{st}$) (Prigogine 1945, 1996). λ is some control parameter.

The next question is: What about systems *far* from equilibrium (also called *nonlinear* nonequilibrium systems)?

Let λ denote the set of control parameters which determines the evolution of a nonlinear dynamical system. The time-evolution of such a system, described by a set of state parameters $\{\chi_i\}$, can be written as (Nicolis and Prigogine (1989)):

$$\partial \chi_i / \partial t = F_i(\{\chi_i\}, \lambda) \tag{A5.10}$$

When there are no constraints on the system, these equations must describe the state of equilibrium:

$$F_i(\{\chi_{i,eq}\}, \lambda_{eq}) = 0 \tag{A5.11}$$

For a nonequilibrium steady state, this generalizes to

$$F_i(\{\chi_{i,s}\}, \lambda_s) = 0 \tag{A5.12}$$

For a linear system, if χ is the unique state-variable, Eq. A5.10 takes the form

$$d\chi / dt = \lambda - k\chi, \tag{A5.13}$$

where k is some parameter of the system. This yields a stationary-state solution:

$$\lambda - k\chi_s = 0 \tag{A5.14}$$

A plot of χ against λ is a straight line for a linear system.

χ_s can serve as a *reference state* for examining the effects of perturbations or fluctuations, under the effect of which the stationary state gets modified to

$$\chi(t) = \chi_s + x(t), \tag{A5.15}$$

where x denotes the perturbation.

There are four possible ways in which the system may respond to the deviations created by $x(t)$:

Case 1. The simplest possibility is that, after some transient jitter or adjustment, the system comes close to the reference state χ_s.

Case 2. $\chi(t)$ may approach χ_s asymptotically as time progresses. χ_s is then said to be an *asymptotically stable* state. Asymptotic stability implies irreversibility, so we are dealing with a dissipative structure here. *Such a system can approach a unique attractor reproducibly because it can eliminate the effects of perturbations, wiping out all memories of them.* Asymptotic stability in a dissipative system is a very beneficial effect of irreversibility in Nature (Nicolis and Prigogine 1989; Prigogine 1996). By contrast, conservative systems retain a memory of the perturbations. A conservative system cannot enjoy asymptotic stability.

Case 3. In this category perturbations have a strong destabilizing effect, so that, as time passes, $\chi(t)$ does not remain near χ_s. We speak of a *point instability* of χ_s. Such a situation is possible for both conservative and dissipative systems.

Case 4. It can happen that a system is stable against small initial perturbations, but unstable against large initial perturbations. The χ_s point is then said to correspond to a system that is *locally stable* and *globally unstable*. If, on the other hand, there is stability against *any* initial value of the perturbation, we have *global stability*. In the latter case, χ_s is said to be a *global attractor.* An example is that of thermodynamic equilibrium in an isolated system.

A5.5 Pitchfork bifurcation

Consider a variable x, controlled by a parameter λ through the following rate equation:

$$dx/dt = f(x, \lambda) = -x^3 + \lambda x \qquad (A5.16)$$

The *fixed points* (steady states) are given by

$$-x_s^3 + \lambda x_s = 0 \qquad (A5.17)$$

This equation has three solutions: x_0 and $x_\pm$. Apart from the solution $x_0 = 0$, the other two are given by

$$-x_s^2 + \lambda = 0 \qquad (A5.18)$$

Only positive λ gives meaningful solutions of this equation:

$$x_\pm = \sqrt{\lambda} \qquad (A5.19)$$

The solution $x_0 = 0$ is independent of λ, and the other two solutions correspond to two distinct branches of the plot of x_s against λ. Fig. 3.5 in Chapter 3 shows a plot of the three solutions for the case wherein x has been identified with the concentration X for the chemical reaction discussed there. Thus, for $\lambda \geq 0$ the horizontal curve corresponding to the solution $x_0 = 0$ bifurcates into two branches. This is known as *pitchfork bifurcation.*

The solution $x_0 = 0$ can be identified with *the thermodynamic branch* (which is independent of the control parameter λ), and the other two solutions correspond to dissipative structures.

The control parameter λ can be viewed as playing a role akin to that of the parameter k in the logistic equation in chaos theory. Chaotic behaviour can set in at sufficient distances from equilibrium, and then even tiny fluctuations can have unpredictably large consequences.

A5.6 Extension of Newton's laws

It is necessary to incorporate instability and nonintegrability into conventional Newtonian mechanics to account for the occurrence of irreversible processes associated with entropy production. As we have seen above, nonintegrability arises because of the rampant occurrence of Poincaré resonances, and these resonances have a nonlocal character. Therefore the local-trajectory description, so typical of Newtonian mechanics, must be replaced by a statistical description. Thus *indeterminism appears in classical mechanics also*, and not only in the realm of quantum mechanics. This is to be contrasted with the fact that Newton's laws of dynamics are deterministic.

Poincaré investigated the Hamiltonian $H = H_0(p) + \lambda V(q)$, where the first term on the right-hand side is the *free* Hamiltonian, and λ is a scaling factor which is a measure of the degree of coupling of the potential energy V with H_0. Following Prigogine (1996), let us consider '*large Poincaré systems*' (LPSs): Poincaré resonances are associated with frequencies corresponding to various modes of motion, and LPSs are those in which the frequency varies *continuously* with wavelength. For such systems the volume occupied is large enough to make the surface effects insignificant; that is why they are called 'large'.

The time-evolution of the density function $\rho(q, p, t)$ can be derived from the canonical equations of motion. The evolution operator is the well-known Liouville operator L, and the evolution equations are:

$$i\partial\rho / \partial t = L\rho \qquad\qquad\qquad (A5.20)$$

An operator acting on a function transforms it to another function. An *eigenfunction* of the operator is a function which, when operated on by the operator, remains the same, except that it gets multiplied by a number called the *eigenvalue*. An operator can be expressed in terms of its eigenfunctions and eigenvalues, both defined in a specified *function space*. Of particular interest is the well-known *Hilbert space*, except that this space is not adequate for dealing with the singular functions (associated with Poincaré resonances) needed for introducing irreversibility into the statistical description of dynamics. [Hilbert space is the space of functions for which the integral of the square of the functions is well-defined and finite.]

One has to leave Hilbert space (which is adequate only for localized 'nice' functions) and work with a new space for dealing with unstable dynamical systems involving *persistent* interactions and thus requiring a nonlocal approach. For some details about this, see Prigogine (1996) and the references therein.

A6. Quantum Theory and Particle Physics

But quantum physics agrees with observation. It has never failed a test, and it has been tested more than any other theory in science.

Hawking and Mlodinow (2010)

A6.1 Introduction

All phenomena are governed by the laws of quantum mechanics. Quantum theory has been remarkably successful in explaining accurately a vast range of observations. It is also highly counterintuitive. The electron microscope provides a good real-life example of the counterintuitive behaviour of fundamental particles like electrons (for purposes of imaging, a probing beam of electrons does in an electron microscope what a light beam does in an optical microscope). The layperson may think that an electron is a particle, and not a wave. But the fact is that there is indeed a wavelength associated with a moving electron. Suppose the momentum of an electron is p. Then its wavelength (the so-called *de Broglie wavelength*) λ is given by the equation

$$\lambda = h / p \tag{A6.1}$$

Here h is a fundamental constant of Nature, called the Planck's constant.

But an electron is also a particle, with a definite 'rest mass'. This *wave-particle duality*, though counterintuitive, is an important feature of quantum theory. Similarly, a beam of light (in fact, electromagnetic radiation of any wavelength) is not just a set of waves, but also has a particle aspect. Particles (or *quanta*) of light are called *photons*. Any wave has some frequency v and some wavelength λ, and the two are related through the velocity of propagation of the wave. For the case of electromagnetic radiation the relation is $c = v\lambda$. The energy of a photon is

$$E = hv = hc / \lambda \tag{A6.2}$$

Now, a wave is not something you can specify in terms of location at a point in space (unlike a particle). A wave has *nonlocal* character, with an amplitude and a phase at *every* point in space. That leads us to another counterintuitive aspect of quantum mechanics: Since a particle has a wave aspect also, *it can be everywhere in space at the same time* (with varying probabilities, of course). Thus, in general we cannot specify the position of a particle with complete certainty. In quantum mechanics, we generally speak in terms of *probabilities*, and not certainties.

A6.2 The Heisenberg uncertainty principle

Suppose we want to determine the position and the momentum of an object. As a simple case, suppose that the object is at rest, so its momentum is zero. To observe or to image any object, we have to shine some probing radiation on it, and then analyse how the object responds to (or 'scatters') the radiation impinging on it. Suppose the object is an electron. The act of impinging on it even one quantum of the probe (e.g. a photon) will impart it some momentum, and also disturb its initial position.

Suppose we want to determine the position very accurately. That would require a probing photon of very small wavelength. But such a photon would also have more momentum and energy compared to a longer-wavelength photon (in accordance with Eq. A6.2), so the disturbance or uncertainty in the momentum will be larger. The converse is also true: If we try to reduce the uncertainty in our knowledge of the momentum by using a larger-wavelength (or smaller energy) probe, the uncertainty in the measurement of the position would become larger. This tradeoff between the uncertainty in position and the uncertainty in momentum is captured neatly by the celebrated Heisenberg uncertainty principle of quantum mechanics. Suppose the uncertainty in position is Δx, and the uncertainty in momentum in the direction of the x-axis is Δp_x. Then the principle says that the product $\Delta x \Delta p_x$ cannot be arbitrarily small; it must always be such that it is greater than a certain minimum, governed by the relation

$$\Delta x \Delta p_x \geq \hbar / 2 = \frac{h}{4\pi} \qquad\qquad\qquad (A6.3)$$

Here $\hbar = h / 2\pi$. Thus we cannot determine *both* x and p_x simultaneously with arbitrarily high precision. There is a lower limit to the product of the uncertainties of these two parameters. The uncertainties in position and momentum mean that *unpredictable quantum fluctuations* can occur in their values within the limits permitted by the Heisenberg principle.

Δx and Δp_x in Eq. A6.3 are an example of what are called *conjugate parameters*. The Heisenberg uncertainty principle holds for other pairs of conjugate parameters as well. Time and energy are another such pair, and therefore:

$$\Delta t \Delta E \geq \hbar / 2 \qquad\qquad\qquad (A6.4)$$

A6.3 The Schrödinger equation

> *Where did we get that [Schrödinger's equation] from? It's not possible to derive it from anything you know. It came out of the mind of Schrödinger.*
> The Feynman Lectures on Physics

In conventional quantum mechanics we use wave functions, ψ, to represent quantum states. And the Schrödinger equation describes how the wave function of a quantum system evolves with time:

$$i\hbar \frac{\partial}{\partial t}\psi = -\frac{\hbar^2}{2m}\nabla^2\psi + V\psi \qquad\qquad\qquad (A6.5)$$

This equation predicts a smooth and *deterministic* time-evolution of the wave function, with no discontinuities or randomness. Just as trajectories in classical mechanics describe the evolution of a system in phase space from one time step to the next, the Schrödinger equation transforms the wave function at a particular time (corresponding to a specific point in phase space) to its value $\psi(t)$ at another time t. The physical interpretation of the wave function is that of a probability: $|\psi|^2$ is a probability: It is the probability of finding the particle at a time t at a given location. It is usually denoted by the symbol ρ. The same symbol is used commonly in classical physics for the density function. In quantum mechanics it is better known as the *density matrix*:

$$\rho = |\psi|^2 \qquad\qquad\qquad (A6.6)$$

The quantum state of an elementary particle can be a superposition of two or more alternative quantum states. Suppose its energy can take two values, E_1 and E_2. Let u_1 and u_2 denote the corresponding wave functions. Then the following *linear superposition* is also a solution of the Schrödinger equation:

$$\psi = c_1 u_1 + c_2 u_2 \qquad\qquad\qquad (A6.7)$$

The interpretation is that the system exists in both the states simultaneously, with u_1^2 and u_2^2 as the respective probabilities. Thus our description moves from that of a pure state to a mixture or ensemble of states. This is sometimes referred to as *the quantum paradox*.

A6.4 The Copenhagen interpretation

Something interesting happens when we humans observe, say, an electron, with an instrument. At the moment of observation, the wave function appears to *collapse* into only one of the possible alternative states, the superposition of which was described by the wave function before the event of measurement. That is, a quantum state becomes *decoherent* when measured or monitored by the environment. This amounts to the introduction of a discontinuity in the smooth evolution of the wave function with time.

This *apparent* collapse of the wave function does not follow from the mathematics of the Schrödinger equation. It was introduced 'by hand' by Bohr and Heisenberg as an *additional postulate* or *axiom*. That is, they *chose* to introduce the interpretation that there is a collapse of the wave function to the state actually detected by the measurement in the 'real' world, to the exclusion of other states represented in the original wave function.

This dualistic interpretation of quantum mechanics for dealing with the measurement problem was suggested by Bohr and Heisenberg at a conference in Copenhagen in 1927, and is known as *the Copenhagen interpretation* (Brooks 2007; Groblacher *et al.* 2007; Folger 2009). But introducing more axioms than necessary is not a good idea in science (see Section A6.6 below).

A6.5 Time asymmetry

Another basic fact in quantum mechanics is that of time asymmetry. In classical mechanics we often make the reasonable-looking assumption that, once we have formulated the Newtonian (or equivalent) equations of motion for a system, the future states are determined by the initial conditions. In fact, in classical mechanics we can not only calculate the future conditions from the initial conditions, we can even calculate the initial conditions in the past if the present conditions or states are known. This is time *symmetry*. In quantum mechanics, the uncertainty principle destroys time symmetry. There can now be a one-to-many relationship between initial and final conditions. Two identical particles, in identical initial conditions, may not be observed to be in any particular final state(s) at a later time. Further, this time asymmetry of quantum mechanics may seem to indicate that measurements only have consequences *after* they are performed, i.e., in the future. However, it can be shown that new information obtained from measurements is also relevant for the *past* of any quantum system (see Aharonov and Tollaksen 2007).

A6.6 Multiple universes

Hugh Everett, during the mid-1950s, expressed total dissatisfaction with the Copenhagen interpretation (see Seigel 2009): 'The Copenhagen Interpretation is hopelessly incomplete because of its *a priori* reliance on classical physics … as well as a philosophic monstrosity with a "reality" concept for the macroscopic world and denial of the same for the microcosm'. Byrne (2007) has given an update of Everett's life and work. The Copenhagen interpretation implied that equations of quantum mechanics apply only to the microscopic world, and cease to be relevant in the macroscopic or 'real' world. Everett offered a different interpretation, which presaged the modern ideas of quantum decoherence. He simply let the mathematics of the quantum theory show the way for understanding logically the interface between the microscopic world and the macroscopic world. He made the observer an integral part of the system being observed, and introduced a *universal wave function* that applies comprehensively to the totality of the system being observed and the observer. This means that even macroscopic objects exist (as they should) as quantum superpositions of all allowed quantum states. There is thus no need or reason for the discontinuity of a 'wave-function collapse' when a measurement is made on the microscopic quantum system in a macroscopic world.

Everett examined the question: What would things be like if no contributing quantum states to a superposition of states are banished artificially after seeing the results of an observation? He showed that the wave function of the observer would then *bifurcate* at each interaction of the observer with the system being observed. Suppose an electron can have two possible quantum states A and B, and its wave function is a linear superposition of the two. The evolution of the composite or universal wave function describing the electron *and* the observer would then contain two branches corresponding to each of the states A and B. Each branch has a copy of the observer, one which sees state A as a result of the measurement, and the other which sees state B. In accordance with the very basic *principle of linear superposition* in quantum mechanics, the branches do not influence each other, and each embarks on a different future, or a different 'universe', independent of the other. The copy of the observer in each universe is oblivious to the existence of other copies of itself and other universes, although the 'full reality' is that each possibility has actually happened.

This reasoning can be made even more abstract and general by removing the distinction between the observer and the observed, and saying that, at each interaction among the components of the composite system, the total or universal wave function would bifurcate as described above, giving rise to *multiple universes* or *many worlds* (DeWitt and Graham 1973).

A modern and somewhat different version of this interpretation of quantum mechanics introduces the term *quantum decoherence* to rationalize how the branches become independent, and how each turns out to represent our classical or macroscopic reality.

A6.7 Feynman's sum-over-histories formulation

Richard Feynman (1967) spoke in terms of multiple or *alternative histories* of the universe (rather than multiple worlds or universes). This work has turned out to be far more important than was realised at that time (Hawking and Mlodinow 2010). Feynman, whose *path integrals* are well known in quantum mechanics, pointed out that when a quantum particle goes from a point P in phase space or probability space to a point Q, it does not have just a single unique trajectory or history; because of its quantum nature it can be everywhere, so all possible trajectories must be considered. [It should be noted that, although we normally associate the world 'history' with *past* events, history in the present context can refer to both the past or the future. 'A history is merely a narrative of a time sequence of events – past, present, or future' (Gell-Mann 1994).]

Feynman emphasized that every possible path or trajectory from P to Q is a candidate history, with an associated probability. The wave function for every such trajectory has an amplitude and a phase. The path integral for going from P to Q is obtained as the weighted vector sum or integration over all such individual paths or histories. Although an infinity of histories are considered and summed up, long-winded or zigzag histories contribute almost nothing to the vector sum. For example, a trajectory in which the particle goes one way and then goes almost the opposite way will contribute almost nothing because the phases of the two vectors will be almost exactly opposite, so they will largely cancel out. The particle is imagined to sample all possible histories *simultaneously*. When the sum over histories is carried out, the effects of all except the one actually measured for a *macroscopic* object get nearly cancelled out. For sub-microscopic particles, of course, the cancellation is far from complete, and there are indeed competing histories and interference effects.

The important point I wish to emphasize here is that this approach is applicable, not just to electrons and other particles, but also to the entire universe, except that now we cannot talk about both point A and point B in spacetime: We know only point B, namely the present. We have no information about the initial conditions represented by point A in phase space. As proposed by Stephen Hawking, *the present point B determines which alternative histories of the universe get included in the sum over histories*. A highly nontrivial statement indeed.

A6.8 Quantum Darwinism

A different resolution of the problem of interfacing the microscopic quantum description of reality with macroscopic classical reality is offered by what has been called quantum Darwinism (Blume-Kohout and Zurek 2007; also see Merali 2007). This formalism does not require the existence of a 'sentient' observer for the universe to exist (a very reasonable stipulation indeed). Instead, the environment becomes a witness. A *selective* witness at that; rather like in Darwin's theory of evolution by natural selection of the fittest. The environment determines which quantum properties are the fittest to survive (to be observed by humans). Many copies of the fitter quantum property get created in the entire environment ('redundancy'). When humans make a measurement, there is a much greater chance that they would all observe and measure the fittest solution of the Schrödinger equation, to the exclusion (or near exclusion) of other possible outcomes of the measurement experiment.

In a computer experiment, Blume-Kohout and Zurek (2007) demonstrated quantum Darwinism in zero-temperature quantum Brownian motion (QBM). A harmonic-oscillator system (S) is made to evolve in contact with a bath (ε) of harmonic oscillators. The question asked is: How much information about S can an observer extract from ε? ε consists of subenvironments ε_i: $\varepsilon = \varepsilon_1 \otimes \varepsilon_2 \otimes \varepsilon_3...$ Each observer has exclusive access to a fragment F consisting of m subenvironments. The mutual information entropy (see Section A4.3) is calculated from the quantum mutual information between S and F:

$$I_{SF} = S_S + S_F - S_{SF} \qquad (A6.8)$$

An important result obtained is that substantial redundancy appears in the model for QBM; i.e., multiple redundant records get made in the environment. As the authors state, this redundancy accounts for the objectivity and the classicality; the environment is a witness, holding many copies of the evidence. When humans make a measurement, it is most likely that they would all interact with one of the stable recorded copies, rather than directly with the actual quantum system, and thus observe and measure the classical value, to the exclusion of other possible outcomes of the measurement experiments.

A6.9 Gell-Mann's coarse-graining interpretation

To discuss this interpretation, let us first understand the difference between fine-grained and coarse-grained histories of the universe. 'Completely fine-grained histories of the universe are histories that give as complete a description as possible of the entire universe at every moment of time' (Gell-Mann 1994). Consider a simplified universe in which elementary particles have no attributes other than positions and momenta, and in which the indistinguishability among particles of a given type is ignored. Then, one kind of fine-grained history of the simplified universe would be that in which the positions of all the particles are known at all times. Unlike classical mechanics which is deterministic, quantum mechanics is probabilistic. One might think that we can write down the probability for each possible fine-grained history. But this is not so. Let us go back to Eq. A6.7 and take squares of both sides:

$$\psi^2 = c_1{}^2 u_1{}^2 + c_2{}^2 u_2{}^2 + 2c_1 c_2 u_1 u_2 \qquad\qquad (A6.9)$$

Here we have assumed, for the sake of simplicity, that the wave function is real, rather than complex. The last term on the right-hand side is the so-called *interference term*. It turns out that the interference terms between fine-grained histories do not usually cancel out, and therefore we cannot assign probabilities to the fine-grained histories.

One has to resort to coarse-graining to be able to assign probabilities to the histories. Gell-Mann and coworkers (see Gell-Mann 1994) applied this approach to a description of the quantum-mechanical histories of the universe. It was shown that the cross term in the above equation (i.e., the interference term) gets cancelled out on coarse-graining, and what is left is the result that the probability expressed by the left-hand side of this equation is simply equal to the sum of the probabilities on the right-hand side. This means that we can work directly with wave functions, rather than having to work with wave-function amplitudes, and there is no problem regarding the interfacing of the microscopic description with the macroscopic world of measurements etc.

Gell-Mann also made the point that the term 'many worlds or universes' should be substituted by '*many alternative histories of the universe*', with the further proviso that the many histories are not 'equally real'; rather they have different probabilities of occurrence.

A6.10 Poincaré resonances and quantum theory

Measurement is an irreversible process. Therefore irreversibility must be incorporated into quantum theory for a satisfactory resolution of its apparently dualistic structure (i.e., microscopic description *vs.* macroscopic measurement). This was achieved by Prigogine and coworkers (see Prigogine 1996/1997). As discussed in Section A5.2, Poincaré resonances can result in strong instabilities. Prigogine *et al.* incorporated these resonances into a statistical description, and derived *diffusive terms* that can be obtained in quantum mechanics in terms of wave functions. Working with the density matrix ρ, rather than with wave functions, and invoking Poincaré resonances, they were able to achieve the transition from wave-function amplitudes to wave functions proper, without having to introduce any nondynamical assumptions by hand.

Poincaré resonances occur in both classical and quantum physics. Therefore, as in classical physics, one has to go beyond Hilbert space for formulating a statistical theory applicable to the quantum version of large Poincaré systems (LPSs; see Appendix A5).

In quantum mechanics there are very few systems for which the eigenvalues of the Hamiltonian operator can be worked out exactly. Therefore a perturbational approach has to be used. Let H_0 be the free part of the Hamiltonian, i.e., when the potential energy is taken to be zero, and let us assume that the eigenvalues of H_0 can be calculated exactly. That is, we know the solutions of the eigenvalue equation

$$H_0 u_n^{(0)} = E_n^{(0)} u_n^{(0)} \tag{A6.10}$$

The full Hamiltonian operator is

$$H = H_0 + \lambda V \tag{A6.11}$$

The parameter λ couples the potential-energy operator V to the free Hamiltonian operator. V is treated as a perturbation. In the standard recursive perturbational method, the eigenvalues and the eigenfunctions are expanded in powers of λ. Equations for each order of λ have to be solved, and terms like $1/(E_n^{(0)} - E_m^{(0)})$ or $1/((h/2\pi)\omega_n^{(0)} - (h/2\pi)\omega_m^{(0)})$ are encountered. This is reminiscent of Poincaré resonances because singularities arise whenever the denominator is zero. For classical systems the LPSs may possibly have a *discrete* spectral distribution (making the problem integrable), but in quantum mechanics the spectrum of an LPS is *always* continuous. Therefore we are dealing with a nonintegrable system always. An example is that of a particle of energy, say, $E_1^{(0)} = (h/2\pi)\omega_1^{(0)}$ coupled with a field. Resonances occur whenever the frequency ω_1 is equal to a frequency ω_k associated with the field.

The solution to the problem of dealing with such singularities is the same as that I mentioned in Appendix A5 for classical physics, namely to abandon the Hilbert space, and work with a suitably chosen, more general, function space. The quantum Liouville eigenvalue problem is solved in this space for LPSs. Introduction of delocalized distribution functions leads to singularities, and introduction of Poincaré resonances leads to new dynamical processes. I give a mere outline of the approach here (Prigogine 1996).

The quantum Liouville equation is

$$i(\partial\rho/\partial t) = L\rho = H\rho - \rho H \tag{A6.12}$$

Taking note of the time-asymmetry of the problem (engendered by Poincaré resonances), Prigogine (1996) wrote its solution as

$$\rho(t) = e^{-iLt}\rho(0) \tag{A6.13}$$

This solution describes a single time sequence in Liouville space (instead of two time sequences, one for the future and the other for the past): *The presence of Poincaré resonances breaks the time symmetry*. The interactions are described as a succession of events separated by free motion.

In classical physics, Poincaré resonances couple the creation and destruction of correlations, introduce diffusion in phase space, break determinism, and break time-symmetry. In the extended quantum theory something similar happens, but with one important difference. In quantum mechanics the coordinate and momentum operators q_{op} and p_{op} do not commute, a fact which forms the basis of the Heisenberg uncertainty principle. Therefore we can either have a *coordinate representation* $\psi(q)$ for the wave function, or a *momentum representation* $\psi(p)$. One writes the density matrix either in terms of $\psi(q)\psi^*(q')$ or $\psi(p)\psi^*(p')$. The equation for the time evolution of the density matrix is the quantum Liouville equation written above.

I cannot go into details here. Some salient features of the extended quantum theory are as follows:

1. Since the Liouville operator and the Hamiltonian operator do not commute, the eigenvalues of the Liouville operator are no longer just the differences between the eigenvalues of the Hamiltonian.

2. The quantum superposition principle associated with the linearity of the Schrödinger equation *is violated*.

3. And, most significantly, the eigenfunctions of the Liouville operator are not expressed in terms of the wave-function amplitudes or probability amplitudes, but in terms of probabilities themselves.

The Copenhagen interpretation was introduced artificially in quantum mechanics because measurement breaks time-symmetry. It is therefore not in conformity with the time-symmetric Schrödinger equation. It is a happy situation that the LPSs considered here already break time-symmetry, and thus blur the distinction between microscopic and macroscopic quantum physics.

A6.11 Model-dependent realism, intelligence, existence

> *The universe is comprehensible because it is governed by scientific laws;*
> *that is to say, its behaviour can be modelled.*
>
> Hawking and Mlodinow (2010)

I introduced Hawking and Mlodinow's (2010) idea of model-dependent realism in Section 2.9. For all practical purposes, reality exists only to the extent that we can model it in an elegant, simple, and successful manner, the success of our model being measured by agreement with experiment. At the present moment, that model is most 'real' which has the maximum agreement with the available experimental data. If tomorrow a different model turns out to be more successful, then our perception of reality would change accordingly.

The model-dependent-realism approach helps provide satisfactory answers to several basic questions about ourselves and about our universe (Hawking and Mlodinow 2010):

What is human intelligence?

I described Hawkins' (2004) model of human intelligence in Section 41.3. It is completely in line with the model-dependent-realism approach described above. According to his model, 'the brain uses vast amounts of memory to create *a model of the world*. Everything we know and have learnt is stored in this model. The brain uses this memory-based model to make continuous predictions of future events. It is the ability to make predictions about the future that is the crux of intelligence'.

What is the meaning of existence?

You leave this book on your study table and go out of the room. Does the book exist when you are not in the room? If you say 'no', there are problems to sort out. If the book is not there when you are not watching it, and reappears when you come back, nothing can happen to it in your absence. Now suppose somebody comes to the room in your absence and steals the book. But you will say that this is not possible because how can a book be stolen when it is not there. And yet, when you come back, the book is not there; it *has* been stolen.

I think the way out of this conundrum is to postulate a model according to which the book exists, whether you are there or not. Your new model says that the book has a 'material' existence, independent of whether somebody sees it or not. Is it a better model? The answer is a resounding 'yes' because it is the best model for explaining all the observations in a simple, efficient, elegant, and convenient manner. Model-dependent realism is all we can hope to have, and nobody should complain. What else is possible for us?

Wave-particle duality

Hawking and Mlodinow (2010) go beyond their suggestion of model-dependent realism. They also argue that the same system may well be described by *different* models under different situations, so long as the models agree in the overlapping domains of application. For example, small objects like electrons can be described by a particle model and a wave model. Ditto for photons. But when we are dealing with a macroscopic object, the particle model is more effective and convenient for describing its 'reality'.

A6.12 The principle of conservation of quantum information

As indicated by Eq. A6.6, if we want to determine the probability of finding a particle in a given volume V in phase space, we have to sum up or integrate $|\psi|^2$ over that volume. This means that if we integrate over *all* spacetime, and not only over that specific volume, the result for the probability should be unity, or 100%; the particle has to be *somewhere* in the universe. This result is attributed to Bohr, and is referred to as *the unitarity principle of quantum mechanics*, the word 'unitarity' coming from the value unity for the probability. In fact, the particle can be everywhere in the universe, with appropriate probabilities for each spacetime point. A rider here is that we have only this universe, and there are no multiple universes (in line with Bohr's stance on this issue). So the particle must be somewhere in *this* universe.

There is another, very important, implication of this result. Since the wavefunction carries all the information about the particle, it follows that the overall information about the particle cannot be destroyed. This is the all-important *principle of conservation of quantum information*. No information can be lost in any quantum-mechanical process.

There is a more esoteric aspect of this principle, involving quantum field theory. I quote from a university website for a particularly compact description (http://van.physics.illinois.edu/qa/listing.php?id=24045):

'The conservation of information is derived from quantum field theory via the quantum Liouville theorem. Quantum field theory works both forward and backward in time, so the conservation of entropy (or information) works both ways. If quantum field theory is correct (as it so far seems to be) then information, in the abstract, is neither created nor destroyed. Pure states remain pure states. A probabilistic combination of pure states keeps the same set of probabilities.

'This may sound very strange to anyone familiar with the second law of thermodynamics, which says that entropy generally increases and never decreases. How can these claims be consistent?

'The general idea is that the entropy described by the second law is the sum of the entropies of many local objects, such as the Earth, the Sun, the radiation in the nearby space, etc. These things keep getting more and more quantum-entangled. Entanglement means that only some of the possible states of one part can accompany particular states of another. Thus the number of states for the collection is not really the product of the numbers for the parts. Therefore the conserved entropy of the collection is not the sum of the entropies of the parts. The growth of the entropies of the parts is cancelled by the growing negative entanglement entropy.

'In practice, that entanglement entropy has no measurable consequences in ordinary circumstances. If half the possible states of the Earth, all looking very similar to the other half, can only be paired with half the possible states of Jupiter, which look just like the other half, we see no special consequences. The entropy described by the second law is the one we can monitor.

'There may be some more interesting consequences of large-scale entanglement for the problem of what happens at the surface of a black hole. There has been a lot of recent discussion of that. The key words for a search would be "black hole firewall". To try to understand the overall picture may also require understanding the effects of cosmic horizons, discussed in some papers by Lenny Susskind and collaborators.'

A6.13 Particle physics

Laws of nature involve four types of fundamental interactions: gravitational; electromagnetic; weak nuclear; and strong nuclear. The gravitational interaction was the first to be formulated in the form of a mathematical law of Nature through the ground-breaking work of Newton. The electromagnetic interaction was given a firm mathematical footing by Maxwell through his well-known Maxwell equations.

Before Maxwell did the consolidation work regarding the electromagnetic interaction (through his celebrated equations, which also resulted in the identification of 'displacement current'), it was Faraday who first recognized the presence of a *force field* for any interaction or 'action at a distance' between two or more entities. It is now accepted that *all* forces are transmitted through fields. Electromagnetic fields propagate through space at the speed of light. In fact, light itself is an electromagnetic wave.

Maxwell's theory of the electromagnetic interaction was a *classical* theory. So also was Einstein's theory of the gravitational interaction ('the general theory of relativity'). But we need a quantum theory for sub-atomic length scales. The nascent universe was also at a sub-atomic length scale. We therefore need *quantum field theories*.

The electromagnetic interaction was the first of the four interactions to be reformulated in a quantum version, and the result is known as *quantum electrodynamics* (QED). This was mainly the handiwork of Richard Feynman, an all-time genius of science.

In QED and other quantum field theories, the force fields are envisaged as comprising of various fundamental particles called *bosons*. And the particles among which the quantum force fields operate are called *fermions*. Electrons are an example of fermions. In QED the electromagnetic interaction between two electrons is mediated by the bosons we call *photons*. One electron emits a photon, with the accompanying recoil, and the other electrons receives it, again with due recoil. The exchange of photons between the electrons comprises the electromagnetic interaction. Similarly, the *graviton* is the postulated boson responsible for the gravitational interaction.

Feynman further stipulated that when we consider the electromagnetic interaction between two charged particles, we must consider all possible ways in which the photons can be exchanged. Even the number of photons exchanged can be more than one. In other words, the quantum requirement is that a *sum over all alternative histories* must be performed for calculating the net interaction. And there are infinitely many histories conceivable (with different probabilities, of course).

When Feynman and others were developing the theoretical framework for QED, *infinities* started appearing in the final results. For example, it emerged that the electron should have infinite mass and infinite charge, a clearly absurd result. This difficulty was overcome by a process called *renormalization*. It amounts to subtracting infinite quantities which are so defined that the positive infinities and negative infinities almost cancel out, leaving as remainder the correct values for the charge and the mass of the electron.

Admittedly, this is not an elegant way of developing a model for explaining natural phenomena. The theory does not *predict* the mass and the charge of the electron. But the saving grace is that, once the charge and the mass have been 'cooked up' correctly, the theory is able to make a large number of predictions which have excellent agreement with experiment. An example is the correct prediction of what is called *Lamb shift*, a small change in one of the energy states of the hydrogen atom.

Why are there four fundamental interactions, and not just one? Actually, there was indeed only one basic high-symmetry interaction at the very high energies prevailing at and soon after the Big Bang. As the universe expanded and cooled, it underwent symmetry-breaking phase transitions and, as a result, other interactions appeared. Therefore physicists have been trying both experimentally and theoretically to achieve a unification of the four interactions. Experimentally, it amounts to creating in the laboratory the conditions of the high energies prevailing in the early universe when the symmetry-breaking transitions occurred. The idea is to achieve the reverse (i.e., symmetry-enhancing) transitions.

The first successful theoretical attempt at unification was made in 1967 (by Abdus Salam, and independently by Steven Weinberg) for the electromagnetic interaction and the weak interaction. It so happened that attempts at building a quantum theory of the weak interaction on the lines of QED had failed because it was not possible or easy to do the renormalization to get rid of the infinities. But in the theory *unifying* the two interactions it was possible to do the renormalization in a satisfactory manner.

The quantum mechanical theory which does the renormalization for the strong force is called *quantum chromodynamics* (QCD). In this theory the proton and the neutron, as also several other particles, are conceived as made up of a more fundamental set of particles called *quarks*. Quarks come in three 'colours': red, green, and blue, along with the respective anticolours. They do not exist as free particles. Only those combinations of them can exist as free particles which do not have a net colour. For example, a colour and its anticolour cancel, giving a neutral net colour. Composite particles in which this occurs are the unstable *mesons*. Another possibility is that all three colours (one each), or all three anticolours, occur in a composite particle. The name for such a composite particle is *baryon*. Protons and neutrons are examples of baryons.

QCD has the *asymptotic freedom* feature: The quarks interact via the strong force, which behaves like a piece of stretched rubber: The force is small at shorter distances, and becomes stronger (like a taut rubber) when the distance between the interacting quarks increases.

During the 1970s, physicists formulated a number of *grand unification theories* (GUTs) which aimed to unify the electroweak interaction and the strong interaction. Most of them predicted that the proton should decay with a half-life of $\sim 10^{32}$ years. Experimental verification of proton decay has not been achieved, so the GUTs have not been an unqualified success.

In the so-called *standard model* of particle physics, the electromagnetic interaction and the weak nuclear interaction are one single interaction, namely the electroweak interaction. In addition we have the strong nuclear interaction, as also the gravitational interaction, all four interactions still to be unified into a single interaction.

The gravitational interaction has not even been cast in terms of a quantum theory yet. One reason for this is the uncertainty principle. As mentioned above in Section A6.2, the principle holds for various pairs of conjugate parameters. One such pair is the value of a field and its rate of change. The gravitational field is extremely weak, meaning that a very large uncertainty in its rate of change is readily permitted by the uncertainty principle. This has a very important consequence: *There is no such thing as empty space.* This is because a really empty space would mean that both the value of the gravitational field and its rate of change are zero. Since a large uncertainty in the rate of change is permitted by the uncertainty principle (for regions where the gravitational force is very weak), an empty space would not be able to remain empty for a significant duration. There is a minimum-energy state called the *vacuum*, but the vacuum is subject to *quantum fluctuations* permitted by the uncertainty principle: 'Virtual' particles and fields get formed and annihilated spontaneously.

The virtual pairs of particles have energies, and they add up to an infinite value. Thus there is once again the problem of infinities, but what is apparently an insurmountable problem here is that it is not possible to apply the renormalization trick. This is because there are not enough adjustable parameters in this case (unlike charge and mass in the case of the electromagnetic interaction).

The only hope can be that somehow the various infinities get cancelled out, without our having to resort to renormalization. This has been achieved by postulating the presence of *supersymmetry*. An important feature of supersymmetry is that matter particles (fermions) and force particles (bosons) have a supersymmetric relationship: They are two aspects of the same thing. This means that each matter particle, say a quark, must have a force counterpart, and each force particle, say a photon, must have a matter counterpart. This can solve the infinities problem because it turns out that the infinities associated with force particles have a positive sign and those for matter particles have a negative sign, so they can cancel each other out to a large extent. But the calculations are so complicated that it has not yet been possible to substantiate such a theory of *supergravity*.

Another problem is that it has not been possible to find the partners expected to pair with the known fundamental particles. The supersymmetry theory predicts that the partner particles should be ~1000 times as massive as the proton, perhaps even more. Such massive particles have not been observed yet.

The idea of supersymmetry actually had its origin in *string theory*. Some details were discussed in Chapter 10. As explained there, there are, in fact, five different string theories. The present tentative view is that the five different string theories, as also supergravity, are just different approximations to a more fundamental theory, each approximation being valid in different situations. The more fundamental theory is called *M-theory*, with M standing for 'master', 'miracle', or 'mystery'. Some features of M-theory were described in Chapter 10.

A7. Theory of Phase Transitions and Critical Phenomena

There is no theoretical reason to expect evolutionary lineages to increase in complexity with time, and no empirical evidence that they do so. Nevertheless, eukaryotic cells are more complex than prokaryotic ones, animals and plants are more complex than protists, and so on. This increase in complexity may have been achieved as a result of a series of major evolutionary transitions. These involved changes in the way information is stored and transmitted.

Szathmáry and Maynard Smith (1995)

Familiarity with the basics of phase transitions is necessary for a proper understanding of complex systems. This is because one of the most important results in the theory of complexity is that, by and large, complexity in a system evolves with time in such a way as to take the system close to an *edge-of-chaos* configuration. There is order on one side of the edge of chaos, and chaos on the other, and complexity thrives best in this 'phase-transition' region. The jargon used in complexity science for describing this situation draws substantially from the theory of phase transitions.

A7.1 A typical phase transition

Water exists as ice at low temperatures, as steam at high temperatures, and as liquid water at intermediate temperatures. The basic chemical species is H_2O. Its boiling point at atmospheric pressure is 100°C, and its freezing point is 0°C. We say that H_2O can exist in three *phases*: vapour, liquid, and solid. We can go from one phase of water to another by changing some control parameter; typically temperature. Suppose the temperature is above 100°C, so we have steam. As we cool the system, the steam condenses to liquid water at 100°C. That is, it makes a transition from the vapour phase to the liquid phase, so we speak of a *phase transition*. If we go on cooling the system, another phase transition occurs at 0°C, when the liquid phase changes to ice, which is a crystalline solid phase.

Why do phase transitions occur? Why does anything happen? Just invoke the second law of thermodynamics. The law says that all processes occur so as to minimize the overall free energy. At any particular temperature, that phase of H_2O (or of any other system) is favoured (is more stable) for which the free energy is the least. Suppose the temperature is, say, 30°C. At this temperature, the intensity of thermal agitation of the molecules is such that the liquid phase of water is more stable (i.e., has a lower free energy), compared to the other two competing phases, namely the vapour phase and the solid phase. As we cool the system, the degree of thermal agitation goes on decreasing, and at 0°C a different phase of water (namely ice) becomes a stronger contender for existence because the system can lower its free energy by a substantial amount by making a transition to the ice phase.

Any phase transition occurs because the system can lower its free energy by undergoing the phase transition.

A7.2 Liberal definitions of phase transitions

The term 'phase transition' in complex systems tends to be used in a more liberal manner than what it means when inanimate matter changes over to a different, lower-free-energy, phase under the action of some control parameter. For example, the transition from chaos to order at the edge of chaos is viewed as a 'phase transition' from a chaotic state to a state of self-organized partial order.

Here is an example of an even more liberal definition of a phase transition, taken from the literature on complexity (Achlioptas, D'Souza and Spencer 2009): 'A large system is said to undergo a phase transition when one or more of its properties change abruptly after a slight change in a controlling variable. Besides water turning into ice or steam, other prototypical phase transitions are the spontaneous emergence of magnetization and superconductivity in metals, the epidemic spread of disease, and the dramatic change in connectivity of networks and lattices known as percolation'.

'Bifurcations in phase space' would be a better description of the phenomena mentioned in the above paragraph, rather than 'phase transitions'.

A7.3 Instabilities can cause phase transitions

It is indeed true that any process (including a phase transition) is favoured if it results in a lowering of the free energy. But this is a sort of blanket statement, and one would like to know more about how a phase transition occurs, say, in a crystal as a function of temperature.

In a crystal the atoms interact with one another, particularly with their nearest neighbours. They are also vibrating around their mean positions ('thermal vibrations'), and we can describe the interactions of an atom with its neighbours as *couplings*. We can go further and notice that the strengths of the couplings change with temperature. At any particular temperature the crystal has a structure determined by a balance between the various attractive and repulsive couplings or links. There are weak links and there are strong links, and like the proverbial strength of a chain being determined by its weakest link, the structure may collapse (i.e., undergo a phase transition) to a different structure if one of the links breaks or weakens too much. An analogy will help understand this.

Consider a rectangular tent supported on four vertical poles, dug a little bit into the ground. As everyone knows, such a tent can collapse rather easily, unless there are ropes used for making the structure sturdy, as shown schematically in Fig. A7.1(a); typically, four ropes are used, one for each pole.

Suppose one of the ropes snaps. The tent would distort, but will be prevented from collapsing completely because enough number of ropes have been used. The mechanical tension among these remaining ropes adjusts to a new set of values, but the tent now has a *less symmetrical* shape.

Something similar happens to a crystal at a structural phase transition. The interactions among the atoms can be viewed as a set of 'force constants', and one of them becomes very small, rather like the breaking of one of the four ropes in Fig. A7.1, in the vicinity of the phase transition. As a result, the crystal structure starts to collapse a bit, but soon finds a new (though somewhat distorted) stable configuration. In other words, it makes a transition to a structure of lower symmetry.

This weak force constant can be viewed as an *instability* ('the weakest link in the chain'). We say that the phase transition was *driven* by this instability. This instability interpretation of a phase transition has generalizations in various fields. For example, an inherent instability in the management of an economy may result in a collapse of the stock market. Or a whole civilization may head towards upheaval or extinction because of some inherent instability in the way it has been organized and governed.

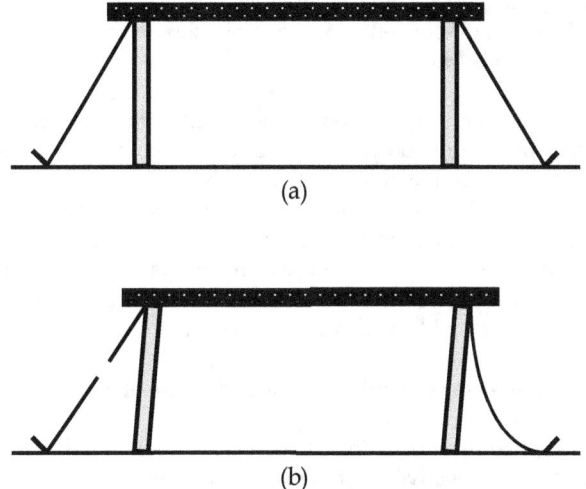

Fig. A7.1 A symmetrical tent on four legs (a) distorts to a less symmetrical shape if one of the four ropes keeping it in shape gives way (b).

A7.4 Order parameter of a phase transition

In a crystal we have positively charged ions (cations), and either negatively charged ions (anions) or free electrons, or both, and the crystal as a whole is charge-neutral, meaning that there is no excess positive or negative charge. A material is called a dielectric or an insulator if there are practically no free electrons present. An example is that of barium titanate ($BaTiO_3$).

Above 130°C, a crystal of $BaTiO_3$ has cubic symmetry. This means that in it one can identify a cube-shaped volume called the *unit cell*, which when repeated endlessly along the three edges of the cube will generate the entire crystal. The unit cell contains one molecule of $BaTiO_3$. The unit cell is, of course, only an imaginary construct. The actual bonding among the atoms cuts across the unit cells, and is shown schematically in Fig. A7.2.

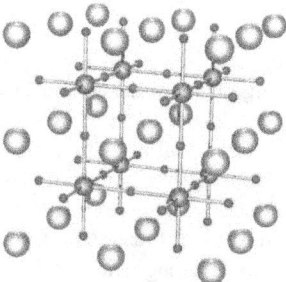

Fig. A7.2 The crystal structure of $BaTiO_3$. The smallest spheres are oxygen ions (O^{2-}), and six of them (lying on the corners of an imaginary octahedron) bond to a Ti^{4+} cation lying at the centre of the octahedron. The largest spheres are the Ba^{2+} cations. Image credit:
http://en.wikipedia.org/wiki/File:Perovskite.jpg

So this phase transition results in the emergence of a *spontaneous* polarization, and the underlying charge separation occurs along one of the edges of the cubic unit cell. Suppose we take any corner of the cube as the origin of coordinates, and the three edges that meet there as pointing along the three coordinate axes x, y, z. Then the spontaneous polarization can point along any of the six directions: $+x, -x, +y, -y, +z, -z$. There is nothing to choose one over the other. So, in the tetragonal phase, some portions of the crystal will have the polarization pointing along one of these, and some other portions will choose another of these possibilities. Each such portion is called a *domain*.

What has happened is that, as the crystal was being cooled from a temperature above 130°C, the various force constants were undergoing changes. But one of them became vanishingly small at the phase transition. We identify it as the instability that is primarily responsible for driving the transition. The result is the emergence of spontaneous polarization (usually denoted by the symbol P), pointing along any of the six equivalent directions (equivalent in the cubic phase). We can also describe the process in a different way by saying that spontaneous polarization is the *order parameter* of the phase transition. Its emergence at the phase transition lowers the symmetry of the crystal from cubic to tetragonal.

A7.5 The response function corresponding to the order parameter

If an electric field is applied to a collection of charges, the positive charges move along the direction of the field, and the negative charges move in the opposite direction, thus creating a *field-induced* dipole moment. The magnitude of this response is a measure of how efficiently and strongly this electric polarization occurs, and is called *the response function corresponding to the polarization*. It is also called the dielectric susceptibility, usually denoted by the symbol χ. In the phase transition in $BaTiO_3$, since the genesis of the emergence of the polarization is the fact that a force constant became zero, it follows that even a vanishingly small external electric field, applied along the direction of the polarization, can invoke a very large response. Thus the response function χ becomes very large in the vicinity of this *ferroelectric* phase transition.

A7.6 Phase transitions near thermodynamic equilibrium

A *phase* of a material is characterized by its atomic structure and composition. The phase a material is in depends on parameters like temperature (T), environmental pressure (p), composition (x), electric field ($\mathbf{E}$), magnetic field ($\mathbf{H}$), and directed or uniaxial mechanical stress (σ). For the sake of concreteness, let us consider here the effect of temperature alone as the control parameter.

The free energy of the system changes as we vary the control parameter. At and below a certain temperature T_c a different competing phase of the system may have a lower free energy than the phase the system is in. It will therefore make a *transition* to the new phase (see Fig. 3.3 in Chapter 3). For example, suppose we are dealing with a large crystal, and above T_c the crystal is in a paramagnetic phase, characterized by a zero magnetic moment in the absence of any applied magnetic field. Below T_c the disordering thermal fluctuations become weak enough to be overpowered by the ordering tendency of the crystal, so that there exists a nonzero average magnetization even when no external magnetic field is present. Thus a *spontaneous* magnetization emerges below T_c, and we speak of a *ferromagnetic phase transition*.

The simplest and the best-known model for understanding a ferromagnetic phase transition is the *Ising model* (see Puri and Wadhawan 2009). In it the relevant microscopic variables are the magnetic spins $\mathbf{S}$, which can take only two values: 'up' and 'down.' The long-range ferromagnetic ordering, resulting in the emergence of a macroscopic spontaneous magnetic moment, is mediated by the magnetic exchange interaction among the spins. For spins $\mathbf{S}_i$ and $\mathbf{S}_j$ on sites i and j of the crystal lattice, the spin Hamiltonian describing the interaction is written as

$$H_{ij} = -J_{ij}S_iS_j \tag{A7.1}$$

In this equation for the so-called *Ising Hamiltonian*, summation over repeated indices is implied for all the lattice sites; and only nearest-neighbour interactions are assumed to be nonzero. The spin in this example may play the role of an order parameter of a phase transition. The term 'order parameter' was first used by Landau, and I outline his basic theory of phase transitions next.

A7.7 The Landau theory of phase transitions

Landau (1937) introduced the concept of an order parameter for formulating a general theory of an important class of phase transitions in crystals, namely *continuous phase transitions*. The order parameter is a thermodynamic quantity, the emergence of which at the phase transition results in a lowering of the symmetry of the parent phase of the crystal. The order parameter, like other thermodynamic parameters, is subject to thermal fluctuations. It is envisaged as having a zero mean value, or expectation value, for temperatures above the phase-transition temperature T_c. As the temperature is lowered from values above T_c, the mean value of the order parameter becomes nonzero below T_c. A *continuous* phase transition is defined as one for which the mean value of the order parameter approaches zero continuously as T_c is approached from below, i.e., as we approach the parent phase. [We assume that the phase transition to the ordered phase occurs on cooling, rather than on heating; this is usually the case.]

In the Landau theory of phase transitions, the Gibbs free energy density, g, of the crystal is expressed as a Taylor series in powers of the order parameter, and minimized with respect to it:

$$g = g_0 + \alpha\eta + (a/2)\eta^2 + \beta\eta^3 + (c/6)\eta^6 + ... \tag{A7.2}$$

Here g_0 is the free energy per unit volume in the parent phase. The pre-factors for the various powers of the order parameter η depend on temperature, pressure and certain other parameters.

For minimizing g with respect to η, we equate its first derivative to zero:

$$\partial g / \partial \eta = \alpha + a\eta + 3\beta\eta^2 + b\eta^3 + c\eta^5 + ... = 0 \tag{A7.3}$$

In addition, the second derivative of g with respect to η must be positive (so that the free energy is minimum, rather than maximum):

$$\partial^2 g / \partial \eta^2 = a + 6\beta\eta + 3b\eta^2 + ... > 0 \tag{A7.4}$$

If Eq. A7.4 is not satisfied, the system is unstable. It is therefore a *stability condition*.

Since Eq. A7.3 must hold even when $\eta = 0$, we must have $\alpha = 0$. Eq. A7.3 can be therefore rewritten as

$$\eta(a + 3\beta\eta + b\eta^2 + ...) = 0 \tag{A7.5}$$

Thus there are two solutions. One of them, namely $\eta = 0$, corresponds to the parent (or the disordered) phase. The other, with a nonzero mean value of the order parameter, corresponds to the daughter phase or the ordered phase.

Suppose X is the field 'conjugate' to the order parameter. It is defined by

$$\partial g / \partial \eta = X \qquad \qquad (A7.6)$$

The stability condition, Eq. A7.4, can therefore be written as

$$\partial^2 g / \partial \eta^2 = \partial X / \partial \eta \equiv 1 / \chi > 0 \qquad \qquad (A7.7)$$

Here χ is a *generalized susceptibility*.

Thus, *for a phase to be stable, its inverse generalized susceptibility must be positive.*

The disordered or higher symmetry phase is characterized by $\eta = 0$. Putting this in Eq. A7.4, we get the following as the condition for the stability of the disordered or parent phase:

$$\chi^{-1} = a > 0 \qquad \qquad (A7.8)$$

The inverse susceptibility is a function of temperature. Let T_p be the temperature below which the parent phase (the disordered phase) stops being stable:

$$\chi^{-1}(T = T_p) = 0 \qquad \qquad (A7.9)$$

In the vicinity of T_p, we can write

$$a = (\partial a / \partial T)(T - T_p) \equiv a'(T - T_p) \qquad \qquad (A7.10)$$

Substituting this into Eq. A7.8 we get

$$\chi^{-1} = a'(T - T_p) \qquad \qquad (A7.11)$$

or

$$\chi = C / (T - T_p) \qquad \qquad (A7.12)$$

for $T > T_p$. This is the familiar *Curie-Weiss law* for the parent phase.

Likewise, the stability limit of the ordered daughter phase can also be worked out. Let us denote it by T_d. It turns out that this stability limit coincides with T_p if we are dealing with a *continuous* (or 'second-order') phase transition. We can then replace T_p in Eq. A7.12 by T_c, the temperature of the phase transition:

$$\chi = C / (T - T_c) \qquad \qquad (A7.13)$$

For *discontinuous* phase transitions, by contrast, the stability limits of the two phases do not coincide; i.e., $T_p \neq T_d$ (see, for example, Wadhawan (2000)). As the parent phase is cooled, its inverse susceptibility (or the inverse of the response function corresponding to the order parameter) becomes zero at T_p, and below this temperature the parent phase is unstable. Similarly, as the daughter phase is heated towards the phase-transition temperature T_c (*which is the temperature at which the free energies of the two phases become equal*), it becomes unstable above the temperature T_d. It turns out that $T_p < T_c < T_d$. This means that, for a first-order or discontinuous phase transition, *there is a range of temperatures in which the parent phase and the daughter phase can coexist.* And the order parameter, instead of decreasing continuously all the way to zero as the daughter phase is heated, drops *suddenly* to zero at the temperature T_c. Similarly, as the parent phase is cooled, the order parameter for a discontinuous phase transition suddenly becomes nonzero by a finite amount at T_c.

Another important result of the theory of phase transitions is embodied in Eq. A7.13 (or, for that matter, in Eq. A7.12). It tells us that *the response function χ corresponding to the order parameter becomes extremely large and nonlinear in the vicinity of a continuous phase transition.*

A7.8 Spontaneous breaking of symmetry

The order parameter can be visualized as a vector in a certain phase space. It can have two or more allowed orientations, each corresponding to the free-energy minimum. For simplicity, let us assume that there are only two configurations, corresponding to up and down spins (if we are dealing with a ferromagnetic phase transition). I have to skip details here, but it can be shown that the two orientations of the order parameter are separated by *an energy barrier*, which must be overcome by an external force if the crystal is to switch from one allowed configuration to the other. And the barrier between the two free-energy minima is proportional to the volume V of the crystal. Therefore, in the *thermodynamic limit*, namely $V \to \infty$ (corresponding to the extrapolation of *microscopic* results to *macroscopic* dimensions), the barrier separating the up and down spins becomes infinitely high. This is an instance of *ergodicity breaking* at and below T_c: The phase space of the system is now divided into two equal valleys, with an infinitely high hill separating them, and therefore not all the microscopic states are available for 'visiting' and ensemble-averaging; either spin-up or spin-down states are available for *observable* thermodynamics. In fact this is the reason why, in spite of the fact that the Ising Hamiltonian (Eq. A7.1) is symmetric (i.e., remains the same) with respect to a global change of the signs of all the spins, we end up with *two* ground states, one for each global sign of spin. This is *spontaneous breaking of symmetry* (Strocchi 2005).

A7.9 Field-induced phase transitions

Equation A1.6 in Appendix A1 defines the Gibbs free energy as

$$G = U + pV - TS + \sum_j \mu_j N_j \tag{A7.14}$$

In this equation, N_j is the number of molecules of type j in the system, and μ_j is the *chemical potential* (chemical free energy per unit volume) for molecules of type j. Let G_1 and G_2 denote the free energies of two competing phases, say Phase 1 and Phase 2. Since the phases are different, the variation of their respective free energy with temperature is also different. Suppose Phase 1 is more stable compared to Phase 2 at high temperatures; and Phase 2 is more stable at low temperatures (Fig. 2.3). That is, $G_1 < G_2$ at high temperatures, and $G_1 > G_2$ at low temperatures. Naturally, there is some intermediate temperature, T_c, at which $G_1 = G_2$. This is the temperature (*the phase-transition temperature*) at which Phase 1 can lower its free energy by making a transition to Phase 2 on cooling.

Eq. A7.14 for G tells us that temperature need not be the only control parameter for effecting a phase transition. We can well use pressure p instead. Also, the last term in this equation tells us we can even use *chemical concentration* as a control parameter. A phase transition can occur when the chemical concentration of a component of a chemical system reaches a critical value. So far we have presumed that the system is at some temperature T, under an isotropic pressure p. Now suppose there is also an electric field **E** present, as also a magnetic field **H**, and a directed (or anisotropic) mechanical stress **σ**. Then a more general form of Eq. A7.14 for the Gibbs potential can be written as follows:

$$G = U + pV - TS + \sum_j \mu_j N_j - \sum_j E_j D_j - \sum_j H_j B_j - \sum_{i,j} \sigma_{ij} e_{ij} \qquad (A7.15)$$

Two distinct types of control parameters can be identified in this equation: *scalars* (T, p, or μ), and directed quantities or *nonscalars* (**E**, **H**, or **σ**). Each of them influences the free-energy, and can cause a phase transition at a certain critical value. Phase transitions caused by variation of the scalars (temperature, pressure or composition) are more familiar, and better investigated. Phase transitions caused by any of the three nonscalars listed above are called *field-induced phase transitions*.

Some of the most spectacular effects of practical importance in complex materials are caused by field-induced phase transitions: colossal magnetoresistance (CMR); shape-memory effect (SME); giant photoelastic effect; etc. Relaxor ferroelectrics are a prominent example of complex materials in which electric field induces an unusually large response, including the modification of the phase-transition behaviour (Wadhawan 2007).

A7.10 Ferroic phase transitions

Ferromagnetic phase transitions considered above, characterized by the emergence of spontaneous magnetization (which is a macroscopic tensor property), are an example of a more general class of phase transitions called *ferroic phase transitions* (FPTs). Another example of FPTs are ferroelectric phase transitions, in which there emerges a spontaneous polarization (rather than a spontaneous magnetization) when the crystal is cooled to a temperature below T_c (Wadhawan 2000).

Let us consider all phase transitions involving a change of the space-group symmetry of a crystal. Space-group symmetry has a translational part and a directional or point-group part. The translational part is described by the Bravais group, and the directional part by one of the crystallographic point groups. If there is a change of the point-group symmetry at the phase transition, we have an FPT. This is the formal definition of an FPT. If only the translational part of the symmetry changes and the point-group symmetry remains the same, we speak of a nonferroic phase transition. A *ferroic material* is one which can undergo at least one FPT (Wadhawan 2000).

Across an FPT, the parent phase has a higher point-group symmetry than the daughter phase or ferroic phase. Group-theoretic considerations demand that the orders of the two point groups must differ by at least a factor of 2 (see Wadhawan 2000). This means that the ferroic phase would have *domains*. For example, if there is a ferromagnetic phase transition, one type of domains will have spin up, and the other spin down. Similarly, in the case of the ferroelectric phase transition which occurs in a crystal of $BaTiO_3$ at $T_c = 130°C$, six ferroelectric *domain-types* or *variants* can arise. This is because the point symmetry above T_c is cubic ($m\bar{3}m$) (with no preferred or 'polar' direction), and the symmetry in the ferroelectric phase below T_c is $4mm$, which is a polar group belonging to the tetragonal crystal system. The 4-fold axis is the polar axis, and it can point along any of six possible directions: $+x, -x, +y, -y, +z, -z$. Each of these possibilities gives rise to a distinct ferroelectric domain type (Wadhawan 2000).

A7.11 Prototype symmetry

The concept of *prototype symmetry* is crucial for a proper description of FPTs. A crystal may undergo a sequence of symmetry-lowering phase transitions, say on cooling. All but one phase in this sequence has a 'parent' phase, from which it arose on cooling. Prototype symmetry is not just the symmetry of the next higher 'parent' phase. It is the *highest* symmetry conceivable for that crystal structure, so that all the daughter phases can be considered as derived from it by the loss of one or more symmetry operators. For example, $BaTiO_3$ undergoes a sequence of FPTs on cooling: cubic to tetragonal to orthorhombic to rhombohedral. For each of the lower phases, it is the cubic phase which has the prototypic symmetry. For a rigorous definition of prototype symmetry, see Wadhawan (1998, 2000).

A7.12 Critical phenomena

Consider a loosely wound spring made from a metallic wire. If you stretch it and then release it, it not only springs back to its original length but overshoots that point and then goes the other way, and so on: It starts vibrating about its average original length, at a certain 'natural' frequency. Now suppose you thin down the wire of the spring a bit by dissolving it in some acid. If you repeat the above experiment you would find that the natural frequency of vibration is less than before. What has happened is that the thinned-down wire exerts a smaller restoring force (it has a smaller *force constant*) when it is twisted by the stretching of the spring, so it takes a little longer to do the restoring. What is more, the same stretching force as in the above experiment now causes a longer stretching. So the frequency of vibration has gone down, and the amplitude of vibration has gone up.

Now imagine that you go on reducing the force constant of the spring. Ultimately the natural frequency of vibration will tend towards zero, and the amplitude of vibration will become arbitrarily large. This is also what happens when a crystal makes a structural phase transition. We can imagine the attractive and repulsive interactions among the atoms as mediated by a set of springs of a variety of force constants. Each such spring is responsible for a particular types of vibration of the atoms, and the net vibrations are just the superposition of these component vibrations. When one of the springs or force constants becomes weak and finally gives way at a phase transition (like the snapping of a rope in Fig. A7.1), the corresponding vibrations have arbitrarily large amplitudes, extending over many unit cells of the crystal. These are described as *critical fluctuations*. There is an inherent instability, with a corresponding set of critical fluctuations.

There is an analogy with upheavals in complex social networks. When there is a revolution underway in a society, even a tiny rumour can snowball into something terrible, simply because of the inherent instability of the overall situation.

The first derivative of a function is its rate of change, and its second derivative defines its curvature. The degree of positivity of the second derivative of free energy (Eq. A7.4) changes with temperature, and it approaches the value zero as the stability limit T_c is approached for the case of a continuous phase transition. Thus, at T_c the free-energy *vs.* η curve acquires a *flat bottom* (i.e., a zero curvature, or an infinite radius of curvature) (Fig. A7.3). This means that the order parameter η can make large excursions around the value zero, without costing any changes of free energy. Thus, there are large, *scale-free*, fluctuations in η. Since T_c is called the *critical point*, these are called *critical-point fluctuations*, or just *critical fluctuations*.

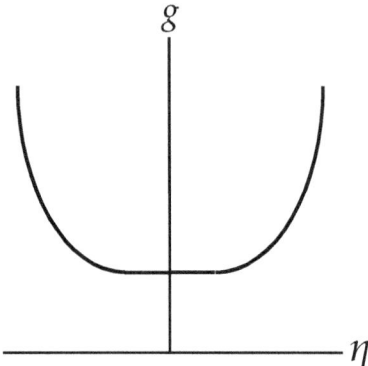

Fig. A7.3 A schematic and exaggerated depiction of free energy density g vs. order parameter η at and very near the critical point $T = T_c$ for a continuous phase transition. The curvature of the curve is zero at $\eta = 0$, and almost zero near $\eta = 0$ ('flat bottom' potential). This means that in the vicinity of the critical point the order parameter can acquire nonzero values at practically all length scales (scale-free 'critical fluctuations') without costing any energy.

Critical fluctuations of the order parameter at and near T_c are just one of the manifestations of a host of complex *critical phenomena*, which occur when the crystal is very close to the critical point (Ball 1999b).

A7.13 Universality classes and critical exponents

> *Nature normally hates power laws. In ordinary systems all quantities follow bell curves, and correlations decay rapidly, obeying exponential laws. But all that changes if the system is forced to undergo a phase transition. Then power laws emerge - nature's unmistakeable sign that chaos is departing in favour of order. The theory of phase transitions told us loud and clear that the road from disorder to order is maintained by powerful forces of self-organization and is paved by power laws. It told us that power laws are not just another way of characterizing a system's behaviour. They are the patent signatures of self-organization in complex systems.*
>
> A.-L. Barabási, *Linked*

Another manifestation, which is actually a consequence of the critical fluctuations of the order parameter, is the occurrence of *universality classes*: There can exist a whole class of phase transitions, involving a variety of interatomic interactions, which exhibit a common or universal temperature dependence of the order parameter and several other physical properties (see Wadhawan 2000). This temperature dependence is conveniently described in terms of the so-called *reduced temperature t*:

$$t = (T - T_c)/T_c = T/T_c - 1 \tag{A7.16}$$

In terms of the reduced temperature, Eq. A7.13 for the response function χ has the following form:

$$\chi = t^{-1} = t^{-\gamma} \tag{A7.17}$$

Here γ is the so-called *critical exponent* for the response function corresponding to the order parameter. We say that the Landau theory predicts that $\gamma = 1$. For a given system, if this critical exponent is found to be very different from this prediction, then the Landau theory is not a good theory for that system. Notice that this prediction of the theory is independent of the exact nature of the interactions driving the phase transition.

If there are several systems (involving different types of atomic interactions), all exhibiting the same (or nearly the same) set of critical exponents, then they are said to belong to the same *universality class*. One example of such a class are systems involving long-range interactions, irrespective of the nature of the interaction (e.g., whether electrical or mechanical). The Landau theory, being a mean-field theory, provides a good description of such a universality class. [In a mean-field theory one replaces the actual local configurations of the order parameter by their mean value, ignoring the fluctuations from the mean value.]

The critical exponents are largely determined by three factors:

1. The effective spatial dimension of the system.
2. The symmetry and dimensionality of the order parameter.
3. The symmetry and the range of the interaction.

For example, phase transitions that are well-described by the Ising Hamiltonian (Eq. A7.1) have nearly the same set of critical exponents. Ferromagnetic transitions in uniaxial magnetic systems, liquid-gas transitions, and some order-disorder transitions are examples of this.

A8. Chaos Theory

Chaos, which is regarded as an information channel, has characteristics in which information mixes in the binary space of state variables and self-similar structures appear in the information. This characteristic establishes sufficient conditions correctly to transmit arbitrary external information and is common to chaos discovered so far in living systems.

Kaneko and Tsuda (2000)

A8.1 The logistic equation

The excellent introductory book by Williams (1997) on chaos theory illustrates many of its basic features by recourse to the very simple *logistic equation* (also see Addison 1997). This equation was formulated as early as in 1845 to model the long-term population dynamics of a species. How does the population of a species, confined to a certain geographic area, vary from year to year? Obviously, the population in a year $t+1$ will depend on the population in the previous year t: $x_{t+1} = kx_t$ (this equation is an example of a *difference equation*). But there are other factors to consider. For example, if the population in the year t becomes too large, there can be an increased decimation of the population, either from predators, or due to shortage of food, or due to the increased competition in the reproduction dynamics. Therefore a more realistic logistic equation is as follows:

$$x_{t+1} = kx_t(1 - x_t) \tag{A8.1}$$

A plot of x_{t+1} against x_t gives an inverted parabola. It is convenient to divide both sides of the above equation by the largest value, N, that the population can attain. Then x varies between 0 and 1. The population is zero for both $x = 0$ and $x = 1$, and has the maximum value ($k/4$) for $x = 0.5$. It follows that, since $x \leq 1$, the largest value that the control parameter k can have is 4. Thus, $0 \leq k \leq 4$.

If we take $k < 1$ and carry out the iteration by starting with some initial population value x_0, and calculate x_1 from it for the next year, and then calculate x_2 from x_1, and so on, we find that the population eventually becomes zero (Fig. A8.1).

The eventual or final value of x, denoted by x^*, is an *attractor* in phase space. There is a *basin of attraction* such that every starting value x_0 in this case is eventually drawn or 'attracted' towards the attractor $x^* = 0$. Thus for $k < 1$ we have a *fixed-point attractor* at the zero value of the population.

A different population dynamics is predicted by the logistic equation for $1 < k < 3$. Now x^* is not zero; rather it increases from zero to ~0.667 as k is increased from 1 to 3. Fig. A8.2 shows the plot for $k = 2.9$.

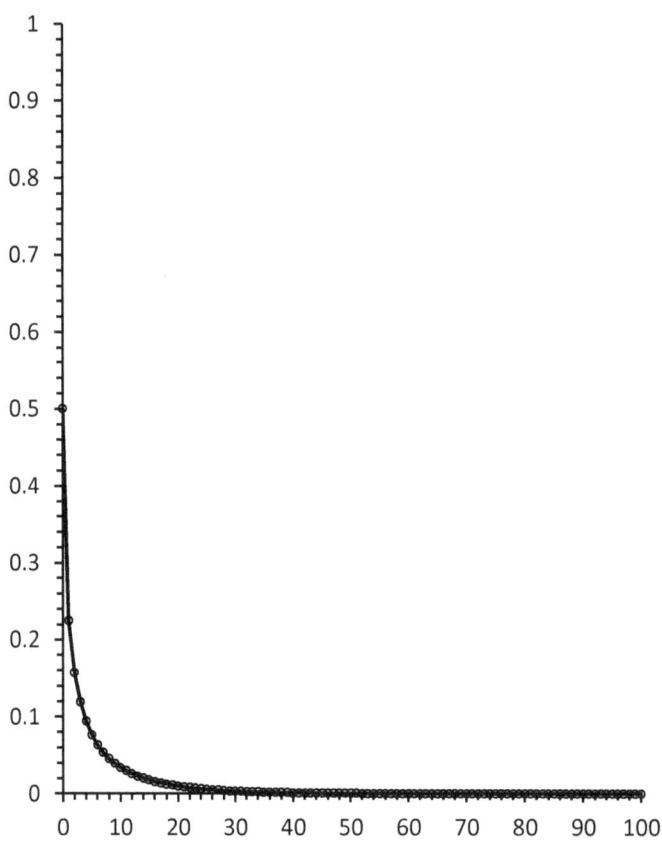

Fig. A8.1 Plot of the time dependence of the population x in successive 100 generations for the case when the control parameter has the value $k = 0.9$. The staring population is taken as 50% of the maximum possible population, also known as the *biotic potential*; i.e., $x_0 = 0.5$.

A fundamentally different dynamics emerges for $3 < k < 4$: The population trajectory no longer converges to a single fixed-point attractor. Further, the trajectory becomes increasingly sensitive to the value of k. For example, for $k = 3.3$ (Fig. A8.3) the trajectory has not one but two fixed points: at $x_1{}^* \approx 0.479$ and $x_2{}^* \approx 0.824$. This means that, from year to year, the population oscillates between ~48% and ~82% of the maximum possible value. There is a *two-point attractor* now. We describe such an oscillating system as having a *period* 2.

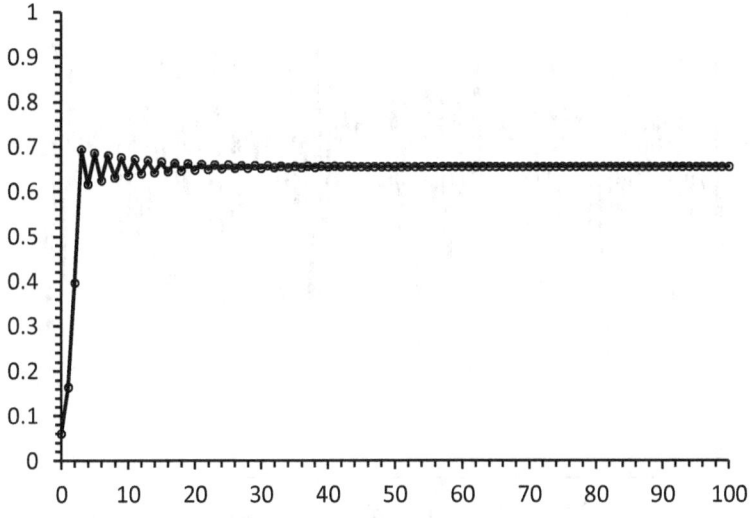

Fig. A8.2 Population variation from year to year for $x_0 = 0.06$ and $k = 2.9$.

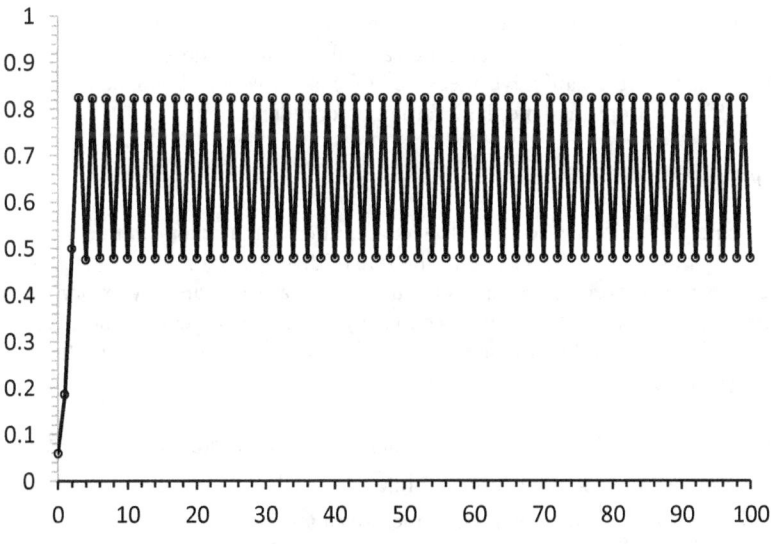

Fig. A8.3 Population variation for $x_0 = 0.06$ and $k = 3.3$.

As the value of k is increased further, it is found that there is a critical value $k \approx 3.4495$ beyond which we get a *four-point attractor*. For example, for $k = 3.5$ we get $x_1^* \approx 0.875$, $x_2^* \approx 0.383$, $x_3^* \approx 0.827$, $x_4^* \approx 0.501$. Such successive *bifurcations* of each attractor into two, such that there are 4, 8, 16, 32, .. etc. fixed points, occur with smaller and smaller increases in k. And we move into the *chaotic regime* of complexity for values of k above ~3.57. The periods now double every time k is increased by even an infinitesimally small amount. The number of points that comprise the *chaotic attractor* is practically infinite, and the trajectory looks erratic (Fig. A8.4), although there may be some ranges of k values for which there is apparent stability.

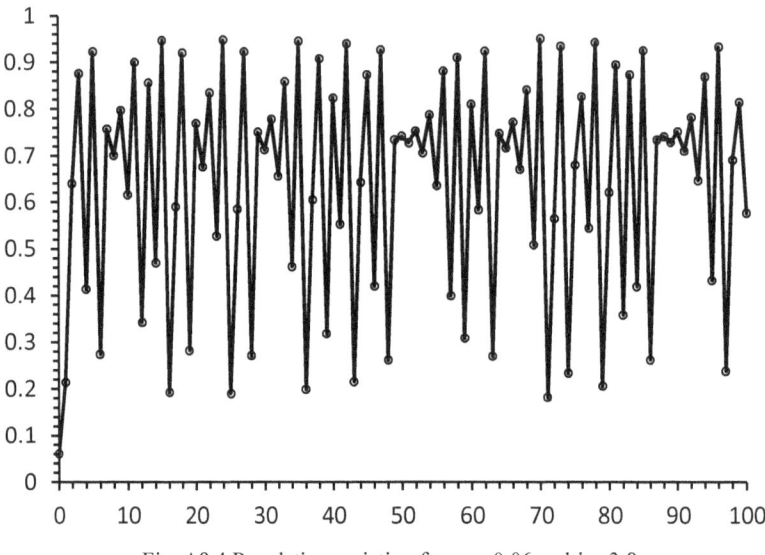

Fig. A8.4 Population variation for $x_0 = 0.06$ and $k = 3.8$.

One would like to have quantitative measures of how chaotic a given system is. This is not easy to specify in practical systems, but one can still attempt to do so, at least theoretically. There are at least three quantitative measures of chaos, namely Lyapunov exponents, the Kolmogorov-Sinai entropy, and mutual information or redundancy.

A8.2 Lyapunov exponents

Consider two neighbouring phase-space trajectories at or near a chaotic attractor. A Lyapunov exponent is a number that is a measure of the rate of divergence or convergence of two such trajectories, averaged over the entire attractor. A negative Lyapunov exponent signifies convergence, and if it is zero, there is neither convergence nor divergence. A positive Lyapunov exponent is an indicator of the presence of chaos in a system, and its magnitude reflects how chaotic the system is.

Nearby trajectories usually diverge exponentially on chaotic attractors. Let ε_0 denote the separation between two such trajectories at time zero, and ε_t its value at a later time t. Since this separation evolves exponentially with time, we can write

$$\varepsilon_t = \varepsilon_0 e^{\lambda t} \tag{A8.2}$$

Here λ is a Lyapunov exponent:

$$\lambda \equiv \frac{1}{t} \ln \left[\frac{\varepsilon_t}{\varepsilon_0} \right] \tag{A8.3}$$

We can treat the first trajectory as the reference trajectory or the *fiducial* trajectory, and the second as the test trajectory. Since different selections of the test trajectory can give somewhat different values of λ, we must average over the various selections over a portion of the attractor cycle time. If we test N nearby trajectory segments, the average Lyapunov exponent can be calculated from

$$\bar{\lambda} = \frac{1}{N} \sum_{i=1}^{N} \frac{1}{t} \ln \left[\frac{\varepsilon_{t(i)}}{\varepsilon_{0(i)}} \right] \tag{A8.4}$$

The bar over λ is normally omitted in practice.

Nearby trajectories on chaotic attractors diverge. Trajectories near, but not on, chaotic attractors converge because they are all drawn towards the attractor. Thus there is divergence in some directions, and convergence in others. Therefore, to characterize fully a chaotic attractor, one requires a *set* of Lyapunov exponents ($\lambda_1, \lambda_2, ... \lambda_n$), where n is the dimensionality of the phase space: There is one exponent for each orthogonal direction in phase space. The set of these exponents constitutes *the Lyapunov spectrum*. The first exponent, λ_1 , is taken to be the largest positive exponent, and the rest are written in order of decreasing value, so that λ_n is the largest negative exponent. The magnitude of λ_1 is an indicator of the degree of chaos in the system.

Chaotic attractors have at least one finite positive Lyapunov exponent (i.e., there is at least one direction in phase space along which there is a divergence of trajectories).

Random or noisy attractors have an infinite positive Lyapunov exponent, representing the fact that there is no correlation whatsoever between the time evolution of two neighbouring trajectories: The divergence is instantaneous.

For stable periodic attractors the Lyapunov exponents are either zero or negative.

A8.3 Divergence of neighbouring trajectories

Divergence of neighbouring trajectories is a characteristic feature of chaotic systems. Let us hark back to the logistic equation to gain some more understanding of this divergence. We have seen above that for $k = 3.8$ the equation describes a chaotic regime. When the iterations are performed successively, initially there may be some 'transients' or 'jitter' before the chaotic attractor is reached. Let us suppose that we are working in this chaotic-attractor regime when we choose an initial value of x_0 on our reference or fiducial trajectory. We next choose a neighbouring point ($x_0 + \varepsilon_0$), and carry out the iterations for the trajectory passing through this point. The difference between the fiducial trajectory and this neighbouring trajectory can then be plotted as a function of the number of iterations. Fig. A8.5 shows a typical result for $\varepsilon_0 = 0.002$ (see Williams (1997)). This plot was obtained as follows: As seen from Fig. A8.4, when we start with $x_0 = 0.06$ (for $k = 3.8$), there is some jitter for a few generations or cycles, and then the yearly variation is random because the system is in the chaotic attractor regime. After 100 generations we can be sure to have reached the attractor regime. We find that $x_{100} = 0.57627200$. For generating Fig. A8.5, I first took this as the new value for x_0, and then repeated the calculation for $x_0 = 0.57827200$. That is, I took $\varepsilon_0 = 0.002$. What is plotted in Fig. A8.5 is the natural logarithm of the absolute difference between the two trajectories (as a function of x_t).

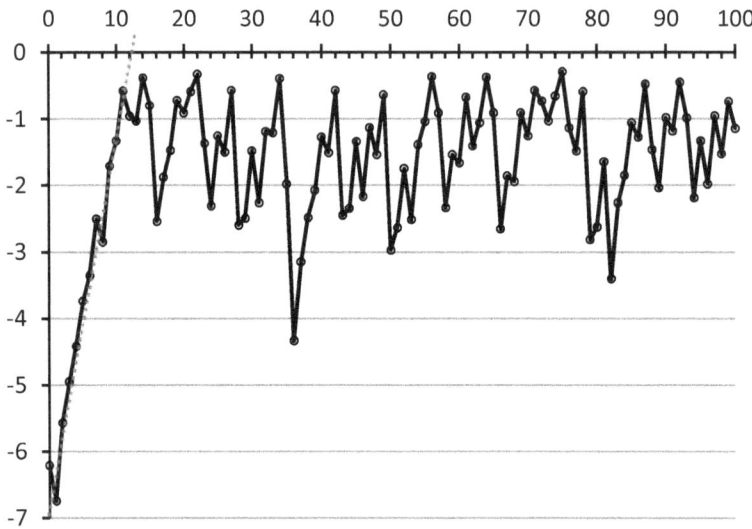

Fig. A8.5 A plot of the natural logarithm of the evolution of the separation of two neighbouring trajectories with time for the logistic equation for k = 3.8 and ε_0 = 0.002.

We see that initially, i.e., for the first few (~12) iterations, there is a systematic divergence of the trajectories (although there is some scatter also), but soon the behaviour becomes totally erratic. For the initial 'systematic divergence' part we can fit a straight line through the iteration points.

This reflects the very important fact about chaotic systems, namely that we can make only short-term *predictions about their evolution with time.*

Fig. A8.5 shows that, if the accuracy with which we can measure the population x is 0.002 or 0.2%, then we can make correct predictions about x up to ~12 generations.

We can get a feel for the meaning of 'short-term predictions' by repeating the computer experiment depicted in Fig. A8.5 for a substantially smaller value of ε_0. The result for $\varepsilon_0 = 0.00000001$ is shown in Fig. A8.6.

We now see a substantially larger initial portion (i.e., it encompasses a larger number of iteration points) corresponding to the systematic divergence of trajectories, with a smaller scatter than in the corresponding portion of the plot in Fig. A8.5. As before, a smaller ε_0 can be interpreted as corresponding to data with higher accuracy than that for Fig. A8.5.

What Fig. A8.6 is telling us is that if the experimental data are of higher accuracy and reliability, our short-term predictions about the chaotic system can extend a little more into the future (in this case up to 28 generations).

An example is that of weather forecasts. *The better the quality of the data on which a prediction is made for a complex system, the higher is the reliability of the short-term predictions about the weather, and the larger is the period about which the weather can be predicted with high reliability.*

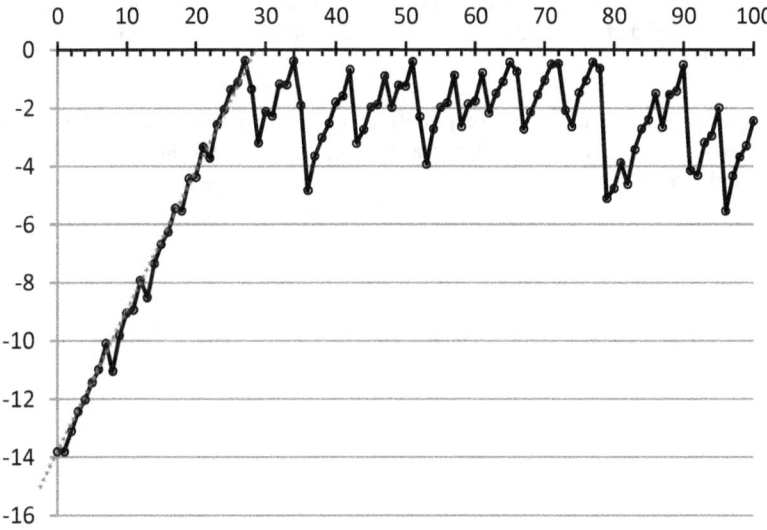

Fig. A8.6 Same as Fig. A8.5, but with $\varepsilon_0 = 0.00000001$.

A8.4 Chaotic attractors

The complexity of atmospheric phenomena is well-illustrated by our inability to make long-term predictions about weather with a high degree of reliability. The chaotic nature of these phenomena was first discovered by Lorenz when he was investigating a coupled system of ordinary differential equations, using a simplified model of 2-dimensional thermal convection, namely Rayleigh-Bénard convection. We considered Rayleigh-Bénard convection (also called just Bénard convection) in Chapter 1.

The equations Lorenz formulated for weather phenomena are now called *the Lorenz equations*:

$$\dot{x} = -\sigma(x - y) ; \tag{A8.5}$$

$$\dot{y} = x(r - z) - y ; \tag{A8.6}$$

$$\dot{z} = xy - bz . \tag{A8.7}$$

Here σ, r, and b are control parameters, like the control parameter k in the logistic equation: σ is the Prandtl number, connecting the energy losses due to viscosity to those due to thermal conduction; r is the Rayleigh number, a dimensionless parameter related to $T_b - T_t$; and b is related to the ratio of the height of the fluid layer and the horizontal dimension of the convective rolls in the fluid. And x, y, z denote, respectively, the convective overturning, the horizontal temperature variation, and the vertical temperature variation.

Suppose we take $\sigma = 10.00$ and $b = 2.67$, and let r be the variable control parameter. It is found that there is a critical value of r, namely $r_c = 24.74$, at which there is a sudden change of behaviour. Below this value the system decays to a steady non-oscillating state, i.e., a stable fixed point (see Addison (1997) for some instructive illustrations). For $r > r_c$, continuous oscillatory behaviour is observed. And $r = 28.00$ produces aperiodic behaviour or '*deterministic nonperiodic flow*'; in other words, chaos. The corresponding chaotic attractor is better known as a *strange attractor*. The phase-space trajectories evolve around two distinct lobes. After an unpredictable number of revolutions around one lobe, the trajectory switches to the other lobe, and then back again. Fig. 15.1 in Chapter 15 is a plot of the Lorenz attractor for $r = 28$, $\sigma = 10$, and $b = 8/3$.

The Lorenz strange attractor has an extremely complex structure. A further increase in the value of r causes a series of reverse bifurcations, i.e., the system moves back from chaotic orbits to periodic orbits.

A9. Network Theory and Complexity

The mystery of life begins with the intricate web of interactions, integrating the millions of molecules within each organism. The enigma of the society starts with the convoluted structure of the social network. The unpredictability of economic processes is rooted in the unknown interaction map behind the mythical market. Therefore, networks are the prerequisite for describing any complex system, indicating that complexity theory must inevitably stand on the shoulders of network theory.

A.-L. Barabási, *Linked*

In physics one can, in principle, work out the properties of a noncomplex system in terms of the structure of the system and the distance-dependent interactions among its constituents. For instance, one can understand the emergence of macroscopic magnetism in a crystal in terms of the distance-dependent interactions among the spins. But there are a large number of complex systems in Nature in which the physical distances among the constituents are largely irrelevant, and in which there is ambiguity about whether or not there is interaction between a chosen pair of constituents. Social systems are one such example, as are ecological systems, neurons in the brain, investors in a stock market, and participants in the Internet. Network theory and statistical mechanics provide powerful tools for dealing with such systems (Strogatz 2001).

A9.1 Graphs

The construction and structure of graphs or networks is the key to understanding the complex world around us. Small changes in the topology, affecting only a few of the nodes or links, can open up hidden doors, allowing new possibilities to emerge.

A.-L. Barabási, *Linked*

A graph is a collection of points (called nodes or vertices) together with a collection of lines (called edges or links) that connect certain pairs of these points. A *directed graph* is a graph in which the edges are *ordered* pairs of vertices; that is, every edge in a directed graph has a direction.

A *network* is a directed graph in which every edge is given a label. The term network is used in several different contexts, and can have different, context-dependent, meanings. Electrical networks, social networks, communication networks, and neural networks are some examples. Networks with a large number of nodes and edges generally exhibit complex behaviour, and are called *complex networks* (Albert and Barabási 2002). Complex systems embody organizing principles encoded in the topology of the underlying network.

I introduce a few concepts and formal definitions in this appendix (see Bondy and Murthy (1976) and Faudree (1992) for more details).

A graph G is a nonempty finite set $N(G)$ together with a finite set $E(G)$ of distinct unordered pairs of elements of $N(G)$. Each element in $N(G)$ is called a *node* or a *vertex*, and each element in $E(G)$ is called an *edge*. The *order* $|N|$ of G is the number of elements in N, and the *size* $|E|$ of G is the number of elements in E.

In practical applications of graph theory, it is often necessary to assign a *weight* to each edge.

For example, if the vertices in a graph represent cities, and the edges the roads between the cities, then a weight proportional to the length of a road can be assigned to the edge representing that road.

Consider two vertices u and v of G which are joined by an edge e. The edge $e = \{u, v\}$ is often represented as just uv (or $u-v$). We say that u is *adjacent* with v. And e is said to be incident to u and v. Two edges incident to a common vertex are said to be *adjacent with each other*.

The *neighbourhood* $n(u)$ of a vertex u is the set of vertices in the graph that are adjacent with u. The number of vertices in the set $n(u)$ is called the *degree* of u. It is denoted by $d(u)$, and is the number of edges of the graph incident to u.

Fig. A9.1(a) illustrates these definitions. For the graph G shown in it, $N = \{u, v, w, x, y\}$; $E = \{ux, uy, vy, xy\}$; order $|N| = 5$; size $|E| = 4$. The degrees of the vertices are: $d(u) = d(x) = 2$, $d(v) = 1$, $d(y) = 3$, $d(w) = 0$. The vertex w is said to be an *isolated* vertex.

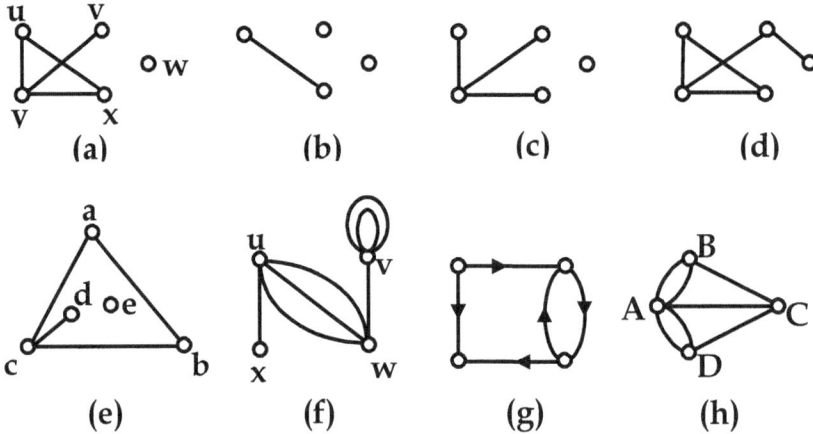

(a)   (b)   (c)   (d)

(e)   (f)   (g)   (h)

Fig. A9.1. (a) Graph G. (b) Subgraph $G - y$. (c) Spanning graph $G - ux$. (d) Spanning supergraph $G + vw$. (e) Graph H isomorphic to the graph G. (f) Multigraph. (g) Directed graph. (h) Königsberg graph, depicting the Königsberg-bridges problem posed and solved by Euler.

A graph H is a *subgraph* of a graph G if $N(H)$ is a subset of $N(G)$ and $E(H)$ is a subset of $E(G)$. G is called a *supergraph* of H. In Fig. A9.1(a), if we delete the vertex y and also all the edges adjacent with y, then we obtain the subgraph $G-y$ shown in Fig. A9.1(b).

If H and G have the same set of vertices, then H is called a *spanning subgraph* of G. For example, Fig. A9.1(c) has the same set of vertices as Fig. A9.1(a), but the edge ux is missing; consequently, $G-ux$ is a spanning subgraph of G.

Fig. A9.1(d) depicts a *spanning supergraph* of G. In it, the vertices are the same as for G, but an edge vw has been added. It is denoted by $G+vw$.

Two graphs G and H are said to be *isomorphic* if there is a one-to-one mapping from the vertices of one graph onto the vertices of the other graph such that the edges are preserved. For example, the graph shown in Fig. A9.1(e) is isomorphic to that in Fig. A9.1(a).

A graph or subgraph in which every pair of vertices is adjacent is called a *complete graph or subgraph*. That is, each vertex of the graph or subgraph is connected to every other vertex of the graph or subgraph. If it is of order *n*, then each vertex in it is of degree (*n*-1), and its size is $n(n-1)/2$. A complete subgraph in a graph is said to constitute a *cluster* in the graph.

If the vertices of a graph G can be divided into two disjoint subsets G_1 and G_2 such that every edge in the graph G connects a vertex in G_1 only to a vertex in G_2, it is called a *bipartite graph*. If every vertex in G_1 is connected to every vertex in G_2, then G is a *complete bipartite graph*.

If all vertices of a graph have the same degree r, then it is called an *r-regular graph*, or regular graph.

A *random graph* is one in which the edges are distributed randomly.

If a graph has multiple edges between two vertices, or if it has edges in the form of *loops* (e.g. *vv*), it is called a *multigraph* (see Fig. A9.1(f)).

A *walk W* in a graph G from a vertex v to a distinct vertex v is a finite sequence of alternating vertices and edges ($W = v_0 e_1 v_1 ... v_k e_k v_{k+1}$, with $v_0 = u$ and $v_{k+1} = v$) such that each edge in the sequence is incident to each of the two vertices on either side of it in the sequence. The *length* of the walk is the number k, i.e., the number of edges in it.

If all the edges in W are distinct, then it is called a *trail*. A trail in which all the vertices are distinct is called a *path*.

A walk, trail, or path is said to be *closed* if the initial and the final vertex are the same ($u = v$). A closed path with $k \geq 3$ is called a *cycle*.

If there is a path between two vertices of a graph, then the vertices are said to be *connected*. If each pair of vertices in a graph is connected, then the graph is said to be connected. If only a subset of the vertices in a graph are connected, and if they are connected to a particular vertex, then this subset constitutes a *connected subgraph* (or a cluster). A graph that is not a connected graph can be partitioned into *maximal connected subgraphs* called the *components* or clusters of the graph. The graph in Fig. A9.1(a) has two components: $\{u, v, x, y\}$ and $\{w\}$.

We define the *distance* between any two vertices u and v of a graph as the length of the shortest path between them. It is denoted by $d(u,v)$. If the edge connecting these two vertices has been assigned a weight w, the weighted distance is $wd(u,v)$.

A *tree* is a connected graph that does not contain any cycles. It is readily argued that any connected graph has a spanning tree.

The maximal distance between any pair of its nodes defines the *diameter* of the graph. A *disconnected graph* is one that is made of two or more isolated clusters. Strictly speaking, for such a graph the diameter is infinite, so a modified definition of diameter is used by defining it as the maximum diameter of its clusters.

Partitioning of the vertices of a graph or network

Consider a set V of vertices. A partition of V is a set of its disjoint nonempty subsets such that the union of all the subsets is the set V. The subsets or 'elements' of a partition are called *cells*.

If the partition has only one cell, it is called a *unit partition*. If each cell of a partition contains only one vertex, it is called a *discrete partition*.

Naturally, a variety of partitions are possible, depending on the criteria used. For example, members of a subset or cell can be chosen to be those vertices which have the same degree, with different subsets having vertices with different degrees. This is called a *degree partition*. Consider two different partitions P and Q of the set V of vertices:

$$P: V = P_1 \cup P_2 \cup P_3 ... \tag{A9.1}$$

$$Q: V = Q_1 \cup Q_2 \cup Q_3 ... \tag{A9.2}$$

If every cell or subset of P is a subset of some cell of Q, then P is said to be a *finer* partition of V compared to Q. Or Q is *coarser* than P.

Eulerian graphs

The first paper in graph theory was published in 1836 by Euler (see Barabási 2002). Euler solved the famous Königsberg bridges problem in that paper (see Biggs, Lloyd and Wilson (1976) for an account of the early days of graph theory). There were seven bridges in Königsberg over a river, which connected pairs of regions from among the four which we may denote by, say, A, B, C, D. The problem posed was whether it is possible to traverse the seven bridges without crossing any bridge twice. If we represent the tour regions as the vertices in a graph, and the bridges as the edges connecting pairs of vertices, we obtain the multigraph shown in Fig. A9.1(h). Euler established that the problem posed had an answer in the negative. In other words, he showed that the graph in Fig. A9.1(h) does not have a trail that uses all the edges in the graph.

An *Eulerian trail* of a graph is defined as a trail that uses all the edges of the graph. A graph that contains a closed Eulerian trail is called an *Eulerian graph*. Euler showed that there is no Eulerian trail in the graph shown in Fig. A9.1(h). For such a trail to exist, a graph must be a *connected* graph, and all its vertices (except possibly the first and the last vertices of the trail) must be of even degree.

Euler proved the theorem that *a connected graph or multigraph G has a closed Eulerian trail if and only if each vertex has even degree, and that G has an open Eulerian trail if and only if precisely two of its vertices are of odd degree.*

Hamiltonian graphs

A graph is said to be Hamiltonian if it contains a spanning cycle. The spanning cycle is called a *Hamiltonian cycle*. The objective in Euler's problem discussed above was to visit each bridge or *edge* exactly once. In the case of a Hamiltonian graph, the objective is to visit each *vertex* exactly once.

The necessary and sufficient conditions for a graph to be Hamiltonian are not known. However, many sufficient conditions are known, as also some necessary conditions. Here is an example of one such sufficient condition: If for each pair of non-adjacent vertices u and v of a graph of order $n \geq 3$ we have $d(u) + d(v) \geq n$, then the graph is Hamiltonian.

Directed graphs

A directed graph or *digraph D* is a graph in which every edge is an arrow, i.e., it has a direction (Fig. A9.1(g) depicts an example). This means that an edge *uv* (meaning $u \rightarrow v$) in *D* is distinct from an edge *vu* (meaning $v \rightarrow u$). Consequently, the phrase 'adjacent with' in the context of a graph *G* must be replaced by 'adjacent to' or 'adjacent from.' Similarly, one must make a distinction between 'incident to' and 'incident from.' The term *arc* is sometimes used for a directed edge.

For a digraph the degree $d(v)$ of a vertex comprises of two parts: $d(v) = d^+(v) + d^-(v)$, where $d^+(v)$ is the number of edges incident *from v*, and $d^-(v)$ is the number of edges incident *to v*.

For every digraph *D* there is an underlying graph *G* obtained by removing the direction part of every edge, and replacing any multiple edges connecting two vertices by a single edge.

A digraph is connected if the underlying graph is connected. A vertex *v* is *reachable* from a vertex *u* if there is a directed path from *u* to *v*. A digraph is *strongly connected* if all its distinct pairs of vertices are reachable from each other.

A9.2 Networks

> *If we want to understand life – and ultimately cure disease – we must think networks.*
>
> A.-L. Barabási, *Linked*

A network is a directed graph in which every edge or arc has been assigned a label. Networks may be *isotropic*, or otherwise. Isotropic networks do not have well-defined inputs and outputs. By contrast, neural networks are examples of networks with well-defined inputs, which they are supposed to convert into certain desired outputs.

Networks can be of different types. A network with a well-defined input and output may be formally defined as follows: A network is a directed graph *D* in which there is vertex *s* (called the *source*) and a vertex *t* (called the *sink*), and for which a function *c* (called the capacity) is defined which assigns to each edge of the network a nonnegative number. Such a network is relevant, for example, for building a model for maximizing the flow of goods from one point to another in a transportation system. Although we are considering a single-input single-output network here, such considerations can be generalized to multiple sources and/or multiple sinks.

A *flow* on *D* is a function *f* from *E(D)* into nonnegative numbers such that the following two conditions are satisfied. The first is that the flow cannot exceed the capacity of an edge; i.e., for an edge *e*, $f(e) \leq c(e)$ for all *e* belonging to *E*. The second condition is that the input into any intermediate vertex must be equal to the output from that vertex. That is, for all vertices *v* belonging to the subgraph $D - \{s,t\}$, we must have $\sum_{uv \in E} f(u,v) = \sum_{vw \in E} f(v,w)$.

A *cut* in a network is a partition of its nodes *N* into two subsets S_1 and S_2 such that the source node *s* belongs to S_1 and the sink node *t* belongs to S_2. The *capacity of the cut* is the sum of the capacities of the edges from S_1 to S_2.

The maximin theorem

An important theorem in the theory of network-flows states that *in a network the value of a maximum flow is equal to the capacity of a minimum cut*. It is known as *the max-flow min-cut theorem*, or *the maximin theorem*.

r-regular networks

The notion of an *r*-regular or regular graph was defined above. An *r-regular network* is one which has an underlying *r*-regular graph. All vertices of an *r*-regular network have the same degree *r*. In other words, in a regular network all nearest-neighbour nodes are connected in a regular manner. Such networks have local groups of highly interconnected nodes. For them the number of edges per node has an approximately Gaussian distribution, with a mean value that is a measure of the 'scale' of the network (Bray 2003).

We consider an example from systems biology to illustrate the notion of a network. The processes occurring in a biological cell can be modelled in terms of a variety of molecular networks (Alon 2003, 2008; Bray 2003). *Transcription networks* are an example. In a transcription network the nodes or vertices are the genes, and the edges are the transcriptional interactions. Consider two nodes *u* and *v*. If there is a transcription interaction between them, it is represented by a directed edge or arrow: $u \rightarrow v$. In general, more than one arrows may point at a particular node in some networks, in which case a suitable input function has to be defined (e.g. AND or OR). Similarly, two nodes may be simultaneously involved in more than one types of interaction (e.g. on different time scales). One can then use different *colours* for the arrows joining them. The strength of the interaction between two nodes is represented by assigning a weight to the arrow joining them.

Network motifs

In spite of the fact that evolutionary processes involve a fair amount of random tinkering, Nature has zeroed-in again and again on certain recurring circuit elements, called *network motifs* (NMs) (Milo *et al.* 2004). Each such motif in the molecular network performs a specific information-processing job. Examples of such jobs in the case of the transcriptional networks of the cell comprising the bacterium *E. coli* are (see Alon 2003): filtering out spurious input fluctuations; generating temporal programs of gene expression; and accelerating the throughput of the network. Similar NMs have also been found in the transcription networks of yeast (Milo *et al.* 2002). Operational amplifiers and memory registers in human-made electronic circuits are some of the equivalents of NMs.

NMs are formally defined as follows (Milo *et al.* 2002). Given a network with *N* nodes and *E* edges, one can construct a *randomized network* or *random network* from it which has the same single-node characteristics as the actual network (i.e., the numbers *N* and *E* are unchanged), but where the connections between nodes are made at random. In other words, we construct a spanning random network from the actual network. NMs are those patterns that occur in the actual network significantly more frequently than in the spanning random network.

There are a large number of possibilities by which a given set of nodes can be interconnected. We shall consider some of them below, after considering the 'travelling postman problem' as an illustration of some of the graph-theoretical concepts described above.

A9.3 The travelling-salesman problem

Suppose a salesman wants to visit every city in his route exactly once and then return home. The travelling-salesman problem (TSP) is to determine the route for which the distance travelled (or the *cost*) is the least. In the language of graph theory, the TSP amounts to finding a minimum-weight Hamiltonian cycle in a complete graph having weighted edges, the weight of an edge representing the distance or cost of travel between the two cities connected by the edge.

The TSP amounts to enumerating the possible number of distinct itineraries. Let the order N of the graph be, say, 6. That is, the salesman has to visit five cities and then return home. There are five ways of choosing the first destination city. For each of these, there are four different ways of choosing the second city. Having chosen any of these, there are only three ways of picking up the next city (because the salesman cannot touch any city twice). And so on. Thus the total number of possible itineraries is 5 x 4 x 3 x 2 = 120, or 5! (factorial 5), or $(N-1)!$. For each of these 120 options, one computes the total distance to be covered. The option with the least distance is the best solution.

This is a brute-force way of solving the TSP. Suppose the number of cities to be visited is 50, rather than 5. The search space now comprises of 50! possibilities; i.e., $\sim 10^{64}$. This is a very large number indeed, and algorithms for reducing the size of the search space must be found. At present it is not known whether such a *good algorithm* actually exists which always gives a minimum solution. In fact, it is an NP-complete problem. But the situation is not totally hopeless. Algorithms are known that give reasonable solutions most of the time, or give solutions that are close to the best solution (Applegate *et al.* 2006).

The website http://en.wikipedia.org/wiki/Traveling_salesman_problem gives useful information about the TSP.

A9.4 Random networks

> *Erdős and Rényi acknowledged for the first time that real graphs, from social networks to phone lines, are not nice and regular. They are hopelessly complicated. Humbled by their complexity, the two assumed that these networks are random.*
>
> A.-L. Barabási, *Linked*

A random network is a network in which the edges are distributed randomly. Such networks are *easily traversed*; this means that the path length or distance (i.e., the number of edges) between any two nodes is typically quite small. For random networks, as also for regular networks, the number of edges per node has an approximately Gaussian distribution, with a mean value that is a measure of the *scale* of the network.

Historically, complex graphs with no discernible design principles were modelled by Erdős and Rényi (ER) in the 1950s as random graphs (see Barabási (2003) for a highly readable historical account of graph theory and network theory). In this simplest possible model (*the ER model*) of a complex graph, if there are N nodes, and if every pair of nodes is connected with a probability p, then there are $\sim pN(N-1)/2$ edges, distributed randomly. The ER model amounted to equating complexity with randomness and was therefore, naturally, superseded by more sophisticated models of complex networks. But it still provides a very useful benchmark for understanding the degree of complexity of real-life networks.

The degrees of different nodes in a graph or network are different. One can define a *degree-distribution function* $P(k)$ which gives the probability that any particular node has a degree k, i.e. it is connected to k other nodes. Suppose $<k>$ is the average degree of a network. Most of the nodes in a random network have nearly the same degree, which is close to $<k>$. The degrees of the various nodes in a random graph or network follow a Poisson distribution, with a peak at $P(<k>)$. The degree distribution $P(k)$ in most of the real networks deviates significantly from the Poisson distribution. In fact, it generally follows a *power law* (see below).

The notion of *clustering coefficient* has a particularly simple meaning for a random network. Consider a particular node in a random network. The clustering coefficient C_i of a node i is a number which is a measure of how many other nodes are connected to it. For a random network, this number is nothing but the average degree $<k>$ of the network, which is also the probability p that any two nodes of the network are connected. All nodes in a random network have the same clustering coefficient:

$$C_{random} = p = <k> / N \tag{A9.3}$$

Evolution of a graph or network as more and more nodes or edges are added is of direct relevance to the study of evolution in complex systems. Historically, such investigations in graph theory were first carried out for random graphs. One starts with a set of N nodes, and introduces random edges in it successively. As the number of edges grows, the connection probability p for any pair of nodes increases. Eventually one obtains a fully connected graph, i.e., $p \rightarrow 1$, and the number of edges is the maximum possible, namely $N(N\text{-}1)/2$.

One aim of such studies is to see what happens when $N \rightarrow \infty$. Another is to investigate how new properties emerge as the value of p increases gradually. An important result obtained by Erdős and Rényi was that *many properties of random graphs appear quite suddenly*. Usually there is a *critical or threshold connection probability* $p_c(N)$ such that if $p(N)$ grows slower than $p_c(N)$ as $N \rightarrow \infty$, then almost every random graph with connection probability $p(N)$ does not have the property which would have otherwise appeared in the graph if $p(N)$ were greater than $p_c(N)$.

A9.5 Percolation transitions in random networks

> *Random network theory tells us that as the average number of links per node increases beyond the critical one, the number of nodes left out of the giant cluster decreases exponentially.*
>
> A.-L. Barabási, *Linked*

> *Recognizing that passing a critical threshold is the prerequisite for the spread of fads and viruses was probably the most important conceptual advance in understanding spreading and diffusion.*
>
> A.-L. Barabási, *Linked*

For a random network there exists a critical connection probability p_c below which only isolated clusters exist, but above which a *giant cluster* emerges which spans a large portion of the network, or the entire network. We then speak of a percolation transition.

Significant analogies or correspondences exist between percolation transitions in networks and the regular thermodynamic phase transitions in materials. In particular, the notion of continuous (or second-order) and discontinuous (or first-order) phase transitions in materials has its parallel in network theory (Achlioptas, D'Souza and Spencer 2009). The critical connection probability p_c in a random network can be interpreted as the equivalent of the critical point in a thermodynamic phase transition. Suppose, in the spirit of the ER model, we start with a set of N isolated vertices and introduce edges randomly, one by one. At any stage in the evolution of this random network, we can identify components or connected subgraphs in the network. The larger the size of a component, the greater is the chance that it would merge with another component as more and more edges are added, rather like the gravitational attraction between objects.

Suppose rN edges have been introduced randomly at a given stage. Let C denote the size of a component. For $r < 1/2$ the largest value of C is quite small, as it scales as log$N$. But, for $r > 1/2$, something very different happens. C scales as $(4r-2)N$ when r is slightly greater than ½. That is, C scales linearly with N, starting with the value 0 for $r = 1/2$. This is very similar to how the order parameter of a continuous phase transition varies as a function of the control parameter, say temperature: It is zero above the critical temperature T_c, and rises continuously as the temperature is lowered gradually from T_c. Thus the percolation transition in a random network marks a distinct change in the evolution of the connection topology.

Can we have a percolation transition in a random network that is the equivalent of a *discontinuous* or first-order phase transition? The answer according to Achlioptas, D'Souza and Spencer (2009) is 'Yes,' provided we change the rule for the evolution of the network. Several such rules are conceivable, and even small changes may be effective. In one such rule, we begin by choosing *two* edges randomly, and then discard one of them according to some criterion. Of course, if the edge to be discarded is selected randomly from the pair of edges, we get nothing new. However, if a non-random selection rule is operative for deciding the edge to be discarded, a discontinuous or first-order percolation transition can become possible. Small changes in edge-formation rules can sometimes alter drastically the nature of percolation transitions in complex networks. When an edge is introduced, it connects two vertices, each of which has a certain component size. One example of a selection rule for obtaining first-order percolation transition behaviour is that, from the pair of edges chosen randomly we keep the one which minimizes the product of the sizes of the two components which will get merged by this choice, and discard the other candidate edge. Alternatively, instead of minimizing the product of the two component sizes, we could choose to minimize their sum. In either case, what would have been a continuous percolation transition becomes a discontinuous, sudden, or *explosive* percolation transition (Achlioptas, D'Souza and Spencer 2009).

A9.6 Small-world networks

It is observed that in most of the large networks, the distance or path-length between any two nodes does not increase in proportion to the number of nodes in the network. Perhaps the best known example of this is the *six degrees of separation* feature discovered by Stanley Milgram (1967). He found that between most pairs of people in the USA there is typically a network distance of six acquaintances. This is commonly referred to as a small-world feature in a network. It can occur even in random networks or graphs, for which Erdős and Rényi had shown that the typical distance between any two nodes scales as the logarithm of the total number N of nodes.

Milgram's idea stood discredited for some time, but has got extensive confirmation recently. A Microsoft team (E. Horvitz and J. Leskovec) studied the addresses of 30 billion Microsoft Messenger instant messages sent during a single month (June) in 2006. Two people were taken to be acquaintances if they had sent each other an instant message. It was found that any two persons on average are linked by seven or fewer acquaintances.

Another feature of social and many other networks is the occurrence of *cliques*. In social networks, cliques are groups of friends or acquaintances in which everyone knows every other member of the group. The *clustering coefficient* of a network serves as a good measure of this tendency to form clusters or cliques (Wasserman and Faust 1994; Watts 1999). Consider a particular node i. Suppose it is connected to k_i other nodes. If these other nodes are part of a clique (which naturally includes the node i also), there would be $k_i(k_i - 1)/2$ edges among them. Suppose the actual number of edges among them is E_i. Then the clustering coefficient of the node i is defined as

$$C_i = \frac{2E_i}{k_i(k_i - 1)} \tag{A9.4}$$

And the clustering coefficient C for the entire network is the average of all the C_i's.

As we have seen above, for a random network, $C = p$. For practically all real networks, C is much larger than p, indicating that the small-world characteristic is indeed very common in complex networks.

Small-world networks have short path lengths like random networks, and they have clustering coefficients that are quite independent of network size, like regular networks. This last feature of regular networks or graphs is well illustrated by a crystal lattice, in which the clustering coefficient is fixed, no matter what the size of the crystal is. Consider, for simplicity, a ring lattice, i.e. a 1-dimensional lattice with periodic boundary conditions. Let each node be connected to k other nodes nearest to it. It has a high clustering coefficient because most of the immediate neighbours of any lattice point are also neighbours of one another. Its clustering coefficient can be shown to be

$$C = \frac{3(k - 2)}{4(k - 1)} \tag{A9.5}$$

A low-dimensional lattice like this does not have short path lengths. For a d-dimensional cubic lattice, the average distance between two nodes scales as $N^{1/d}$; this is a much faster increase with N than the logarithmic increase typical of random networks, as also of most of the real-life networks.

The Watts-Strogatz model

Real small-world graphs or networks have short path lengths unlike ordered or regular networks, and they have clustering coefficients that are quite independent of network size, unlike random networks. Watts and Strogatz (WS) (1998) proposed a one-parameter model that explored the regime intermediate between random graphs and regular graphs. They started with ring lattice of order N, in which each node is connected to its nearest k nodes, $k/2$ on either side. For ensuring that they were dealing with a sparse but connected network, they assumed that $N \gg k \gg \ln(N) \gg 1$. Then they introduced randomness by rewiring each conceivable edge with a certain probability p. Thus even distant nodes had a chance of being connected. In fact, the procedure introduces $pNk/2$ long-distance edges connecting nodes which would have been of parts of different neighbourhoods otherwise.

Connection probability $p = 0$ corresponds to perfect order or regularity, and $p = 1$ to randomness. *Somewhere between these two limits a transition occurs from order to randomness (or chaos).* This transition is of great interest in the context of complex systems, because this 'edge of chaos' is where complexity thrives. It is in the vicinity of the edge of chaos that the network exhibits a coexistence of clustering and short path lengths.

For a ring lattice, i.e., when $p = 0$, the average path length is

$$L(0) \approx N / 2k \gg 1 \tag{A9.6}$$

And the average clustering coefficient is

$$C(0) \approx 3/4 \tag{A9.7}$$

Thus the average path-length scales linearly with order, and the average clustering coefficient is large. At the other extreme in the WS model, i.e., when $p \to 1$, the model describes a random graph, for which

$$C(1) \approx C_{random} \sim k/N << 1 \tag{A9.8}$$

and

$$L(1) \approx L_{random} \sim \ln(N)/\ln(k) \tag{A9.9}$$

Thus the clustering coefficient decreases with increasing order of the net, and the path length increases logarithmically with order.

The WS model has the feature that there is a substantial range of probabilities p for which large C values coexist with small L values, in agreement with what is observed in a variety of real networks. However, it is not able to reproduce the power-law degree-distribution exhibited by many real networks for large k. Next we focus attention on this power-law feature.

A9.7 Scale-free networks

> *This unparalleled license of expression, coupled with diminished publishing costs, makes the Web the ultimate forum of democracy; everybody's voice can be heard with equal opportunity. Or so insist constitutional lawyers and glossy business magazines. If the web were a random network, they would be right. But it is not. The most intriguing result of our Web-mapping project was the <u>complete</u> absence of democracy, fairness, and egalitarian values on the Web. We learned that the topology of the Web prevents us from seeing anything but a mere handful of the billion documents out there.*
>
> A.-L. Barabási, *Linked*

> *Hubs appear in most large complex networks that scientists have been able to study so far. They are ubiquitous, a generic building block of our complex, interconnected world.*
>
> A.-L. Barabási, *Linked*

A scale-free network is one in which a few of the nodes have a much larger number of connections than others. For regular networks and for random networks, the number of edges per node has an approximately Gaussian distribution, with a mean value that is a measure of the scale of the network. In a scale-free network, by contrast, there are a few strongly connected nodes and a large number of weakly connected ones. Typically, the degree distribution follows a power law:

$$P(k) \sim k^{-\gamma} \tag{A9.10}$$

The exponent γ generally lies between 2 and 3. Most of the complex networks in Nature have a power-law degree distribution.

Since the distribution function $P(k)$ does not show a characteristic peak (unlike the Gaussian peak for random networks and regular networks), the network is described as scale-free (Barabási and Albert 1999; Barabási 2009). For it, the average distance between any two nodes is almost as small as for a random network, and yet its clustering coefficient is practically as large as for a regular network. In other words, scale-free networks exhibit *cliquishness*, like small-world networks.

Generalized random networks

Often one can bring a random-network model closer to reality by introducing a power-law degree distribution into it. Such a topologically modified network is still random in the sense that the edges still connect randomly selected nodes. But a constraint is introduced that the degree distribution function $P(k)$ must follow a power law. For more details, see Albert and Barabási (2002).

A9.8 Evolution of complex networks

> *We are witnessing a revolution in the making as scientists from all different disciplines discover that complexity has a strict architecture. We have come to grasp the importance of networks.*
>
> A.-L. Barabási, *Linked*

Networks evolve with time. It has been observed in a wide variety of complex networked systems that their topology evolves to achieve higher and higher degrees of *robustness* against attacks and failures (Albert and Barabási 2002; Barabási 2003). How does this happen? *How does the power-law degree distribution arise*? What mechanisms result in the coexistence of large clustering coefficients and small path lengths?

We have seen above that the Watts-Strogatz model captures some of the important features of many real networks, even though it does not reproduce their power-law or scale-free feature. The generalized random-graph model, on the other hand, *is* able to introduce power-law degree-distribution behaviour. The ad hoc ways in which these models are introduced do not shed light on the underlying mechanisms operative in the evolution of the topology of real networks. What is needed is that, instead of modelling the network topology, we should try to model the assembly of the network and its evolution. This was first done by Barabási and Albert (1999).

The Barabási-Albert model

The BA model captured some essential features of network dynamics, and could reproduce the scale-free connectivity exhibited by many real networks, by doing away with two assumptions made in the models we have discussed so far. We have assumed that the order N of a network is constant. We have also assumed that the connection probability p does not depend on which two nodes are selected for connection. What usually happens in the real world, however, is that, firstly, networks *grow* with time by the addition of new nodes (e.g. the World Wide Web grows exponentially with time by the continuous addition of new web pages); and secondly, there is likelihood of *preferential attachment* to certain nodes (e.g. when a new manuscript is prepared for publication, there is a greater chance that well-known papers will be cited, rather than less frequently cited average-quality papers).

The growth feature in the BA model is introduced as follows: One starts with a small number

m_0 of nodes, and at every time step a new node having $m(\le m_0)$ edges is added. Thus there are $N = m_0 + t$ nodes after t time steps. And the preferential attachment of new nodes is carried out as follows: It is assumed that the probability Π that the new node will be connected to an existing node i is linearly related to the degree k_i of the node i:

$$\Pi(k_i) = \frac{k_i}{\sum_i k_i} \qquad\qquad (A9.11)$$

Thus there are mt edges in the network after t time steps.

Numerical simulations demonstrate that the BA network evolves to be a scale-free network, with the probability that a node has k edges given by a power law with exponent $\gamma_{BA} = 3$.

But the properties of real networks can differ substantially from what is predicted by the BA model. For example, the power-law exponent for the actual degree distribution may lie anywhere between 1 and 3. Even non-power-law features like an exponential cutoff, or saturation for small k, are observed sometimes. Evolving networks in Nature can develop both power-law and exponential degree distributions. Certain preferential attachments, aging effects, and growth constraints can make a network crossover from power-law to exponential distribution of degree. Various refinements of the BA model have been investigated (see Albert and Barabási 2002).

Robustness of networks

Networked complex adaptive systems display a remarkable degree of error tolerance. Naturally, this robustness aspect has been investigated widely by computer simulation, as well as by analytical studies. The error tolerance of real networks has a dynamic aspect (Watts 1999; Lago-Fernández et al. 2000; Garlaschelli, Capocci and Caldarelli 2007) and a topological aspect. The topological aspect is easier to simulate on a computer. One may focus either on edges or on vertices, or on both. *A network is defined as robust if it contains a giant cluster consisting of most of the nodes even after a fraction of its nodes are removed.*

The progressive removal of edges randomly is like an inverse edge-percolation problem. The removal of a node, on the other hand, has a more damaging effect on the network topology, since all the edges adjacent with that node also get removed.

Simulation studies have shown a strong correlation between robustness and topology. Scale-free networks are found to be generally more robust than random networks. However, if the most connected nodes of a scale-free network are the targets of attack, then they are much more vulnerable than random networks (Albert, Jeong and Barabási 2000).

A9.9 Emergence of symmetry in complex networks

Several networks are rich in symmetry, and exploring the origins of that symmetry can provide insights useful for modelling the dynamics and topology of the network (MacArthur and Anderson 2006; MacArthur, Sánchez-García and Anderson 2007; Holme 2006; Xiao et al. 2008a, b).

In network theory, the symmetry transformations under which a network remains invariant are usually taken to be the appropriate permutations of vertices. And by 'invariance' we mean the invariance of the *adjacency* of the vertices of the network. Thus a network structure is said to possess symmetry if certain permutations of its vertices do not change the adjacency of the vertices.

Symmetry analysis of networks has led to the important result that *similar linkage pattern* is the underlying factor responsible for the manifestation of symmetry by networks (Xiao *et al.* 2008b).

I first introduce some formal definitions in the context of symmetry of graphs or networks. Let G be the graph underlying a network. A one-to-one mapping from the vertices of G onto itself is called a *permutation*. Let $S(V)$ be the set of all the permutations possible for a set V of vertices comprising the graph G. In the set $S(V)$ there can be some permutations which preserve the adjacency of each vertex of the set V. These are called the *automorphisms* acting on the set V of vertices of the graph G. The set of all automorphisms of the graph G is denoted as $\text{Aut}(G)$.

A network is said to be symmetric if its underlying graph G has at least one automorphism which is not an identity or trivial permutation. If the identity permutation is the only automorphism of G, then the graph and the network are asymmetric.

Since permutations are mappings, one can define successive permutations or *products* of permutations. A product of two permutations f and g (written as fg) is another permutation h ($h = fg$). h is the mapping which takes the vertex set V onto itself, and this mapping has the same effect on a vertex x as if we first applied f on x (getting x^f), and then applied g on x^f (getting $(x^f)^g$). Thus, $x^h = (x^f)^g$. It can be shown that the set of automorphisms under the product of permutations forms a group, called the *automorphism group* (Godsil and Royle 2001).

Given the automorphism group $\text{Aut}(G)$ for the vertex set V, we can create a partition $P = \{V_1, V_2, ... V_k\}$ such that a vertex x is equivalent to a vertex y if and only if $\text{Aut}(g)$ contains a mapping g such that $x^g = y$. Such a partition is called an *automorphism partition*. And each cell of this partition is called an *orbit* of $\text{Aut}(g)$. A *trivial* orbit contains only a single vertex. Otherwise it is nontrivial orbit.

Fig. A9.2 illustrates some of these ideas (Xiao *et al.* 2008b). There are 7 vertices, so the possible number of permutations is 7!, or 5040. Clearly, the adjacency of vertex 3 is different from that of 4. So a permutation that interchanges 3 and 4, and does nothing else, is not a symmetry transformation. So it is not an automorphism of this graph. A permutation that *is* an example of an automorphism is that in which only vertices 1 and 2 are interchanged. Thus, vertices 1 and 2 belong to the same orbit. Similarly, the vertices 5, 6 and 7 form a subset of the graph in which any interchange is an automorphism, so these three vertices all belong to another orbit. We can find all the orbits, and write the automorphism partition as

$$P = \{\{1,2\},\{3\},\{4\},\{5,6,7\}\} \tag{A9.12}$$

Measures of symmetry of networks

If a graph is shown to be symmetric, the next question is: How symmetric is it? I describe two measures of such symmetry here (Xiao *et al.* 2008).

The order $|\text{Aut}(G)|$ or α_G of the automorphism group of a graph is clearly a measure of the extent of symmetry in the graph. We should normalize it to be able to compare the symmetries of graphs of different orders N. The following is one such normalized measure of symmetry

(MacArthur and Anderson 2006):

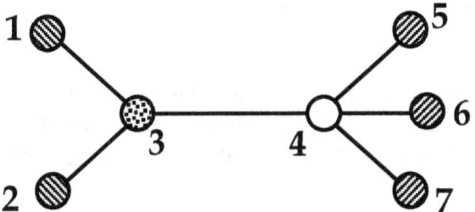

Fig. A9.2 Example of a symmetric graph (from Xiao *et al.* 2008b).

$$\beta_G = (\alpha_G / N!)^{1/N} \tag{A9.13}$$

It is intuitively clear from Fig. A9.2 and Eq. A9.10 that a network or graph with more nontrivial orbits is more symmetric. Xiao *et al.* (2008b) have used a measure of symmetry which accounts for this:

$$\gamma_G = \frac{\sum\limits_{1 \le i \le k, |V_i| \ge 1} |V_i|}{N} \tag{A9.14}$$

Here V_i is the ith orbit in the automorphism partition, and k is the number of cells in the partition.

A9.10 Chua's cellular nonlinear networks as a paradigm for emergence and complexity

Cellular nonlinear networks (CNNs) were introduced and analysed by Chua (1998) as new a paradigm for emergence and complexity. A typical CNN has a spatially discrete collection of cells, and interactions among the cells. The cells are nonlinear dynamical systems wherein information can be encrypted by three independent variables: input, threshold, and initial state.

The interactions among the cells are defined via *coupling laws*. A coupling law relates one or more relevant variables of each cell C_{ij} to all neighbouring cells C_{kl} which are within a prescribed sphere of influence $S_{ij}(r)$ of radius r, centred at C_{ij}.

If a CNN has only a homogeneous array of cells, and the cells have no inputs, thresholds, or outputs, and if the influence of any cell extends only to the nearest neighbours, i.e., $r = 1$, the CNN is nothing but a *nonlinear lattice*.

Let x_{ij}, y_{ij}, u_{ij}, and z_{ij} denote scalars which are, respectively, the state, the output, the input, and the threshold of a cell C_{ij}. The standard CNN equation is then written as (Chua 1998):

$$
\begin{aligned}
\dot{x}_{ij} &= -x_{ij} + \sum_{k,l \in S_{ij}(r)} a_{kl} y_{kl} + \sum_{k,l \in S_{ij}(r)} b_{kl} u_{kl} + z_{ij} \\
y_{ij} &= f(x_{ij}) = \tfrac{1}{2}(|x_{ij}+1| - |x_{ij}-1|) \\
i &= 1,2,\dots,M; \, j = 1,2,\dots,N
\end{aligned}
\tag{A9.15}
$$

Here a_{kl} and b_{kl} are scalars called *synaptic weights*.

When the radius of influence is unity ($r = 1$), a standard CNN is uniquely defined by 19 real numbers: a uniform threshold $z_{kl} = z$, nine feedback synaptic weights a_{kl}, and nine control synaptic weights b_{kl}. Such a CNN is called a *gene* because it determines completely the properties of the CNN.

A universe of all CNN genes is called a *genome*. A large number of applications, ranging from brain science to image processing, can be implemented by a CNN 'program', defined by a string of CNN genes, called a CNN *chromosome* (like in an implementation of a genetic algorithm).

Here is a sampling of new results and insights obtained by employing the CNN paradigm (Chua 1998):

• Theorem: Every Boolean function of the neighbouring-cell inputs can be explicitly synthesized by a CNN chromosome. This implies that every cellular automaton with binary states is a CNN chromosome.

• In particular, the Game-of-Life CA can be realized by a CNN chromosome made of only three CNN genes. This means that the Game-of-Life CNN chromosome is a universal Turing machine, and is capable of self-replication in the von Neumann sense (Berlekamp, Conway and Guy 1982).

• The new concept of *generalized cellular automata* (GCA) is introduced. It is distinct from the von Neumann CA in that it cannot be defined by local rules. Rather, it is defined by iterating a CNN gene or chromosome in a 'CNN DO LOOP'. The GCA include not only *global* Boolean maps, but also *continuum-state CA* wherein the initial state configuration and its iterates are *real* numbers, and not just a finite number of states as in classical CA.

• Successful implementation of the so-called *CNN universal chip* is another new result described in Chua (1998). It is an analogue-input analogue-output CNN universal machine on a single silicon chip. It is a complete dynamic-array, stored-program, computer in which a CNN algorithm or flow-chart can be programmed and executed on the chip at the extremely high speed of 10^{12} analogue instructions per second. It is based entirely on nonlinear dynamics.

• If there are no inputs to a CNN, we are dealing with an *autonomous* CNN. It is found that this subclass of CNNs 'can exhibit a great variety of complex phenomena, including pattern formation, Turing patterns, knots, autowaves, spiral waves, scroll waves, and spatiotemporal chaos. It provides a unified paradigm for complexity, as well as an alternative paradigm for simulating nonlinear partial differential equations (PDE's)' (Chua 1998).

• Chua (1998) advocates that nonlinear PDEs should be regarded as mere idealizations of CNNs. This is a much stronger stance compared to saying that autonomous CNNs should be regarded as approximations of PDEs. It is reminiscent of a similar point made by Wolfram (2002) regarding his 'new kind of science' (NKS).

• *The local-activity dogma*: Chua (1998) asserts that 'all of the phenomena described in the complexity literature under various names and headings (e.g., synergetics, dissipative structures, self-organization, cooperative and competitive phenomena, far-from-thermodynamic equilibrium phenomena, edge of chaos, etc.) are merely qualitative manifestations of a more fundamental and quantitative principle called the local-activity dogma'. According to the dogma, in order for a non-conservative system or model to exhibit any form of complexity, the associated CNN parameters must be chosen so that either the cells or their couplings are *locally active*.

Local activity: the genesis of complexity

As stated above, Chua has introduced the central concept of *local activity of cells* in a CNN which, he asserts, is at the heart of a variety of manifestations of complexity. The CNN paradigm is a universal Turing machine, of which CA and lattice-dynamical systems are special cases. The CNN approach is able to trace the origin of emergence and complexity to what Chua calls *local activity* of the cells of the CNN. Complex phenomena like 'order from disorder' (Schrödinger 1944), 'cooperative phenomena' (Haken 1975, 1983), 'self-organization' (Nicolis and Prigogine 1977), 'dissipative structures' (Prigogine 1980, 1996), 'synergetics' (Haken 1983), 'slaving principle' (Haken 1983), 'edge of chaos' (Langton 1990), etc., are explained in terms of a CNN operating near the edge of chaos, where the cells of the CNN are not only locally active, but also *linearly asymptotically stable* (Dogaru and Chua 1998). The local-activity idea has the advantage of being explicit and mathematically definable. One can check mathematically whether or not a cell in a CNN is locally active. The idea of local activity, though of universal applicability, has been discussed in substantial detail by Chua (1998) by considering the case of *the reaction-diffusion CNN*.

The possible presence of locally active cells in a CNN is the equivalent of *non-homogeneous networks* in which local decrease of entropy is possible. Like in the edge-of-chaos picture, such cells are locally stable but potentially unstable. By contrast, locally passive cells are the equivalent of 'lifeless' systems.

Local activity is only a *necessary* condition for complexity. To obtain sufficient conditions, one must bring in the coupling circuits. The main idea is that if the cells are not locally active, no passive coupling circuits of any complexity can be found which can lead to pattern formation in the CNN.

The CNN formalism provides a generalization and unification of a number of fields of research coming under the umbrella term *computational intelligence*, which covers the artificial evolution of complexity also. In fact, it does more than that by coming up with a number of new, mathematically precise, results about complexity, some of which I have stated above.

The CNN formalism is universal in the sense that it is system-independent, having been abstracted mathematically from the law of conservation of energy. Therefore it provides a mathematically precise and universal measure of the degree of complexity.

A10. Game Theory

Life is not a game. Still, in this life, we choose the games we live to play.
J. R. Rim

A10.1 Introduction

Game theory is the mathematical study of situations of conflict of interest. It is the analysis of individual and group decision-making in situations involving a diversity of individual goals and differing levels of strategic control over the environment (Owen 1982, 1992; Shubik 1993; Turocy and von Stengel 2002). It provides a mathematical description of certain interactive phenomena among two or more members of a population.

A *game* is a model of an interactive situation. The formal definition of a game specifies the players, their preferences, their information, the strategic actions available to them, and how these actions may influence the outcome. Naturally, game theory is very useful for understanding certain kinds of biological and cultural evolutions of complexity.

The publication of the book by von Neumann and Morgenstern (1944) on game theory created a major and still expanding interest in this subject. This book included for the first time a proof of the so-called *Fundamental Theorem* in the theory of games (namely *the minimax theorem*), which I shall state below. Classical game theory provides a mathematical formalism of strategy as an extension of individual *rational* behaviour. Since human beings are not exactly famous for rational behaviour, classical game theory is complementary to social psychology and other behavioural sciences which model human actions in terms of limited rationality and nonconscious behaviour. The initial work in game theory was directed towards economics, but was soon diversified to a theory of competitive games relevant not only to competitive economics, but also combat and warfare.

Game theory is, by and large, based on the assumption of *rational choice*. Game theory is widely employed in the social sciences, and the rational-choice assumption is kept intact for modelling many of the real-life situations. This is something unrealistic, but the assumption is made nevertheless, mainly because it allows for the deductive form of reasoning for drawing conclusions (Axelrod 2006).

The main alternative to rational choice is *adaptive behaviour*. And this is what happens in complex adaptive systems (CASs). Members of a CAS have to optimize their interaction not only with the ever-changing environmental conditions, but also with the survival strategies adopted by fellow members of the CAS, as well as other interacting species (Dawkins and Krebs 1979; Axelrod 1984, 2006). Both competition and cooperation play a role in the survival strategies adopted by interacting species. Particularly notable in this context is the pioneering work done by Maynard Smith (1974, 1976) on evolutionary dynamics. He had to modify the game theory existing at that time for analysing the nature of competition and cooperation among species. The important idea he introduced was that of *evolutionarily stable strategies* (ESS) (see Chapter 30).

Abandoning the rational-choice paradigm of game theory makes it very difficult to deduce analytically the consequences of the adaptive processes involved (because of the complex nonlinear interactions among the agents), and computer simulation ('agent-based modelling') is then the only way out for understanding the phenomena involved.

The first defining feature of a game is the number of players involved. In general, this number can be n, and *the set of players* can be represented by $N = \{1,2,...n\}$.

If $n = 0$, we speak of a *zero-player game*, an example being a cellular automaton. Once an automaton starts, it keeps going, without any decision-making imposed on it by a person.

In game theory one generally assumes that the players are *rational*. A rational player is defined as one who always chooses an action which gives the outcome he most prefers, given that he expects his opponents to be rational too.

When $n = 1$, game theory becomes *decision theory*. Games of solitaire are examples of one-person games.

$n = 2$ games, or *two-player games* are the best investigated; the concepts and the conclusions are clearer for them.

For modelling in macroeconomics, the number of players can be extremely large, and sometimes it is assumed to be infinite. One even speaks in terms of a *continuum* of players if the influence any one player has on the game is infinitesimally small.

Apart from the number of players in a game, a characteristic feature that can distinguish one form of game from another is the *level of detail* considered important or relevant for playing the game. One can distinguish three models or forms of games by this criterion: *extensive* form of games; *normal or strategic form* of games; and *coalitional* form of games.

Extensive form of games

Maximum detail is available in the extensive form of games, or *extensive games*. One can speak of a *position* in the game, and of a *move* in the game. A move takes the game from one position to another. This later position can depend on which player's turn it was to make a move. A player may even make random moves (e.g., rolling of dice, or shuffling of cards before dealing them in a game of cards).

Any game is played according to certain *rules*. For example, the rules may specify the probabilities of the outcome of random moves.

In an extensive model of a game, players may have *information* before making a move. A *perfect-information game* is that in which each player knows about all the past moves by the players and the results of all such moves, as also the results of all the past random moves. Two-person perfect-information games, with no chance moves, and with complete knowledge of win or lose outcomes, are called *combinatorial games*. Such games may be either *impartial* or *partisan*. In an impartial game, the two players can make the same set of valid moves from each position. If this is not the case, it is a partisan game.

In the language of graph theory, an extensive game can be represented by a *tree* which depicts the order in which the players make moves, and the information each player has at each decision point.

Normal or strategic form of games

Unlike in the extensive form of games, normal or strategic form of games does not involve details like 'positions' and 'moves.' Rather, we speak in terms of *strategy* and *payoff*. We specify a *strategy set* (also called *action space*) A_i for each player i ($i = 1, 2, ..., n$). Thus, each player can choose a strategy from this possible set of strategies, after considering all the players and their strategies. All players make their choices of strategy simultaneously, and the choices are revealed to all of them. At the end of the game, each player receives some payoff. The choice of strategy made by each player may influence the final outcome for all the players.

If the players are assumed to be rational, they make choices which result in the outcome that they prefer the most, given what their opponents do. For example, a player may have two strategies A and B such that, given any combination of strategies of the other players, the outcome from strategy A is better than the outcome from B. Then A is said to *dominate B*. A rational player will always play a dominant strategy if he can.

The payoffs are generally modelled as having numerical values. For player 1, the strategy set is A_1. Suppose he/she chooses strategy a_1 from this set. Similarly, any player i may choose a strategy a_i from the set A_i. Then the payoff to a player j ($j = 1, 2, 3, ..., n$) is a function $f_j(a_1, a_2, ... a_n)$, called the *payoff function* for player j.

The *strategic form of a game* is defined by the set N of players, the sequence $A_1, A_2, ... A_n$ of the set of strategies of the players, and the sequence $f_1(a_1, a_2, ... a_n)$, $f_2(a_1, a_2, ... a_n)$, ... $f_n(a_1, a_2, ... a_n)$ of the set of payoff functions for the n players.

A game in strategic form is called a *zero-sum game* if the sum of payoffs to all the players is always zero. Two-player zero-sum games are the best investigated. Each such game has a *value*, and the two players have *optimal strategies* that guarantee the value. Because of the minimax theorem (see below), such games have clear-cut solutions.

Two-person non-zero-sum games can be tricky business. Usually such games do not have values or optimal strategies for the players. Two types of theoretical approaches can be used for dealing with such games: *Noncooperative theory*, and *cooperative theory*.

In the noncooperative form of the game, the players may not form binding agreements. This can happen, for example, during negotiations between sovereign nations when there is no overseeing body to enforce agreements. Similarly, in business dealings, trading companies may be forbidden by law to enter into certain types of agreements. John Nash, John Harsanyi and Reinhard Selten were awarded the Nobel Prize in economics in 1995 for work in this area. In their formalism, the concept of *strategic equilibrium* or *Nash equilibrium* replaces value and optimal strategy.

In cooperative game theory (the other approach to non-zero-sum games), the players are permitted or willing to form binding agreements. Thus there is a strong incentive to cooperate for maximizing the total payoff. But now the problem is about how to split the payoffs between the players. There are two kinds of cooperative games. The game may have *transferable utility*, or it may have *non-transferable utility*. Transferable utility is relevant when the players measure utility of the payoff in the same units and there is a means of exchange of utility, such as *side payments*.

The payoffs in the strategic form of a game can be presented in the form of a table, each cell of which represents a distinct strategy combination. Such a tabular presentation becomes possible because the players can choose their strategies simultaneously.

Coalitional or cooperative form of games

When the number of players is large, the available information is usually inadequate for either an extensive-form approach or a strategic-form approach for analysis. Instead, one uses the notions of a *coalition*, and the *value* or *worth* of a coalition. In a many-player scenario, the players are likely to form coalitions for mutual benefit. So the notion of strategies of individual players is replaced by that of coalitions, and payoffs are replaced by the value of the coalition. It is expected that the coalition can guarantee its members a certain amount of benefit, called the *value of the coalition*.

In a coalitional form of a game, the *process* by which the coalition forms is not specified. For example, the players may be a number of parties in parliament. Each party has a strength determined by the number of seats it has in the parliament. The game describes which coalitions of parties can form a majority, but does not specify, for example, the negotiation process through which an agreement to vote en bloc is achieved.

Coalitional game theory is a subset of *cooperative game theory*, with transferable utility. There is a grand coalition involving all the members of the coalition, with some rule(s) for distribution among members the total payoff received by the grand coalition. For example, in the context of an economy, there is the important notion of the *core* of the economy. It is a set of payoffs to the players such that each coalition (as a part of the grand coalition) receives at least its value. There are principles that can result in a unique distribution of the payoff from the grand coalition.

Noncooperative game theory, on the other hand, is concerned with the analysis of strategic choices. Here the details of the ordering and timing of players' choices are crucial to determining the outcome of a game. In a noncooperative model of bargaining, one defines a specific process in which it is stated who gets to make an offer at a given time. Even cooperation can arise in a noncooperative model of a game when players find it in their own interests to cooperate rather than compete.

A10.2 Dual or two-player games

Prisoner's Dilemma

> *There is something rational about choosing not to be rational when playing Traveller's Dilemma.*
>
> Kaushik Basu (2007)

There are two prisoners suspected of a serious crime. There is no judicial evidence for the crime except if one of the prisoners testifies against the other. The two are interrogated separately. If one of them incriminates the other, he will be let off without any punishment (amounting to a payoff of, say, 4), whereas the other prisoner will be awarded a long prison sentence (payoff 0). If both testify against each other, their punishment will be somewhat less severe (payoff 2 for each). However, if they both 'cooperate' with each other by not testifying at all (i.e. by choosing to remain silent), they will only be given a very short term in jail (due to lack of strong evidence), for example for possession of illegal weapons (payoff 3 for each, with a total payoff 6). This is a nonzero-sum strategic game: The total payoff is 4 if both incriminate each other. It is 4 if any one of them testifies against the other; the defecting prisoner is rewarded with immunity from prosecution, and the other gets a long term in jail. If they cooperate with each other (unknowingly, of course), both get only a very short term of imprisonment, and the total payoff is 6. *To defect or to cooperate, that is the prisoner's dilemma (PD).*

In this two-player game, let the first prisoner be called player I, and the other one is player II. And let C or c stand for cooperation, and D or d for defection: C and D are the options for player I, and c and d those for player II. The payoff table for the PD game is shown in Fig. A10.1. Player I can choose either row C or row D. At the same time, player II chooses column c or column d. The possible strategy combinations for the two players are: (C, c), (D, d), (C, d), and (D, c). For (C, c) the payoff for each player is 3, and that for (D, d) it is 2 for each player.

II→ I↓	c	d
C	3 3	4 0
D	0 4	2 2

Fig. A10.1 Payoff matrix for the prisoner's dilemma game.

If only player I defects, i.e., strategy (D, c), the payoff to player I is 4, and that to player II is 0. If only player II defects (strategy (C, d)), he gets a payoff of 4, and player I gets 0.

Two important features of this payoff matrix should be noted. One is that it is *symmetric* about the diagonal from top left to bottom right. The strategic game is symmetric between the two players. This means that the payoff matrix remains the same if the roles of player I and player II are interchanged. This symmetry becomes possible because there is no order between the two players; they act simultaneously, not knowing the action of the other player.

The other feature to note in Fig. A10.1 is that the 'defect' strategy dominates over the 'cooperate' strategy. Consider player I. If II chooses c, the payoff for I is 4 for D and 3 for C. If II chooses d, the payoff for I is 2 for D and 0 for C. In either case, D dominates over C for player I. And the symmetry of the payoff matrix implies that for player II also, 'defect' dominates over 'cooperate'. Thus the selfishness of the independently acting players in the PD game makes defection a dominant strategy, and cooperation a dominated strategy, resulting in a lower total payoff for the two players. Cooperation would have given them the best total payoff of 6. There are many real-life versions of the PD game wherein individual defections at the expense of others lead to outcomes which are, on the whole, less profitable. Here are some examples: arms races; litigation instead of settlement; environmental pollution. Treaties or laws are introduced sometimes to enforce cooperation.

The PD game gets fundamentally changed if played more than once by the same players. In such an *iterated game*, patterns of cooperation emerge as rational behaviour. The fear of punishment in the future outweighs the gain from immediate defection (Axelrod 1984, 2006). What happens when the players are not humans, but 'non thinking' organisms? How does cooperation, rather than only 'tit for tat' or competition, evolve in Nature? Axelrod's answer was the genetic-algorithm (GA) approach developed by his friend John Holland. Several examples of the actual models developed using GAs can be found in his book *The Complexity of Cooperation* (Axelrod 2006). I shall skip the details here.

As Axelrod (2006) said, the PD game has become the E. *coli* of the social sciences, providing a common framework for a large variety of studies. It is an extremely simple model that captures a fundamental feature of many interactions, namely the tension between the advantages of selfishness in the short run and the need to elicit cooperation from the other players to be successful in the long run.

Traveller's Dilemma

A generalization of the PD game is the traveller's dilemma (TD) game, crafted in 1994 by Basu (2007). Lucy and Pete go to a remote island in the Pacific, where they purchase two identical antiques. The antiques get damaged during their return journey, and the airline manager offers to compensate, accept that he has no idea about the actual price of these objects. To make sure that the airline does not have to pay more than what is necessary, he comes up with a game plan that ensures that the travellers do not get away by quoting an inflated figure. He asks each of them to write down simultaneously and without consulting each other the price of each antique as any integral number of dollars between 2 and 100, with the following proviso: If both write the same number, he will pay that amount to each of them. But if they write different numbers, he would conclude that the lower price is the correct one, and that the other person is trying to cheat. Therefore, the person writing the lower price will be rewarded for honesty with an extra two dollars, and the other person will be punished with a payment that is two dollars less than the lower of the two prices submitted by the travellers. For example, if Lucy writes $54 and Pete writes $100, then Lucy will be paid $56, and Pete will get $52. What will the two travellers write to maximize their gains? That is the traveller's dilemma.

Although most people will choose a number 100 or one close to 100, logic demands that both players should choose 2. The actual price of each antique was considerably lower than $100, but let us first suppose that Lucy decides to write $100, hoping that Pete would be similarly greedy. Lucy's logic is that if they both wrote $100, they would both get $100. But she soon realizes that she can get a little more money by writing 99, because she would then get $101 if Pete has written $100. But the same thought can occur to Pete also, in which case Lucy would be better off by writing 98. She would then get $100, if Pete wrote 99 or 100. But once again, if this idea occurs to Pete also, Lucy should be writing $97. This kind of logic (called *backward induction*) can go on till both should write $2. So if both players in the TD game are rational, their joint payoff should be (2, 2). Fig. A10.2 gives the complete payoff matrix for this two-player strategic game.

	2	**3**	**4**	**...**	**98**	**99**	**100**
2	2 2	4 0	4 0	...	4 0	4 0	4 0
3	0 4	3 3	5 1	...	5 1	5 1	5 1
4	0 4	1 5	4 4	...	6 2	6 2	6 2
...	...	...	...	...	...	...	...
98	0 4	1 5	2 6	...	98 98	100 96	100 96
99	0 4	1 5	2 6	...	96 100	99 99	101 97
100	0 4	1 5	2 6	...	96 100	97 101	100 100

Fig. A10.2. Payoff matrix for the traveller's dilemma game between Lucy and Pete (Basu 2007). The leftmost column in this table gives Lucy's choices in dollars, and the top row those of Pete. The square where a chosen row and a chosen column intersect shows the payoffs of Lucy and Pete respectively. For example if Lucy chooses 4 and Pete chooses 98, the payoff to Lucy is $6 and that to Pete is $2. The payoffs (2, 2) when both players choose 2 represent Nash equilibrium: Any unilateral deviation from the Nash choices gives a lower payoff. For example, if Lucy stays at the equilibrium by choosing 2, then Pete does worse by choosing any number other than 2.

The existence of the Nash equilibrium in the TD presents a logical paradox. Although it is a rationally correct solution of the TD, most of us would choose a number much higher than 2. It appears that our intuition contradicts the tenets of game theory. Classical economic theory was based on the axiom that people make choices that game theory can predict. Classical game theory assumes that people are selfish *and rational*. But in practice, as demonstrated by several experiments in which real money was used for participating persons, and in which a variety of magnitudes of rewards and punishments were tested (Basu 2007), *people do not play the Nash strategy on average*. For low rewards, the average choice of numbers was high, and it fell somewhat when the rewards were increased. Repeated playing of the TD threw up some surprises as well. For high rewards, the play converged over time down to the Nash-equilibrium value. But for low rewards, the repeated play went in the other direction, i.e., more and more away from Nash equilibrium. Such experiments come under the relatively new field of research called *experimental economics*.

Why do people behave in such an illogical irrational manner? Why do they not follow the rational path to Nash equilibrium? Several reasons have been advanced (Basu 2007):

1. Many people are not capable of deductive reasoning, and end up making illogical, impulsive, irrational moves. But then why do even game-theorists, who presumably do not belong to this category of people, and who know how to reason deductively, end up making irrational choices? It turned out in real experiments that they chose high numbers because they expected other participating game-theorists to choose high numbers!

2. Perhaps there is a tussle between altruism and selfishness in our brains, with simple-minded Darwinism favouring selfishness (Axelrod 2006).

3. Perhaps many of us do not like to 'let down' our fellow traveller in the TD game just for the sake of getting rewarded an additional dollar. So we may choose 100, knowing fully well that choosing 99 can give us $101.

This is really an unsolved problem. Even if we eliminate factors such as faulty reasoning, altruism, and socialization, it is still doubtful if people would play logically and ruthlessly. *People do not always make selfish rational choices* (Ball 1999a).

The ultimatum game

The real-life experience with the TD results in two conclusions:

1. The most 'efficient' game for the two players to play, that would give the maximum total payoff, is that each chooses 100, giving a payoff (100, 100); i.e., 200. This goes against the grain of classical economics, which was based on the 'libertarian presumption' that if individuals are left to their own selfish devices, the economy would run efficiently.

2. People do not always tend to play the Nash-equilibrium strategy, which in the TD game will give a payoff (2, 2). This violates the assumption of classical economy that people take rational decisions.

Thus people do not always make selfish rational choices. The ultimatum game (UG) is another game to illustrate this point.

Imagine Mr. A and Mr. B, not known to each other, and in separate rooms, unable to communicate with each other. Somebody plays the ultimatum game (UG) with them. He offers $100, which must be shared between them according to certain strict rules. The rules are known to both Mr. A and Mr. B. One of them, chosen at random, has to decide how the money is to be divided and makes this offer to the other person. If the other person accepts the offer, the deal is closed. If he rejects the offer, none of them gets anything. What amount will be offered by the person who has been chosen at random to make the offer?

Our first response may be: 50%. But is this what really happens in practice? Suppose Mr. A has been chosen to make the offer to Mr. B. He may very well think that even if he offers 40%, Mr. B is going to accept it, because $40 is much better than getting nothing at all. But then why not offer just $30? How about $20, 10, … , or even $1? Will Mr. B accept $1 dollar, which is after all better than getting nothing at all, even though Mr. A walks away with $99?

Experimentally, it is found that two thirds of the offers are between 40 and 50% (Sigmund, Fehr and Nowak 2002). Only 4% people offer less than 20% money; offering such a small share is considered risky because the offer may get rejected. More than 50% of all responders reject offers that are less than 20%. But the question is: Why should anyone reject an offer as 'too small'? Obviously, factors such as sense of fair play and/or sense of outrage are at play.

The economics of fair play

Analyses based on Darwinian evolution ideas demonstrate that fairness evolves if the proposer has some information about what deals the responder has accepted in the past (Nowak, Page and Sigmund 2000). Thus, evolution of a sense of fair play, like the evolution of cooperation (Maynard Smith and Price 1973), is linked to reputation. The opposite is also true. If an individual has the reputation of being content with low offers, he is more likely to be offered small shares.

Theoretical classical economics modelled a human being as *Homo economicus*: a rational and purely selfish individual. Studies on the UG and its variants reveal, however, that humans actually are a cross between *H. economicus* and *H. emoticus*. Emotions like generosity, revengefulness, altruism *can* override purely logical and selfish pursuits sometimes. How did this trait evolve in the Darwinian sense? There must have been some evolutionary advantage in this for such genes to survive and propagate. Yes indeed. Unfair play arouses acts of moral indignation and revenge, which can be costly to the social group or the species. Fair play and altruism may not appear to be of immediate benefit to the individual, but can have long-term benefits to the group or the species. Over several generations, those genotypes can evolve for which the phenotype is such that, in pairwise encounters, a purely self-centred strategy is not adopted; instead, due account is taken of the other player's conduct. A player is not interested only in his own payoff, but compares it with that of the other player, and demands fair play.

Such departures from the game-theoretic predictions of the UG have been rationalized in various ways. The chief among them is the premise that the strict rules prescribed by the UG are far too removed from what actually happens when real people deal with one another. For example, haggling is not permitted in the UG, whereas it must have been a common theme in the social life of our ancestors. Therefore, real players are likely to forget or ignore that the UG is a one-off game (the deal is either on or off), with no bargains or second chances.

Another factor to consider is the existence of rather strongly knit groups within a species. Such groups provide a sense of belonging, as also the assurance of protection against enemies or rival groups. Naturally, it makes sense not to outcompete or injure other members within the group to such an extent that successful contests with *other* groups are jeopardised. But while this line of reasoning does explain why proposers offer large sums, it cannot explain why low offers are rejected by responders.

One more explanation softens the condition of the UG that there is strict secrecy, and that a contestant has no access to information about the other contestant (Sigmund, Fehr and Nowak 2002). The fact is that our social evolution has been such that we expect that others will notice our actions and decisions. For example, suppose a player is known to become angry and reject a low offer, others are likely to make him high offers. Therefore, emotional responses to low offers must have been favoured by evolution. In due course, self-esteem entered the picture, making it necessary for a player to reject a low offer 'as a matter of principle,' making it a matter of dignity.

Other factors which influence the outcome of the UG, and also of some of the 'Public Good' games which include a punishment rule, are emotions like a sense of revenge, and the satisfaction of punishing selfish action (Sigmund, Fehr and Nowak 2002). Such responses have evolutionary underpinnings which ensure fitness advantage for the social group. In times of war, pestilence, or famine, an above-average proportion of punishers helps the group to survive as a whole. *Moral guidelines are an offshoot of such tendencies, and they apply to the economic affairs of the community as well.*

Zero-sum two-player games

In such games a single number, i.e., the amount won by the first player, and therefore the amount lost by the second player (or vice versa) determines the payoff. If it is a finite game, its normal form is a matrix of payoffs, such as that shown in Fig. A10.3.

$$\begin{bmatrix} 1 & 3 & 5 \\ 2* & 4 & 3 \\ 0 & 1 & -3 \end{bmatrix}$$

Fig. A10.3. A typical payoff matrix for a two-player zero-sum game (Owen 1992). The entry with a star is a *saddle point*.

Here each row corresponds to the strategies of player I, and each column to those of player II. Suppose player I chooses a strategy corresponding to row 2. The payoffs are 2, 4, or 3, depending on the strategy chosen by player 2. Thus, player I can be certain of winning at least 2 units. By contrast, if player I chooses row 1, what he can win at least is 1 unit. Similarly, if he chooses row 3, what he can win at least is 1 unit. Thus the row 2 strategy is his *maximin strategy* (it maximizes his minimum winnings), and is called the *security level* of this player. The security level of a player is the largest amount he can be sure of winning, no matter what games other players play.

Now consider player 2. For columns 1, 2, and 3, his maximum payoffs are 2, 4, and 5, respectively. Thus column 1 corresponds to his *minimax strategy* (it gives the minimum of the three maximum payoffs).

In this example, 2 units is both the maximin and the minimax payoff. By choosing row 2, player I is sure to win at least 2 units, and by choosing column 1, player II is sure to lose no more than 2 units. It is a zero-sum game, and row 2 and column 1 are *optimal strategies*. A game of this type in which the maximin of one player coincides with the minimax of the other player has a *saddle point* in the payoff matrix; the saddle-point payoff is 2 units in our example.

Suppose we make a 3-dimensional plot of the payoff to player I as a function of his two strategies plotted along the other two axes. The maximin strategy will be seen at the valley in this plot. Now suppose we plot the payoff for player II as a function of his two strategies. The minimax strategy will show as a peak in this plot. When the maximin and the minimax coincide, we see a saddle point in the plot. If the two players can be presumed to be rational, they would both hit the saddle point *without any need to communicate with each other*, except through the consequences of the games played by each. This is a notable result from the point of view of emergence of swarm intelligence and complex behaviour. Each bee in a beehive is obviously 'rational' in the sense that it has no emotions to speak of. Its actions are determined by the innate tendencies (*simple rules*) dictated by its genes. Although there is an apparent visual 'communication' with other bees (the waggle dance), the fact remains that even the response to the dance is nothing more than an operation of simple local rules in a rational manner. The existence of the saddle point, although arrived at above in the context of a zero-sum two-player game, points to the existence of a sustainable level of survival of the bee species, each bee playing the role of a rational player at the individual level.

It was von Neumann who showed in his pioneering work on game theory that for two-person zero-sum games, the idea of rational behaviour can be extended by using the concept of *individual rationality*.

In general, the maximin and minimax strategies may not give the same payoff, and this is illustrated in Fig. A10.4. It is the payoff matrix for a rudimentary version of poker (Owen 1992). Player 1 gets to choose from four possible strategies, and player II must choose from two possible strategies. Here the maximin is -0.6 in row 2, and the minimax is -0.2 in column 1. The difference is not zero, and represents an *indeterminacy* of the game. In such situations it is advisable for the two players to adopt *mixed strategies*. The idea is to prevent the other player from guessing the strategy by choosing a row or a column according to some randomizing scheme. For example, suppose player 1 plays row 1 with probability 1/8, row 2 with probability 7/8, and does not play rows 3 and 4 at all. The four choices can be represented by a vector $x = (\frac{1}{8}, \frac{7}{8}, 0, 0)$, and give player 1 the winnings of at least -0.4. This means that he expects to lose no more than 0.4 units, irrespective of what game the opponent plays. Similarly, player 2 may be advised to play either column with a probability ½ ($y = (\frac{1}{2}, \frac{1}{2})$). The expected payoff for this approach is at the most 0.4.

With this mixed strategy, player 1 expects to lose no more than 0.4, and player 2 expects to win at least 0.4. The number -0.4 is *the value of the game*.

$$
\begin{bmatrix}
-1.8 & 1 \\
-0.2 & -0.6 \\
-2.2 & 1 \\
-0.6 & -0.6
\end{bmatrix}
$$

Fig. A10.4. Payoff matrix representing the normal form of a poker game (Owen 1992). Here each row corresponds to the strategies of player 1, and each column to those of player 2.

We can generalize. Consider a game in which the payoff matrix **A** is an $m \times n$ matrix. The mixed-strategy for player 1 is a vector $x = (x_1, x_2, ..., x_m)$. The components of this vector, which are probabilities, are nonnegative, and add up to 1. For player 2 there is a similar mixed-strategy vector $y = (y_1, y_2, ..., y_n)$. The expected payoff for mixed strategies used by the two players can be written as

$$P(x,y) = xAy^t \tag{A10.1}$$

The maximin in this general case is

$$V_I = \max_x \min_y xAy^t \tag{A10.2}$$

And the minimax is

$$V_{II} = \min_y \max_x xAy^t \tag{A10.3}$$

The minimax theorem

The minimax theorem says that, for all matrix games, $V_I = V_{II}$. The common value of the payoff is called the *value* of the game. The maximization of **x** and the minimization of **y** are *optimal mixed strategies*.

Linear programming is the most common technique used for the computation of optimal strategies in a matrix game.

Chess is an example of a zero-sum two-player game. There is a theorem in game theory according to which *any finite, zero-sum, two-player game like chess has an optimal solution.* That is, there is an optimal way of choosing moves in chess that will make a player do better than any other set of moves. The total possible number of moves in chess is 10^{120}. The problem is that no one knows what that optimal solution is. One never plays the same game twice.

Next, we shall take up games involving more than two players, i.e., *plural games.* In any game, whether dual or plural, the total payoff may be zero (zero-sum games) or nonzero (nonzero-sum games). For the latter category, the interests of the players may not be opposite always; i.e., both noncooperation and cooperation are conceivable.

A10.3 Noncooperative games

In a noncooperative game, communication, correlated strategies, side payments, binding contracts etc. are not available to the players. Let us first consider a two-player game of this type. One usually looks for equilibrium points. Any such point is a pair of strategies such that neither player gains by a unilateral change of strategy. Consider the payoff matrix shown in Fig. A10.5.

$$\begin{bmatrix} (2,3) & (4,4)* \\ (5,2)* & (3,1) \end{bmatrix}$$

Fig. A10.5. A payoff matrix illustrating the presence of *two* equilibrium points (Owen 1992).

The second-row first-column payoff (5, 2) is an equilibrium point. If player 1 unilaterally chooses the first row instead of the second, his payoff goes down from 5 to 3. Similarly, if player 2 unilaterally chooses the second column instead of the first, his payoff goes down from 2 to 1.

Interestingly, Fig. A10.5 has one more equilibrium pair of strategies, namely at first row second column, with a payoff (4, 4). This possibility of more than one equilibrium points is characteristic of nonzero-sum games. It does not occur in zero-sum games.

We have seen earlier that the equilibrium-point scenario is not always in the interests of the players. For example, in the case of the prisoner's dilemma (Fig. A10.1) and the traveller's dilemma (Fig. A10.2), the equilibrium-point strategy, even though unique, gives a payoff (2, 2), which is not in the best interests of the players.

Some improvements over the equilibrium-point concept are the concepts of *undominated equilibrium*, and *perfect equilibrium* (Owen 1992).

A10.4 Nash equilibrium

For plural games in strategic form, the most well-known solution is in terms of the noncooperative equilibrium, better known as the Nash equilibrium (Nash 1950, 1951, 1953). Suppose there are n players in a game, with the set of players represented by $N = \{1,2,...n\}$. Player i selects a strategy s_i from his set of pure strategies S_i. Any vector $s = (s_1, s_2,..., s_n)$ represents a possible set of strategies chosen by the n players. Suppose we replace the ith component of this vector by a strategy $s_i *$. The Nash theorem states that *there exists a pure-strategy noncooperative equilibrium vector* $s* = (s_1 *, s_2 *,..., s_n *)$ *which represents an* optimal *strategy or response, meaning that, given the strategies of all the other players, no individual player i can improve his payoff by unilaterally selecting a strategy s_i other than $s_i *$.*

The theorem can be extended to the case of mixed strategies by allowing each player to select a probability distribution over his set of pure strategies.

The Nash equilibrium situation is one of *self-fulfilling expectations*. Suppose each player is a strictly rational individual, and knows or expects all other players to be rational as well. There is then a mutually consistent set of expectations such that if each player optimizes his expectations, then the predictions of each player will be fulfilled.

But even in the perfect-rationality scenario, the problem with Nash equilibrium is that, as seen in Fig. A10.5, there may be more than one Nash equilibria. Moreover, as seen above with the prisoner's dilemma and the traveller's dilemma problems, the Nash equilibrium choice of strategies in not always the most efficient and profitable way of making strategic moves (Dubey 1986; Dubey and Rogawsky 1990).

The Nash equilibrium solution has been mostly applied to the strategic form of games. Applying it to the extensive form can be illuminating, as the extra detail of information can highlight many additional aspects of the game. An example is the introduction of the concept of *perfect equilibrium point* in this context by Selten (1975). Such an equilibrium has the property that it is an equilibrium point not only for the game as a whole, but also in every subgame.

A10.5 Cooperative games

When cooperation between players is possible, the focus shifts from strategies to *bargaining* about how the payoff will be divided among the cooperating players. Let us consider two-player cooperative games first. Let us plot along the x-axis the payoffs to player 1, and along the y-axis the payoffs to player 2. One can represent the possible payoffs to the players by a subset S of points in the xy-plane (Fig. A10.6). A popular theoretical model assumes that there

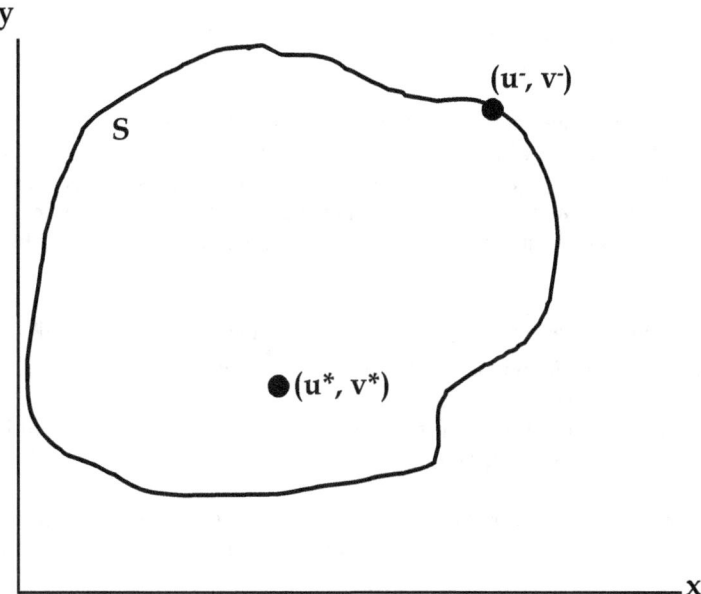

Fig. A10.6 The bargaining problem in a two-player cooperative game (Owen 1992).

is a special point ($u*,v*$) in the subset S, called the *conflict point*. If the two players cooperate, they can reach a fair point (u^-,v^-) in S as a mutually acceptable solution, but if they fail to agree they must receive the conflict-point payoff ($u*,v*$) (Owen 1992).

Several axiomatic solutions to this problem have been suggested. Nash's (1950, 1953) axioms for (u^-,v^-) were that:

1. It must lie in the feasible set S.

2. It must be *Pareto-optimal*, meaning that there can be no point in S that is better for both players.

3. It must be independent of irrelevant alternatives. This means that the elimination of a less desirable alternative cannot change the solution of the problem.

4. It must be covariant with linear changes of the utility scale. There are two kinds of cooperative games. The game may have transferable utility, or it may have non-transferable utility. Transferable utility is relevant when the players measure utility of the payoff in the same units and there is a means of exchange of utility, such as side payments.

5. It must be at least as symmetric as the set $(S,u*,v*)$.

It was shown that, with these axioms, (u^-,v^-) is necessarily that feasible point (u,v) for which the product $(u-u*)(v-v*)$ of the increment in utilities is a maximum, subject to $u \geq u*$, $v \geq v*$.

For *n*-player cooperative games, coalitions are usually the focus of attention, as also the bargaining within the coalitions. Such games are normally investigated in the *characteristic function form*. This function plays a role equivalent to that played by the payoff matrix in the characteristic form of games.

A *coalition* can be any nonempty subset S of the set N. The characteristic function V of the game is defined as one which assigns to each coalition S the set of outcomes that the members of that coalition can obtain by coming together, even against the concerted action of other players. If it can be assumed that the utility is transferable among members of a coalition, then $V(S)$ is the maximum amount of utility that the coalition S can obtain and then distribute among its members.

An *imputation* is a vector $x = (x_1, x_2, ..., x_n)$ such that $x_i \geq V(\{i\})$ for all i, and $\sum x_i = V(N)$. An imputation is an individually rational way of dividing the utility $V(N)$.

Given two imputations x and y, x is said to *dominate* y if there is some coalition S which prefers x and is strong enough to enforce x.

In plural games the main problem is to choose some reasonable set of outcomes, preferably a unique outcome, from the set of all imputations. *Stability* and *fairness* are two possible criteria for making the choices.

The core. One can look for the set of all undominated imputations. This set is known as the core. It is somewhat like the competitive equilibrium of classical economic theory, and illustrates countervailing power.

The stable set. The core may not always be nonempty. von Neumann and Morgenstern (1944) therefore introduced the notion of stable sets. A set of imputations is said to be *internally stable* if no imputation in the set dominates another. A set is *externally stable* if any imputation not in the set is dominated by some imputation in the set. A stable set, also called a *solution*, is any set which is both internally and externally stable. This solution illustrates a form of social stability.

The value. The value form of a solution illustrates fair division. It is a single-point solution for which one considers the combinatorics of all possible ways in which an individual could join a coalition of any size. We calculate the marginal worth of his contribution to any coalition, and then average over all the contributions. A value or expected worth is assigned to every player. Shapley (1953) stated four axioms for calculating the value:

1. *Symmetry*. If two players make the same contribution when any of them joins any coalition, they each obtain the same value.

2. *Dummy player*. The value of a player is zero if he contributes nothing, no matter which coalition he joins.

3. *Efficiency*. The sum of the values assigned to all the players is equal to the available utility $V(N)$; no less, no more.

4. *Additivity*. If two separate games are considered jointly as if it is a single game, the values in the joint single game are simply the sums of the values in the separate games.

Bibliography

Achlioptas, D., R. M. D'Souza and J. Spencer (13 March 2009). 'Explosive percolation in random networks.' *Science*, 323: 1453.

Adamatzky, A. *et al.* (2005). 'Reaction-diffusion computing'. In *Handbook of Natural Computing*, pp. 1897-1920. Springer.
http://link.springer.com/referenceworkentry/10.1007/978-3-540-92910-9_56#page-1

Adams, J. (1998). 'On the application of DNA based computing'.
http://publish.uwo.ca/~jadams/dnaapps1.htm

Adams, W. M. (2006). 'The Future of Sustainability: Re-thinking Environment and Development in the Twenty-first Century'. Report of the IUCN Renowned Thinkers Meeting, 29-31 January 2006.
http://cmsdata.iucn.org/downloads/iucn_future_of_sustanability.pdf

Addison, P. S. (1997/2001). *Fractals and Chaos: An Illustrated Course*. Bristol: IOP Publishing.

Ade, P. A. R. *et al.* (BICEP2 Collaboration) (2014). 'Detection of B-mode polarization at degree angular scales by BICEP2'. *Phys. Rev. Lett.*, 112: 241101.
http://physics.aps.org/articles/v7/64

Aharonov, Y. and J. Tollaksen (8 June 2007). 'New insights on time-symmetry in quantum mechanics.' www.arXiv:0706.1232vl [quant-ph].

Albert, R. and A.-L. Barabási (2002). 'Statistical mechanics of complex networks.' *Reviews of Modern Physics*, 74: 47.

Albert, R., H. Jeong and A.-L. Barabási (2000). 'Error and attack tolerance of complex networks.' *Nature*, 406: 378.

Ali, A. F. and S. Das (2015). 'Cosmology from quantum potential.' *Phys. Lett. B.*, 741: 276–279.

Alon, U. (26 September 2003). 'Biological networks: The tinkerer as an engineer'. *Science*, 301: 1866.

Alon, U. (2007). *An Introduction to Systems Biology: Design Principles of Biological Circuits*. London: Chapman and Hall/CRC.

Ananthaswamy, A. (August 2003). 'March of the motes'. *New Scientist*, 279: 26.

Anderson, P. (1999). 'Complexity theory and organization science'. *Organization Science*, 10: 216.

Anderson, P. W. (4 August 1972). 'More is different.' *Science*, 177: 393.

Anderson, P. W. (1988, 1989). *Physics Today*. A series of six articles: January 1988; March

1988; June 1988; September 1988; July 1989; September 1989.

Angell, C. A. (31 March 1995). 'Formation of glasses from liquids and biopolymers.' *Science*, 267: 1924.

Applegate, D. L., R. E. Bixby, V. Chvátal and W. J. Cook (2006). *The Travelling Salesman Problem: A Computational Study.* Princeton: Princeton University Press.

Arkin, R. C. (1998). *Behaviour-Based Robotics.* Cambridge MA: The MIT Press.

Arthur, W. B. (February 1990). 'Positive feedbacks in the economy'. *Scientific American*, p. 92.

Ashby, W. Ross (1962). 'Principles of the self-organizing System'. In Heinz Von Foerster and George W. Zopf, Jr. (eds.), *Principles of Self-Organization* (Sponsored by Information Systems Branch, U.S. Office of Naval Research). Republished as a pdf in Emergence: Complexity and Organization (E:CO) Special Double Issue Vol. 6, Nos. 1-2 2004, pp. 102–126.

Atkins, P. (1984/1994). *The 2nd Law: Energy, Chaos, and Form*. Scientific American Library.

Aumann, R. J. and M. Maschler (1964). 'The bargaining set for cooperative games.' In M. Drescher, L. S. Shapley and A. W. Tucker (Eds.), *Advances in Game Theory*. Princeton, NJ: Princeton University Press.

Avery, J. (2003). *Information Theory and Evolution*. Singapore: World Scientific.

Axelrod, R. (1984). *The Evolution of Cooperation*. New York: Basic Books.

Axelrod, R. (2006). *The Complexity of Evolution: Agent-Based Models of Competition and Collaboration*. Princeton University.

Baddeley, A. D. and G. J. Hitch (1994). 'Developments in the concept of working memory'. *Neuropsychology* 8:485–93.

Bak, P. and K. Chen (1989). 'The physics of fractals'. *Physica D*, 38: 5.

Bak, P. (1996). *How Nature Works: The Science of Self-Organized Criticality*. New York: Springer.

Bak, P., C. Tang and K. Wiesenfeld (1987). 'Self-organized criticality: An explanation of 1/f noise.' *Phys. Rev. Lett.* 59: 381.

Bak, P., C. Tang and K. Wiesenfeld (1988). 'Self-organized criticality.' *Phys. Rev. A*, 38: 364.

Baker, D., G. Church, J. Collins, D. Endy, J. Jacobson, J. Keasling, P. Modrich, C. Smolke and R. Weiss (June 2006). 'Engineering life: Building a FAB for biology.' *Scientific American*, 294(6): 44.

Ball, P. (1999a). *The Self-Made Tapestry: Pattern Formation in Nature*. Oxford: Oxford University Press.

Ball. P. (2 December 1999b). 'Transitions still to be made.' *Nature*, 402 (Supplement): C73.

Bar-Yam, Y. (1997). *Dynamics of Complex Systems*. Colorado: Westview Press (Perseus Books).

Barabási, A,-L. (2003). *Linked: How Everything is Connected to Everything Else and What It Means for Business, Science, and Everyday Life*. Cambridge, MA: Perseus.

Barabási, A.-L. (24 July 2009). 'Scale-free networks: A decade and beyond.' *Science*, 325: 412.

Barabási, A.-L. and R. Albert (15 October 1999). 'Emergence of scaling in random networks.' *Science*, 286: 509.

Barkhow, J. H. (1989). 'The elastic between genes and culture'. *Ethology and Sociobiology*, 10:111.

Barrow, J. D. and F J Tipler (1986). *The Anthropic Cosmological Principle*. New York: Oxford University Press.

Baumann, O. (2015). 'Models of complex adaptive systems in strategy and organization research'. *Mind and Society*, special issue on 'Complexity Modelling in Social Science and Economics'.

Basu, K. (June 2007). 'The traveller's dilemma'. *Scientific American*, 296(6): 90.

Battram, A. (1998). *Navigating Complexity: The Essential Guide to Complexity Theory in Business and Management*. Industrial Society, University of Michigan.

Berry, E. (2010). 'The Scientific Method'. http://edberry.com/blog/ed-berry/2-the-scientific-method/.

Biggs, N. L., E. K. Lloyd and R. J. Wilson (1976). *Graph Theory: 1736-1936*. Oxford: Clarendon Press.

Bray, D. (26 September 2003). 'Molecular networks: the top down view'. *Science*, 301: 1864.

Bennett, C. H. (1982). 'The thermodynamics of computation - A review.' *International Journal of Theoretical Physics*, 21: 905.

Bennett, C. H. (1987a). 'Demons, engines and the second law'. *Scientific American*, 257(5): 108.

Bennett, C. H. (1987b). 'Dissipation information, computational complexity, and the definition of organization.' In D. Pines (Ed.), *Emerging Syntheses in Science*. Addison Wesley.

Bennett, C. H. (1988). 'Logical depth and physical complexity.' In R. Herken (Ed.), *The Universal Turing Machine: A Half-Century Survey*. Oxford: Oxford University Press.

Bennett, C. H. and R. Landauer (July 1985). 'The fundamental physical limits of computation'. *Scientific American*, 253(1): 38.

Berlekamp, E. R., J. H. Conway and H. K. Guy (1982). *Winning Ways for Your Mathematical Plays*. New York: Academic Press.

Bialynicki, I. and L. Rudnicki (2011). 'Entropic uncertainty relations in quantum physics'. http://arxiv.org/abs/1001.4668 [quant-ph]

Blackmore, S. (1999). *The Meme Machine*. Oxford: Oxford University Press.

Blackmore, S. (October 2000). 'The power of memes.' *Scientific American*, 283(4): 52.

Blume-Kohout, R. and W. H. Zurek (2007). 'Quantum Darwinism in quantum Brownian motion.' www.arxiv.org/abs/0704.3615

Bonabeau, E., M. Dorigo and G. Theraulaz (2000). 'Inspiration for optimization from social insect behaviour'. *Nature*, 406: 39.

Bodenschatz, E., J. R. de Bruyn, G. Ahlers, and D. S. Cannell (1991). *Phys. Rev. Lett.*, 67: 3078.

Bonabeau, E. and Meyer, C. (May 2001). 'Swarm intelligence: A whole new way to think about business.' *Harvard Business Review*, 5: 107.

Bondy, J. A. and U. S. R. Murthy (1976). *Graph Theory with Applications*. New York: Elsevier.

Bradley, A. S. (December 2009). 'Expanding the limits of life.' *Scientific American*, 301(6): 38.

Bray, D. (26 September 2003). 'Molecular networks: the top down view.' *Science*, 301: 1864.

Brooks, M. (23 June 2007). 'Reality check.' *New Scientist*, p. 30.

Brooks, R. A. (1986). 'A robust layered control system for a mobile robot. *IEEE Journal of Robotics and Automation*, 2(1): 1986.

Brooks, R. A. (1990). *Elephants Don't Play Chess: Designing Autonomous Agents: Theory and Practice from Biology to Engineering and Back*. The MIT Press. ISBN 978-0-262-63135-8.

Byrne, P. (December 2007). 'The many worlds of Hugh Everett.' *Scientific American*, p. 98.

Cairns-Smith, A. G. (1982). *Genetic Takeover and the Mineral Origins of Life*. Cambridge: Cambridge University Press.

Callender, C. (June 2010). 'Is time an illusion?' *Scientific American*, 302(6): 41.

Calvin, M. (1969). *Chemical Evolution Towards the Evolution of Life, on Earth and Elsewhere*. Oxford: Oxford University Press.

Camazine, S. (June 2003). 'Patterns in Nature'. Page 34.

Carroll, S. M. (June 2008). 'The cosmic origin of time's arrow.' *Scientific American,* p. 26.

Cepelewicz, J. (31 August 2017). 'Seeing emergent physics behind evolution'. *Quanta Magazine*: https://www.quantamagazine.org/seeing-emergent-physics-behind-evolution-20170831/.

Chakrabarti, B. K., A. Chakraborti and A. Chatterjee (2006). *Econophysics and Sociophysics*. Berlin: Wiley-VCH.

Chaikin and Lubensky (1996). *Principles of Condensed Matter Physics*. Cambridge: Cambridge University Press.

Chaisson, E. (2002). *Cosmic Evolution: The Rise of Complexity in Nature*. Harvard: Harvard University Press.

Chaitin, G. (1987a). *Algorithmic Information Theory*. Cambridge: Cambridge University Press.

Chaitin, G. (1987b). *Information, Randomness and Incompleteness*. Singapore: World Scientific.

Chaitin, G. (2001). *Exploring Randomness*. Berlin: Springer.

Chaitin, G. (March 2006). 'The limits of reason.' *Scientific American*, 294(3): 74.

Chandler, D. (29 September 2005). 'Interfaces and the driving force of hydrophobic assembly.' *Nature*, 437: 640.

Choi, C. Q. (2016). '7 Theories on the Origin of Life'. https://www.livescience.com/13363-7-theories-origin-life.html

Chua, L. O. (1998). *CNN: A Paradigm for Complexity*. Singapore: World Scientific.

Clifton, T. and P. G. Ferreira (April 2009). 'Does dark energy really exist?' *Scientific American*, 300(4): 32.

Crawford, V. P. (1989). 'Learning and mixed strategy equilibria in evolutionary games.' *Journal of Theoretical Biology*, 140: 537.

Crosby, A. (1997). *The Measure of Reality: Quantification and Western Society*. Cambridge: Cambridge University Press.

Cross, M. C. and P. C. Hohenberg (1993). 'Pattern formation outside of equilibrium.' *Reviews of Modern Physics*, 65: 851.

Crothers, S. J. (2014). 'The rise and fall of black holes and big bangs'. http://www.principia-scientific.org/the-rise-and-fall-of-black-holes-and-big-bangs.html.

Crutchfield, J. P., J. D. Farmer, N. H. Packard and R. S. Shaw (December 1986). 'Chaos.' Scientific American, 254(12): 46.

Dagotto, E. (8 July 2005). 'Complexity in strongly correlated electronic systems'. *Science*, 309: 257.

Davies, P. (5 March 2005). 'The sum of the parts.' *New Scientist*, 185: 34.

Dawkins, R. (1979). 'The value judgements of evolution.' In M. A. H. Dempster and D. J. McFarland (Eds.), *Animal Economics*. London: Academic Press.

Dawkins, R. (1986). *The Blind Watchmaker: Why the Evidence of Evolution Reveals a Universe Without Design*. New York: Norton.

Dawkins, R. (1976/1989/2006). *The Selfish Gene*, revised edition. Oxford: The Oxford University Press.

Dawkins, R. (1998). *Unweaving the Rainbow*. Allen Lane: Penguin.

Dawkins, R. (2007). *The God Delusion*. London: Transworld Publishers (paperback edition).

Dawkins, R. and J. R. Krebs (1979). 'Arms races between and within species' *Proceedings of the Royal Society of London B*, 205: 489.

Deacon, T. W. (1997). *The Symbolic Species: The Coevolution of Language and the Brain*. New York: W. W. Norton & Co.

Denbigh, K. G. and J. S. Denbigh (1985). *Entropy in Relation to Incomplete Knowledge*. Cambridge: Cambridge University Press.

Dennett, D. C. (1991). *Consciousness Explained*. New York: Back Bay Books.

Dennett, D. C. (1995). *Darwin's Dangerous Idea: Evolution and the Meanings of Life*. London: Penguin Books.

Dennett, D. C. (2006) in J. Brockman (Ed.), *What We Believe but Cannot Prove*. New York: Harper.

de Rosnay, J. (1979). The Macroscope: A New World Scientific Instrument. New York: Harper and Row Publishers. First published in French in 1975.

DeWitt, B. S. and N. Graham (Eds.) (1973). *The Many-Worlds Interpretation of Quantum Mechanics*. Princeton: Princeton University Press.

Diamond, A. (2013). 'Executive functions'. Ann. Rev. Psychol., 64: 135.

Distin, K. (2005). *The Selfish Meme: A Critical Reassessment*. Cambridge: The Cambridge University Press.

Dogaru, R. and L. O. Chua (1998). 'Edge of chaos and local activity domain of Fitz-Hugh-Nagumo equation.' *International Journal of Bifurcation and Chaos*, 8(2).

Dooley, K. J. (1997). 'A complex adaptive systems model of organization change'. *Nonlinear Dynamics, Psychology, and Life Sciences,* 1: 69.

Dorigo, M., M. Birattari and T. Stutzle (November 2006). 'Ant colony optimization: Artificial ants as a computational intelligence technique'. *IEEE Computational Intelligence Magazine*, 1(4): 28.

Dorigo, M. and M. Colombetti (1997). *Robot Shaping: An Experiment in Behaviour*

Engineering. Bradford.

Douglas, K. (9 April 2005a). 'Language.' *New Scientist*, p. 30.

Douglas, K. (9 April 2005b). 'Superorganisms.' *New Scientist*, p. 34.

Drossel, B. (2008). 'Random Boolean networks.' In H. G. Schuster (Ed.), *Reviews of Nonlinear Dynamics and Complexity*, Vol. 1. Berlin: Wiley-VCH.

Dubey, P. (1986). 'Inefficiency of Nash equilibria.' *Mathematical Operations Research*, 11: 1.

Dubey, P. and J. D. Rogawsky (1990). 'Inefficiency of smooth market mechanisms.' *Journal of Mathematical Economics*, 19: 285.

Duke, C. (July 2006). 'Prosperity, complexity and science.' *Nature Physics*, 2: 426.

Dunbar, R. (1998). *Grooming, Gossip, and the Evolution of Language.* Harvard University Press.

Durand, Fábio Renan, Larissa Melo, Lucas Ricken Garcia, Alysson José dos Santos and Taufik Abrão (2013). 'Optical network optimization based on particle swarm intelligence'. http://www.intechopen.com/books/search-algorithms-for-engineering-optimization/optical-network-optimization-based-on-particle-swarm-intelligence

Dyson, F. (1985). *Origins of Life.* Cambridge: Cambridge University Press.

Dyson, F. (2008) in J. Brockman (Ed.), *Life: What a Concept!* http://www.edge.org/documents/life/life_index.html

Dyson, G. B. (1997). *Darwin Among the Machines: The Evolution of Global Intelligence.* Cambridge: Perseus Books.

Eigen, M. (1987/1996). *Steps Towards Life: A Perspective on Evolution.* Oxford: Oxford University Press.

Einstein, A. (1911). 'On the influence of gravitation on the propagation of light'. Translated from *Über den Einfluss der Schwercraft auf die Ausbreitung des Lichtes*, in Das Relativitätsprincip, 4th edition. Originally from *Annalen der Physik* [35], 1911.

Eldredge, N. and S. J. Gould (1972). "Punctuated equilibria: an alternative to phyletic gradualism". In T.J.M. Schopf, ed., *Models in Paleobiology*. San Francisco: Freeman Cooper. pp. 82-115. Reprinted in N. Eldredge *Time frames*. Princeton: Princeton Univ. Press, 1985, pp. 193-223.

England, J. L. (2013). 'Statistical physics of self-replication'. *J. Chem. Phys.*, 139, 121923; doi: 10.1063/1.4818538. Also see:
https://www.scientificamerican.com/article/a-new-physics-theory-of-life/
https://www.quantamagazine.org/a-new-thermodynamics-theory-of-the-origin-of-life-20140122/
https://futurism.com/new-theory-for-life-suggests-it-was-not-an-accident-of-biology-it-was-physics/
https://www.wired.com/story/controversial-new-theory-suggests-life-wasnt-a-fluke-of-

biologyit-was-physics?mbid=social_fb
https://www.livescience.com/60250-did-life-emerge-from-physical-laws.html
https://futurism.com/abiogenesis-theory-origins-life/

Farías, O. J., C. L. Latune, S. P. Walborn, L. Davidowich and P. H. S. Ribeiro (12 June 2009). 'Determining the dynamics of entanglement.' *Science*, 324: 1414.

Farmer, J. D., S. A. Kauffman and N. H. Packard (1986a). 'Autocatalytic replication of polymers'. *Physica D*, 22: 50.

Farmer, J. D. *et al.* (Eds.) (1986b). *Evolution, Games, and Learning.* Amsterdam: North-Holland. Also available in *Physica D*22, Nos. 1-3.

Farmer, J. D. and A. Belin (1992). 'Artificial Life: The Coming Evolution'. In C. Langton, ed., *Artificial Life II*. Redwood City, California, Addison-Wesley.

Faudree, R. (1992). 'Graph theory.' In R. A. Meyers (Ed.), *Encyclopedia of Physical Science and Technology*, Vol. 7. New York: Academic Press.

Ferreira, C. (2006). *Gene Expression Programming: Mathematical Modelling by an Artificial Intelligence.* Springer.

Feynman, R. (1967). *The Character of Physical Law.* Cambridge: MIT Press.

Feynman, R. P. (1985). *Surely You're Joking, Mr. Feynman!* London: Vintage Books.

Fisher, L. (2009). *The Perfect Swarm: The Science of Complexity in Everyday Life*: New York: Basic Books.

Fisher, M. A. (1998). 'Renormalization group theory: its basis and formulation in statistical physics.' *Reviews of Modern Physics*, 70: 653.

Fisher, R. A. (1930). *The Genetical Theory of Natural Selection.* Oxford: Oxford University Press.

Folger, T. (19 June, 2009). 'Is quantum mechanics tried, true, wildly successful, and wrong?' *Science*, 324: 1512.

Foley, J. (April 2010). 'Boundaries for healthy planet.' *Scientific American India*, 5(4): 42.

Frampton, P. H. (2010). *Did time Begin? Will Time End? Maybe the Big Bang Never Occurred.* Singapore: World Scientific.

Renard, F. (2000). 'What is self-organization?' http://www.viten.com/nyviten/renard.htm

Fritzsch, H. (2009). *The Fundamental Constants: A Mystery of Physics.* Singapore: World Scientific.

Garlaschelli, D., A. Capocci and G. Caldarelli (November 2007). 'Self-organized network evolution coupled to extremal dynamics.' *Nature Physics*, 3(11): 813.

Gates, B. (January 2007). 'A robot in every home'. *Scientific American*, 296(1): 58.

Gefter, A. (29 May 2014). 'Complexity on the horizon'. *Nature*, 509, 552.

Gell-Mann, M. (1994). *The Quark and the Jaguar: Adventures in the Simple and the Complex*. New York: Freeman.

George, D. (2008). *How the Brain Might Work: A Hierarchical and Temporal Model for Learning and Recognition*. (PhD dissertation, Stanford University, June 2008)].
https://en.wikipedia.org/wiki/Hierarchical_temporal_memory

Gibbs, W. W. (2004). 'Synthetic life'. *Scientific American*, April 26, 2004 issue.

Gibson, D. G. *et al.* (2010). 'Creation of a bacterial cell controlled by a chemically synthesized genome.' Sciencexpress/www.sciencexpress.org/20 may 2010/Page 1/10.1126/1190719.

Gielen, S. and N. Turok (2016). 'Perfect quantum cosmological bounce'. Phys. Rev. Lett., 117: 021301.

Gilder, L. (2008). *The Age of Entanglement: When Quantum Physics Was Reborn*. New York: Alfred A. Knopf.

Giro, R., M. Cyrillo and D. S. Galvao (2002). 'Designing Conducting Polymers with Genetic Algorithms'. *Chem. Phys. Letters*, 366 (1-2): 170.
http://www.scielo.br/scielo.php?script=sci_arttext&pid=S1516-14392003000400017

Glandsdorff, P. and I. Prigogine (1971). *Thermodynamic Theory of Structure, Stability and Fluctuations*. New York: Wiley Interscience.

Gödel. K. (1931). *Über formal unentscheidbare Sätze der Principia Mathematica und verwandter Systeme, I. Monatshefte für Mathematik und Physik 38*: 173-98.

Gödel, K. (1964). 'Russell's mathematical logic, and what is Cantor's continuum problem?' In P. Benacerraf and H. Putnam (eds.), *Philosophy of Mathematics*. New Jersey: Prentice-Hall.

Godsil, C. and G. Royle (2001). *Algebraic Graph Theory*. Berlin: Springer.

Goldberg, D. (1989). *Genetic Algorithms in Search, Optimization and Machine Learning*. Addison-Wesley.

Goldenfeld, N. (2014). 'Looking in the right direction: Carl Woese and evolutionary biology'. *RNA Biology*, 11: 3.

Goldenfeld, N. and L. P. Kadanoff (2 April 1999). 'Simple lessons from complexity.' *Science*, 284: 87.

Goldenfeld, N. and C. Woese (2011). 'Life is Physics: Evolution as a collective phenomenon far from equilibrium.' *Ann. Rev. Cond. Matt. Phys.*, 2: 375-399.
http://www.annualreviews.org/doi/abs/10.1146/annurev-conmatphys-062910-140509

Goodwin, C. A. P. *et al.* (2017). 'Molecular magnetic hysteresis at 60 kelvin in dysprosocenium'. Nature, 548 (7668):439 DOI: 10.1038 / nature23447.

Grant, A. (2014). 'The mysterious boundary'.
https://www.sciencenews.org/article/mysterious-boundary

Green, K. L. (May 2011). 'Complex adaptive systems in military analysis'. IDA document D-4313.

Groblacher, S., T. Paterek, R. Kaltenbaek, C. Brukner, M. Zukovski, M. Aspelmeyer and A. Zeilinger (19 April 2007). 'An experimental test of non-local realism.' *Nature*, 446: 871.

Guerra, F. (2006). 'Spin glasses', in J.-P. Francoise, G. L. Naber and T. S. Tsun (eds.), *Encyclopaedia of Mathematical Physics*, Vol. 4, p. 655. Amsterdam: Elsevier.

Gunawardana, G. N. (2015). 'Complexity leadership theory'.
http://www.dailymirror.lk/88693/complexity-leadership-theory

Guth, A. H. *et al.* (2017). 'A Cosmic Controversy'.
https://blogs.scientificamerican.com/observations/a-cosmic-controversy/?WT.mc_id=SA_TW_SPC_BLOG&sf77843181=1

Haken, H. (1975). 'Cooperative phenomena in systems far from thermal equilibrium and in nonphysical systems.' *Reviews of Modern Physics*, 47(1): 67.

Haken, H. (1983). *Synergetics*, 3rd edition. New York: Springer-Verlag.

Hameroff, S. (2012). 'How quantum brain biology can rescue conscious free will'. *Frontiers in Integrative Neuroscience*, 6, 93.
http://www.ncbi.nlm.nih.gov/pmc/articles/PMC3470100/

Hamilton, W., R. Axelrod and R. Tanese (1990). 'Sexual reproduction as an adaptation to resist parasites (A review)'. *Proc. Nat. Acad. Sci. (USA)*, 87: 3566.

Hansmann, U. (2003). *Pervasive Computing: The Mobile World.* Springer.

Yuval Noah Harari (2016). *Homo Deus: A Brief History of Tomorrow*. Harvill Secker.

Harris, S. (2012). *Free Will*. New York: Free Press.

Harrison, P. (1999/2013). *Elements of Pantheism: A Spirituality of Nature and the Universe*.
www.pantheism.net.

Hartnett, J. (2011). 'Does observational evidence indicate the universe is expanding? – part 2: the case against expansion'. http://creation.com/expanding-universe-2

Hauser, M. D., N. Chomsky and W. T. Fitch (22 November 2002). 'The faculty of language: What is it, who has it, and how did it evolve?'. *Science, 298:* pp. 1569-1579.
http://science.sciencemag.org/content/298/5598/1569

Hawking, S. W. (1988). *A Brief History of Time: From the Big Bang to Black Holes*. New York: Bantam Books.

Hawking, S. W. (2001). *The Universe in a Nutshell*. London: Bantam Press.

Hawking, S. W. and T. Hertog (2002). 'Why does inflation start at the top of the hill? *Phys.*

Rev., D66: 123509. www.arXiv:hep-th/0204212.

Hawking, S. W. and T. Hertog (10 February 2006). 'Populating the landscape: A top down approach'. www.arXiv:hep-th/0602091

Hawking, S. W. and L. Mlodinow (2010). *The Grand Design: New Answers to the Ultimate Questions of Life*. London: Bantam Press.

Hawkins, J. (2004). *On Intelligence*. New York: Times Books (Henri Holt and Company).

Hebb, D. O. (1949). *The Organization of Behaviour: A Neuropsychological Theory.* New York: Wiley.

Heudin, J.-C. (1999). *Virtual Worlds: Synthetic Universes, Digital Life, and Complexity*. Reading Massachusetts: Perseus Books.

Hofstadter, D. R. (1979). *Gödel, Escher, Bach: An Eternal Golden Braid*. New York: Basic Books.

Hogan, J. (2007). 'Unseen universe: Welcome to the dark side.' *Nature*, 448(7151): 240.

Holland, J. H. (1975). *Adaptation in Natural and Artificial Systems*. Ann Arbor: University of Michigan Press.

Holland, J. H. (1995). *Hidden Order: How Adaptation Builds Complexity*. Reading, Massachusetts: Addison Wesley.

Holland, J. H. (1998). *Emergence: From Chaos to Order*. Cambridge, Massachusetts: Perseus Books.

Holland, J. H., K. J. Holyoak, R. E. Nisbett and P. R. Thagard (1986). *Induction: Processes of Inference, Learning and Discovery*. Cambridge, MA: MIT Press.

Holldobler, B. and E. O. Wilson (1990). *The Ants*. Berlin: Springer-Verlag.

Holme, P. (2006). 'Local symmetries in complex networks.' *Physical Review E*, 74: 036107.

Hooker, B. (2012). 'Quantum computing: The next information revolution'. http://www.aaas.org/blog/depth/quantum-computing-next-information-revolution

Hooper, R. (7 January 2006a). 'Men inherit hidden costs of dad's vices.' *New Scientist*, p. 10.

Hooper, R. (27 May 2006b). 'Mendel would turn in his grave...' *New Scientist*, p. 16.

Horgan, J. (June 1995). 'From complexity to perplexity.' *Scientific American*, 272: 74.

Horowitz, J. M. and J. L. England (2017). 'Spontaneous fine-tuning to environment in many-species chemical reaction networks'. *PNAS*, USA, 114 (29): 7565.

Hosokawa, K., I. Shimoyama and H. Miura (1995). 'Dynamics of self-assembling systems: Analogy with chemical kinetics.' *Artificial Life*, 1: 413.

Houser, M. (September 2009). 'Origin of the mind.' *Scientific American*, 301(3): 30.

Ijjas, A. *et al.* (25 June 2013). 'Inflationary paradigm in trouble after Planck 2013'. *Phys. Lett.*, 723: 261.

Ijjas, A., P. J. Steinhardt and A. Loeb (1 February 2017). 'Pop goes the universe'. *Scientific American,*

Jayens, E. T. (1957). 'Information theory and statistical mechanics.' *Physical Review*, 106: 620.

Jenkins, A. and G. Perez (January 2010). 'Looking for life in the multiverse.' *Scientific American*, 302(1): 28.

Jogalekar, A. (23 April 2013). 'Stephen Hawking's advice to twenty-first century grads: Embrace complexity'. https://blogs.scientificamerican.com/the-curious-wavefunction/stephen-hawkings-advice-for-twenty-first-century-grads-embrace-complexity/

Johnson, N. (2009). *Simply Complexity: A Clear Guide to Complexity Theory*. Oneworld Publishers; reprint edition.

Kachman, T., J. A. Owen and J. L. England (21 July 2017). 'Self-organized resonance during search of a diverse chemical space'. Phys. Rev. Lett., 119: 038001.

Kaku, M. (1998). *Visions: How Science Will Revolutionize the 21st Century*. Published by Anchor.

Kaneko, K. and I. Tsuda (2000). *Complex Systems: Chaos and Beyond (A Constructive Approach with Applications in Life Sciences)*. Berlin: Springer.

Kauffman, S. (11 October 1969). 'Homeostasis and differentiation in random genetic control networks.' *Nature*, 224: 177.

Kauffman, S. (11 October 1969). 'Homeostasis and differentiation in random genetic control networks.' *Nature*, 224: 177.

Kauffman, S. A. (1991). 'Antichaos and adaptation.' *Scientific American*, 265: 78.

Kauffman, S. A. (1993). *Origins of Order: Self-Organization and Selection in Evolution*. Oxford: Oxford University Press.

Kauffmann, S. A. (1995a). *At Home in the Universe: The Search for Laws of Self-Organization and Complexity*. Oxford: Oxford University Press.

Kauffman, S. A. (1995b). 'Order for free.' In J. Brockman (Ed.), *The Third Culture: Beyond the Scientific Revolution*. New York: Simon & Schuster. http://www.edge.org/documents/ThirdCulture/d-Contents.html

Kauffman, S. A. (2000). *Investigations*. Oxford: Oxford University Press.

Kauffman, S. A. (2006). 'Beyond reductionism: Reinventing the sacred.' http://www.edge.org/3rd_culture/kauffman06/kauffman06_index.html

Keller, M. A., A. V. Turchyn and M. Ralser (2014). 'Non-enzymatic glycolysis and pentose phosphate pathway-like reactions in a plausible Archean ocean.' *Molecular Systems Biology*, 10: 725.

Kelly, K. (1994). *Out of Control: The New Biology of Machines, Social Systems, and the Economic World*. Cambridge: Perseus Books.

Kelly, S. and M. Allison (1999). *The Complexity Advantage: How the Science of Complexity Can Help Your Business Achieve Peak Performance*. McGraw-Hill, New York, NY.

Kennedy, J. (2006). 'Swarm intelligence'. In Zomaya, A. Y. (ed.), *Handbook of Nature-Inspired and Innovative Computing: Integrating Classical Models with Emerging Technologies*. New York: Springer, p. 187.

Kingsley, D. M. (January 2009). 'From atoms to traits.' *Scientific American*, 300(1): 38.

Klein, R. G. and B. Edgar (2002). *The Dawn of Human Culture: A Bold New Theory on what Sparked the 'Big Bang' of Human Consciousness*. New York: Wiley.

Klosin, A. et al. (April 2017). 'Transgenerational transmission of environmental information in C. elegans'. *Science*, 356: 320.

Kodandaramaiah, S. B. *et al.* (2012). 'Automated whole-cell patch-clamp electrophysiology of neurons *in vivo*'. *Nature Methods*, 9: 585.

Kolmogorov, A. N. (1965). 'Quantity of information defined (algorithmic approach which uses recursive functions for defining concept and quantity of information)'. *Probability and Information Transactions*, 1: 1.

Konar, A. (2005). *Computational Intelligence: Principle, Techniques, and Applications*. Berlin: Springer-Verlag.

Koshland, D. E. (2002). 'The seven pillars of life.' *Science*, 295: 2215.

Koza, J. R. *et al.* (1997). 'Automated synthesis of analogue electrical circuits by means of genetic programming.' *IEEE Transactions on Evolutionary Computation*, 1(2): 109.

Krauss, L. M., S. Dodelson and S. Meyer (21 May 2010). 'Primordial gravitational waves and cosmology.' *Science*, 328: 989.

Krauss, L. (2012). *A Universe from Nothing: Why There is Something Rather than Nothing*. New York: Free Press.

Krauss, L. (2014). 'Peering back to the beginning of time'. *Physics*, 7: 54.

Krauss, L. (2017). *The Greatest Story Ever Told – So Far: Why Are We Here*. Simon & Schuster, U.K.

Krauss, L. and F. Wilczek (2014). 'From B modes to quantum gravity and the unification of forces'. http://arxiv.org/abs/1404.0634v3

Kruse, R., C. Borgelt, F. Klawonn, C. Moewes, M. Steinbrecher and P. Held (2013). *Computational Intelligence: A Methodological Introduction*. London: Springer-Verlag.

Krori, K. D. (2010). *Fundamentals of Special and General Relativity*. New Delhi: PHI Learning Pvt. Ltd.

Kurzweil, R. (1998). *The Age of Spiritual Machines: When Computers Exceed Human Intelligence*. New York: Viking Penguin.

Kurzweil, R. (2001). 'The coming merging of mind and machine'. http://www.kurzweilai.net/the-coming-merging-of-mind-and-machine

Kurzweil, R. (2005). *The Singularity is Near: When Humans Transcend Biology*. London: Penguin Books.

Kurzweil, R. (2012). *How to Create a Mind: The Secret of Human Thought Revealed*. London: Penguin Books.

Lago-Fernández, L. F., R. Huerta, F. Corbacho and J. A. Sigüenza (2000). 'Fast response and temporal coherent oscillations in small-world networks.' *Physical Review Letters*, 84: 2758.

Landau, L. D. (1937). *Phys. Z. Sowjet.* 11: 26 (in Russian). For an English translation, see D. ter Haar (ed.) (1965), *Collected Papers of L. D. Landau*. New York: Gordon and Breach.

Langton, C. G. (1986). 'Studying artificial life with cellular automata'. *Physica D*, 22: 120.

Langton, C. G. (ed.) (1989). *Artificial Life*. Santa Fe Institute Studies in the Sciences of Complexity, Proceedings Vol. 6. Redwood City, CA: Addison Wesley.

Langton, C. G. (1990). 'Computation at the edge of chaos.' *Physica D*, 42: 12.

Lawlor, L. R. and Maynard Smith, J. (1976). 'The coevolution and stability of competing species.' *American Naturalist*, 110, 79.

Layzer, D. (December 1975). 'The arrow of time.' *Scientific American*, 233(6): 56.

Lederman, L. M. and C. T. Hill (2008). *Symmetry and the Beautiful Universe*. New York: Prometheus Books.

Lehn, J.-M. (1995). *Supramolecular Chemistry: Concepts and Perspectives*. New York: VCH.

Lehn, J.-M. (1999). In R. Ungaro and E. Dalcanale (eds.), *Supramolecular Science: Where It Is and Where It Is Going*. Dordrecht: Kluwer.

Lehn, J.-M. (16 April 2002). 'Toward complex matter: Supramolecular chemistry and self-organization'. *PNA, USA*, 99(8): 4763.

Lehrer, J. (2009). *How We Decide?* New York: Houghton, Mifflin, Harcourt.

Lerner, E. J., R. Falomo and R. Scarpa (2014). 'UV surface brightness of galaxies from the local universe to z ~ 5.'. *Int. J. Mod. Phys. D*, 23, 1450058. DOI: http://dx.doi.org/10.1142/S0218271814500588

Lin, S. K. (2001). 'The nature of the chemical process: 1. Symmetry evolution – Revised

information theory, similarity principle and ugly symmetry'. *Int. J. Molecular Sciences*, 2:
10.

Lin, S. K. (2008). 'Gibbs paradox and the concepts of information, symmetry, similarity and
their relationship'. *Entropy*, 10:1.

Liscovitch-Brauer, N. *et al.* (6 April 2017). 'Trade-off between transcriptome plasticity and
genome evolution in cephalopods'. *Cell*: www.cell.com/cell/fulltext/S0092-8674(17)30344-
6.

Litvin, D. B. and V. K. Wadhawan (2001). 'Latent symmetry and its group-theoretical
determination.' *Acta Crystallographica,* A57: 435.

Litvin, D. B. and V. K. Wadhawan (2002). 'Latent symmetry.' *Acta Crystallographica,* A58:
75.

Litvin, D. B., V. K. Wadhawan and D. M. Hatch (2003). 'Latent symmetry and domain
average engineered ferroics.' *Ferroelectrics*, 292: 65.

Lloyd, S. (August 2001). 'Measures of complexity: A nonexhaustive list.' *IEEE Control
Systems Magazine*, 21(4): 7.

Lloyd, S. (2006). *Programming the Universe*. New York: Alfred A. Knopf, Random House.

Lloyd, S. and Y. J. Ng (November 2004). 'Black hole computers.' *Scientific American*,
291(5): 30.

Lloyd, S. and H. Pagels (1988). 'Complexity as thermodynamic depth.' *Annals of Physics*,
188: 186.

Lotka, A. J. (1925). *Elements of Physical Biology*. Baltimore: Williams and Wilkins.

Lovat, G. *et al.* (2017). 'Room-temperature current blockade in atomically defined single-
cluster junctions'. *Nature Technology*, DOI: 10.1038/nnano.2017.156.

Lovelock, J. E. (1979). *Gaia, A New Look at Life on Earth*. Oxford: Oxford University Press.

LPPhysics (2014). 'The growing case against the big bang'.
http://lawrencevilleplasmaphysics.com/cosmic-connection/plasma-cosmology/the-growing-
case-against-the-big-bang/

MacArthur, B. D. and J. W. Anderson (2006). 'Symmetry and self-organization in complex
systems.' e-print arXiv:cond-mat/0609274.

MacArthur, B. D., R. J. Sánchez-García and J. W. Anderson (2007). 'On automorphism
groups of networks.' e-print arXiv:0705.3215v2 [physics.soc-ph] .

Mainzer, K. (2005). *Symmetry and Complexity: The Spirit and Beauty of Nonlinear Science*.
Singapore: World Scientific.

Mainzer, K. (2007). *Thinking in Complexity: The Computational Dynamics of Matter, Mind,
and Mankind* (5th edition). Berlin: Springer-Verlag.

Mandelbrot, B. B. (1977). *Fractals: Form, Chance and Dimension*. San Francisco: Freeman.

Mandelbrot, B. B. (1982). *The Fractal Geometry of Nature*. San Francisco: Freeman.

Mange, D. *et al.* (2004). 'Embryonic machines that divide and differentiate'. Volume 3141 of the series Lecture Notes in Computer Science, pp. 201-216.
http://link.springer.com/book/10.1007/b101281

Margulis, L. and D. Sagan (2002). *Acquiring Genomes: A Theory of the Origins of Species*. New York: Basic Books.

Maynard Smith, J. (1974). 'The theory of games and the evolution of animal conflicts.' *Journal of Theoretical Biology*, 47: 209.

Maynard Smith, J. (1976). 'Evolution and the theory of games.' *American Scientist*, 64: 41.

Maynard Smith, J. (1977). 'Parental investment – a prospective analysis.' *Animal Behaviour*, 25: 1.

Maynard Smith, J. (1978). 'Optimization theory in evolution.' *Annual Reviews of Ecological Systems*, 9: 31.

Maynard Smith, J. (1982). *Evolution and the Theory of Games*. Cambridge, MA: Cambridge University Press.

Maynard Smith, J. and G. R. Price (1973). 'The logic of animal conflict.' *Nature*, 246: 15.

Maynard Smith and E. Szathmary (12 February1998). *The Major Transitions in Evolution*. Oxford University Press, paperback reprint edition.

Mayr, E. (1942). *Systematics and the Origins of Species.* Cambridge Massachusetts: Harvard University Press.

McCulloch, W. and W. Pitts (1943). 'A logical calculus of the ideas immanent in nervous activity'. *Bull. Math. Biophys.*, 5: 115.

McDowell, N. (17 October 2002). 'Ecological footprint forecasts face sceptical challenge'. Nature, 419: 656.

McKibben, B. (April 2010). 'Breaking the growth habit.' *Scientific American India*, 5(4): 49.

McMahon, T. A. and J. T. Bonner (1983). *On Size and Life*. San Francisco: W. H. Freeman.

Menger, F. M. and K. D. Gabrielson (1995). 'Cytomimetic organic chemistry: Early developments.' *Angew. Chem. Int. Ed. Engl.* 34: 2091.

Merali, Z. (30 June 2007). 'Quantum reality, Darwinian style.' *New Scientist*, 194(2610), 18.

Michaelian, K. (2011). 'Thermodynamic dissipation theory for the origin of life.' *Earth System Dynamics*, 2: 37-51. https://doi.org/10.5194/esd-2-37-2011, 2011.

Milgram, S. (1967). *Psychology Today*, 1: 60.

Miller, G. A. (1956). 'The magical number seven, plus or minus two: Some limits on our capacity for processing information'. *The Psychological Review*, 63: 81-97.
http://www.musanim.com/miller1956/

Milo, R., S. Shen-Orr, S. Itzkovitz, N. Kashtan, D. Chklovskii and U. Alon (2002). 'Network motifs: Simple building blocks of complex networks'. *Science*, 298: 824.

Minsky, M. and S. Papert (1969/1972). *Perceptrons: An Introduction to Computational Geometry*. The MIT Press, Cambridge MA.

Minsky, M. (1981). *Jokes and Their Relation to the Cognitive Unconscious*. In Vaina and Hintikka (Eds.), *Cognitive Constraints on Communication*. Reidel.

Minsky, M. (1986/1988). *The Society of Mind*. New York: Simon & Schuster.

Minsky, M. (1990). 'Logical vs. analogical or symbolic vs. connectionist or neat vs. scruffy'. In P. H. Winston (Ed.), *Artificial Intelligence at MIT, Expanding Frontiers*, Vol. 1.MIT Press. Reprinted in AI Magazine in 1991. Also see Singh (2003).

Minsky, M. (2006). *The Emotion Machine: Commonsense Thinking, Artificial Intelligence, and the Future of the Human Mind*. New York: Simon & Schuster.

Mirsky, S. (January 2009). 'What's good for the group.' *Scientific American*, 300(1): 37.

Mitra, A. (2013). 'Einsteinian revolution's wrong turn: lumpy interacting cosmos assumed as smooth perfect fluid: no dark energy'. *Proceedings of the XXIX Workshop on High Energy Physics: New Results and Actual Problems in Particle and Astroparticle Physics and Cosmology*, 26-28 June, 2013, Protvino, Moscow region, Russia.
http://indico.cern.ch/event/211539/session/5/contribution/41

Mitra, A. (2014). 'Einsteinian revolution's misinterpretation: no true black holes, no information paradox: just quasi-static balls of quark gluon plasma'. *Proceedings of the XXIX Workshop on High Energy Physics: New Results and Actual Problems in Particle and Astroparticle Physics and Cosmology*, 26-28 June, 2013, Protvino, Moscow region, Russia.

Moalem, S. (2007). *Survival of the Sickest*. New York: Harper Collins.

Montegna, R. N. and H. E. Stanley (2000). *An Introduction to Econophysics: Correlations and Complexity in Finance*. Cambridge: Cambridge University Press.

Moravec, H. (1999). *Robot: Mere Machine to Transcendent Mind*. Oxford: Oxford University Press.

Moravec, H. (1999b). 'Rise of the robots'. *Scientific American*, December issue, pp.124-135.
http://www.frc.ri.cmu.edu/~hpm/project.archive/robot.papers/1999/SciAm.scan.html

Moravec, H. (2000). 'Robots, re-evolving mind'.
http://www.frc.ri.cmu.edu/~hpm/project.archive/robot.papers/2000/Cerebrum.html.

Moravec, H. (2003). 'Robotics'.
http://www.frc.ri.cmu.edu/~hpm/project.archive/robot.papers/2003/robotics.eb.2003.html

Moravec, H. (23 March 2009). 'Rise of the robots - The future of artificial intelligence'. *Scientific American*. https://www.scientificamerican.com/article/rise-of-the-robots/

Morrison, P. (1964). 'A thermodynamic characterization of self-reproduction.' *Reviews of Modern Physics*, 36: 517.

Moskowitz, C. (2016). 'Did the universe boot up with a "Big Bounce"?'. https://www.scientificamerican.com/article/did-the-universe-boot-up-with-a-big-bounce/

Mountcastle, V. B. (1978). 'An organizing principle for cerebral function: The unit model and the distributed system.' In G. M. Edelman and V. B. Mountcastle (eds.), *The Mindful Brain*. Cambridge, Massachusetts: MIT Press.

Murray, J. D. (1993). *Mathematical Biology*, 2nd edition. Berlin: Springer-Verlag.

Murthy, K. P. N. (2009). *Excursions in Thermodynamics and Statistical Mechanics*. Hyderabad: The Universities Press.

Mydosh, J. A. (1993). *Spin Glasses: An Experimental Introduction*. London: Taylor and Francis.

Myhrvold, Nathan and M. Bilet (2011). *Modernist Cuisine*: *The Art and Science of Cooking*. New York: The Cooking Lab; Slp, Spi Ha edition.

Nakahara, M. and T. Ohmi (2008). *Quantum Computing: From Linear Algebra to Physical Realizations*. New York: CRC Press.

Nash, J. F. (1950). 'The bargaining problem.' *Econometrica*, 18: 155.

Nash, J. F. (1951). 'Non-cooperative games.' *Annals of Mathematics*, 54: 289.

Nash, J. F. (1953). 'Two person cooperative games.' *Econometrica*, 21: 128.

Neumann, von J. (1963). *Collected Works*. Oxford: Pergamon Press.

Neumann, von J. (1966). *Theory of Self-Reproducing Automata*. Completed and edited by A. W. Burks. Champaign-Urbana: University of Illinois Press.

Newnham, R. E. and S. E. Trolier-McKinstry (1991a). 'Crystals and composites' *Journal of Appl. Crystallography*, 23: 447.

Nicolis, G. and I. Prigogine (1977). *Self-Organization in Nonequilibrium Systems: From Dissipative Structures to Order through Fluctuations*. New York: Wiley.

Niele, F. (2005). *Energy: Engine of Evolution*. Amsterdam: Elsevier.

Nöbauer, T. *et al*. (2017). 'Video rate volumetric Ca2 imaging across cortex using seeded iterative demixing (SID) microscopy'. *Nature Methods*, DOI: 10.1038/nmeth.4341. Also see: http://www.kurzweilai.net/how-to-capture-videos-of-brains-in-real-time?utm_source=KurzweilAI+Weekly+Newsletter+Plain+Text&utm_campaign=dd3b6962 37-UA-946742-1&utm_medium=email&utm_term=0_43205bd0ea-dd3b696237-282140345

Nolfi, S. and D. Floreano (2000). *Evolutionary Robotics: The Biology, Intelligence, and*

Technology of Self-organizing Machines. MIT Press.

Nowak, M. A., K. M. Page and K. Sigmund (8 September 2000). 'Fairness versus reason in the ultimatum game.' *Science*, 289: 1773.

O'Hare, G. M. P. and N. R. Jennings (Eds.) (1996). *Foundations of Distributed Artificial Intelligence.* New York: Wiley.

Orr, H. A. (January 2009). 'Testing natural selection.' *Scientific American*, 300(1): 30.

Owen, G. (1982). *Game Theory.* New York: Academic Press.

Owen, G. (1992). 'Game theory.' In R. A. Meyers (Ed.), *Encyclopedia of Physical Science and Technology*, Vol. 7. New York: Academic Press.

Packard, N. H. (1988). 'Adaptation toward the edge of chaos.' In J. A. S. Kelso, A. J. Mandell and M. F. Shlesinger (Eds.), *Dynamic Patterns in Complex Systems.* Singapore: World Scientific.

Pagels, H. R. (1988). *The Dreams of Reason: The Computer and the Rise of the Sciences of Complexity.* New York: Simon & Schuster.

Pajares, F. (2002). 'Overview of social cognitive theory and of self-efficacy'. Retrieved on January 15, 2017 from http://www.emory.edu/EDUCATION/mfp/eff.html.

Paulos, J. A. (2008). *Irreligion: A Mathematician Explains Why the Arguments for God Just Don't Add Up.* New York: Hill & Wang.

Pennisi, E. (21 May 2010). 'Synthetic genome brings new life to bacterium.' *Science*, 328: 958.

Peters, E. E. (1999). *Patterns in the Dark: Understanding Risk and Financial Crisis with Complexity Theory.* New York: Wiley.

Phister, P. W. (2010). 'Cyberspace: The ultimate complex adaptive system'. *The International C2 Journal*, 4: 1.

Plischke, M. and B. Bergersen (1994). *Equilibrium Statistical Mechanics*, second edition. Singapore: World Scientific.

Poincaré, H. (1993). In D. Goroff (ed.), *New Methods of Celestial Mechanics.* New York: American Institute of Physics.

Politi, A. (2006). 'Complex systems.' In G. Fraser (ed.), *The New Physics for the Twenty-First Century.* Cambridge, U. K.: Cambridge University Press.

Popper, K. R. (1982). *The Open Universe: An Argument for Indeterminism.* Cambridge: Routledge.

Popper, K. (2005). *The Logic of Scientific Discovery.* Taylor & Francis e-Library edition, London and New York: Routledge / Taylor & Francis e-Library.

Prigogine, I. (1945). *Bull. Acad. Roy. Belgique*, 31: 600.

Prigogine, I. (1980). *From Being to Becoming: Time and Complexity in the Physical Sciences.* San Francisco: W. H. Freeman.

Prigogine, I. (1996/1997). *The End of Certainty: Time, Chaos, and the New Laws of Nature.* New York: The Free Press.

Prigogine, I., G. Nicolis and A. Babloyantz (1972). 'Thermodynamics of evolution.' *Physics Today*, 25(11): 23, and 25(12): 38.

Puri, S. and V. K. Wadhawan (Eds.) (2009). *Kinetics of Phase Transitions.* Boca Raton: CRC Press.

Ramachandran, V. S. (2010). *The Tell-Tale Brain: Unlocking the Mystery of Human Nature.* Noida: Random House India.

Raychaudhuri, A. K. (1955). 'Relativistic cosmology I.' *Phys. Rev.,* 98 (4): 1123.

Rees, M. (1999). *Just Six Numbers.* London: Weidenfeld & Nicolson.

Renshaw, E. (1991). *Modelling Biological Populations in Space and Time.* Cambridge: Cambridge University Press.

Ricardo, A. and J. W. Szostak (September 2009). 'Origin of life on Earth.' *Scientific American*, 301(3): 38.

Robson, B. (2011). 'A brief history of the brain'. *New Scientist.* 21 September issue. https://www.newscientist.com/article/mg21128311.800-a-brief-history-of-the-brain?full=true#.UqHS9Sey6L8

Rochester, N. *et al.* (1956). *IRE Trans. Info. Theory*, IT2: 80-93.

Rochester, N. *et al.* (2004) in Copper, L. N., N. Intrator, B. S. Blais and H. Z. Shouval (2004). *Theory of Cortical Plasticity.* Singapore: World Scientific.

Ross, W. Ashby (1947). 'Principles of the self-organizing dynamic system'. *J. General Psychology*, 37(2): 125.

Rubi, J. M. (November 2009). 'The long arm of the second law.' *Scientific American India*, 3(11): 40.

Rudnicki, K. (2014). 'The generalized Copernican cosmological principle'. http://southerncrossreview.org/51/rudnicki4.htm

Sagan, C. (1995). *The Demon-Haunted World: Science as a Candle in the Dark.* Ballantine Books.

Sanders, T. I. (2003). 'What is complexity?'. http://www.complexsys.org/downloads/whatiscomplexity.pdf

Sapp, J. (1994). *Evolution by Association: A History of Symbiosis.* New York: Oxford University Press.

Schmeidler, D. (1969). 'The nucleolus of a characteristic function game.' *SIAM Journal of Applied Mathematics*, 17: 1163.

Schneider, E. D. and Kay, J. J. (1994). 'Life as a manifestation of the second law of thermodynamics.' *Mathematical and Computer Modelling*, 19(6-8): 25.

Schneider, J. J. and F. Dettenwanger (2007). 'Avenues into complex materials: Cooperative projects of the natural, engineering, and biosciences'. *Small*, 3(6): 907.

Schrödinger, E. (1944). *What is Life?* Cambridge: Cambridge University Press.

Seeley, T. D., P. K. Visscher and K. M. Passino (May-June 2006). 'Group decision making in honey bee swarms.' *American Scientist*, 94(3): 220.

Seeman, N. C. (23 January 2003). 'DNA in a material world'. *Nature*, 421: 427.

Seeman, N. C. (June 2004). 'Nanotechnology and the double helix'. *Scientific American*, p. 64.

Segre, D. et al. (2001). 'The lipid world'. *Origins of Life and Evolution of the Biosphere*, 31, 119.

Seigel, E. (2009): 'A tale of two slits'. http://scienceblogs.com/startswithabang/2009/06/01/a-tale-of-two-slits/

Selten, R. C. (1975). 'Reexamination of the perfectness concept for equilibrium points in extensive games.' *International Journal of Game Theory*, 4: 25.

Selten, R. C. (1980). 'A note on evolutionary stable strategies in asymmetric animal conflicts.' *Journal of Theoretical Biology*, 84: 101.

Shannon, C. E. (October 1948). 'A mathematical theory of communication.' *Bell Systems Technical Journal*, 27: 379-423, 623-656.

Shannon, C. E. and W. Weaver (1949). *The Mathematical Theory of Communication.* Urbana-Champaign: University of Illinois Press.

Shapley, L. S. (1953). 'A value for *n*-person games.' In H. Kuhn and A. W. Tucker (Eds.), *Contributions to the Theory of Games.* Princeton, NJ: Princeton University Press.

Sharma, A. S., et al. (2012). 'Complexity and Extreme Events in Geosciences: An Overview', in *Extreme Events and Natural Hazards: The Complexity Perspective* (eds. A. S. Sharma, A. Bunde, V. P. Dimri and D. N. Baker), American Geophysical Union, Washington, D. C.. doi: 10.1029/2012GM001233

Sheftal, N. N. (1966a). 'The definition of symmetry.' In A. V. Shubnikov and N. N. Sheftal (Eds.), *Growth of Crystals*, Vol. 4. New York: Consultants Bureau.

Sheftal, N. N. (1966b). 'The physical meaning of symmetry.' In A. V. Shubnikov and N. N. Sheftal (Eds.), *Growth of Crystals*, Vol. 4. New York: Consultants Bureau.

Sheftal, N. N. (1976). 'A crystal as a medium that orders phenomena.' In N. N. Sheftal (Ed.), *Growth of Crystals*, Vol. 10. New York: Consultants Bureau.

Shenoy, S. R., T. Lookman and A. Saxena (2005). 'Spin, charge, and lattice coupling in multiferroic materials', in A. Planes, L. Manosa and A. Saxena (eds.), *Magnetism and Structure in Functional Materials*. Berlin: Springer.

Shermer, M. (2006). In J. Brockman (Ed.), *What We Believe but Cannot Prove*. New York: Harper.

Shubik, M. (1993). 'Game theory'. In G. L. Trigg (ed.), *Encyclopedia of Applied Physics*, Vol. 7, p. 69. New York: VCH Publishers.

Sigmund, K., E. Fehr and M. A. Nowak (January 2002). 'The economics of fair play.' *Scientific American*, 286(1): 80.

Simon, H. A. (1970). 'The axiomatization of physical theories.' *Philosophy of Science,* 37: 16 26.

Simon, H. A. (1996). *The Sciences of the Artificial*. Cambridge: The MIT Press.

Singh, P. (2003). 'Examining the society of mind'. Computers and Artificial Intelligence, 22.6: 521. http://www.jfsowa.com/ikl/Singh03.htm

Sipper, M. (2002). *Machine Nature: The Coming of Age of Bio-Inspired Computing*. McGraw-Hill.

Smil, V. (2003). *The Earth's Biosphere: Evolution, Dynamics, and Change*. Cambridge: The MIT Press.

Smith, K. (31 August 2011). 'Neuroscience vs. philosophy: taking aim at free will'. *Nature*, 477: 23. http://www.nature.com/news/2011/111001/full/477023a.html

Smith, E. E. and J. Jonides (1999). 'Storage and executive processes in the frontal lobes'. *Science*, 283:1657–61.

Smolin, L. (January 2004). 'Atoms of space and time.' *Scientific American*, 290(1): 56.

Smuts, J. C. (1929). 'Holism.' *Encyclopaedia Britannica,* 14th ed., vol. 11, p. 640.

Solé, R. and B. Goodwin (2000). *Signs of Life: How Complexity Pervades Biology*. New York: Basic Books.

Stanford Encyclopedia of Philosophy (2009). 'Compatibilism'. http://plato.stanford.edu/entries/compatibilism/

Stewart, J. (2000). *Evolution's Arrow: The Direction of Evolution and the Future of Humanity*. Canberra: Chapman Press.

Stillinger, F. H. (31 March 1995). 'A topographic view of supercooled liquids and glass formation.' *Science*, 267: 1935.

Stix, G. (January 2010). 'Darwin's living legacy.' *Scientific American*, 300(1): 24.

Storhoff, J. J. and C. A. Mirkin (1999). 'Programmed materials synthesis with DNA.'

Chemical Reviews, 99: 1849.

Strocchi, F. (2005). *Symmetry Breaking*. Berlin: Springer.

Strogatz, S. H. (8 March 2001). 'Exploring complex networks.' *Nature*, 410: 268.

Sugden, A., B. Hanson, E. Pennisi and E. Cullota (9 January 2009). 'A celebration and a challenge.' *Science*, 323: 185.

Szent-Györgyi, A. (1957). *Bioenergetics*. New York: Academic Press.

Szilard, L. (1929). 'Uber die entropieverminderung in einem thermodynamischen system bei eingriffen intelligenter wesen'. *Zeitschrift fur Physik*, 53: 840.

Szathmáry, E. and Maynard Smith, J. (16 March 1995). 'The major evolutionary transitions.' *Nature*, 374: 227.

Szilard, L. (1964). 'On the decrease of entropy in a thermodynamic system by the intervention of intelligent beings'. *Behavioural Science*, 9: 301.

Taber, M. (1969). *Chaos and Integrability in Nonlinear Dynamics*. New York: Wiley.

Tegmark, M. (2017). *Life 3.0: Being Human in the Age of Artificial Intelligence*. Knopf Doubleday Publishing, A Division of Penguin Random House LLC.

Thich Nhat Hanh (1991). *Peace in Every Step*. Bantam, New York.

Toffoli, T. (1994), "Neural networks". In G. L. Trigg (Ed.), *Encyclopedia of Applied Physics*, Vol. 11. VCH Publishers, New York.

Tribus, M. and E. C. McIrvine (September 1971). 'Energy and information.' *Scientific American*, 225(3): 179.

Tsagas, C. G. (2011). 'Peculiar motions, accelerated expansion, and the cosmological axis.' *Physical Review D*, 84: 063503.

Tsallis, C. (1988). 'Possible generalizations of Boltzmann-Gibbs statistics.' *J. Stat. Phys.* 52: 479.

Tsallis, C. (1995a). 'Some comments on Boltzmann-Gibbs statistical mechanics.' *Chaos, Solitons and Fractals*, 6: 539.

Tsallis, C. (1995b). 'Non-extensive thermostatistics: brief review and comments.' *Physica A*, 221: 277.

Tsallis, C. (July 1997). 'Levy distributions.' *Physics World*: p. 42.

Tsallis, C. (2009). *Introduction to Nonextensive Statistical Mechanics*. Springer, New York.

Turing, A. M. (1950). 'Computing machinery and intelligence'. *Mind: A Quarterly Review of Psychology and Philosophy*, 59 (236): 433.
http://www.amazon.in/Alan-Turing-His-Work-Impact/dp/0123869803

Turing, A. M. (1952). 'The chemical basis of morphogenesis'. *Philosophical Transactions of the Royal Society of London*, Series B, 237: 37.

Turner, M. S. (September 2009). 'Origin of the universe.' *Scientific American*, 301(3): 22.

Turner, M. S., D. Baumann and P. Steinhardt (2014). 'Instantaneous cosmic growth: Have we found the smoking gun?'.
http://www.kavlifoundation.org/science-spotlights/theory-cosmic-inflation-bicep2#.U4s0MXbm4mh

Turocy, T. L. and B. von Stengel (2002). 'Game theory.' In *Encyclopedia of Information Systems*. New York: Academic Press.

Uhl-Bien, M., R. Marion, and B. McKelvey (2007). 'Complexity Leadership Theory: Shifting leadership from the industrial age to the knowledge era'. *The Leadership Quarterly*, 18: 298. *Leadership Institute Faculty Publications.* Paper 18.
htp://digitalcommons.unl.edu/leadershipfacpub/18
Additional works available at http://digitalcommons.unl.edu/leadershipfacpub

Varadan, V. K., X. Jiang and V. V. Varadan (2001). *Microstereolithography and Other Fabrication Techniques for 3D MEMS.* Wiley.

Varadan, V. K., K. J. Vinoy and S. Gopalakrishnan (2006). *Smart Material Systems and MEMS: Design and Development Methodologies.* Wiley.

Villasenor, J. and W.H. Mangione-Smith (June 1997). "Configurable Computing". *Scientific American*, pp. 66-71. Also see:
http://authors.library.caltech.edu/1632/1/MANcompute97.pdf

Vogelsang, J. (2002). 'Futuring: A complex adaptive systems approach to strategic planning'. *OD Practitioner*, 34: 8.

Volterra, V. (1926). 'Variations and fluctuations of the number of individuals in animal species living together.' The original paper was in Italian. An English translation was published in R. N. Chapman (1936), *Animal Ecology*. New York: McGraw-Hill.

von Neumann, J. (1963). *Collected Works.* Oxford: Pergamon Press.

von Neumann, J. (1966). *Theory of Self-Reproducing Automata.* Completed and edited by A. W. Burks. Champaign-Urbana: University of Illinois Press.

von Neumann, J. and O. Morgenstern (1944). *Theory of Games and Economical Behaviour.* Princeton N.J.: Princeton University Press.

Wadhawan, V. K. (1987). 'The generalized Curie principle, the Hermann theorem, and the symmetry of macroscopic tensor properties of composites'. *Materials Research Bulletin*, 22: 651.

Wadhawan, V. K. (1998). 'Towards a rigorous definition of ferroic phase transitions'. *Phase Transitions,* 64: 165.

Wadhawan, V. K. (2000). *Introduction to Ferroic Materials.* Amsterdam: Gordon and Breach.

Wadhawan, V. K. (July 2002). 'Ferroic materials: A primer'. *Resonance*, p. 15.

Wadhawan, V. K. (2007a). *Smart Structures: Blurring the Distinction Between the Living and the Nonliving*. Oxford: Oxford University Press.

Wadhawan, V. K. (August 2007b). 'Robots of the future'. *Resonance*, p. 61.

Wadhawan, V. K. (2010). *Complexity Science: Tackling the Difficult Questions We ask about Ourselves and about Our Universe*. Saarbrücken: LAP Lambert Academic Publishing.

Wadhawan, V. K. (2011, 2014). *Latent, Manifest, and Broken Symmetry: A Bottom-up Approach to Symmetry, with Implications for Complex Networks*. Printed by CreateSpace, USA. ISBN: 978-1463766719 / 1463766718.

Wadhawan, V. K. (2014). 'Science, scientists, and scientific temper in society'. http://vinodwadhawan.blogspot.in/2014/05/science-scientists-and-scientific.html

Wadhawan, V. K. (2015). 'Climate justice and what we eat'. http://vinodwadhawan.blogspot.in/2015/09/climate-justice-and-what-we-eat.html

Wadhawan, V. K., P. Pandit and S. M. Gupta (2005). 'PMN-PT based relaxor ferroelectrics as very smart materials'. *Materials Science & Engineering B*, 120: 199.

Wagner, G. P. and Lee Altenberg (1996). 'Perspective: Complex Adaptations and the Evolution of Evolvability'. *Evolution*, 50(3): 967.

Wahl, D. C. (May 2016). *Designing Regenerative Cultures*. Triarchy Press.

Waldrop, M. M. (1992). *Complexity: The Emerging Science at the Edge of Order and Chaos*. New York: Simon and Schuster.

Wasserman, S. and K. Faust (1994). *Social Network Analysis: Methods and Applications*. Cambridge: Cambridge University Press.

Watkins, N. W., Pruessner, G., Chapman, S. C. *et al*. (2016). *Space Sci Rev*, 198: 3. https://doi.org/10.1007/s11214-015-0155-x https://link.springer.com/article/10.1007/s11214-015-0155-x

Watts, D. J. (1999). *Small Worlds: The Dynamics of Networks between Order and Randomness*. Princeton, NJ: Princeton University Press.

Watts, D. J. and S. H. Strogatz (4 June 1998). 'Collective dynamics of "small-world" networks.' *Nature*, 393: 440.

Weigend, A. S. and N. A. Gershenfeld (1993) (Eds.). *Time-Series Prediction: Forecasting the Future and Understanding the Past*. Reading, Massachusetts: Addison-Wesley.

Weiner, N. (1948). *Cybernetics; or Control and Communication in the Animal and the Machine*. New York: Wiley (The Technology Press).

Whitehead, A. N. and B. Russell. *Principia Mathematica*, three volumes. Cambridge: Cambridge University Press, 1910, 1912, and 1913. Second edition, 1925 (Vol. 1), 1927

(Vols. 2, 3).

Whitesides, G. M. and M. Boncheva (16 April 2002). 'Beyond molecules: Self-assembly of mesoscopic and macroscopic components'. *PNA, USA*, 99(8): 4769.

Whitesides, G. M. and B. Grzybowski (29 March 2002). 'Self-assembly at all scales'. *Science*, 295: 2418.

Williams, G. P. (1997). *Chaos Theory Tamed.* Washington: Joseph Henry Press.

Williams, N. (25 July 1997). 'Biologists cut reductionist approach down to size'. *Science*, 277: 476.

Wilson, E. O. (1975). *Sociobiology: The New Synthesis*. Cambridge, Massachusetts: Harvard University Press.

Williams, G. P. (1997). *Chaos Theory Tamed*. Washington DC: Joseph Henry Press.

Wolchover, N. (2011). '"Accelerating universe" could be just an illusion'. http://www.nbcnews.com/id/44690771/ns/technology_and_science-science/t/accelerating-universe-could-be-just-illusion/#.U5Rvs3bm4mh

Wolfram, (1983). 'Statistical mechanics of cellular automata'. *Reviews of Modern Physics*, 55: 601.

Wolfram, S. (1984a). 'Universality and complexity in cellular automata'. *Physica D*, 10: 1.

Wolfram, S. (4 October 1984b). 'Cellular automata as models of complexity'. *Nature*, 311: 419.

Wolfram, S. (2002). *A New Kind of Science*. Champaign, Illinois: Wolfram Media Inc. (for more details about this book, see www.wolframscience.com).

Wolfram, S. (2012a). 'It's been 10 years: What's happened with a new kind of science?'. http://blog.stephenwolfram.com/2012/05/its-been-10-years-whats-happened-with-a-new-kind-of-science/

Wolfram, S. (2012b). 'Living a paradigm shift: Looking back on reactions to a *New Kind of Science*'. http://blog.stephenwolfram.com/2012/05/living-a-paradigm-shift-looking-back-on-reactions-to-a-new-kind-of-science/

Wolfram, S. (2012c). 'Looking to the future of a *New Kind of Science*'. http://blog.stephenwolfram.com/2012/05/looking-to-the-future-of-a-new-kind-of-science/

Wright, J. (2014). 'Roll over, Boltzmann'. *Physics World*, May 2014 issue. http://physicsworld.com/cws/article/indepth/2014/may/29/roll-over-boltzmann

Wudka, J. (1998). 'What is the 'scientific method'?' http://physics.ucr.edu/~wudka/Physics7/Notes_www/node6.html.

Xiao, Y., W-T. Wu, H. Wang, M. Xiong and W. Wang (2008a). 'Symmetry-based structure entropy of complex networks.' *Physica A*, 387: 2611.

Xiao, Y., M. Xiong, W. Wang and H. Wang (2008b). 'Emergence of symmetry in complex networks.' *Physical Review E*, 77: 066108.

Yates, F. E. (Ed.) (1987). *Self-Organizing Systems*. New York: Plenum.

Zadeh, L. A. (1965). *Fuzzy Sets, Information, and Control*, 8: 338.

Zadeh, L. A. (1975). 'The concept of a linguistic variable and its applications to approximate reasoning, Parts, I, II, III'. *Information Sciences*, 8, 9: 199-249, 301-357, 43-80. Also see: http://www.seattlerobotics.org/encoder/mar98/fuz/flindex.html

Zimmer, C. (2001). *Evolution: The Triumph of an Idea*. London: Harper Collins.

Zomaya, A. Y. (ed.) (2006). *Handbook of Nature-Inspired and Innovative Computing: Integrating Classical Models with Emerging Technologies*. New York: Springer.

Index

1/f noise system, 173
80-20 law, 173
8-fold way, 9

A New Kind of Science (NKS), 149
a principle of self-organization, 46
action potential, 325
activation barrier, 166, 187
activation barrier, 29
adaptive computation, 154, 276
adenine (A), 193
adiabatic processes, 358
agrian period, 242
agrocultural regime, 253
agro-energy revolution, 242
algorithmic information content, 66, 79, 141
algorithmic information content, 70
algorithmic probability, 69, 185, 374
algorithmically incompressible, 339
Alice in Wonderland, 134
alleles, 236
altruism and natural selection, 81
ambiguity, 373
AMP, 217
AMP (adenosine monophosphate), 215
animal and human migrations, 296
answer engine, 350
ant logic, 160
anthropic principle, 317, 348
anthropogenic fire, 249
anthroposphere, 249
anthroposystem, 249
antiferromagnetic coupling, 166
apparent laws, 130
apparent laws of physics, 111
apparent laws of a universe, 127
application-specific ICs (ASICs), 284
applied NKS, 150, 153
approximate reasoning, 261
Arcadian Man, 254
Arcadian view, 253
archaea, 226
archaebacteria, 225
Archeozoic and Proterozoic eras, 217
arms races, 290

arrow of time, 117, 118, 348, 383
artificial evolution, 143, 287, 289, 297
artificial intelligence, 287, 300
artificial intelligence (AI), 271, 293, 317, 349
artificial life (AL), 276
artificial neural networks (ANNs), 263
artificial pattern-recognition systems, 333
association areas, 314
asymmetric arms race, 237
asymmetric bond, 189
asymmetric conflict, 236
asymmetric contest, 235
ATP, 216, 242, 244, 246
ATP (adenosine triphosphate), 215
attractor, 221, 413
autoassociation and invariance, 333
autocatalysis, 200, 202, 204, 206
autocatalytic reactions, 212, 222
autocatalytic sets of molecules, 202-204
autocatalytic-set model, 279, 280
automated patch-clamp robot, 328
autonomous adaptability, 284
autonomous mobile robots, 288
autonomous robots, 287
axion, 106, 123
axis of evil, 112, 114
axon, 263

back propagation method, 326
bacteria, 225, 226, 228, 273
bacteriophage, 212
bacterium, 312
Barabási-Albert model, 432
baryon, 399
basic building blocks of information, 264
basin of attraction, 39, 46, 138, 205, 413
beehive, 218
behaviour-based robotics, 287
Belousov-Zhabotinsky (BZ) reactions, 205
Bénard cells, 7
bifurcations in phase space, 42, 341, 343, 367
Big Bang, 3, 14, 44, 54, 184, 348, 350
Big Bounce, 115

Big Crunch, 104
Big Questions, 347
bilayer sheets, 200
bino, 107
biological symbiosis, 225
biotic potential, 76, 414
BioWatch, 284
black holes (BHs), 112
Blue Brain Project, 128, 327
blue-green algae, 225, 244, 258
Boltzmann factor, 363
bootstrap problem, 289
broken symmetry, 86
building block (BB), 89-90, 93, 180, 266
bush robot, 299
BZ reaction, 285

C. elegans nematodes (roundworms), 230
Cambrian explosion, 297
Cantor's diagonal argument, 58
Carbian Period, 242, 254
carbo-energy revolution, 242
carbon footprint, 253, 259
catalysis, 203
causal networks, 153
causality principle, 21
CCR (concurrence and coordination run-time), 293
cell assemblies, 264
cell differentiation, 193, 219, 220, 221, 222
cells that fire together wire together, 326
cellular automata (CA), 178, 229
cellular nonlinear networks (CNNs), 435
central dogma of molecular biology, 194, 229, 261, 297
Chaitin's halting probability Ω, 67
chaotic attractor, 139, 415
chaotic inflation, 109
chaotic systems, xx
chemical adaptation, 200, 201
chemical complexity, 211
chemical Darwinism, 204
chemical evolution, 187, 203, 211, 212, 217, 351
chemical natural selection, 190
chemical potential, 359
chemical self-assembly, 197
Chess, 449
chloroplasts, 193, 226
chromatin, 195
chromosomal crossover, 230

chromosomes, 195
chronology of cosmic events, 123
Church-Turing thesis, 150, 151
Class 1 CA, 146
Class 2 CA, 146
Class 3 CA, 146
Class 4 CA, 147
classifiers, 274
cliques, 429
cliquishness, 432
closed system, 379
CLT, 345
CMB, 123
coacervate' droplets, 211
coalition, 452
coalitional or cooperative form of games, 442
COBE satellite, 112
codon, 195
coevolution of species, 290
columnar organization of the neocortex, 327
combinatorial games, 440
compatibilism, 19
competing ground states, 164
competing schemata, 352
complementary charge distributions, 197
complex adaptive organizations, 342
complex adaptive superorganism, 269
complex adaptive system (CAS), 4, 70, 77, 120, 141, 151, 157, 159, 179, 247, 273, 276, 279, 299, 301, 311, 319, 341, 343, 349, 351-352, 439
complex materials, 163
complex system, xix, 4
complexity leadership theory (CLT), 345
complexity metric, 183
complexity transitions, 343
comprehension is compression, 59
compression of information, 66
computational complexity, 71
computational equivalence, 151
computational intelligence (CI), 261, 289, 437
computational irreducibility, 149, 151, 156
computational universality, 150
computational universe, 149, 154
computationally irreducible, 6
computing with words, 261
concurrency, 293
conditional entropy, 376
conditional probability, 366

conditional strategy, 238
conducting polymers, 267
conflict point, 451
connectionist ideas, 326
connectionist movement, 326
connections-in-waiting, 285, 329, 333
consciousness, 320, 323
constructionism, 13
controlled-NOT operation, 132
cooperative games, 450
Copernican principle, 51, 112, 114
cortical column, 327, 331
Cosmic Background Explorer (COBE),
    103, 115
cosmic background radiation (CBR), 103,
    112
cosmic inflation, 87, 98, 108
cosmological constant, 107
Cosmological principle, 51, 112, 114
critical exponent, 411
critical fluctuations, 409
critical phenomena, 410
critical RBNs, 221
cuckoo hosts, 237
cultural assemblies, 308
cultural evolution, 199, 303
cultural trigger, 242
culture, 304
cumulative natural selection, 76
cyanobacteria, 212, 225, 244
cybernetic information, 199
cybernetics, 287
cyborg, 254, 300
cytosine (C), 193

Darwinian evolution, 266, 279, 296, 349,
    353, 446
Darwinian natural selection, 273, 307
decentralized software services (DSSs),
    296
decision problem, 60
decision theory, 440
declarative memory, 315, 323
Deep Blue, 297
degree of complexity, 57
degree of logical randomness, 67
dendrite, 263
descent with modification, 77
deterministic nonperiodic flow, 419
development and evolution of
    evolvability, 290
difference equation, 137, 413

digital cortical brain, 350
digital programmable universe, 143
dipole moment, 189
dissipation-driven adaptive organization,
    xvii, 80
dissipative path, 241
dissipative structures, 42-44
dissipative system, 38, 136
distal description, 288
distributed intelligence, 300
distributed perceptive networks, 295
distributed superintelligence, 300
DNA (deoxyribonucleic acid), 193, 220
domains, 408
double-slit experiment, 52
drug design, 267

E. coli, 220, 221, 225, 426
early life, 218
ecological dominance, 252
ecological footprint, 253-254
ecological niches, 344
economics of fair play, 446
edge of chaos and order, xx, 4, 172, 177,
    203, 214, 222, 231, 278, 352, 401, 430,
    437
edge-of-chaos existence, 271
effective complexity, 70, 79, 141, 184
effective theory, 23
effectively calculable, 61
efficient dissipation of gradients, 80
efficient-market hypothesis (EMH), 338
electric dipole, 189
electroactive gel, 285
embodiment, 288
embryonic electronics, 284
emergence, xviii, 41, 90, 175, 177, 179,
    197, 202, 211, 216, 228, 267, 272, 276,
    289, 319, 345, 347, 435, 448
emergence of a word-speaking species,
    305
emergence of life out of nonlife, 32, 348
emergent behaviour, 78, 265, 344
emergent phenomena, 169, 383
emergent property, 288, 295, 298
energy regime, 242
energy-dissipating pathway, 243
ensemble average, 361
entropic effect, 192
enzymes, 203
epigenetic factors, 237
epigenetic inheritance, 229

epigenetics, 229
equation of state, 106
equivalence principle, 50
ergodicity breaking, 407
ergodicity hypothesis, 361
error catastrophe, 213
ESS, 232, 233, 236, 238
eucarya, 226
eukaryotic cell, 193, 245, 258
evolution of biochemical complexity, 191
evolution of biological complexity, 225
evolution of cellular complexity, 226, 229
evolution of strategies, 267
evolution of the empirical sciences, 251
evolutionarily stable strategies (ESS),
    231-232, 439
evolutionary arms races, 231, 236
evolutionary computation, 276, 289
evolutionary or adaptive robotics, 287
evolutionary robotics, 288
evolvable hardware, 283
examples of complex systems, xix
expert system, 263, 273, 297
exploitation vs. exploration, 275
extensive form of games, 440
extracting supervision from the
    environment, 290

f.g-generalization, 263
falsifiable statement, 10
feedforward mechanism, 230
ferroelastic phase transitions, 164
ferroelectric phase transitions, 164, 408
ferroic phase transition, 89, 164
ferromagnetic coupling, 166
ferromagnetic phase transition, 164, 404
Feynman's sum-over-histories
    formulation, 55
field-programmable gate arrays (FPGAs),
    284, 330
fire economy, 250
firing of the synapse, 263
fittest individuals, xviii
fixed point, 176
fixed-point attractor, 138, 413
Flatness Problem, 107, 108
fMRI (functional magnetic resonance
    imaging), 326
force field, 398
four-point attractor, 139, 415
FPGAs, 284, 330
free energy, 28
free expansion of gas, 25, 33

free will, 152, 310, 320
free-energy rate density, 183
Friedmann equation, 102
functional areas, 314
functional composites, 163
functional neuronal assemblies, 329
functional polar groups, 192
fungi, 225
fuzzy logic, 262
Fuzzy-rule-based systems (FRBSs), 262
Gaia, 4
Game of Life, 277, 278, 436
gametes, 229
garbage bag model, 216
gauge boson, 99
gene, 194
gene doping, 254
gene expression, 229, 230
gene pool, 224
gene regulation, 220
generalized cellular automata (GCA), 436
generalized random networks, 432
genetic algorithms (GAs), 289
genetic mutations, 224
genetic regulatory network (GRN), 219,
    221, 279
genetic-algorithm (GA), 443
genotype, 223, 228, 229, 231, 271, 290,
    308, 318, 446
giant cosmic void, 106
Gibbs potential, 358
glassy behaviour, 165
Gödel sentence, 60
Goldilocks zone, 128
GPS technology, 48
grand unification theories (GUTs), 98,
    399
gravitational potential energy, 358
graviton, 99
green biofuels, 259
Green Valley, 254
grooming, 305
group selection, 81, 82, 235
GTYPE, 271, 272, 276
guanine (G), 193
Gutenberg-Richter law, 171
H. economicus, 446
H. emoticus, 446
halting probability, 374
haplodiploidy, 238
hardware you can store in a bottle, 285
Hebb's (1949) neural-network model, 273
Hebbian assemblies, 264, 328

Hebbian learning, 326
Hebbian response theory, 326
Heisenberg uncertainty principle, 347
heliocultural energy regime, 254, 260
heliocultural energy revolution, 259
HGT, 218, 229
hidden Markov models, 335
hierarchical hidden Markov models, 317, 335
hierarchical thinking, 318
hierarchy of structures, 165
Higgs boson, 99
Higgs field, 87, 98
Higgsino, 107
Hilbert space, 387
hippocampus, 315
homeostasis, 221, 222
Homo economicus, 446
Homo heidelbergensis, 304, 305
Homo sapiens, 304, 307
Horizon Problem, 108
horizontal gene transfer (HGT), 218, 228
how do arms races end?, 239
Hubble expansion, 102
Human Connectome Project, 330
Human Development Index (HDI), 256
hurricane, 202
hydrocarbons, 191
hydrophilic end, 192
hydrophobic end, 192
hydrophobic interaction, 191, 201
hyperthermophiles, 242, 244

IBM's Deep Blue computer, 291
IBM's Watson, 331
ideomotor apraxia, 20
imperial approach, 254
Imperial Man, 255
imperial view, 253
imputation, 452
incompatibilism, 19
incompatible propositions that imply one another, 58
incremental evolution, 289
induction, 276
industrial robots, 287
inflation, 348
information irreducibility, 149
Information Paradox, 113
inorganic life (intelligent robots), 298
integrable systems, 381
integrated genomes, 228
integrated sensors, 282

interspecific asymmetric arms races, 237
interspecific symmetric arms races, 239
into what is the universe expanding?, 104
intraspecific asymmetric arms races, 238
intraspecific symmetric arms races, 239
irreversible processes, 117
Ising model, 404
joint entropy, 376
joint probability, 366
KAM theory, 382
kingdoms, 225
Koch's snowflake, 139
Kolmogorov complexity, 374
Kolmogorov-Sinai (K-S) entropy, 152, 376
Königsberg bridges problem, 424
K-S entropy, 152, 376
lac operon, 220
Lamarckian evolution, 261, 297
Lamarckian process, 303
Lamarckism, 229
language of thought, 333, 335
laser, 201
last universal common ancestor (LUCA), 218, 225
law of accelerating returns (LOAR), 299, 327
law of diminishing returns, 339
law of increasing returns, 159, 160, 161, 205, 339
law of requisite complexity, 345
lead magnesium niobate (PMN), 167
learning, 333
Lego-like building blocks of knowledge for perception, 328
Leibniz's principle of sufficient reason, 59
LEOGER, 244
leptons, 98
Life 3.0, 301
life originated twice, 215
life-dinner principle, 237
lightest supersymmetric particle (LSP), 106
limited war strategy, 234, 235
linear no threshold principle, 257
linear nonequilibrium situations, 384
linguistic variables, 261, 262
lipid bilayers, 200
lipids-first model, 214, 216
liposomes, 200, 201
liquid logic gates, 285
liquid robot brain, 285

local activity: the genesis of complexity, 437
local interactions, 6
local rules, xviii, 22, 178, 276
lock-and-key-like shapes, 197
logic of animal conflict, 233, 234
logical randomness, 72
logistic difference equation, 138
London dispersive interaction, 190
long-term memory, 315
Lorenz attractor, 137
Lotka-Volterra equations, 238
LUCA, 218, 225, 228
machine intelligence, 268
macroscope, 253
Macroscopical Signal, 242, 253, 254, 258
Magnetospheric Eternally Collapsing Objects (MECOs), 113
Malthusian ideas, 349
marginal probability, 366
mass extinctions, 172
massive compact halo objects (MACHOs), 106
massive compact objects (MCOs), 113
maximin strategy, 447
maximin theorem, 425
meaning of entropy, 375
mechanical procedure, 61
meme, 307, 320
meme pool, 297, 307
memetic DNA, 308
memory and prediction theory, 323
MEMS, 283
Mendel's laws of genetics, 223
messenger RNA (mRNA), 194
metacognition, 321-322
metamaterials, 163
metametarules, 320
metarules, 320
methylation, 230
Mexican-hat potential, 86
micelles, 200
microelectromechanical systems (MEMS), 283
mind children, 247, 287, 350
mindbody, 311
minimax procedure, 61
minimax strategy, 232, 447
minimax theorem, 439, 441, 449
MIPS, 291
mitochondria, 193, 226
mitosis, 195
mixed-strategy game theory, 233

model dependent realism, 23, 120
molecular complementarity, 197
molecular Darwinism, 204
molecular recognition, 197, 198, 201
Moore's law, 353
'More is Different', 15
morphogenesis, 205
morphotropic phase boundary (MPB), 167
mote, 295, 297
Mountcastle's hypothesis, 323, 327
mRNA, 221
M-theory, 127, 348, 400
multicore processors, 292
multiferroics, xx
multifunctional polymers, 283
multi-hop networking approach, 295
multiple universes, 392
multiverse, 109
mutations, 224
mutual entropy, 377
mutual information, 132, 377
naked RNA molecules, 213
nanocomposite, 163
Nash equilibrium, 441, 444, 445, 450
natural selection, 45, 200, 307
natural spaceship, 241
Nature abhors gradients, 312, 351
negative entropy, 93, 208, 349, 372
negative feedback, 160, 178, 339
negentropy, 208, 372
neocortex, 263, 313
neo-Darwinism, 224, 225
network motifs, 426
Neumann universe, 145, 276, 278
Neumann's self-reproducing cellular automata (CA), 276
neural networks, 425
neurotransmitter, 263
neutralino, 106, 107, 123
niche separation, 239
NKS, 152
NKS way of thinking, 150
NMs, 426
'no such thing as empty space', 110
Noether's theorem, 84
nonadaptive complex systems, 163
nonbiological natural composites, 164
noncooperative equilibrium, 234
non-homogeneous networks, 437
nonintegrable dynamical systems, 381
nonlinear causality, 159
nonlinear dynamics, 135

nonlinear nonequilibrium systems, 384
nonlinear response, xx, 165
non-random origin of life, 222
normal or strategic form of games, 440
NP-class problem, 71
NP-complete-class problem, 71, 427
nuclear fusion, 255
nucleation site, 168
nucleic acids, 193
nucleocultural energy regime, 255
nucleotide, 193, 216
nude replicating RNA molecules, 213
nude ribozyme polymerase, 214
Ockham's razor, 59, 185
old brain, 313
omnipresent superintelligence, 300
one-way SME, 168
ontogenetic adaptation, 290
ontogeny, 284
open-source heredity, 217
operator DNA sequence, 220
opposable thumb, 184, 318
optimal mixed strategies, 449
optimal strategies, 447
order for free, 77
organelles, 193, 245
orientation states, 88
'ours is a flat-geometry universe', 128
oxian period, 242
oxo-energy revolution, 241
PAMPS, 285
parasites, 267
parasitism, 226, 229
Pareto distribution, 173
Pareto principle, 173
Pareto-optimal, 451
path dependence in phase space, 340
pattern formation, 41, 83, 178, 204, 205,
    296, 436, 437
pattern recognition modules, 285, 329,
    330
Pattern Recognition Theory of the Mind
    (PRTM), 300, 316, 331
pattern recognizer (PR), 303, 331, 333-
    334
pattern-recognition capability, 331, 350
pattern-recognition circuits, 300
pattern-recognition module, 332, 334
pattern-recognition scheme, 316
Pauli exclusion principle, 96
payoff, 233
p-branes, 111
PCE, 149-152

P-class problem, 71
perceptron, 265
percolation transition, 429
permutation symmetry, 83, 92, 93
perpetual novelty, 4, 78, 159, 177, 280,
    289, 352
personal robots (PRs), 296
pervasive computing, 296
phase-space trajectory, 38
phenotype, 223, 224, 228, 229, 230, 231,
    271, 290, 308, 318, 446
pheromone, 157
phospholipids, 214
photian period, 242
photino, 107
photo-energy revolution, 241
photosynthesis, 244
phototrophy, 244
phyllotaxis, 205
phylogenetic adaptation, 290
Planck time, 102
pleuromona, 214
plural games, 449
PMN, 167
PMN-PT, 167
Poincaré resonances, 382
polar clusters, 167
polar liquid, 189, 191
polymer, 202, 211
polymerase enzyme, 212
population dynamics, 137
positive feedback, 158, 159, 160, 177,
    204, 264, 267, 274, 299, 339
post-biological world, 298
potential energy, 358
power law, 431
power of gradual change, 349
power-law degree distribution, 432
power-law dependence, 363
power-law feature, 431
predator-prey dynamics, 238
predictability or incremental redundancy,
    378
premonitory phenomena, 278
primary bonds, 188
primary structure of a DNA molecule,
    193
primordial soup, 203
Principle of Computational Equivalence
    (PCE), 155
Principle of Computational Equivalence
    (PCE), 149-152

principle of conservation of quantum
    information, 397
principle of general relativity, 50
principle of inheritance of acquired
    characteristics, 229
principle of relativity, 48
principle of use and disuse, 229
Prisoner's Dilemma, 442, 450
processing element (PE), 265
profit motive, 275
progenitive cosmic gradient, 215
progenote, 228
programmable living machines, 254
program-size complexity, 67, 373
prokaryotic cell, 193, 212
proteinoids, 212
protobionts, 212
protoctists, 225
proton acceptor, 190
proton donor, 190
proximal description, 288
PRTM, 331, 332, 335
PTYPE, 271, 272, 276
punctuated equilibrium, 169, 172, 231
pure NKS, 150
pyrian period, 242
pyrocultural regime, 253
pyro-energy revolution, 241, 249

Quantificational Signal, 242, 251, 253
quantum chromodynamics (QCD), 399
quantum electrodynamics (QED), 398
quantum parallelism, 133
quark, 96, 123, 399
quasi-isolated subsystem, 12
quasi-isolated system, xix
qubit, 131

R symmetry, 106
random Boolean networks (RBNs), 221
random network, 219
random system, 121
RBN, 221
R-brain, 313, 327
RCC (reinforced cement concrete), 163
reaction-diffusion equations, 205
reconfigurable computers (RCs), 284
reconfigurable hardware, 284
recursion, 304
recursive functions, 61
recursive procedure, 61
recursive process, 318

redundancy, 377
regulatory genes, 220
reinforced learning, 289
relative complexity, 374
relativity of simultaneity, 48
renormalization-group approach, 176
replication errors, 224
replication template, 196
representation instinct, 309
representation problem, 290
representational content, 308
representational system (RS), 309
repressor protein, 220
reptilian brain, 313
resistance to antibiotics, 218
resonances among the degrees of
    freedom, 381
restructuring phase of problem-solving,
    321
ribosomal RNA (rRNA), 213
ribosome, 193, 213, 217
ribozymes, 213
RNA (ribonucleic acid), 193, 212
RNA invented viruses, 217
RNA species, 213
Robo sapiens, 254
robot shaping, 289
robustness of networks, 433
Rule 110, 148, 150
Rule 30, 147
Rule 90, 147, 148

saddle point in the payoff matrix, 447
sbaryons, 106
scale invariance, 139
schema, 78, 351
schema theorem, Holland's, 267
scientific method, 254, 300, 347, 350
search engine, 350
search space, 229
second law of thermodynamics, xvii, 348
secondary chemical interactions, 189
seeding, 31
selective destruction of information, 265
self-assembly, 46, 201
self-healing machines, 284
self-information, 377
selfish gene, 230, 231, 309
selfish meme, 309
selfish replicator, 230

self-organization, 77, 81, 83, 91, 165, 169, 172, 177, 178, 197, 199, 200, 201, 222, 225, 279, 326, 343, 347, 352, 436, 437
self-organize, 172, 192
self-organized criticality' (SOC), 203, 231
self-referential question, 62
self-reproducing CA, 276
self-reproducing machine, 145
self-similar structure, 175
self-similarity, 139
semiotic information, 199
sequence probability, 366-367
sequential switching automata, 221
sex ratio, 233
sexual crossover, 266
sexual signalling, 305
Shannon entropy, 372
Shannon formula, 64, 370
shape-memory effect (SME), 168
short-term memory, 314, 326
similar linkage pattern, 92, 434
single-self concept, 320
singularity, 300
situatedness, 288
six degrees of separation, 429
smart dust, 295
SOC, 203, 231
social progress, 199
societies of agents, 319
solution space, 229
solvation shell, 192
some truths are unprovable, 60
spacetime continuum, 49
spatio-temporal complexity, 165
speciation, 217, 224, 227-228
species, 227
specific free energy rate, 183
spontaneous breaking of symmetry, 407
spontaneous magnetization, 88, 404
spontaneous polarization, 89, 403
spontaneous strain, 89
stasis, 169, 231
steady state, 380
strange attractor, 137, 146, 419
strange loops, 321
strategic equilibrium, 441
strong emergence, 17
structural composite, 163
structural phase transitions, 164
structure of a pattern (in the brain), 332
subsumption architecture, 287

supergravity, 110
superhuman intelligence, 315
superintelligence, 351
superintelligence created by humans, 350
supernova explosion, 125
superorganism, 81, 157, 160, 218, 258, 349
superstrings, 111
supersymmetry (SUSY), 110, 400
supervised learning, 265, 289
supramolecular aggregates, 197, 198
surface energy, 30
surface of last scatter, 104, 123
survival of the fittest, 76, 200
sustainable energy, 259
swarm intelligence, 78, 159-160, 178, 287, 295, 297, 311, 341, 346, 448
Symbian approach, 259
Symbian Man, 258, 259
Symbian partnerships, 260
symbiogenesis, 226, 227, 228
symbiont, 216, 225
symbiont-integration evolutionary route, 245
symbiosis, 226, 228-229, 231, 258, 260
symbiosis in action, 225
symbolic representation, 304
symbolic thinking, 250
Symbolisational Signal, 242, 250, 253
symmetric contest, 235
symmetry-breaking bifurcation in phase space, 351
symmetry-breaking phase transitions, 399
synapse, 263
synthetic life, 254
System Earth, 241-242, 258
systems biology, 213

technological singularity, 286
tentacle patterns on Hydra, 205
theory of cognition, 276
thermion period, 242
thermodynamic potentials, 358
thermodynamically large system, 173, 362
thermodynamically small system, 173
thermodynamic-limit, 383
thermophilic energy regime, 212
thermophilic organisms, 242
'This statement is unprovable', 60, 68
thymine (T), 193
time irreducibility, 149

time paradox, 383
totalistic rule, 152
total-war strategy, 235
transcription network, 426
transcription of a stretch of DNA, 194
transfer RNA (tRNA), 194
transition-metal oxides, 164
Traveller's Dilemma, 444, 450
tree structure, 269
triplet code, 207
Tsallis entropy, 362
Tsallis thermodynamics, 173
Turing machine, 62
Turing's universal computer, 61
two-person non-zero-sum games, 441
two-point attractor, 139, 414
two-way SME, 168

UCA, 148
ultimatum game, 445
uncertainty principle, xviii, 53, 400
undecidability theorem of computer
    science, 271
unitarity principle of quantum mechanics,
    397
universal cellular automata (UCA), 148,
    150
universal computer, 61, 148
universal constructor, 145
universal infinite set, 58
universal robots, 293, 297
universal Turing machine, 436
unstable evolutionary progressions, 236

unsupervised learning, 289
uracil (U), 193

vacuoles, 193
vagueness, 373
valence electron, 188
vertical gene transfer, 218
vesicles, 201, 214
viruses, 193
vivisystem, 78, 167, 168, 287

waggle dance, 158
weak emergence, 16
weakly interacting massive particles
    (WIMPs), 106
weather forecasting, 296
what is complexity science, xviii
what is the meaning of existence?, 396
what was there before the Big Bang?, 104
whole-cell patch-clamp electrophysiology
    of neurons, 328
wild fire, 249
WMAP satellite, 112
Wolfram Alpha, 350
Wolfram's four classes of CA, 276
working memory, 321
World Wide Web, 351

zero-sum game, 441
zero-sum two-player games, 447
Zipf's law, 171, 173
zygote, 195, 206, 284, 285